MICROBIAL ENZYMES AND BIOTECHNOLOGY

2ND EDITION

Other titles in the Elsevier Applied Biotechnology Series:

M. Y. Chisti. *Airlift Bioreactors*
T. J. R. Harris (ed.). *Protein Production by Biotechnology*
E. J. Vandamme (ed.). *Biotechnology of Vitamins, Pigments and Growth Factors*

MICROBIAL ENZYMES AND BIOTECHNOLOGY

2ND EDITION

Edited by

WILLIAM M. FOGARTY & CATHERINE T. KELLY

Department of Industrial Microbiology,
University College, Dublin 4, Republic of Ireland

ELSEVIER APPLIED SCIENCE

LONDON AND NEW YORK

ELSEVIER SCIENCE PUBLISHERS LTD
Crown House, Linton Road, Barking, Essex IG11 8JU, England

Sole distributor in the USA and Canada
ELSEVIER SCIENCE PUBLISHING CO., INC
655 Avenue of the Americas, New York, NY 10010 USA

First edition 1983

WITH 76 TABLES AND 80 ILLUSTRATIONS

© 1990 ELSEVIER SCIENCE PUBLISHERS LTD

Softcover reprint of the hardcover 2nd edition 1990

British Library Cataloguing in Publication Data

Microbial enzymes and biotechnology.—2nd ed.
  1. Industrial microbiology. Use of enzymes
  I. Fogarty, William M.   II. Kelly, Catherine T.
  660.62
ISBN-13: 978-94-010-6830-7   e-ISBN-13: 978-94-009-0765-2
DOI: 10.1007/978-94-009-0765-2

Library of Congress Cataloging-in-Publication Data

Microbial enzymes and biotechnology/edited by William M. Fogarty &
Catherine T. Kelly.—2nd ed.
    p.   cm.

    1. Microbial enzymes—Biotechnology.   2. Microbial enzymes—
—Industrial applications.   I. Fogarty, William M.   II. Kelly,
Catherine T.
TP248.65.E59M53   1990
660′.634—dc20   90-3015
                                                                  CIP

PREFACE

Biotechnology is now one of the major growth areas in science and engineering and within this broad discipline enzyme technology is one of the areas earmarked for special and significant developments.

This publication is the second edition of *Microbial Enzymes and Biotechnology* which was originally published in 1983. In this edition the editors have attempted to bring together accounts (by the relevant experts) of the current status of the major areas of enzyme technology and specifically those areas of actual and/or potential commercial importance. Although the use of microbial enzymes may not have expanded at quite the rate expected a decade ago, there is nevertheless intense activity and considerable interest in the whole area of enzyme technology.

Microbial enzymes have been used in industry for many centuries although it is only comparatively recently that detailed knowledge relating to their nature, properties and function has become more evident. Developments in the 1960s gave a major thrust to the use of microbial enzymes in industry. The commercial success of alkaline proteases and amyloglucosidases formed a bed-rock for subsequent research and development in the area.

This book is a collection of chapters covering the most important areas in enzyme technology. It is intended primarily as an update of recent developments and should provide an insight into advances in specific areas. One notable feature of the current edition, which should greatly widen its usefulness, is that the number of contributing authors has been increased and broadened in comparison with the first edition. The range of topics covered includes a number of emerging or expanding areas, such as enzymes in organic synthesis (S. M. Roberts), enzymes in antibiotics, steroid and other conversions (J. O'Sullivan), microbial lipases (S. E. Godtfredsen), cellulases (fungal—M. P. Coughlan; bacterial—F. Stutzenberger), glucose transforming enzymes (A. and W. Crueger), alkalophilic enzymes (K. Horikoshi), and microbiosensors and immunosensors (I. Karube, A. Seki and K. Sode).

One omission from this edition is that of microbial hemi-cellulases/xylanases—an area of considerable interest. Unfortunately, the contributor in question was unable, in the end, to provide a manuscript.

The editors are greatly indebted to the contributors who have given so generously of their time and expertise.

W. M. FOGARTY
C. T. KELLY

CONTENTS

LIST OF CONTRIBUTORS

C. O. L. BOYCE
Novo-Nordisk A/S, Novo Allé, DK-2880, Bagsvaerd, Denmark

M. P. COUGHLAN
Department of Biochemistry, University College, Galway, Republic of Ireland

A. CRUEGER
Verfahrensentwicklung Biochemie, Bayer AG, Postfach, 5600 Wuppertal 1, FRG

W. CRUEGER
WV Umweltschutz, Bayer AG, 5090 Leverkusen, Bayerwerk, FRG

A. L. DEMAIN
Fermentation Microbiology Laboratory, Department of Biology, Massachusetts Institute of Technology, Cambridge, Massachusetts 02139, USA

W. M. FOGARTY
Department of Industrial Microbiology, University College, Belfield, Dublin 4, Republic of Ireland

S. E. GODTFREDSEN
Novo-Nordisk A/S, Novo Allé, DK-2880, Bagsvaerd, Denmark

K. HORIKOSHI
Department of Bio-engineering, Tokyo Institute of Technology, O-okayama, Meguro-ku, Tokyo 152, Japan
Present address: Department of Applied Microbiology, The Institute of Physical and Chemical Research, Wako-shi, Saitama 351-01, Japan

I. KARUBE
Research Center for Advanced Science and Technology, University of Tokyo, 4-6-1 Komaba, Meguro-ku, Tokyo 153, Japan

C. T. KELLY
Department of Industrial Microbiology, University College, Belfield, Dublin 4, Republic of Ireland

J. O'SULLIVAN
The Squibb Institute for Medical Research, Princeton, New Jersey 08543-4000, USA

H. OUTTRUP
Novo-Nordisk A/S, Novo Allé, DK-2880, Bagsvaerd, Denmark

L. W. POWELL
Beecham Pharmaceuticals, Clarendon Road, Worthing, West Sussex BN14 8QH, UK

S. M. ROBERTS
Department of Chemistry, University of Exeter, Stocker Road, Exeter EX4 4QD, UK

A. SEKI
Research Center for Advanced Science and Technology, University of Tokyo, 4-6-1 Komaba, Meguro-ku, Tokyo 153, Japan

K. SODE
Research Center for Advanced Science and Technology, University of Tokyo, 4-6-1 Komaba, Meguro-ku, Tokyo 153, Japan

F. STUTZENBERGER
Department of Microbiology, Clemson University, Clemson, South Carolina 29634-1909, USA

J. R. WHITAKER
Department of Food Science and Technology, University of California, Davis, California 95616, USA

Chapter 1

CELLULOSE DEGRADATION BY FUNGI

MICHAEL P. COUGHLAN

Department of Biochemistry, University College, Galway, Republic of Ireland

CONTENTS

1 INTRODUCTION

Cellulose is the most abundant organic macromolecule on earth. It has been estimated that total biomass (fossil fuels excepted) amounts to about 1.8×10^{12} tonnes, 1×10^{11} tonnes being replenished each year by photosynthesis (Bassham, 1975; Stephens & Heichel, 1975). Since 40% of this biomass consists of cellulose (Brown, 1983), one may calculate that 7×10^{11} tonnes of this material exist, mainly in higher plants, and that annual productivity is about 4×10^{10} tonnes. The magnitude of these figures can be appreciated by noting that the rate of cellulose synthesis is equivalent to 70 kg per person per day (Lützen *et al.*, 1983) or to 50 000 barrels of oil per second in energy terms (Sienko & Plane, 1976). Indeed, current terrestrial biomass, at 640 billion tonnes of oil equivalent, is equal to total proven fossil fuel reserves (De Montelambert, 1983). But, unlike the latter it is constantly being renewed. Lignocellulosic wastes or residues of forest, agriculture, industrial or domestic origin, are generated in great quantities especially in the more developed countries. Many reports have suggested that much of the demand for fuels and chemical feedstocks, currently met by oil, could be met by appropriate exploitation of such wastes (Avgerinos & Wang, 1980; Chartier, 1981; Eveleigh, 1982; Brown, 1983; Grohman & Villet, 1983; Hall, 1983; Sheppard & Lipinsky, 1983; Sinskey, 1983; Soltes, 1983; Eveleigh, 1984; Lloyd, 1984).

As yet there is little industrial-scale realization of this potential of cellulose as a source of food, fuels and chemical feedstocks. One is reminded of King Henry's words, 'Such are the rich, That have abundance and enjoy it not' (Shakespeare, 1600). This lack of realization is attributable to the recalcitrance of cellulose itself, exacerbated by its association with other polysaccharides and lignin *in vivo*. Thus, worthwhile saccharification of cellulosic biomass requires pretreatment to render it amenable to conversion and subsequent use of large amounts of enzyme. This involves costs that render the overall process uneconomical (Ladisch *et al.*, 1983; Mandels, 1985; Ladisch & Tsao, 1986). Nevertheless, the study of cellulases continues. These enzymes have other applications, including extraction of juices and flavours from appropriate plants and fruits, improvement of the quality of animal fodder and facilitation of brewing processes (Coughlan & Folan, 1979; Mandels, 1985). Moreover, there is an inherent scientific challenge in the study of the cooperative attack on an insoluble substrate by multicomponent systems of enzymes with subtly-overlapping substrate specificities.

Several excellent books and symposia proceedings are available to readers interested in the development of research on cellulose and cellulolytic enzyme systems. These include those by Gascoigne & Gascoigne (1960), Hajny & Reese (1969), Wilke (1975), Gaden *et al.* (1976), Ghose (1978) and Aubert *et al.* (1988). All of the relevant papers published between 1972 and 1982 are accessible on database (Anon., 1982). Several reviews provide an alternative means of bringing oneself up-to-date (Brown, 1983; Enari, 1983; Gilbert & Tsao, 1983; Coughlan, 1985; Eriksson & Wood, 1985; Ljungdahl & Eriksson, 1985; Marsden & Gray, 1986; Enari & Niku-Paavola, 1987; Eveleigh, 1987; Fan *et al.*, 1987). The emphasis in this chapter will be on fungal endo- and exoglucanases and their rôle in the degradation of cellulose. Bacterial cellulases are reviewed in the following chapter in this volume.

2 CELLULOSE—THE SUBSTRATE

Cellulose is a linear homopolymer of β-1,4-linked glucose residues in the chair configuration. The number of such residues, i.e. the degree of polymerization, is approximately 14 000 in native plant cellulose (Marx-Figini & Schultz, 1966), about 3500 in that produced by bacteria, such as *Acetobacter xylinum*

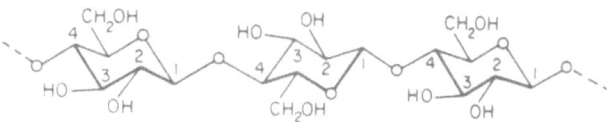

Fig. 1. Part of a cellulose chain showing the β-1,4-linked glucosyl residues rotated 180° with respect to neighbouring residues. Because of this rotation, adjacent linkages will be sterically different to an enzyme approaching the crystalline substrate.

(Marx-Figini, 1982), while that of commercial celluloses ranges from 50 to 5000 (Ljungdahl & Eriksson, 1985). Each glucose residue is rotated 180° with respect to its neighbours along the main axis of the chain (Fig. 1). X-ray diffraction and infrared spectral data indicate that anhydrocellobiose, $C_{12}H_{20}O_{10}$, is the basic recurring unit (Tonnesen & Ellefsen, 1971). The crystal unit cell is $10.5 \text{ Å} \times 8.35 \text{ Å} \times 7.9 \text{ Å}$ (Blackwell, 1982; Eveleigh, 1987). The rotation of adjacent residues and the β-1,4 linkages give cellulose chains a flat ribbon-like structure that is stabilized by intrachain hydrogen bonding (see below). Sixty to seventy adjacent chains, having the same polarity, associate with one another through interchain hydrogen bonding and van der Waals interactions to form highly-ordered crystalline microfibrils (Blackwell, 1982; Rees *et al.*, 1982). Bundles of such microfibrils aggregate to form the strong, insoluble fibres characteristic of the primary and secondary cell walls of higher plants (Rees *et al.*, 1982).

Four crystalline forms (polymorphs) of cellulose, designated as I to IV, differing from one another in their X-ray and/or electron diffraction patterns, have been described in the literature (Rees *et al.*, 1982; Atalla, 1983, 1984). The two most common polymorphs are celluloses I and II. The former is akin to that found in nature while cellulose II is obtained by treatment of type I with concentrated (>15%) solutions of sodium hydroxide. On the basis of solid-state ^{13}C-NMR and Raman spectroscopic studies, Atalla (1983, 1984) and Atalla & VanderHart (1984) concluded that all native celluloses, i.e. celluloses of type I, are composites of two distinct crystalline forms, I_α and I_β. Cellulose types I and II differ from one another with respect to the equivalence, or otherwise, of the glycosidic linkages and of the primary hydroxyl group at C-6. In cellulose I, the C-6 hydroxyl on alternate residues participates, with the ring oxygen of the same residue and the hydroxyl at C-3 of the adjacent residue, in bifurcated intramolecular bonding similar to that observed in methyl-β-cellobioside (Fig. 2). This conformation, designated k_I, is the major factor stabilizing the glycosidic linkages in cellulose I. In this form, the C-6 hydroxyls on adjacent anhydroglucose units are non-equivalent as are the glycosidic linkages between the successive anhydroglucoses. I_α and I_β are two different crystalline lattices containing cellulose molecules in the same k_I conformation. One or other form predominates depending on the source of the cellulose. Bacterial (*Acetobacter xylinum*) and algal (*Valonia macrophysa*) celluloses are comprised of 60–70% I_α whereas cotton, and presumably other plant

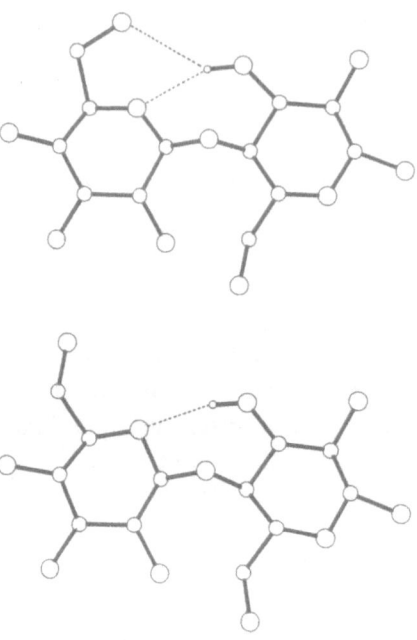

Fig. 2. Schematic representation of the anhydrocellobiose repeat units in the two ordered states, k_I (upper figure) and k_{II} (lower figure), based on the stable conformations of methyl β-cellobioside, with a bifurcated hydrogen bond (Ham & Williams, 1970), and cellobiose with one isolated hydrogen bond (Chu & Jeffrey, 1968), respectively. The computer-drawn figures were generated using molecular modelling (CHEMX, Chemical Design Ltd, Oxford, UK). Adapted from Atalla (1983).

celluloses, are 60–70% I_β. In cellulose II, on the other hand, non-equivalence is centred at the glycosidic linkages whereas the C-6 hydroxyls on adjacent anhydroglucose units are equivalent and exist in the conformation, designated k_{II}, characteristic of cellobiose (Fig. 2).

The resistance of native cellulose to enzymic hydrolysis is due in part to its intrinsic properties and to the fact that its association with lignin and the polysaccharides, hemicellulose and pectin, hinder access by cellulolytic enzymes (Crawford, 1981; Eriksson, 1981; Beldman et al., 1984; Coughlan et al., 1985; Marsden & Gray, 1986; Saddler, 1986; Schwald et al., 1988). The degradability of cellulose by mixed rumen flora was shown to be inversely proportional to the crystallinity of the substrate (Dunlap et al., 1976). Treatments that decreased substrate crystallinity enhanced degradability by *Trichoderma viride* and *Aspergillus niger* cellulases (Sasaki et al., 1979). However, crystallinity indices are not of overriding importance in determining susceptibility of cellulose to enzymic degradation. The degree of swelling, fibrillar structure and pore volume may be more important factors (Marchessault & St-Pierre, 1980; Marchessault et al., 1983; Grethlein, 1985; Puls et al., 1985; Weimer & Weston, 1985; see also Section 4.1.4.).

Total pore volume of water-swollen cellulose is the sum of the volumes of the large number of small cell capillaries (up to 200 Å in diameter) between adjacent microfibrils and by the fewer but larger gross capillaries (pit apertures ranging from 200 Å to 1000 Å in diameter) along the external surfaces of microfibrillar bundles (Cowling, 1975). Treatments that effect an increase in pore volume, hence providing access by cellulases to the inner microfibrils, and/or treatments that effect a decrease in particle size (i.e. increase in surface area) or degree of polymerization are considered to be the most appropriate ways of increasing the extent of saccharification (Marchessault & St-Pierre, 1980; Marchessault *et al.*, 1983; Puri, 1984; Grethlein, 1985; Puls *et al.*, 1985; Saddler, 1986). These disparate conclusions may not be contradictory, however, since it has been repeatedly stressed that it is difficult, if not impossible, to alter one property of native cellulose without modifying others. The physicochemical properties of the substrate that limit its degradation and the pretreatments that may be used to overcome these constraints have been extensively reviewed by Marsden & Gray (1986). The intrinsic properties of the enzymes may also limit conversion. Cellulolysis, in general, is inhibited by cellobiose, an endproduct of the process (Saddler *et al.*, 1986), and the enzymes may be unproductively adsorbed on lignin (Deshpande & Eriksson, 1984) or on the recalcitrant cellulose (Saddler *et al.*, 1986) remaining even after extensive saccharification. Reutilization of the adsorbed or the desorbed enzymes may be practical and economically worthwhile (Pourquié *et al.*, 1988) although the ratio of the individual components after desorption may differ from that of the unreacted enzyme mixture (Stutzenberger, 1988).

3 CELLULOSE-DEGRADING FUNGI

Much of the carbon dioxide fixed by photosynthesis is found in the cellulose of plants. As biomass undergoes turnover it is degraded and oxidized to carbon dioxide which is returned to the atmosphere. Ljungdahl & Eriksson (1985) state that while most (*c.* 90%) of the cellulose is degraded to carbon dioxide by aerobic microorganisms (eqn (1)), as much as 10% is converted by anaerobic microorganisms to methane and carbon dioxide (eqn (2)).

$$C_6H_{12}O_6 + 6O_2 \rightarrow 6CO_2 + 6H_2O \qquad (1)$$

$$C_6H_{12}O_6 \rightarrow 3CH_4 + 3CO_2 \qquad (2)$$

Microorganisms that degrade cellulose are both abundant and ubiquitous in nature. They include fungi, bacteria and actinomycetes, aerobes and anaerobes, mesophiles and thermophiles. In what they described as a partial list of cellulolytic fungi, Ljungdahl & Eriksson (1985) named 60 different species. However, it should be emphasized that while many fungi can grow on cellulose, or produce enzymes that degrade amorphous cellulose, relatively few produce the 'complete' extracellular cellulase systems that degrade crystalline cellulose extensively *in vitro* (Mandels, 1975). *Trichoderma* and *Phanerochaete* species are the best known (Ljungdahl & Eriksson, 1985) but bacteria, such as *Clostridium thermocellum* (Béguin *et al.*, 1987; Tailliez *et al.*, 1989a, b), and the anaerobic rumen fungus, *Neocallimastix frontalis* (Wood *et al.*, 1986), may be more promising for commercial exploitation. It was noted above that production of the large amounts of cellulase required for biomass conversion may account for as much as 60% of the total costs (Mandels, 1985). For this reason much effort and ingenuity has been expended on isolating hyper-cellulolytic mutants and in optimizing fermentation conditions. Productivities of 200–500 filter paper units (FPU) per litre per hour by mutants of *T. reesei* have been achieved using fed-batch techniques and soluble substrates (Watson *et al.*, 1984; Pourquié *et al.*, 1988). However, Eveleigh (1987) considers that productivities of as much as 1000 FPU per litre per hour may be achievable.

Wood decay is caused mainly by aerobic fungi. These microorganisms may be classified as white-rot, brown-rot or soft-rot species depending on the pattern of attack on the major components, viz. cellulose, hemicellulose and lignin, of wood (Eriksson & Wood, 1985; Ljungdahl & Eriksson, 1985). The white-rot fungi, a heterogeneous group, effect substantial degradation of all components, including lignin and leave erosion troughs in the vicinity of the fungal hyphae (Ljungdahl & Eriksson, 1985). Of the white-rot fungi, *Sporotrichum pulverulentum* (*Phanerochaete chrysosporium*) has been the most extensively studied (see Sections 4.1 and 4.2). A sub-group of the white-rot fungi degrade the various components at approximately the same rate and so are called the simultaneous-rot fungi (Liese, 1970). This would indicate that metabolism of all components is integrated and interdependent. Indeed, it is now clear that metabolism of the polysaccharide components provides the energy and the hydrogen peroxide needed to degrade lignin (Kersten & Kirk, 1987; Eriksson, 1988). A second sub-group, called the white-pocket-rot fungi, preferentially degrade lignin leaving substantial amounts of cellulose and relatively sound wood surrounding the pocket (Blanchette, 1980*a, b,* 1982; Otjen & Blanchette, 1982). Such is the pattern of attack by *Phellinus pini* on conifers and by *Ionotus dryophilus* on oak (Blanchette, 1980*a,* 1982; Otjen & Blanchette, 1982).

Brown-rot fungi, generally Basidiomycetes, degrade cellulose and hemi-cellulose extensively but their action against lignin may be limited to removal of methoxyl groups (Cowling, 1961; Kirk & Highley, 1973; Ander & Eriksson, 1978). By contrast with white-rots, the brown-rot fungi degrade cellulose faster than the products are utilized and the point of attack on wood may be at some distance from the fungal cell wall (Blanchette *et al.,* 1978; Eriksson *et al.,* 1980). This would imply the operation of a diffusible factor(s) in the attack on polysaccharide components (see Section 4.2). Typically, the brown-rot fungal hyphae are localized in the wood cell lumen. They penetrate adjacent cells through pre-existing openings or by forming bore holes in the wood. Removal of cell wall substances begins in the S_2 layer of the secondary cell wall—the lignin-rich primary cell wall and middle lamella being resistant to degradation. In the latter stages of decay, i.e. when the bulk of the polysaccharides have been consumed, the cell walls collapse (Eriksson & Wood, 1985; Ljungdahl & Eriksson, 1985).

The soft-rot fungi, so called because they effect the softening of wood, are found among the Ascomycetes and Fungi imperfecti (Ljungdahl & Eriksson, 1985). They catalyse the degradation of polysaccharides but are virtually inactive against lignin. Growth of such fungi begins in the wood cell lumen. This is followed by invasion of the secondary wall where they grow parallel to the fibre axes. Attack by these organisms is characterized by cylindrical boreholes with conical ends in the S_2 layer of wood or by an erosion of the cell walls beginning at the lumen.

Only in the relatively recent past have the rumen flagellates, *Neocallimastix frontalis, Sphaeromonas communis* and *Piromonas communis* been isolated in pure culture and recognized as being species of anaerobic phycomycetous fungi (Orpin 1975; 1977*a, b*; Heath *et al.*, 1983). They degrade cellulose and other polysaccharides (Orpin, 1977*c*; Orpin & Letcher, 1979; Bauchop & Mountfort, 1981). The nature of the cellulose fermentation products *in vitro* depend on whether such organisms are grown in monoculture or in coculture with methanogens, which utilize the endproducts of the fungal fermentation of cellulose (Bauchop & Mountfort, 1981). Electron microscopic studies have shown that in the rumen these anaerobic fungi develop preferentially on lignocellulosic materials to which they adhere (Bauchop, 1981; Orpin, 1981; Akin & Barton, 1983; Akin *et al.*, 1983; Akin, 1987). It is not yet known whether these rumen anaerobic fungi, reviewed recently by Orpin & Joblin (1988), are also engaged in cellulose conversion in anaerobic soils, muds and aquatic habitats.

The microorganisms engaged in cellulose conversion in nature do not operate alone. Whether aerobic or anaerobic, they act in consortium with other microorganisms including non-cellulolytic species. Under these conditions, both the rate and the extent of cellulose degradation is substantially greater than would be the case with a monoculture.

4 FUNGAL CELLULASE SYSTEMS

The oxidative and hydrolytic enzymes known to be associated with cellulose degradation by fungi are cellobiose:quinone dehydrogenase (cellobiose: quinone 1-oxidoreductase; EC 1.1.5.1), cellobiose oxidase/hydrogenase (cellobiose:acceptor 1-oxidoreductase; EC 1.1.99.18), lactonase (D-glucono-1,5-lactonohydrolase, EC 3.1.1.17), endoglucanase (1,4(1,3;1,4)-β-D-glucan 4 glucanohydrolase; EC 3.2.1.4), β-glucosidase (β-D-glucoside glucohydrolase; EC 3.2.1.21), exoglucohydrolase (1,4-β-D-glucan glucohydrolase; EC 3.2.1.74) and cellobiohydrolase (1,4-β-D-glucan cellobiohydrolase; EC 3.2.1.91). To my knowledge cellobiose phosphorylase (EC 2.4.1.20), cellodextrin phosphorylase (EC 2.4.1.49) and cellobiose epimerase (EC 5.1.3.11) have been found only in bacterial systems (Coughlan, 1985; Coughlan & Ljungdahl, 1988). Procedures for the preparation of substrates and for assay, purification and activity staining of all of the relevant fungal and bacterial enzymes have recently been covered in detail by Wood & Kellogg (1988) and will not be repeated here.

4.1 Hydrolytic Components

The 'complete' cellulase systems of white- and soft-rot fungi that actively degrade crystalline cellulose are comprised, in the main, of endoglucanases (EG), exocellobiohydrolases (CBH) and β-glucosidases. The cellulase systems of most individual fungi are not 'complete' in that they lack the exo-cellobiohydrolase component (Wood & Bhat, 1988). However, the distinction between endo- and exo-acting components is not as simple as was considered heretofore and the necessity for development of a more informative nomenclature is urgent (Enari, 1983; Enari & Niku-Paavola, 1987; Coughlan & Ljungdahl, 1988; Knowles *et al.*, 1988*a*; see also discussion in Sections 4.1.3.4, and 4.1.4). Among the better characterized of the 'complete' systems are those of the white-rot fungi *Sporotrichum pulverulentum* (*Phanerochaete chrysosporium*; Eriksson & Pettersson, 1975*a, b*; Eriksson, 1978; Eriksson & Wood, 1985; Ljungdahl & Eriksson, 1985) and *Schizophyllum commune* (Paice *et al.*, 1984; Clarke & Yaguchi, 1985; Willick & Seligy, 1985; Clarke, 1987), of the soft-rots, *Fusarium lini* (Mishra *et al.*, 1983), *Fusarium solani* (Wood, 1971; Wood & McCrae, 1977), *Penicillium funiculosum* (Wood *et al.*, 1980; Wood & McCrae, 1982), *Penicillium pinophilum* (Wood & McCrae, 1986; Wood *et al.*, 1989), *Talaromyces emersonii* (McHale & Coughlan, 1980, 1981*a*, 1982; Moloney *et al.*, 1985), *Thermoascus aurantiacus* (Feldman *et al.*, 1988), *Trichoderma koningii* (Halliwell & Griffin, 1973; Wood & McCrae, 1975, 1978; Halliwell & Vincent, 1981) and *Trichoderma reesei* (formerly *T. viride*; Berghem & Pettersson, 1973; Berghem *et al.*, 1975; Shikata & Nisizawa, 1975; Gum & Brown, 1976, 1977; Anon., 1981; Nummi *et al.*, 1983), and of the phytopathogenic fungus, *Sclerotium rolfsii* (Lachke & Deshpande, 1988).

The brown-rot fungi produce endoglucanases but not cellobiohydrolases (Highley, 1975*a, b*). Oxidative degradation of cellulose appears to be more important to the brown-rots than to the white-rot and soft-rot species, although they also engage in such activity (see Section 4.2). The anaerobic fungus, *Neocallimastix frontalis*, isolated from the rumen of cattle (Pearce & Bauchop, 1985) and sheep (Wood *et al.*, 1986) produces an extracellular cellulase with high activity against crystalline cellulose. This is the only member of this group whose cellulase system has yet been studied, so generalization is premature. It produces endoglucanase and β-glucosidase and probably cellobiohydrolase(s) as do the white-rot and soft-rot fungi (Wood *et al.*, 1986). Unlike the latter systems, however, the system produced by *N. frontalis* exists as a multienzyme complex (Wood *et al.*, 1988).

4.1.1 Substrate Specificities of the Hydrolytic Components—Generalizations
As a rule, endoglucanases are inactive against crystalline celluloses, such as cotton or Avicel, but hydrolyse amorphous celluloses and soluble substrates such as carboxymethylcellulose. Cellooligosaccharides are also substrates, the rate of hydrolysis increasing with increasing chain length (Wood & Bhat,

1988). Attack by these enzymes is characterized by random cleavage of β-glycosidic linkages resulting in a rapid drop in the degree of polymerization relative to the rate of release of reducing sugars. The products of action on amorphous cellulose include glucose, cellobiose and cellodextrins of various lengths. By contrast, true cellobiohydrolases degrade amorphous cellulose by consecutive removal of cellobiose from the non-reducing ends of the substrate. Such enzymes, when acting alone, do not attack cotton extensively (but see Section 4.1.4 for contrary opinions) whereas the extent of degradation of the microcrystalline substrate, Avicel, may be as much as 40% or more (McHale & Coughlan, 1980; Henrissat *et al.*, 1985; Wood & Bhat, 1988). With exo-acting enzymes, the rate of release of reducing sugars is greater than that expected from the decrease in the degree of polymerization or, when appropriate, the decrease in solution viscosity. Endoglucanases and exo-cellobiohydrolases act synergistically in the hydrolysis of crystalline cellulose (see Section 4.1.4).

Exoglucohydrolases have been found in only a few cellulase systems as yet. The enzymes from *Penicillium funiculosum* and *Talaromyces emersonii* catalyse the removal of glucose from the non-reducing ends of cellodextrins, activity decreasing with decreasing chain length (McHale & Coughlan, 1981*a*; Wood & McCrae, 1982). The exoglucohydrolases do not act synergistically with endoglucanases in the hydrolysis of crystalline cellulose (Wood & McCrae, 1982). The β-glucosidases complete the process of cellulose hydrolysis by cleaving cellobiose and by removing glucose from the non-reducing ends of smaller cellodextrins. However, unlike the exoglucohydrolases, their activity increases with decreasing degree of polymerization (McHale & Coughlan, 1981*a*). In a recent report on β-glucosidase, Bock & Sigurskjold (1989) say that high substrate specificity can only be obtained at the expense of the overall rate of reaction. In general, they conclude that the rate of enzymic reaction is optimized by evolving the lowest possible energy in the catalytic step while specificity is optimized by evolving the most stable enzyme–substrate complex. Thus, the need for compromise.

4.1.2 Multiplicity of Hydrolytic Components
Each of the major cellulolytic components synthesized by an individual organism exists in a number of forms. Thus, for example, culture filtrates of *S. pulverulentum* contain five separate endoglucanases (Eriksson & Pettersson, 1975*a*) and two distinct β-glucosidases (Deshpande *et al.*, 1978). The extracellular system of *T. emersonii* contains three β-glucosidases, four endoglucanases and as many as five exo-acting enzymes (McHale & Coughlan, 1981*a, b*; Moloney *et al.*, 1985). Eveleigh (1987) lists several possible causes of multiplicity, including: (a) microheterogeneity due to the complexing of cellulolytic components with other proteins, glycoproteins or polysaccharides; (b) macroheterogeneity based on the formation of multienzyme aggregates; (c) synthesis of variants of a single gene product as a result of infidelity in translation, differential proteolysis or glycosylation, and interaction with components of the culture broth; (d) the occurrence of multiple genes—in

which case one may reason that the gene products have separate rôles to play in cellulolysis. Examples of all possibilities have been reported.

4.1.2.1 MICRO- AND MACROHETEROGENEITY

Cellobiohydrolases from commercial *T. viride* preparations were found to contain various amounts of non-covalently bound carbohydrate even after purification to electrophoretic homogeneity (Alluralde & Ellenreider, 1984). Failure to remove this extraneous material contributes to microheterogeneity. The cellulase systems of anaerobic bacteria, at least in the early stages of cultivation, exist as large multisubunit complexes (Coughlan & Ljungdahl, 1988). So also does that of the anaerobic fungus, *Neocallimastix frontalis* (Wood *et al.*, 1988). By contrast, the extracellular components of aerobic fungal systems have generally been considered to be free and when aggregates were found they were adjudged to be incidental rather than functional (Coughlan, 1985). However, Sprey & Lambert (1983*a, b*) reported the finding of unexpected complexes in culture filtrates of *T. reesei*. One such complex consisted of six proteins and exhibited endoglucanase, β-glucosidase and xylanase activities and appeared to be held together by remnants of the fungal cell wall. This complex appeared as one homogeneous band on preparative isoelectrofocusing and could be partially resolved into its individual components only when subjected to urea–octylglucoside treatment and refocused in the presence of urea. Sprey & Lambert (1983*a, b*) emphasize the fact that homogeneity, after isofocusing under detergent-free conditions, reflects the purity of a complex and not of a single protein. These observations together with the known difficulty of separating cellulolytic components by conventional fractionation procedures and the observed synergistic interactions between components may therefore imply that complex formation in aerobic fungal culture filtrates is functional rather than incidental (see Section 4.1.4).

4.1.2.2 GLYCOSYLATION

Most, if not all, fungal extracellular cellulolytic components are glycoproteins, the extent of glycosylation depending on the enzyme and the fungus in question and ranging from virtually nil to as much as 90% on a weight basis (Coughlan, 1985). Indeed, differential glycosylation was shown to be among the factors leading to heterogeneity in the cellulases of *Schizophyllum commune* (Willick & Seligy, 1985). The same may be true of the four glycoprotein endoglucanases of *Talaromyces emersonii* (Moloney *et al.*, 1985) and of the chromatographically-distinct but immunologically cross-reactive and apparently functionally-identical cellobiohydrolases isolated from culture filtrates of *T. reesei* (Gum & Brown, 1977). The covalently-bound carbohydrate in enzymes generally may be required for stabilization of protein conformation, protection against proteolysis or denaturation, secretion or substrate recognition.

Removal of the carbohydrate from the cellobiohydrolase (26% carbohydrate) and endoglucanase (39% carbohydrate) of the thermophilic fungus

Humicola insolens resulted in a substantial decrease in thermal and pH stability (Hayashida & Yoshioka, 1980). Cloning and expression in yeast of the *T. reesei* gene encoding endoglucanase I resulted in the secretion of a hyperglycosylated enzyme (Van Arsdell *et al.*, 1987). The hyperglycosylated enzyme was substantially more thermostable than the native EGI. Growth of *Schizophyllum commune* in the presence of tunicamycin, an inhibitor of N-linked glycosylation, decreased the secretion of β-glucosidase, and to a lesser extent that of endoglucanase, but did not block completely the secretion of either enzyme (Willick & Seligy, 1985). More surprising was the finding that the M_r values of the mRNA-directed enzymes were the same as those of the enzymes secreted in the presence of the antibiotic. Treatment with tunicamycin also reduced the ability of *T. reesei* QM6a and Rut-C30 to secrete endoglucanases and β-glucosidases (Murphy-Holland & Eveleigh, 1985). In each case, the secreted enzymes had lower carbohydrate contents than the native proteins. However, the modified enzymes, and those prepared by cleavage with endoglycosidase H (which removes N-linked carbohydrate), were not significantly different than the native components with respect to activity or susceptibility to thermal or proteolytic inactivation (Murphy-Holland & Eveleigh, 1985). By contrast, Merivuori *et al.* (1985), who treated *T. reesei* with tunicamycin and 2-deoxyglucose (which blocks both N- and O-linked glycosylation), concluded that *N*-glycosylation is essential for thermal- and pH-stability and protection against proteolysis but that *O*-glycosylation is needed for secretion and activity. On the basis of the effects of endoglycosidase H treatment or β-elimination (which removes O-linked carbohydrate), Kubicek *et al.* (1987) claimed that both EG I and EG II of *T. reesei* contain mainly O-linked neutral carbohydrate (i.e. rather than N-linked) and that this is essential for secretion. *T. reesei* cellulases when expressed in *Escherichia coli* are not glycosylated and yet retain their activity against cellooligosaccharides thus indicating that glycosylation is not required for hydrolysis of soluble substrates (Knowles *et al.*, 1988a). The known affinity of mannans for crystalline cellulose (Chanzy *et al.*, 1979) suggests that the mannose-rich carbohydrate moieties of cellulases may be involved in binding to the native substrate (Chanzy *et al.*, 1984). Further work confirming the fact that the carbohydrate moieties are needed for binding to crystalline cellulose but are not required for hydrolysis of soluble substrates is presented in Section 4.1.3.2.

4.1.2.3 PROTEOLYSIS *IN VIVO*

Treatment of an endoglucanase from filtrates of *T. viride* with a protease from the same source yielded a number of endoglucanase forms with minor alterations in substrate specificity (Nakayama *et al.*, 1976). The pattern of enzyme secretion by *T. reesei* was reported to be 'simple' under certain growth conditions but that multiplicity increased with the age and protease activity of the culture (Gong & Tsao, 1979; Gritzali & Brown, 1979). While some of the observed multiplicity of forms may result from incidental or artefactual proteolysis, other evidence indicates that such modification may be functionally-essential *in vivo*. Extracellular protease production, common among fungi (Aunstrup, 1978), may be required to effect the release of endoglucanases (and presumably other components) from fungal cell walls (Kubicek, 1981; Eriksson & Pettersson, 1982). This may be one of the functions of the two types of acidic protease in culture filtrates of *Sporotrichum pulverulentum* (Eriksson & Pettersson, 1982) and in filtrates of the *T. reesei* mutants, Rut-C30 and RL-P37 (Sheir-Neiss & Montenecourt, 1984). (There are five distinct endoglucanases in *S. pulverulentum* filtrates (Eriksson & Pettersson, 1975a) and six in those of *T. reesei* (Sheir-Neiss & Montenecourt, 1984). High endoglucanase activity is always accompanied by high proteolytic activity when *S. pulverulentum* is cultivated on cellulose and the endoglucanase activity of culture filtrates is enhanced more than 10-fold by treatment with a mixture of the two acidic proteases from the same source (Eriksson & Pettersson, 1982). Moreover, Deshpande *et al.* (1984) showed that the endoglucanase in culture filtrates of *Penicillium janthinellium* is activated by treatment with protease-containing filtrate of *Penicillium funiculosum*. They postulated that the enzyme is secreted as an inactive procellulase and subsequently activated by proteolysis. The endoglucanases, I and II, in filtrates of *Schizophyllum commune* differ from one another only in that an alanine-rich sequence at the N-terminus of form I is absent from form II (Paice *et al.*, 1984). The authors reasoned that this may be a leader sequence required for secretion and that EG II may be derived from EG I by proteolysis at or outside the cell wall (see also Section 4.1.3.3).

Other investigators argue that multiplicity of forms does not result from (fortuitous?) proteolytic modification but that it is an intrinsic property of fungal cellulase components (Dunne, 1982; Labudova & Farkas, 1983). All of the five major endoglucanases in filtrates of *Penicillium pinophilum* appear simultaneously in the early stages of growth (Bhat & Wood, 1989). Moreover, the authors state that a protease from the same source is not active against these enzymes. However, the stage of cultivation at which the enzymes were resistant to proteolysis was not stated (cf. previous paragraph) and it may be argued that the endoglucanases in question had already been cleaved to their mature, and thus resistant, forms at the time of isolation. They concluded that expression of different genes accounted for the observed heterogeneity. Indeed, it has become increasingly clear in the last few years that much of the multiplicity of forms in bacterial (Béguin *et al.*, 1988) and fungal (Knowles *et al.*, 1988a) systems is genetically determined. This had of course been anticipated by the earlier finding of two or more forms of cellobiohydrolase and of endoglucanase differing in immunological properties or in amino acid composition in culture filtrates of *T. reesei* (Berghem *et al.*, 1975; Fägerstam & Pettersson, 1979; Håkansson *et al.*, 1979; Nummi *et al.*, 1983) *P. pinophilum* (Wood *et al.*, 1980; Wood & McCrae, 1986), *Fusarium lini* (Mishra *et al.*, 1983) and *Sclerotium rolfsii* (Lachke & Deshpande, 1988).

Furthermore, Willick & Seligy (1985) had demonstrated that there are two distinct mRNA-directed products for each of the β-glucosidases, cellobiohy-drolases and endoglucanases of *S. commune*. The *T. reesei* genes coding for EG I, EG II, CBH I and CBH II have since been cloned and sequenced (Knowles *et al.*, 1988*a*; see also Section 4.1.3.3).

One of the more curious findings with respect to multiplicity of forms has been that of Niku-Paavola *et al.* (1985). They showed that the most active *T. reesei* endoglucanase, originally M_r 43 000, pI 4, becomes spontaneously mod-ified during purification and storage (and possibly during aging of the culture) to give an enzyme, M_r 56 000, pI 5, with only 20% of the activity of the original. The cause and the reason for this remains to be established.

Whatever the causes, the multiplicity of forms (other than those of adventitious origin) clearly implies that individual isoenzymes perform different functions in the cellulolytic process. Differences between individual endoglucanases (and cellobiohydrolases) from *Trichoderma* sp. with respect to their adsorption/desorption behaviour on crystalline celluloses (Rabinovitch *et al.*, 1982; Beldman *et al.*, 1987; Kyriacou *et al.*, 1988) argues in favour of this. But, what are the different rôles? One plausible answer to this, and as yet only a theoretical one, is that advanced by Wood (1985). He suggested that two endoglucanases and two cellobiohydrolases, in both cases differing in stereo-specificity, would be required to 'handle' the stereospecifically-different glycosidic linkages occurring in cellulose (Fig. 1). This line of reasoning was prompted in part by the need to explain the observed, but then unexpected, synergistic behaviour between cellobiohydrolases in catalysing hydrolysis of crystalline cellulose. This and other possible functions of these enzymes are discussed in more detail in Sections 4.1.4 and 5.

4.1.3 Structure/Function Studies

4.1.3.1 ACTIVE SITES

Over twenty five years ago Vernon & Banks (1963) proposed that all enzymes engaged in cleaving β-1,4 glycosidic linkages might do so using an acid hydrolysis mechanism similar to that of lysozyme—a mechanism that is described in most biochemistry textbooks (e.g. Lehninger, 1982). The large cleft-like active site of lysozyme can accommodate a six-residue portion of the substrate, i.e. the alternating *N*-acetylglucosamine-*N*-acetyl muramic acid backbone of bacterial cell wall peptidoglycan. Two tryptophan residues, viz. 62 and 63, at the active site are involved in substrate binding (Phillips, 1968). Glutamic and aspartic acid residues situated adjacent to the linkage to be cleaved participate in catalysis. These are surrounded by non-polar side chains that enhance proton transfer. The un-ionized Glu-35 donates a proton to the glycosidic oxygen linking an *N*-acetylmuramic acid residue to an *N*-acetylglucosamine residue in the bound substrate. This results in release of the *N*-acetylglucosamine-terminated part of the substrate. Meanwhile, the *N*-acetylmuramic acid carbanion formed as a result of the above reaction is stabilized by the ionized Asp-52 at the active centre. The reaction is completed by water donating a proton to replace that lost by Glu-35 and a hydroxyl to the carbanion, thereby displacing Asp-52 and releasing the muramic acid-terminated part of the substrate.

Active site studies on cellulases have largely confirmed the predictions of Vernon & Banks (1963). Thus, the endoglucanases from *Aspergillus niger* (Hurst *et al.*, 1978), *Penicillium notatum* (Pettersson, 1969), *Myrothecium verrucaria* (Hanstein & Whittaker, 1963), *Thermoascus aurantiacus* (Shepherd *et al.*, 1981), *Trichoderma koningii* (Halliwell & Vincent, 1981) and *T. viride* (Li *et al.*, 1965) all have large active centres that accommodate five or six glucosyl residues of cellulose. Indeed, for optimal activity the substrate must be at least this long.

Hurst *et al.* (1977*a*, *b*) concluded that catalysis by an endoglucanase from *A. niger* must involve a carboxylate anion, pK 4·0–4·5, and a protonated carboxyl group, pK 5·0–5·5. An amino acid sequence in the endoglucanase from *Schizophyllum commune* is homologous with the active site sequence of hen egg-white lysozyme (Yaguchi *et al.*, 1983). On this basis it was speculated that Glu-33 and Asp-50 participate in catalysis by the endoglucanase. Similarly, sequence homology between the catalytic site region of phage T4 lysozyme and a sequence in cellobiohydrolase I from *T. reesei* suggested that Glu-65 and Asp-74 are the catalytic residues in CBH I (Paice *et al.*, 1984). Chemical modification and kinetic studies confirmed the involvement of carboxyl groups in catalysis by the *S. commune* endoglucanase but neither the number nor the location of the residues involved was determined (Clarke & Yaguchi, 1985). Putative active site residues have since been located in a number of such enzymes (see Section 4.1.3.3). However, precise attribution must await site-specific mutation studies. Experimental evidence has also indicated that a tryptophan residue is at the active site of the endoglucanase of *A. niger* (Hurst *et al.*, 1977*b*). Clarke's (1987) studies indicated that there are two essential Trp residues in the endoglucanase of *S. commune*. One, appears to be directly involved in the binding of substrate and the second is thought to constitute an integral part of the catalytic site. On the basis of computer-generated molecular modelling of the interaction of CBH I with cellulose it was suggested that Ser-87 and Trp-85 side-chains might be involved in hydrogen-bonding to rings 1 and 2, respectively, of a cellobiose unit of the substrate (Knowles *et al.*, 1988*a*).

A more recent development has been the finding that not all cellulolytic enzymes effect hydrolysis by way of a lysozyme-type mechanism. Knowles *et al.* (1988*a*) pointed out that some such enzymes appear to operate a β-amylase-like mechanism. One may distinguish between these mechanisms as follows. The lysozyme operates a double displacement, glycosyl–enzyme intermediate mechanism with retention of anomeric configuration at the newly created reducing end. By contrast, a β-amylase mechanism involves a single displacement by a nucleophilic water molecule and results in inversion of configuration at the newly created reducing end. Analysis of the stereochemical course of the reactions catalysed by various cellulases has shown that

CBH I and EG I from *T. reesei* (Knowles *et al.*, 1988*b*; Claeyssens & Tomme, 1989) act with overall retention of configuration—and presumably uses a lysozyme-like mechanism—whereas the reactions catalysed by CBH II from *T. reesei* (Knowles *et al.*, 1988*b*; Saloheimo *et al.*, 1988) and a cellobiohydrolase from *Phanerochaete chrysosporium* (Streamer *et al.*, 1975) proceed with overall inversion of configuration—as is the case with β-amylase. Coughlan (1989) has summarized the properties of those fungal and bacterial cellulases, that have been appropriately analysed, as follows: cellulases, in which the catalytic domain is located at the N-terminus of the protein (see below), can utilize cellobiosides or lactosides as substrates and the stereochemistry of the reaction catalysed is one of retention of configuration. By contrast, those enzymes in which the catalytic domain is at the C-terminus do not use cellobiosides or lactosides as substrates and the reaction catalysed proceeds with inversion of configuration. At present, the underlying reasons for these observed relationships are unclear.

4.1.3.2 DOMAIN STRUCTURES

Perhaps because it accounts for more than half of the total cellulolytic protein produced by *T. reesei*, cellobiohydrolase I (CBH I) was the first to be subjected to detailed structure/function studies. Bhikhabhai & Pettersson (1984*a*), who had elucidated the disulphide bonding pattern of CBH I (it contains 12 disulphide bridges and no free cysteine residues), considered that their results reflected the presence of domain structures within the enzyme. This was subsequently shown to be the case. Limited proteolysis of CBH I (65 kDa) with papain yielded a core protein (56 kDa) and a polypeptide (10 kDa) (Van Tilbeurgh *et al.*, 1986). The latter was identified as the heavily glycosylated C-terminal 59 residues of the native protein. The core protein had no activity against insoluble cellulose such as Avicel, was only 50% as effective as native CBH I in binding to Avicel, but retained all of the activity of the native enzyme against soluble substrates, such as cellodextrins. Accordingly, the authors proposed that CBH I is comprised of two distinct domains: one, corresponding to the carboxy-terminal, acts as a binding site for insoluble cellulose while the other, the core protein, contains the catalytic site.

It has since been shown that CBH II (58 kDa) also has a two domain organization (Tomme *et al.*, 1988). However, in this case, the small, heavily-glycosylated fragment (*c.* 13 kDa) released by papain treatment represents the N-terminal region of the native enzyme. The papain cleavage sites were assigned to residues 431 in CBH I and 82 in CBH II (Tomme *et al.*, 1988). (As illustrated in Fig. 3, these are located in the hinge regions linking the two domains.) In neither case did papain treatment affect activity against soluble substrates. However, comparison of the catalytic and adsorptive properties of these enzymes and their core proteins on insoluble celluloses elicited some interesting differences between CBH I and II. Thus, both with respect to binding to crystalline (Avicel) or amorphous cellulose and activity against such substrates, the CBH II core protein is only 50% as effective as

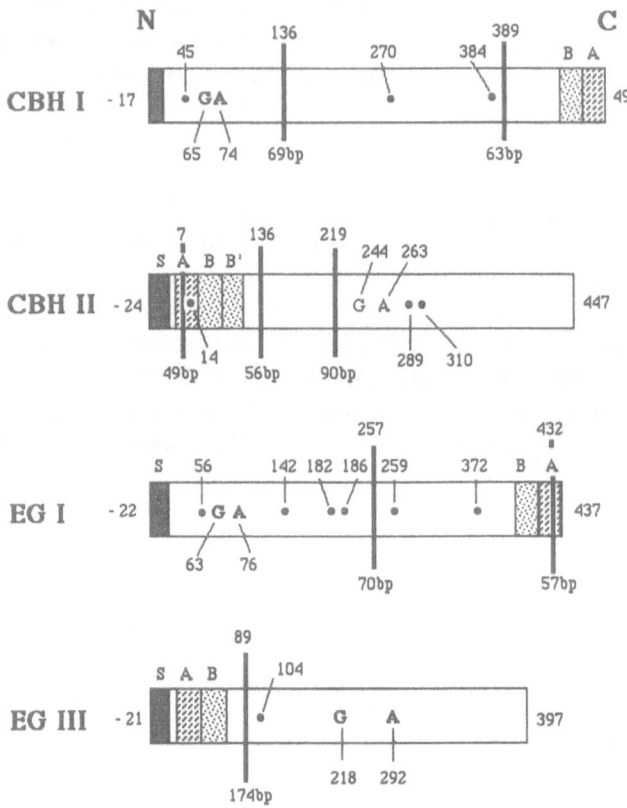

Fig. 3. Diagrammatic representation of the preprotein forms of cellobiohydrolases I and II (CBH I, CBH II) and endoglucanases I and III (EG I and EG III) of *Trichoderma reesei*. The number at the left of each box (i.e. at the amino terminus, N) refers to the number of amino acid residues in the signal sequence (S) while that at the right of each box (i.e. at the carboxy terminus, C) refers to the number of residues in the mature protein. Each mature protein is comprised of a catalytic domain (represented by open boxes) linked via a flexible hinge region, B (dotted boxes) to a tall domain, A (striped boxes). The A regions (which are cysteine- and glycine-rich) and the B regions (which are serine-, threonine- and proline-rich and the site of *O*-glycosylation) are located at the C termini of CBH I and EG I but at (or almost at) the N-termini, and in reverse order, in CBH II and EG III. Each putative *N*-glycosylation site is represented by a small closed circle with the sequence number of the relevant asparagine residue indicated. Putative active site glutamate and aspartate residues are shown as G and A, repectively, and their sequence numbers in the mature protein indicated. Introns in the respective genes are represented by heavy vertical lines. These occur at the position corresponding to the amino acid whose sequence number in the mature protein is given above the line. The number below each such line indicates the length of the intron in base pairs. The figure above was prepared by adaptation of figures in Teeri *et al.* (1987), Knowles *et al.* (1988*a*) and Saloheimo *et al.* (1988), and from data in Shoemaker *et al.* (1983*a*), Penttilä *et al.* (1986) and Chen *et ah* (1987).

native CBH II. Cellobiose, the reaction product, decreased the ability of CBH II or its core protein to adsorb to either Avicel or amorphous cellulose. In the case of CBH I, the core protein had only 35% of the ability of the native protein to bind to Avicel and only 14% of its ability to hydrolyse this substrate. By contrast, the core protein was just as effective as the native enzyme in binding to and hydrolysing amorphous cellulose. Furthermore, cellobiose, a potent inhibitor of the hydrolysis of soluble substrates, did not influence (>5% reduction) the ability of CBH I or its core protein to adsorb to either the crystalline or the amorphous substrate. (Since cellulose must be bound at the active site for hydrolysis of β-glycosidic linkages, a clear distinction is being made here between binding and adsorption.) Tomme *et al.* (1988) considered their findings to indicate that CBH I degrades both crystalline and amorphous regions of cellulose *in vivo* while CBH II is engaged in attacking the amorphous regions. Synergistic interactions between these enzymes and/or their cores were adjudged to be consistent with this hypothesis (see also Section 4.1.4).

4.1.3.3 RECOMBINANT DNA STUDIES

During the time the studies above were underway great strides were also being made in the application of recombinant DNA technology to solving some of the outstanding questions regarding cellulolysis by fungi (Lehtovaara *et al.*, 1986; Knowles *et al.*, 1988a). The *T. reesei* genes encoding CBH I (Shoemaker *et al.*, 1983a; Teeri *et al.*, 1983), CBH II (Chen *et al.*, 1987; Teeri *et al.*, 1987), EG I (formerly EG II) (Penttilä *et al.*, 1986; Van Arsdell *et al.*, 1987) and EG III (formerly II) (Saloheimo *et al.*, 1988) and the corresponding cDNAs have now been cloned and sequenced. Comparison of chromosomal and corresponding cDNA sequences, and of the amino acid sequence derived from the latter with those of the native enzymes, isolated and sequenced by traditional means (Shoemaker *et al.*, 1983b; Bhikhabhai & Pettersson, 1984b; Fägerstam *et al.*, 1984), have been particularly instructive exercises. Each enzyme is secreted as a preprotein with a hydrophobic signal sequence initiated by Met. The signal sequence, which is presumed to direct secretion, is subsequently removed by proteolysis to yield the mature protein with a pyroGlu at the N-terminus. (As discussed in Section 4.1.2.3, the existence of preprotein/mature protein mixtures in culture fluids was thought to contribute to the observed multiplicity of enzyme forms.)

Overall homology between CBH I and EG I is about 45%, although intron positions have not been conserved. CBH II and EG III are unlike one another or either of the other two proteins. By contrast, EG III (from *T. reesei*) shows considerable homology with EG I from *S. commune* (Saloheimo *et al.*, 1988). Also, a gene isolated from *Phanerochaete chrysosporium* codes for a protein showing about 60% homology with CBH I from *T. reesei*, but again neither the number nor the position of introns have been conserved (Sims *et al.*, 1988).

Despite their overall lack of similarity (except that between CBH I and EG I), a short conserved sequence is common to all of the *T. reesei* enzymes.

This polypeptide region, represented as blocks A and B in Fig. 3, is at the N-terminal end of CBH II and EG III but at the C-terminal end of CBH I and EG I. This is the small domain that may be removed from these proteins by treatment with papain (see Section 4.1.3.2). The A region is the best conserved, the overall homology being about 70%. It is approximately 35 residues long and is rich in glycine and cysteine. The number and position of the latter is strictly conserved. Having regard to the earlier work of Bhikhabhai & Pettersson (1984*a*), it was suggested that the A region may fold into a small domain stabilized by disulphide bridging, the glycine residues facilitating close folding (Teeri *et al.*, 1987). The A domain is joined to the rest of the protein via the B region. Homology in this *c.* 33 residue-long region, which is rich in serine, threonine and proline is 50–60%. It is known that the small domains of CBH I and CBH II are rich in carbohydrate (Fägerstam & Pettersson, 1980; Bhikhabhai & Pettersson, 1984*a*; Fägerstam *et al.*, 1984; Van Tilbeurgh *et al.*, 1986; Tomme *et al.*, 1988). Thus, it is concluded that the B region in each of these proteins is the site of *O*-glycosylation, the proline residues facilitating close packing of the glycosylated peptides.

Subtle differences between the substrate specificities of *T. reesei* and other cellulases have been determined by using chromophoric substrates (of low M_r) and unlabelled cellodextrins (Van Tilbeurgh *et al.*, 1982; 1985; Van Tilbeurgh & Claeyssens, 1985; Claeyssens, 1988; Saloheimo *et al.*, 1988). Space does not permit other than noting that specificities overlap and in no case is specificity absolute. However, by judicious choice of substrates, distinction can be made between the various components. For example, cellobiosides and lactosides are hydrolysed by CBH I or EG I but not by CBH II or EG III. This, as Saloheimo *et al.* (1988) point out, is of interest in view of the differential location of the homologous domains (at the C-termini of CBH I and EG I and N-termini of CBH II and EG III). Does it mean that these small domains, although not needed for activity against small molecules as we saw above, can nevertheless modify enzyme specificity? There is a precedent for this possibility. In lactating mammary gland, α-lactalbumin associates with galactosyltransferase (the complex is lactose synthase) to modify the specificity of the latter and ensure that lactose is produced (Lehninger, 1982).

CBH I, CBH II and EG III, as predicted by computer, are similar with respect to the various types of secondary structure they contain (Penttilä *et al.*, 1986; Teeri *et al.*, 1987). In CBH II, for example, β-structures, turns (found mainly at termini) and α-helices account for 34%, 28% and 15% of the total residues, respectively. Small angle X-ray scattering measurements have indicated that CBH I has a 'tadpole-like' shape with an isotropic head (diameter 4·4 nm) and a long tail, the overall length being 18 nm (Schmuck *et al.*, 1986). Tomme *et al.* (1988) subsequently showed the CBH I core protein (see Section 4.1.3.2) to be isotropic and presume it to be the 'tadpole' head. Thus, one may conclude that the small domain that is removed by papain treatment, i.e., the homologous AB regions of these proteins, represents the tail or part thereof. On the basis of the combined structure/function studies discussed above, it was postulated that as cellulases move along the crystalline substrate the A region frees ('unzips') the cellulose chains from the crystal so that the hydrolytic domain (the core protein) can effect hydrolysis (Knowles *et al.*, 1988*a*; see further discussion in Section 5). It has also been suggested that the terminal homologous domains may be removed by proteolysis *in vivo* (i.e. in culture filtrates) when the crystalline cellulose has been degraded to shorter and more accessible cellooligosaccharides (Saloheimo *et al.*, 1988). If this is a common occurrence, it represents another cause of and reason for the observed multiplicity of forms discussed in Section 4.1.2.3.

4.1.4 Synergistic Interactions Between Endo- and Exo-acting Enzymes

Gilligan & Reese (1954) showed that the amount of reducing sugar released from cellulose by the combined fractions of fungal culture filtrate was greater than the sum of the amounts released by the individual fractions. Since that time many investigators, using a variety of fungal preparations, have demonstrated this synergistic interaction between exo- and endo-acting components in the degradation of substrate (Mandels & Reese, 1964; Li *et al.*, 1965; Selby, 1969; Wood, 1969, 1975, 1980; Halliwell & Riaz, 1970; Wood & McCrae, 1972, 1975, 1977, 1978; Eriksson, 1975; Pettersson, 1975; McHale & Coughlan, 1980; Moloney *et al.*, 1985). Cross-synergism between endo- and exo-acting enzymes from filtrates of different aerobic fungi has also been demonstrated several times (Selby, 1969; Wood, 1969, 1975, 1980; Wood & McCrae, 1975, 1977; Coughlan *et al.*, 1987). Synergism between the cellobiohydrolase of *Trichoderma koningii*, an aerobic fungus, and the cellulase of the anaerobic fungus, *Neocallimastix frontalis,* and of the anaerobic bacterium, *Clostridium thermocellum,* has recently been reported (Wood *et al.*, 1988).

For several years endo–exo synergism has been interpreted as indicating that neither type of enzyme, acting alone, can effect extensive hydrolysis of crystalline cellulose. However, in appropriate mixtures of components, the endocellulases effect random scission of glycosidic linkages and so provide chain ends from which the exoglucanases cleave cellobiose. It follows from this theory that chain ends are not available in quantity for cellobiohydrolase action in the absence of endoglucanases. By the same token, because of the high degree of order in crystalline cellulose, bonds cleaved by endoglucanases would reform spontaneously in the absence of cellobiohydrolases. Consistent with the above hypothesis, is the finding by White (1982), using electron microscopy, that hydrolysis of cellulose required the simultaneous presence of both endo- and exo-acting components.

It also follows from the above theory that the degree of synergism should be high with crystalline substrate, low with amorphous cellulose and non-existent with soluble derivatives of cellulose. This was found to be the case (Wood & McCrae, 1979). Henrissat *et al.* (1985) also showed that synergism was not a factor in the hydrolysis of carboxymethyl cellulose. However, synergistic hydrolysis of soluble celluloses by endo- and exo-acting enzymes has since been observed, but it was said to occur only in the early stages of the reaction (Fujii & Shimizu, 1986).

The saga of synergism received an unexpected twist with the demonstration by Fägerstam & Pettersson (1980) that purified CBH I and CBH II from *T. reesei* filtrates cooperated to solubilize cotton. They concluded that this exo–exo synergism reflected the structural asymmetry in the cellulose molecule. Cooperative interaction between the cellobiohydrolases of *T. reesei* has since been confirmed by Henrissat *et al.* (1985) and by Kyriacou *et al.* (1987). The former group found that a 1:4 mixture of CBH I/CBH II is most effective in the hydrolysis of crystalline substrates. CBH I and CBH II from *Penicillium pinophilum* were also found to act synergistically in the hydrolysis of Avicel (Wood, 1985; Wood & McCrae, 1986). In keeping with the suggestion by Fägerstam & Pettersson (1980), Wood (1985) postulated that CBH I attacks only one of the two stereospecifically-different glycosidic linkages in the substrate while CBH II attacks the other (see Fig. 1). Sequential removal of cellobiose from one type of non-reducing chain end by CBH I could expose, on a neighbouring chain, a non-reducing chain end of the other type from which CBH II would remove cellobiose, and vice versa. As will be discussed below, subsequent investigations by Wood and colleagues were to show the observed exo–exo synergism to be artefactual, at least in the case of *P. pinophilum*. However, the theory was attractive in that it may be operative in some cases. At the time it also offered an explanation for the apparent exo–exo synergism and for the fact that the fungus in question produces, and needs to produce, two types of cellobiohydrolase. However, only one type of cellobiohydrolase has been isolated from filtrates of *T. koningii* and *F. solani,* both of which are considered to produce 'complete' systems (Wood & McCrae, 1972, 1977, 1986). One notes that of the two cellobiohydrolases of *P. pinophilum* only CBH II acted synergistically with the CBH of *T. koningii* or *F. solani* to solubilize Avicel (Wood & McCrae, 1986). This would imply that the cellobiohydrolase isolated from *T. koningii* and *F. solani* filtrates is a CBH I-type and that if a second is synthesized by these fungi it would be a CBH II-type.

Kanda *et al.* (1976) reported considerable hydrolysis of crystalline cellulose by the combined actions of two endoglucanases, from *Irpex lacteus,* differing from one another in the manner of their attack on soluble substrates. Similarly, Rabinovitch *et al.* (1986) noted the synergistic hydrolysis of microcrystalline cellulose by two endoglucanases from *T. viride* differing ten-fold in their binding-affinity for substrate. By analogy with the hypothesis above, one might consider that this endo–endo synergism reflects the simultaneous operation of two stereospecifically-different endoglucanases. However, it is the general finding that endoglucanases are inactive against crystalline substrates. An alternative explanation of the observations of Rabinovitch *et al.* (1986) is that the tight-binding endoglucanase in question is identical with CBH II from the same source (Enari & Niku-Paavola, 1987).

Henrissat *et al.* (1985) examined the hydrolysis of a variety of crystalline substrates by purified CBH I, CBH II, EG I and EG II from *T. reesei*, when acting alone and in combination. CBH I readily digested homogenized Avicel and bacterial cellulose but degraded filter paper, Avicel and algal cellulose much more slowly, and CM-cellulose not at all. By contrast, EG I was active against CM-cellulose as expected but was inactive against all crystalline substrates with the exception of homogenized Avicel. CBH I plus EG I acted synergistically to solubilize filter paper, Avicel and homogenized Avicel—a 1:1 mixture of the enzymes being most effective—but did not cooperate in the hydrolysis of algal cellulose or CM-cellulose. Synergistic degradation of Avicel and bacterial cellulose by CBH I plus EG II was also observed—again a 1:1 mix being best. These findings emphasize the fact that the structure of the substrate used is an important factor in saccharification and synergy (see Section 2). No cooperativity was observed with algal cellulose, which has the highest crystallinity index (CI), i.e. approaching 100. However, specific surface area is at least as critical as CI. The relative extent of degradation of the other crystalline celluloses by CBH I plus EG I (or EG II) was roughly proportional to the specific surface areas of the substrates, viz. bacterial cellulose (CI = 76) > homogenized Avicel (CI = 47) > Avicel (CI = 47) > filter paper (CI = 45). The ability of CBH I to hydrolyse soluble $(1,3:1,4)$-β-D-glucan and the failure to hydrolyse CM-cellulose reflect the specificity of the enzyme for unsubstituted segments of the substrate used.

The behaviour of CBH II differs from that of CBH I (Henrissat *et al.*, 1985). CBH II plus EG I, or EG II, catalysed the synergistic degradation of insoluble substrates (e.g. Avicel and bacterial cellulose) but the most effective mixture of CBH II:EG I (or EG II) was 95:1. This is the sort of ratio one would expect of the proposed mechanism of exo–endo synergism defined earlier. A small amount of endo-acting enzyme should provide sufficient chain ends to saturate the exo-enzyme. Clearly, the observed synergism between CBH I and EG I (or EG II), most effective as a 1:1 mixture, does not follow this pattern. More recently, Tomme *et al.* (1988) showed that while the CBH I core enzyme is almost devoid of Avicelase activity, it does act synergistically with CBH II or with the CBH II core enzyme to solubilize Avicel. They concluded that this effect is not related to the hydrolytic activity of CBH I or its core but to the capacity to adsorb to crystalline cellulose and provide new sites for attack by CBH II. This is reminiscent of the endo–exo mechanism of synergism described above.

Henrissat *et al.* (1985) concluded that CBH II is close to being a typical exo-acting enzyme while CBH I is more like an endo-enzyme. For example, CBH II is more specific than CBH I for the penultimate glycosidic linkage of umbelliferyl-cellodextrins (Van Tilbeurgh *et al.*, 1982, 1985; Van Tilbeurgh & Claeyssens, 1985). Moreover, electron microscopic observations show it to catalyse the unidirectional degradation of algal cellulose (Chanzy & Henrissat, 1985). By contrast, CBH I binds randomly to cellulose as would be expected of an endoglucanase (Chanzy *et al.*, 1984) and degrades barley β-glucan in an endo-fashion (Henrissat *et al.*, 1985). Okada & Tanaka (1988) go even further by saying that not only is the predominant Avicel-saccharifying enzyme in *T. reesei* filtrates (i.e. CBH I) not an exo-acting enzyme neither is it a cellobiohydrolase. They claim that it is a 'low endocellulase' active against all cellulosaccharides ranging from cellobiose to cotton. They also claim that cellobiose is the major endproduct in the early stages of reaction with all substrates, but that at high substrate concentration glucose is the major product in the long term.

Having accepted the endo-character of CBH I, Henrissat *et al.* (1985) suggested that it might effect a multiple attack on cellulose in the manner of α-amylase attack on starch. Amylase action may involve initial random adsorption of the enzyme to substrate, cleavage at the point of attack, followed by recurrent removal of maltooligosaccharide without desorption of the enzyme from the substrate. This, as discussed in Section 5 (see also Section 4.1.3.3), would imply that CBH I acts processively as, for example, do DNA polymerases (Lehninger, 1982). Henrissat *et al.* (1985) also say that despite the endo-action of CBH I, the observed synergism cannot be explained by a simple endo–exo cooperation between these enzymes. Taking account of the fact that CBH I exhibits unusually strong and irreversible binding to cellulose (Reese, 1982) and the postulates of Rabinovitch *et al.* (1982) and Ryu *et al.* (1984) they suggest that the CBH I/CBH II interaction may be explained in terms of competitive adsorption that might enhance the turnover of the enzymes. They also say that binary enzyme complex formation at the cellulose surface (in reality this implies a ternary complex between both enzymes and substrate) cannot be ruled out.

If the disparities between the findings of the French (Henrissat and colleagues) and Scottish (Wood & colleagues) groups is confusing, the findings of the Finnish group adds even more confusion. The Finns find that *T. reesei* CBH I, although not a true exo-acting enzyme in all its activities, has more exo-like character than CBH II (Nummi *et al.*, 1983; Niku-Paavola, 1984; Enari & Niku-Paavola, 1987). They also claim that CBH II, while not a true endo-acting enzyme, has considerable endo-like character and is identical to the tight-binding endoglucanase isolated by Rabinovitch *et al.* (1982, 1983, 1986). Kyriacou *et al.* (1987) have also concluded that CBH II is an endo-acting enzyme. These conclusions regarding CBH I and CBH II are essentially the reverse of those arrived at by the French group—yet it does not appear to be a case of simply 'calling the same enzyme by two different names'. The Finns also conclude that hydrolysis of native crystalline cellulose *in vivo* is accomplished by CBH I and CBH II acting in synergy and that EG I and EG II participate in the process only at the stage that soluble cellodextrins become available (Enari & Niku-Paavola, 1987).

More recently, Wood *et al.* (1989) have shown that even electrophoretically homogeneous cellobiohydrolase preparations from *P. pinophilum* and *T. reesei* are contaminated by endoglucanase activity. They also suggest that current concepts of the mechanism of cellulase action may be the result of incomplete resolution of these complexes. The contaminating endoglucanase activity was readily separated from the cellobiohydrolases by affinity chromatography on *p*-aminobenzyl-1-thio-β-D-cellobioside. Contrary to earlier findings with contaminated enzymes, there was no synergistic activity between 'clean' preparations of CBH I and CBH II, nor between either CBH and endoglucanases. Rather, cotton was degraded to a significant degree only when *three* enzymes were present, viz. CBH I, CBH II and some specific endoglucanases. The optimum ratio of the CBH's was 1:1 and only a trace of endoglucanase activity was required to make such a mixture effective. The addition of CBH I or CBH II individually to endoglucanase preparations from other fungi showed little cooperativity. However, a mixture of endoglucanases and *both* cellobiohydrolases was effective.

To some extent the differences between the observations by the various groups above may reflect the use of different fungi, substrates, purification and assay procedures. However, since the issues are of fundamental importance to an understanding of cellulose breakdown, it behoves all investigators to demonstrate unequivocally that their enzyme preparations are uncontaminated.

4.2 Oxidative Components

In their review of cellulose degradation up to the late 1950s, Gascoigne & Gascoigne (1960) state 'many of the older ideas (viz. of the 20s and 30s) on an oxidative mechanism for cellulose breakdown must be discarded and the considerable evidence for a hydrolytic mode of attack must be presented'. The author's conviction may well have reflected popular opinion of the time. Certainly, the emphasis since the 1950s has been on the hydrolytic components. Nevertheless, several reports have clearly shown the involvement of oxidative steps in the degradative process. Brown-rot fungi, as stated earlier, do not produce exocellulases but they do produce endocellulases and significant amounts of peroxide (Highley, 1975*a*, *b*). It was proposed that peroxide in the presence of iron might, by oxidizing the sugar residues, so disrupt the ordered packing of cellulose chains as to allow the endoglucanases to effect hydrolysis (Koenigs, 1975). Others reported that iron must be chelated if the peroxide–iron pretreatment is to be effective (Ljungdahl & Eriksson, 1985). Only then could the endoglucanases operate. A small molecular iron-containing 'microfibril-generating factor' isolated from filtrates of *T. reesei* is said to generate short fibres (but not releasing sugars) from cellulose and to act synergistically with endo- and exoglucanases from the same source in bringing about substantial solubilization of crystalline substrate (Griffin *et al.*, 1984).

Vaheri (1982*a*, *b*, 1983, 1985) reported that the first step in the breakdown of cotton by *T. reesei* is the appearance of short, insoluble fibres. During this stage gluconic and cellobionic acids appear in culture filtrates. Short fibre formation did not occur under strictly anaerobic conditions. He concluded that 'in the initial stages of the breakdown of cotton an oxidative step engenders changes in cellulose that are not reflected in the formation of soluble sugars but that make the substrate more accessible to endoglucanase attack.' One presumes that the immediate products of oxidation would be lactones. Being unstable these would undergo spontaneous hydrolysis to yield the free acids observed. But, the reaction would be speeded up by the presence of lactonases. Indeed, Bruchmann *et al.* (1987) have since isolated lactonase from a commercial (*A. niger*) preparation and deem it to be an integral component of the cellulase system.

Whatever, about the difficulty of accepting the involvement of apparently non-enzymic (e.g. H_2O_2/Fe^{2+}; although one might have expected the operation of a suitable peroxidase) oxidative steps in cellulose breakdown, the involvement of cellobiose:quinone oxidoreductases (CBQs) and cellobiose oxidases/dehydrogenases is well documented. The former is produced by all white-rot fungi when grown on cellulose (Ljungdahl & Eriksson, 1985) and by at least one non-white-rot, i.e. the Ascomycete, *Chaetomium cellulolyticum* (Fähnrich & Irrgang, 1984). The only CBQs to have been characterized to date are those produced by *Phanerochaete chrysosporium* (Westermark & Eriksson, 1974*a, b*, 1975; Morpeth & Jones, 1986). These flavoproteins catalyse the oxidation of cellobiose to cellobionic acid using quinone or phenoxy radicals arising from the breakdown of lignin as acceptor. Thus, they act as a link between lignin and cellulose degradation in the white-rot species (Eriksson, 1978). Moreover, by reducing the laccase-derived phenoxy radical intermediates of lignin degradation they prevent repolymerization (Eriksson, 1988). The CBQs can also use oxygen (yielding superoxide radical and peroxide) and some dyes as electron acceptors (Morpeth & Jones, 1986). In this respect they are akin to the flavin-cytochrome b-containing cellobiose oxidase that is also produced by *Phanerochaete chrysosporium* (Ayers *et al.*, 1978; Morpeth, 1985). This enzyme catalyses the oxidation of cellobiose and cellodextrins to their corresponding -onic acids. One of its functions may be to oxidize the reducing end group formed when a β-glycosidic linkage in cellulose is cleaved by endoglucanase and so to prevent reformation of the linkage (Ljungdahl & Eriksson, 1985). Indeed, such a reaction (with the intermediation of lactonase) could account for the observed appearance of -onic acids during the initial stages of cellulose degradation by *T. reesei* (see above).

Cellobiose oxidases are also considered to be part of the cellulase system of *Myrothecium verrucaria*, *Polyporus adustus* and *T. reesei* since the breakdown of cotton by culture filtrates of these fungi in a nitrogen atmosphere was only half that observed when the reactions were carried out in air or oxygen (Eriksson *et al.*, 1974). By contrast, replacement of air by nitrogen did not affect substrate degradation by filtrates of *F. solani*, *P. funiculosum*, *T. emersonii* or *T. koningii* (Wood & McCrae, 1979; McHale & Coughlan, 1980). It was assumed at the time that oxidation was not involved in cellulose degradation by these organisms. In retrospect, all these experiments showed was that oxygen was not an electron acceptor for any cellobiose-oxidizing enzyme that might be present in these filtrates. However, we have since shown that unconcentrated filtrates of these organisms, and of a variety of other mesophilic and thermophilic fungi, did not effect measurable oxidation of cellobiose using dichlorophenolindophenol as acceptor (P. A. Grogan, T. O. Griffin & M. P. Coughlan, unpublished results). Indeed, of more than twenty species tested, only *Phanerochaete chrysosporium* (*Sporotrichum pulverulentum*), *Myrothecium verrucaria* and *Schizophyllum commune* exhibited cellobiose dehydrogenase activity. Other investigators have shown that cellobiose dehydrogenases are produced by *Monilia* sp. (Dekker, 1980), *Sporotrichum thermophile* (Coudray *et al.*, 1982) and *Sclerotium rolfsii* (Sadana & Patil, 1985). These enzymes cannot transfer electrons to oxygen, quinones, $NAD(P)^+$ or Methylene blue but can utilize dichlorophenolindophenol, ferricyanide, phenol blue and cytochrome c as acceptors. Since cellobiose-oxidizing enzymes are extracellular, one wonders what electron acceptor (other than quinone or oxygen in the appropriate cases) is used by these enzymes *in vivo*. Perhaps, iron or other transition metals present in media or in the environment of the fungi may participate.

It is claimed that the cellobiose dehydrogenase of *S. rolfsii* is devoid of flavin or haem (Lachke & Deshpande, 1988). This is in direct contrast to reports on those enzymes that have been characterized to date (see above) and clearly needs to be confirmed. The authors also state that it does not act cooperatively with the purified hydrolytic components in the degradation of pure cellulose but that it may have an important rôle to play in lignin and cellulose degradation, i.e. when the organism is growing on lignocellulosic substrates. This would seem to imply that *S. rolfsii* degrades lignin in addition to cellulose. It is said 'to cause white-rot of onions and garlic and to be an omnivorous and destructive parasite of wood' (Alexopoulos, 1962).

5 CELLULOSE DEGRADATION—THE OVERALL MECHANISM

What then is the overall mechanism by which fungi effect the degradation of cellulose? In an earlier attempt to reconcile disparate reports in the literature the scheme outlined in Fig. 4 was proposed (Coughlan & Ljungdahl, 1988). It was intended that this scheme be viewed, not as a series of discrete events, but as a continuum in the conversion of the crystalline polymer to smaller molecular weight soluble products. In recognition of the original suggestion by Reese *et al.* (1950), the term 'amorphogenesis' was coined as a 'collective' description of the process(es) by which the recalcitrant substrate is rendered susceptible to the action (or further action) of hydrolytic enzymes (Coughlan, 1985). Whether this process precedes, or is brought about by, short fibre formation, segmentation or destratification is a matter for speculation, as is the extent of the involvement of hydrolases, or oxidative reactions, or both in these early stages.

Vaheri (1982*a, b,* 1983, 1985) showed short fibre formation, with the concomitant production of -onic acids, by filtrates of *T. reesei* to be an

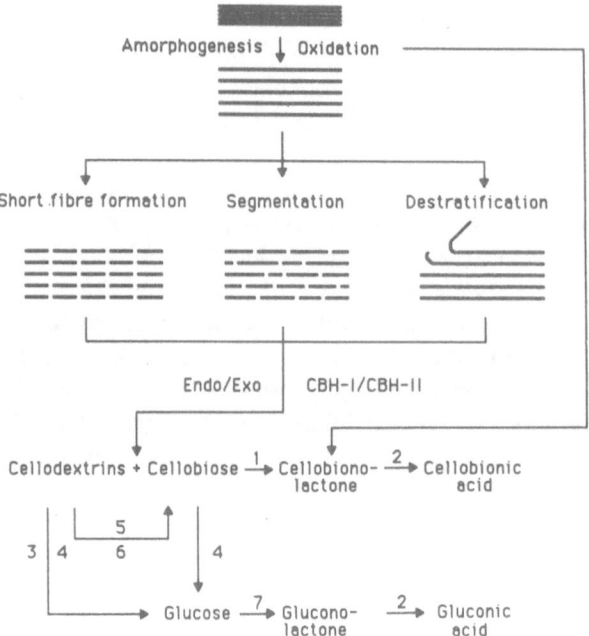

Fig. 4. Proposed mechanism of cellulose degradation by fungi (from Coughlan & Ljungdahl (1988)). The numbered enzymes are: 1, cellobiose oxidase/dehydrogenase; 2, lactonase; exoglucohydrolase; 4, β-glucosidase; 5, endoglucanase; 6, exo-cellobiohydrolase; 7, glucose oxidase.

oxygen-dependent process (see Section 4.2). One presumes that short fibre formation begins at the outermost fibrils and that these are freed for solubilization before inner fibrils come under attack. Short fibre formation, without the appearance of soluble sugars, is also effected by components of the systems produced by *Myrothecium verrucaria*, *T. koningii* (Halliwell & Riaz, 1970; Halliwell, 1975; Halliwell *et al.*, 1985) and *T. reesei* (Pettersson, 1975). However, by contrast with Vaheri's findings, the components involved were reported to be endoglucanases (EG). Indeed, Pettersson (1975) suggested that one of the *T. reesei* endoglucanases was particularly suited to this action because of its small size (M_r, 12 000) and high diffusibility. By way of further contrast, the short fibre-forming ability of *T. reesei* was attributed to CBH I by Enari & Niku-Paavola (1987) and to CBH II by Kyriacou *et al.* (1987). However, both groups claim that the enzyme in question is an endo-acting enzyme or has endo-like characteristics. Had the effects of anaerobiosis on the proposed EG- or CBH I-catalysed reactions been studied, it might have been possible to determine whether both oxidative and hydrolytic reactions are involved in this process. If hydrolytic reactions are involved, are they brought about by endo- or exo-acting enzymes? Unless it can be demonstrated unequivocally that the preparations used in the above studies are uncontaminated (see Section 4.1.4), one must accept the reports with caution.

Segmentation, as we described it, would involve the simultaneous attack on many chains in the cellulose fibre by randomly-acting enzymes (Coughlan & Ljungdahl, 1988). Although not an appealing mechanism *per se*, it was included as a possibility because of the observation (using electron microscopy) by White (1982) that *T. reesei* cellulases, when mixed with cellulose, quickly 'cover' the substrate. Using the same technique, Chanzy *et al.* (1984) showed that colloidal gold-labelled CBH I bound randomly along the cellulose chains. Obviously, for segmentation to work such random binding would have to be productive, i.e. it would have to be accompanied by hydrolysis of glycosidic linkages. Indeed, because of the highly-ordered structure of cellulose, further action by CBH I, or by CBH I in cooperation with other enzymes, would be necessary to prevent spontaneous reformation of these linkages (Eriksson & Wood, 1985). Chanzy *et al.* (1983) did show that CBH I, unaided by other enzymes, solubilized cellulose, but at rates substantially less than that given by the complete *T. reesei* system (see below). Once again, the difficulty of obtaining uncontaminated enzyme preparations requires that caution be exercised in interpreting experimental findings (see Section 4.1.4). The authors in question are clearly aware of this problem since they state that the CBH I preparations used exhibited negligible ($3 \times 10^5 \, \mu$eq glycosidic bonds split per second per mg protein) CMCase activity.

A third possible mechanism operating in the early stages of cellulose conversion was termed 'destratification, i.e. that degradation could be initiated by stripping the cellulose fibre, stratum by stratum, prior to (or concomitant with) solubilization of the chains so released (Coughlan & Ljungdahl, 1988). Knowles *et al.* (1988*a*) have proposed that as cellulases move along the crystalline substrate the terminal domain 'unzips' cellulose chains from the crystal so that the hydrolytic domain (the core protein) can effect hydrolysis (see Section 4.1.3.3). The ability to move along the cellulose fibre would of course imply that these enzymes act processively (i.e. remain associated with the substrate) rather than distributively (i.e. associate, do something, dissociate and subsequently reassociate with substrate). Experiments aimed at proving or disproving such a mechanism have not been reported, but Henrissat *et al.* (1985) suggest that, by analogy with α-amylase CBH I may act in this fashion (see Section 4.1.4). In any event, the proposal is appealing in that it provides another reason for the two domain structure of these enzymes (see Section 4.1.3.3).

What evidence is there to support destratification? White (1982), using electron microscopic techniques, found that the first morphological changes in the structure of *Acetobacter xylinum* cellulose when incubated with *T. reesei* cellulases become apparent when the ribbons 'splay' into smaller bundles of microfibrils. Analogous studies have been carried out by Chanzy & Henrissat (1985) with highly crystalline *Valonia macrophysa* cellulose. They found that the fibrils were eroded only at one end by CBH II. The subsequent addition of EG II, which attacked amorphous regions along the length of the fibrils, allowed CBH II to effect 'surface stripping'. The same investigators also showed that CBH I acting on its own, broke down the algal cellulose by 'splitting lateral hydrogen bonds thereby yielding narrower subfibrillated crystalline elements that retained their original length' (Chanzy *et al.*, 1983). More recently, Hayashida & Mo (1986) reported the isolation of an electrophoretically-homogeneous endoglucanase from *Humicola grisea* var. *thermoidea*. This enzyme exhibited 'intensive' filter paper-disintegrating activity but did not release reducing sugar. Scanning electron microscopic studies showed that it 'disintegrated Avicel fibrils, layer by layer from the surface, yielding thin sections with exposed chain ends.

It should be emphasized that short fibre formation, segmentation and destratification are not necessarily mutually exclusive processes. For example, if the fibrils removed in destratification are cleaved at random points short fibres would be produced. Following, or during, the above processes the various hydrolytic components of complete systems interact to effect further degradation of the cellulose chains. It is universally agreed that much of this interaction is synergistic, although, as discussed in Section 4.1.4, details of the mechanism whereby synergism occurs are in dispute. The products at this stage of degradation are mainly cellodextrins and cellobiose. According to the Finnish group endoglucanases become operative at this stage, but again this is disputed (see Section 4.1.4). Exoglucohydrolases, if present, would remove glucose residues from the non-reducing ends of the cellodextrins as would β-glucosidases, the former showing preference for the longer oligosaccharides and the latter for the shorter substrates. The β-glucosidases, in converting the cellobiose to glucose, not only cater for further metabolic requirements but also relieve the inhibition of cellulase activity by cellobiose. Cellobiose-oxidizing enzymes and lactonases would also have a role to play at this stage, and perhaps earlier stages (Fig. 4).

6 CONCLUSION

Much of our understanding of cellulase systems and of the rôles of the individual components therein has been and will continue to be based on results obtained using enzymes prepared by conventional biochemical procedures. Progress has been hampered by the difficulty in obtaining individual enzymes free from contamination by related activities. Affinity chromatographic procedures will help in this regard. The application of recombinant DNA methodology, which has already provided answers to many of the outstanding questions, may be of even greater use. Individual enzymes, absolutely free from contamination, could be obtained by cloning and expression of the appropriate genes. Cellulase-negative mutants of the same or related strains might be the best recipients. In this way, the expressed gene product should be glycosylated and secreted in the normal way. Reassessment of the nomenclature pertaining to cellulolytic enzymes is urgently required.

ACKNOWLEDGEMENTS

Some of the work carried out in the author's laboratory was funded by EC Contract Nos. RUW/035/EIR, RNW/122/IRL and MA1D/0017/IRL. It is a pleasure to acknowledge the assistance of Drs Patrick McArdle and Angela Savage (Department of Chemistry, University College, Galway) in the preparation of Fig. 2.

REFERENCES

Akin, D. E. (1987). *Animal Feed Science and Technology,* **16,** 273.

Akin, D. E. & Barton, F. E. II. (1983). In *Wood and Agricultural Residues. Research on Use for Feed, Fuels and Chemicals,* ed. E. J. Soltes. Academic Press, New York, p. 33.

Akin, D. E., Gordon, G. L. R. & Hogan, J. P. (1983). *Applied and Environmental Microbiology,* **46,** 738.

Alexopoulos, C. J. (1962). *Introductory Mycology,* 2nd. edn. Wiley, New York.

Alluralde, J. L. & Ellenreider, G. (1984). *Enzyme and Microbial Technology,* **6,** 467.

Ander, P. & Eriksson, K.-E. (1978). *Progress in Industrial Microbiology,* **14,** 1.

Anon. (1981). *Enzymatic Hydrolysis of Cellulose to Glucose. A Report on the Natick Programme.* US Army Natick Research & Development Command, Natick, Mass., USA.

Anon. (1982). *Cellulose Enzymolysis.* 1978–December 1982. (Citations from the Institute of Paper Chemistry Data Base). Available from NTIS no. PB82-855874.

Atalla, R. H. (1983). In *Wood and Agricultural Residues. Research on Use for Feed, Fuels and Chemicals,* ed. E. J. Soltes. Academic Press, New York, p. 59.

Atalla, R. H. (1984). In *Structure, Function and Biosynthesis of Plant Cell Walls,* eds W. M. Dugger & S. Bartnicki-Garcia. Waverly Press, Baltimore, MD, p. 381.

Atalla, R. H. & VanderHart, D. L. (1984). *Science,* **223,** 283.

Aubert, J.-P., Béguin, P. & Millet, J. (eds) (1988). *Biochemistry and Genetics of Cellulose Degradation.* Academic Press, London.

Aunstrup, K. (1978). *Annual Reports of Fermentation Processes,* **2,** 125.

Avgerinos, G. & Wang, D. I. C. (1980). *Annual Reports of Fermentation Processes,* **4,** 165.

Ayers, A. R., Ayers, S. B. & Eriksson, K.-E. (1978). *European Journal of Biochemistry,* **90,** 171.

Bassham, J. A. (1975). *Biotechnology and Bioengineering Symposium,* **5,** 9.

Bauchop, T. (1981). *Agriculture and Environment,* **6,** 339.

Bauchop, T. & Mountfort, D. O. (1981). *Applied and Environmental Microbiology,* **42,** 1103.

Béguin, P., Grépinet, O., Tailliez, P., Joliff, G., Girard, H., Millet, J. & Aubert, J.-P. (1987). Conversion of cellulose into ethanol by *Clostridium thermocellum*: Genetic engineering of cellulases and improvement of ethanol production. In *Proc. Int. Conf. Biomass for Energy and Industry,* ed. G. Grassi, B. Delmoa, J.-F. Molle & H. Zibetta, Elsevier Applied Science, London, p. 346.

Béguin, P., Millet, J., Grépinet, O., Navarro, A., Juy, M., Amit, A., Poljak, R. & Aubert, J.-P. (1988). The cel (cellulose degradation) genes of *Clostridium thermocellum.* In *Biochemistry and Genetics of Cellulose Degradation,* eds J.-P. Aubert, P. Béguin & J. Millet, Academic Press, London p. 267.

Beldman, G., Rombouts, F. M., Voragen, A. G. J. & Pilnik, W. (1984). *Enzyme and Microbial Technology,* **6,** 503.

Beldman, G., Voragen, A. G. J., Rombouts, F. M., Searle-van Leeuwen, M. F. & Pilnik, W. (1987). *Biotechnology and Bioengineering*, **30**, 251.

Berghem, L. E. R. & Pettersson, L. G. (1973). *European Journal of Biochemistry*, **37**, 21.

Berghem, L. E. R., Pettersson, L. G. & Axiö-Frederiksson, U.-B. (1975). *European Journal of Biochemistry*, **53**, 55.

Bhat, K. M. & Wood, T. M. (1989). *Biochemical Society Transactions*, **17**, 104.

Bhikhabhai, R. & Pettersson, G. (1984a). *Biochemical Journal*, **222**, 729.

Bhikhabhai, R. & Pettersson, G. (1984b). *FEBS Letters*, **167**, 301.

Blackwell, J. (1982). In *Cellulose and Other Natural Polymer Systems: Biosynthesis, Structure and Degradation*, ed. R. M. Brown, Jr, Plenum Press, New York, p. 403.

Blanchette, R. A. (1980a). *Journal of Forestry Research*, **78**, 734.

Blanchette, R. A. (1980b). *Canadian Journal of Botany*, **58**, 1496.

Blanchette, R. A. (1982). *Canadian Journal of Forestry Research*, **12**, 304.

Blanchette, R. A., Shaw, C. G. & Cohen, A. L. (1978). *Scanning Electron Microscopy*, **11**, 61.

Bock, K. & Sigurskjold, B. W. (1989). *European Journal of Biochemistry*, **178**, 711.

Brown, D. E. (1983). *Philosophical Transactions of the Royal Society of London*, **B300**, 305.

Bruchmann, E.-E., Schach, H. & Graf, H. (1987). *Biotechnology and Applied Biochemistry*, **9**, 146.

Chanzy, H. & Henrissat, B. (1985). *FEBS Letters*, **184**, 285.

Chanzy, H., Dube, M., Marchessault, R. H. & Revol, J. F. (1979). *Biopolymers*, **18**, 887.

Chanzy, H., Henrissat, B., Vuong, R. & Schülein, M. (1983). *FEBS Letters*, **153**, 113.

Chanzy, H., Henrissat, B. & Vuong, R. (1984). *FEBS Letters*, **172**, 193.

Chartier, P. (1981). In *Energy from Biomass*, 1st EC Conference, eds W. Palz, P. Chartier & D. O. Hall. Elsevier Applied Science, London, p. 22.

Chen, C. M., Gritzali, M. & Stafford, D. W. (1987). *Biotechnology*, **5**, 274.

Chu, S. C. & Jeffrey, G. A. (1968). *Acta Crystallographica*, **B24**, 830.

Claeyssens, M. (1988). In *Biochemistry and Genetics of Cellulose Degradation*, eds J.-P. Aubert, P. Béguin & J. Millet. Academic Press, London, p. 393.

Claeyssens, M. & Tomme, P. (1989). In *Enzyme Systems for Lignocellulose Degradation*, ed. M. P. Coughlan. Elsevier Applied Science, London, p. 37.

Clarke, A. J. (1987). *Biochimica et Biophysica Acta*, **912**, 424.

Clarke, A. J. & Yaguchi, M. (1985). *European Journal of Biochemistry*, **149**, 233.

Coudray, M. R., Canevascini, G. & Meier, H. (1982). *Biochemical Journal*, **203**, 277.

Coughlan, M. P. (1985). In *Biotechnology and Genetic Engineering Reviews*. Vol. 3, ed. G. E. Russell. Interscience, Newcastle upon-Tyne, UK, p. 37.

Coughlan, M. P. (1989). Mechanism of cellulose degradation by fungi and bacteria. In *Proceedings OECD Workshop Cell Wall: Structure, Degradation and Utilization*, ed. J. Delort-Laval. *Journal of Animal Feed Science*. (In press.)

Coughlan, M. P. & Folan, M. A. (1979). *Comparative Biochemistry and Physiology*, **10**, 103.

Coughlan, M. P. & Ljungdahl, L. G. (1988). In *Biochemistry and Genetics of Cellulose Degradation*, eds J.-P. Aubert, P. Béguin & J. Millet. Academic Press, London, p. 11.

Coughlan, M. P., Mehra, R. K., Considine, P. J., O'Rorke, A. & Puls, J. (1985). *Biotechnology and Bioengineering Symposium*, **15**, 447.

Coughlan, M. P., Moloney, A. P., McCrae, S. I. & Wood, T. M. (1987). *Biochemical Society Transactions*, **15**, 263.

Cowling, E. B. (1961). *US Dept Agric. Tech. Bull. 1258*, p. 1.

Cowling, E. B. (1975). *Biotechnology and Bioengineering Symposium*, **5**, 163.

Crawford, R. L. (1981). *Lignin Biodegradation and Transformation*. Wiley, New York.

Dekker, R. F. H. (1980). *Journal of General Microbiology*, **120**, 309.

De Montelambert, M. R. (1983). In *Energy from Biomass*, 2nd EC Conference, eds A. Strub, B. Chartier & G. Schleser. Elsevier Applied Science, London, p. 82.

Deshpande, M. V. & Eriksson, K.-E. (1984). *Enzyme and Microbial Technology*, **6**, 338.

Deshpande, M. V., Eriksson, K.-E. & Pettersson, B. (1978). *European Journal of Biochemistry*, **90**, 191.

Deshpande, V., Rao, M., Keskar, S. & Mishra, C. (1984). *Enzyme and Microbial Technology*, **6**, 371.

Dunlap, C. E., Thomson, J. & Chiang, L. C. (1976). *Association of Industrial and Chemical Engineering Symposium Series*, **72**, 58.

Dunne, C. P. (1982). *Enzyme Engineering*, **6**, 355.

Enari, T.-M. (1983). In *Microbial Enzymes and Biotechnology*, ed. W. M. Fogarty. Elsevier Applied Science Publishers, London, p. 183.

Enari, T.-M. & Niku-Paavola, M.-L. (1987). *CRC Critical Reviews in Biotechnology*, **5**, 67.

Eriksson, K.-E. (1975). In *Symposium on Enzymatic Hydrolysis of Cellulose*, eds M. Bailey, T.-M. Enari & M. Linko. SITRA, Helsinki, p. 263.

Eriksson, K.-E. (1978). *Biotechnology and Bioengineering*, **20**, 317.

Eriksson, K.-E. (1981). In *Trends in the Biology of Fermentations for Fuels and Chemicals*, ed. A. Hollaender. Plenum Press, New York, p. 19.

Eriksson, K.-E. (1988). In *Biochemistry and Genetics of Cellulose Degradation*, eds J.-P. Aubert, P. Béguin & J. Millet. Academic Press, London, p. 285.

Eriksson, K.-E. & Pettersson, B. (1975a). *European Journal of Biochemistry*, **51**, 193.

Eriksson, K.-E. & Pettersson, B. (1975b). *European Journal of Biochemistry*, **51**, 213.

Eriksson, K.-E. & Pettersson, B. (1982). *European Journal of Biochemistry*, **124**, 635.

Eriksson, K.-E. & Wood, T. M. (1985). In *Biosynthesis and Biodegradation of Wood Components*, ed. T. Higuchi. Academic Press, New York, p. 469.

Eriksson, K.-E., Pettersson, B. & Westermark, U. (1974). *FEBS Letters*, **49**, 282.

Eriksson, K.-E., Grünewald, A., Nilsson, T. & Vallander, L. (1980). *Holzforschung*, **34**, 207.

Eveleigh, D. E. (1982). In *Biomass Utilization*, ed. W. A. Côté. NATO Advanced Study Institute, Series A: Life Sciences. Plenum Press, New York, Vol. 67, p. 365.

Eveleigh, D. E. (1984). In *Current Perspectives in Microbial Ecology*, eds M. J. Klug & C. A. Reddy. Amer. Soc. Microbiol., Washington, p. 553.

Eveleigh, D. E. (1987). *Philosophical Transactions of the Royal Society of London*, **A321**, 435.

Fan, L.-T., Gharpuray, M. M. & Lee, Y. H. (1987). *Cellulose Hydrolysis*, Biotechnology Monographs, Vol. 3, Springer-Verlag, Berlin.

Fägerstam, L. G. & Pettersson, L. G. (1979). *FEBS Letters*, **98**, 363.

Fägerstam, L. G. & Pettersson, L. G. (1980). *FEBS Letters*, **119**, 97.

Fägerstam, L. G., Pettersson, L. G. & Engström, J. A. (1984). *FEBS Letters*, **167**, 309.

Fähnrich, P. & Irrgang, K. (1984). *Biotechnology Letters*, **6**, 251.

Feldman, K. A., Lovett, J. S. & Tsao, G. T. (1988). *Enzyme and Microbial Technology*, **10**, 262.

Fujii, M. & Shimizu, M. (1986). *Biotechnology and Bioengineering*, **28**, 878.

Gaden, E. L., Mandels, M. & Reese, E. T. (eds) (1976). *Enzymatic Conversion of Cellulosic Materials: Technology and Applications, Biotechnology and Bioengineering Symposium*, **6**.

Gascoigne, J. A. & Gascoigne, M. M. (1960). *Biological Degradation of Cellulose*. Butterworth, London.

Ghose, T. K. (ed.) (1978). *Bioconversion of Cellulosic Substances into Energy, Chemicals and Microbial Protein*. IIT, New Delhi.

Gilbert, I. G. & Tsao, G. T. (1983). *Annual Report of Fermentation Processes*, **7**, 232.

Gilligan, W. & Reese, E. T. (1954). *Canadian Journal of Microbiology*, **1**, 90.
Gong, C. S. & Tsao, G. T. (1979). *Annual Report of Fermentation Processes*, **3**, 111.
Grethlein, H. E. (1985). *Biotechnology*, **3**, 155.
Griffin, H., Dintzis, F. R., Krull, L. & Baker, F. L. (1984). *Biotechnology and Bioengineering*, **26**, 296.
Gritzali, M. & Brown, R. D. (1979). The cellulase system of *Trichoderma*. Relationships between purified extracellular systems from induced or cellulose-grown cells. In *Hydrolysis of Cellulose: Mechanisms of Enzymatic and Acid Catalysis*, eds R. D. Brown, Jr & L. Jurasek. *Advances in Chemistry Series*, Vol. 18. American Chemical Society, Washington, DC, p. 237.
Grohman, K. & Villet, R. (1983). In *Bioconversion Systems*, ed. D. L. Wise. CRC Press, Boca Raton, FL, p. 1.
Gum, E. K. & Brown, R. D., Jr (1976). *Biochimica et Biophysica Acta*, **446**, 371.
Gum, E. K. & Brown, R. D., Jr (1977). *Biochimica et Biophysica Acta*, **492**, 225.
Hajny, G. J. & Reese, E. T. (eds) (1969). Cellulases and their applications. *Advances in Chemistry Series*, Vol. 95. American Chemical Society, Washington, DC.
Håkansson, O., Fägerstam, L. G., Pettersson, L. G. & Andersson, L. (1979). *Biochemical Journal*, **179**, 141.
Hall, D. O. (1983). In *Energy from Biomass*, 2nd EC Conference, eds A. Strub, B. Chartier & G. Schleser. Elsevier Applied Science, London, p. 43.
Halliwell, G. (1975). In *Symposium on Enzymatic Hydrolysis of Cellulose*, eds M. Bailey, T.-M. Enari & M. Linko. SITRA, Helsinki, p. 319.
Halliwell, G. & Griffin, M. (1973). *Biochemical Journal*, **135**, 587.
Halliwell, G. & Riaz, M. (1970). *Biochemical Journal*, **116**, 35.
Halliwell, G. & Vincent, R. (1981). *Biochemical Journal*, **199**, 409.
Halliwell, G., Wahab, M. N. B. A. & Patel, A. H. (1985). *Journal of Applied Biochemistry*, **7**, 43.
Ham, J. T. & Williams, D. G. (1970). *Acta Crystallographica*, **B26**, 1373.
Hanstein, E. G. & Whittaker, D. R. (1963). *Canadian Journal of Biochemistry and Physiology*, **41**, 707.
Hayashida, S. & Mo, K. (1986). *Applied and Environmental Microbiology*, **51**, 1041.
Hayashida, S. & Yoshioka, H. (1980). *Agricultural and Biological Chemistry*, **44**, 481.
Heath, I. B., Bauchop, T. & Skipp, R. A. (1983). *Canadian Journal of Botany*, **61**, 295.
Henrissat, B., Driguez, H., Viet, C. & Schülein, M. (1985). *Biotechnology*, **3**, 722.
Highley, T. L. (1975a). *Forests Product Journal*, **25**, 38.
Highley, T. L. (1975b). *Wood Fibre*, **6**, 275.
Hurst, P. L., Nielsen, J., Sullivan, P. A. & Shepherd, M. G. (1977a). *Biochemical Journal*, **165**, 33.
Hurst, P. L., Sullivan, P. A. & Shepherd, M. G. (1977b). *Biochemical Journal*, **167**, 549.
Hurst, P. L., Sullivan, P. A. & Shepherd, M. G. (1978). *Biochemical Journal*, **169**, 389.
Kanda, T., Wakabayashi, K. & Nisizawa, K. (1976). *Journal of Biochemistry*, **79**, 997.
Kersten, P. J. & Kirk, T. K. (1987). *Journal of Bacteriology*, **169**, 2195.
Kirk, T. K. & Highley, T. L. (1973). *Phytopathology*, **63**, 1338.
Knowles, J., Teeri, T., Lehtovaara, P., Penttilä, M. & Saloheimo, M. (1988a). In *Biochemistry and Genetics of Cellulose Degradation*, eds J.-P. Aubert, P. Béguin & J. Millet. Academic Press, London, p. 153.
Knowles, J. K. C., Lehtovaara, P., Murray, M. & Sinnott, M. L. (1988b). *Journal of the Chemical Society—Chemical Communications*, 1401.
Koenigs, J. W. (1975). *Biotechnology and Bioengineering Symposium*, **5**, 151.
Kubicek, C. P. (1981). *European Journal of Applied Microbiology and Biotechnology*, **13**, 226.

Kubicek, C. P., Panda, T., Schreferl-Kunar, G., Gruber, F. & Messner, R. (1987). *Canadian Journal of Microbiology*, **33**, 698.

Kyriacou, A., Neufeld, R. J. & MacKenzie, C. R. (1987). *Enzyme and Microbial Technology*, **9**, 25.

Kyriacou, A., Neufeld, R. J. & McKenzie, C. R. (1988). *Enzyme and Microbial Technology*, **10**, 675.

Labudova, I. & Farkas, V. (1983). *Biochimica et Biophysica Acta*, **744**, 135.

Lachke, A. H. & Deshpande, M. V. (1988). *FEMS Microbiological Reviews*, **54**, 177.

Ladisch, M. R. & Tsao, G. T. (1986). *Enzyme and Microbial Technology*, **8**, 66.

Ladisch, M. R., Lin, K. W., Voloch, M. & Tsao, G. T. (1983). *Enzyme and Microbial Technology*, **5**, 82.

Lehninger, A. L. (1982). *Principles of Biochemistry*. Worth, New York.

Lehtovaara, P., Knowles, J., André, L., Penttilä, M., Teeri, T., Salovuori, I., Niku-Paavola, M. J. & Enari, T.-M. (1986). In *Biotechnology in the Pulp and Paper Industry*, Third International Conference, eds K.-E. Eriksson & P. Ander. STFI, Stockholm, p. 90.

Li, H., Flora, R. M. & King, K. W. (1965). *Archives of Biochemistry and Biophysics*, **111**, 439.

Liese, W. (1970). *Annual Review of Phytopathology*, **8**, 231.

Ljungdahl, L. G. & Eriksson, K.-E. (1985). In *Advances in Microbial Ecology*, Vol. 5, ed K. C. Marshall. Plenum Press, New York, p. 237.

Lloyd, A. (1984). *New Scientist*, no. 1047, 21.

Lützen, N. W., Nielsen, M. H., Oxenboell, K. M., Schülein, M. & Stentebjerg-Olesen, B. (1983). *Philosophical Transactions of the Royal Society of London*, **B300**, 283.

Mandels, M. (1975). *Biotechnology and Bioengineering Symposium*, **5**, 81.

Mandels, M. (1985). *Biochemical Society Transactions*, **13**, 414.

Mandels, M. & Reese, E. T. (1964). *Developments in Industrial Microbiology*, **5**, 5.

Marchessault, R. H. & St-Pierre, J. (1980). In *Future Sources of Organic Raw Material—Chemrawn*, eds. L. E. St-Pierre & G. R. Brown. Pergamon Press, Oxford, p. 613.

Marchessault, R. H., Malhatra, S. L., Jones, A. Y. & Perovic, A. (1983). In *Wood and Agricultural Residues. Research on Use for Feed, Fuels and Chemicals*, ed. E. J. Soltes. Academic Press, New York, p. 401.

Marsden, W. L. & Gray, P. P. (1986). *CRC Critical Reviews in Biotechnology*, **3**, 235.

Marx-Figini, M. (1982). In *Cellulose and Other Natural Polymer Systems: Biosynthesis, Structure and Degradation*, ed. R. M. Brown, Jr. Plenum Press, New York, p. 243.

Marx-Figini, M. & Schultz, G. V. (1966). *Naturwissenschaften*, **53**, 466.

McHale, A. & Coughlan, M. P. (1980). *FEBS Letters*, **117**, 319.

McHale, A. & Coughlan, M. P. (1981*a*). *Biochimica et Biophysica Acta*, **662**, 152.

McHale, A. & Coughlan, M. P. (1981*b*). *Biochimica et Biophysica Acta*, **662**, 145.

McHale, A. & Coughlan, M. P. (1982). *Journal of General Microbiology*, **128**, 2327.

Merivuori, H., Siegler, K. M., Sands, J. A. & Montenecourt, B. S. (1985). *Biochemical Society Transactions*, **13**, 411.

Mishra, C., Vaidya, M., Rao, M. & Deshpande, V. (1983). *Enzyme and Microbial Technology*, **5**, 430.

Moloney, A. P., McCrae, S. I., Wood, T. M. & Coughlan, M. P. (1985). *Biochemical Journal*, **225**, 365.

Morpeth, F. F. (1985). *Biochemical Journal*, **228**, 557.

Morpeth, F. F. & Jones, G. D. (1986). *Biochemical Journal*, **236**, 221.

Murphy-Holland, K. & Eveleigh, D. E. (1985). In *Abstracts 85th American Society for Microbiology*, American Society of Microbiology, Washington, DC, K-130.

Nakayama, M. J., Tomita, Y., Suzuki, H. & Nisizawa, K. (1976). *Journal of Biochemistry*, **79**, 955.

Niku-Paavola, M.-L. (1984). In *Soviet-Finnish Seminar on Bioconversion of Plant Raw Materials by Microorganisms*. USSR Academy of Sciences, Puschino, p. 26.

Niku-Paavola, M.-L., Lappalainen, A., Enari, T.-M. & Nummi, M. (1985). *Biochemical Journal*, **231**, 75.

Nummi, M., Niku-Paavola, M.-L., Lappalainen, A., Enari, T.-M. & Raunio, V. (1983). *Biochemical Journal*, **215**, 677.

Okada, G. & Tanaka, Y. (1988). *Agricultural and Biological Chemistry*, **52**, 2981.

Orpin, C. G. (1975). *Journal of General Microbiology*, **91**, 249.

Orpin, C. G. (1977a). *Journal of General Microbiology*, **99**, 107.

Orpin, C. G. (1977b). *Journal of General Microbiology*, **101**, 181.

Orpin, C. G. (1977c). *Journal of General Microbiology*, **98**, 423.

Orpin, C. G. (1981). *Journal of General Microbiology*, **123**, 287.

Orpin, C. G. & Joblin, K. N. (1988). In *The Rumen Microbial Ecosystem*, ed. P. N. Hobson. Elsevier Applied Science, London, p. 129.

Orpin, C. G. & Letcher, A. J. (1979). *Current Microbiology*, **3**, 121.

Otjen, L. & Blanchette, R. A. (1982). *Canadian Journal of Botany*, **60**, 2770.

Paice, M. G., Desrochers, M., Jurasek, L., Roy, C., Rollin, C. F., DeMiguel, E. & Yaguchi, M. (1984). *Biotechnology*, **2**, 535.

Pearce, P. D. & Bauchop, T. (1985). *Applied and Environmental Microbiology*, **49**, 1265.

Penttilä, M., Lehtovaara, P., Nevalainen, H., Bhikhabhai, R. & Knowles, J. (1986). *Gene*, **45**, 253.

Pettersson, G. (1969). *Archives of Biochemistry and Biophysics*, **130**, 286.

Pettersson, L. G. (1975). In *Symposium on Enzymatic Hydrolysis of Cellulose*, eds M. Bailey, T.-M. Enari & M. Linko. SITRA, Helsinki, p. 255.

Phillips, D. C. (1968). In *Proceedings Seventh International Congress of Biochemistry*, International Union of Biochemistry, Vol. 36, p. 63.

Pourquié, J., Warzywoda, F., Chevron, F., Théry, M., Lonchamp, D. & Vandecasteele, J. P. (1988). In *Biochemistry and Genetics of Cellulose Degradation*, eds J.-P. Aubert, P. Béguin & J. Millet. Academic Press, London, p. 72.

Puls, J., Poutanen, K., Körner, H.-U. & Viikari, L. (1985). *Applied Microbiology and Biotechnology*, **22**, 416.

Puri, V. P. (1984). *Biotechnology and Bioengineering*, **26**, 1219.

Rabinovitch, M. L., Van Viet, N. & Klyosov, A. A. (1982). *Biokhimiya*, **47**, 465.

Rabinovitch, M. L., Chernoglazov, V. M. & Klyosov, A. A. (1983). *Biokhimiya*, **48**, 369.

Rabinovitch, M. L., Van Viet, N. & Klyosov, A. A. (1986). *Prikladnaya Biokhimiya I Mikrobiologiya*, **22**, 70.

Rees, D. A., Morris, E. R., Thom, D. & Madden, J. K. (1982). In *Polysaccharides*, Vol. I, ed. G. O. Aspinall. Academic Press, New York, p. 195.

Reese, E. T. (1982). *Process Biochemistry*, **17**, 2.

Reese, E. T., Siu, R. G. H. & Levinson, H. S. (1950). *Journal of Bacteriology*, **59**, 485.

Ryu, D. D. Y., Kim, C. & Mandels, M. (1984). *Biotechnology and Bioengineering*, **26**, 488.

Sadana, J. C. & Patil, R. V. (1985). *Journal of General Microbiology*, **131**, 1917.

Saddler, J. N. (1986). *Microbiological Science*, **3**, 84.

Saddler, J. N., Hogan, C. M. & Mes-Hartree, M. (1986). In *Biotechnology in the Pulp and Paper Industry*, Third International Conference, eds K.-E. Eriksson & P. Ander. STFI, Stockholm, p. 96.

Saloheimo, M., Lehtovaara, P., Penttilä, M., Teeri, T. T., Ståhlberg, J., Johansson, G., Pettersson, G., Claeyssens, M., Tomme, P. & Knowles, J. K. C. (1988). *Gene*, **63**, 11.

Sasaki, T., Tanaka, T., Nanbu, N., Sato, Y. & Kainuma, K. (1979). *Biotechnology and Bioengineering*, **21**, 1031.

Schmuck, M., Pilz, I., Hayn, M. & Esterbauer, H. (1986). *Biotechnology Letters*, **8**, 397.

Schwald, W., Chan, M., Brownell, H. H. & Saddler, J. N. (1988). In *Biochemistry and Genetics of Cellulose Degradation*, eds J.-P. Aubert, P. Béguin & J. Millet. Academic Press, London, p. 303.

Selby, K. (1969). The purification and properties of the C_1 component of the cellulase complex. In *Cellulases and their Applications*, eds G. J. Hajny & E. T. Reese. *Advances in Chemistry Series*, Vol. 95, American Chemical Society, Washington, DC, pp. 34–50.

Shakespeare, W. (1600). *The Second part of Henrie the fourth, continuing to his death, and coronation of Henrie the fift. With the humours of Sir Iohn Falstaffe, and swaggering Pistoll.* Printed by V. S. for Andrew Wise & William Aspley, London. From the title page of the Elizabethan Club copy of the only early Quarto Edition, as illustrated in The Second Part of King Henry the Fourth, ed. S. B. Hemingway, Yale University Press, New Haven and London, 1921, 6th. printing, 1965. The quotation is from act IV, sc. 4, lines 107–108.

Sheir-Neiss, G. & Montenecourt, B. S. (1984). *Applied Microbiology and Biotechnology*, **20**, 46.

Shepherd, M. G., Cole, A. L. & Tong, C. C. (1981). *Biochemical Journal*, **193**, 67.

Sheppard, W. J. & Lipinsky, E. S. (1983). In *Energy from Biomass*, 2nd EC Conference, eds A. Strub, B. Chartier & G. Schleser. Elsevier Applied Science, London, p. 63.

Shikata, S. & Nisizawa, K. (1975). *Journal of Biochemistry*, **78**, 499.

Shoemaker, S., Schweickart, V. L., Ladner, M. B., Gelfand, D. H., Kwok, S., Myambo, K. & Innis, M. A. (1983*a*). *Biotechnology*, **1**, 691.

Shoemaker, S., Watt, K., Tsitovsky, G. & Cox, R. (1983*b*). *Biotechnology*, **1**, 687.

Sienko, M. & Plane, R. (1976). *Chemistry*. McGraw-Hill, New York.

Sims, P., Brown, A., James, C., Raeder, U., Schrank, A. & Broda, P. (1988). In *Biochemistry and Genetics of Cellulose Degradation*, eds J.-P. Aubert, P. Béguin & J. Millet. Academic Press, London, p. 365.

Sinskey, A. J. (1983). In *Organic Chemicals from Biomass*, ed. D. W. Wise. Benjamin Cummings, Menlo Park, CA, p. 1.

Soltes, E. J. (ed) (1983). *Wood and Agricultural Residues. Research on Use for Feed, Fuels and Chemicals*. Academic Press, New York.

Sprey, B. & Lambert, C. (1983*a*). *FEMS Microbiology Letters*, **23**, 227.

Sprey, B. & Lambert, C. (1983*b*). *FEMS Microbiology Letters*, **18**, 217.

Stephens, G. R. & Heichel, G. H. (1975). *Biotechnology and Bioengineering Symposium*, **5**, 27.

Streamer, M., Eriksson, K.-E. & Pettersson, B. (1975). *European Journal of Biochemistry*, **59**, 607.

Stutzenberger, F. (1988). *Letters in Applied Microbiology*, **6**, 1.

Tailliez, P., Girard, H., Longin, R., Béguin, P. & Millet, J. (1989*a*). *Applied and Environmental Microbiology*, **55**, 203.

Tailliez, P., Girard, H., Longin, R., Béguin, P. & Millet, J. (1989*b*). *Applied and Environmental Microbiology*, **55**, 207.

Teeri, T. T., Salovuori, I. & Knowles, J. (1983). *Biotechnology*, **1**, 696.

Teeri, T. T., Lehtovaara, P., Kauppinen, S., Salovuori, I. & Knowles, J. (1987). *Gene*, **51**, 43.

Tomme, P., Van Tilbeurgh, H., Pettersson, G., Van Damme, J., Vandekerckhove, J., Knowles, J., Teeri, T. & Claeyssens, M. (1988). *European Journal of Biochemistry*, **170**, 575.

Tonnesen, B. & Ellefsen, O. (1971). In *Cellulose and Cellulose Derivatives*, Vol. 5, Part IV, eds N. M. Bikales & L. Segal. Wiley, New York, p. 265.

Vaheri, M. P. (1982*a*). *Journal of Applied Biochemistry*, **4**, 153.

Vaheri, M. P. (1982b). *Journal of Applied Biochemistry*, **4**, 356.

Vaheri, M. P. (1983). *Journal of Applied Biochemistry*, **5**, 66.

Vaheri, M. P. (1985). In *Monitoring and Control of Plant Raw Material Bioconversion*, VTI Symp. **60**, Espoo, Finland, p. 166.

Van Arsdell, J. N., Kwok, S., Schweickart, V. L., Ladner, M. B., Gelfand, D. H. & Innis, M. A. (1987). *Biotechnology*, **5**, 60.

Van Tilbeurgh, H. & Claeyssens, M. (1985). *FEBS Letters*, **187**, 283.

Van Tilbeurgh, H., Claeyssens, M. & DeBruyne, C. K. (1982). *FEBS Letters*, **149**, 152.

Van Tilbeurgh, H., Pettersson, L. G., Bhikhabhai, R. & Claeyssens, M. (1985). *European Journal of Biochemistry*, **148**, 329.

Van Tilbeurgh, H., Tomme, P., Claeyssens, M., Bhikhabhai, R. & Pettersson, G. (1986). *FEBS Letters*, **204**, 223.

Vernon, C. A. & Banks, B. E. C. (1963). *Biochemical Journal*, **86**, 7P.

Watson, T. G., Nelligan, I. & Lessing, L. (1984). *Biotechnology Letters*, **6**, 667.

Weimer, P. J. & Weston, W. M. (1985). *Biotechnology and Bioengineering*, **27**, 1540.

Westermark, U. & Eriksson, K.-E. (1974a). *Acta Chemica Scandinavica*, **B28**, 204.

Westermark, U. & Eriksson, K.-E. (1974b). *Acta Chemica Scandinavica*, **B28**, 209.

Westermark, U. & Eriksson, K.-E. (1975). *Acta Chemica Scandinavica*, **B29**, 419.

White, A. R. (1982). In *Cellulose and Other Natural Polymer Systems. Biogenesis, Structure and Degradation*, ed. R. M. Brown, Jr. Plenum Press, New York, p. 489.

Wilke, C. R. (ed) (1975). *Biotechnology and Bioengineering Symposium*, **5.**

Willick, G. E. & Seligy, V. L. (1985). *European Journal of Biochemistry*, **151**, 89.

Wood, T. M. (1969). *Biochemical Journal*, **115**, 457.

Wood, T. M. (1971). *Biochemical Journal*, **121**, 353.

Wood, T. M. (1975). *Biotechnology and Bioengineering Symposium*, **5**, 111.

Wood, T. M. (1980). In *OECD Workshop No. 2 Conversion of Cellulosic Materials*, ed. B. A. Rijkens, IBVL, Wageningen, p. 246.

Wood, T. M. (1985). *Biochemical Society Transactions*, **13**, 407.

Wood, T. M. & Bhat, K. M. (1988). *Methods in Enzymology*, **160**, 87.

Wood, T. M. & McCrae, S. I. (1972). *Biochemical Journal*, **128**, 1183.

Wood, T. M. & McCrae, S. I. (1975). In *Symposium on Enzymatic Hydrolysis of Cellulose*, eds M. Bailey, T.-M. Enari & M. Linko. SITRA, Helsinki, p. 231.

Wood, T. M. & McCrae, S. I. (1977). *Carbohydrate Research*, **57**, 117.

Wood, T. M. & McCrae, S. I. (1978). *Biochemical Journal*, **171**, 61.

Wood, T. M. & McCrae, S. I. (1979). In *Advances in Chemistry Series*, **181**, eds R. D. Brown, Jr & L. Jurasek. American Chemical Society, Washington, DC, p. 181.

Wood, T. M. & McCrae, S. I. (1982). *Carbohydrate Research*, **110**, 291.

Wood, T. M. & McCrae, S. I. (1986). *Carbohydrate Research*, **148**, 331.

Wood, T. M., McCrae, S. I. & Macfarlane, C. C. (1980). *Biochemical Journal*, **189**, 51.

Wood, T. M., Wilson, C. A., McCrae, S. I. & Joblin, K. N. (1986). *FEMS Microbiology Letters*, **34**, 37.

Wood, T. M., McCrae, S. I., Wilson, C. A., Bhat, K. M. & Gow, L. A. (1988). In *Biochemistry and Genetics of Cellulose Degradation*, eds J.-P. Aubert, P. Béguin & J. Millet. Academic Press, London, p. 31.

Wood, T. M., McCrae, S. I. & Bhat, K. M. (1989). *Biochemical Journal*, **260**, 37.

Wood, W. A. & Kellogg, S. T. (eds) (1988). *Biomass—Part A, Methods in Enzymology*, **160**, Academic Press, New York.

Yaguchi, M., Roy, R. C., Rollin, C. F., Paice, M. G. & Jurasek, L. (1983). *Biochemistry and Biophysics Research Communications*, **116**, 408.

Chapter 2

BACTERIAL CELLULASES

FRED STUTZENBERGER

*Department of Microbiology, Clemson University, Clemson,
South Carolina 29634-1909, USA*

CONTENTS

1 INTRODUCTION

The study of bacterial cellulases has, until recent years, lagged badly behind
that of fungal enzymes. While extensive reviews of cellulase production,
localization and activity have been compiled for individual fungal species,
cellulolytic bacterial species have, with few exceptions, escaped such distinction. One result of this inequity, as pointed out recently by Coughlan &
Ljungdahl (1988) is that concepts and conclusions based on fungal studies have
strongly influenced the bacteriologists' perception of what (and where) a
cellulase is 'supposed' to be. Such preconceptions based on other systems,

provide good points from which to start research, but in the final analysis, their contribution may be more that of contrast than similarity.

In the preceding chapter on fungal cellulases, the author has given the reader an excellent background on cellulose as a substrate, mechanisms of cellulase action and of synergistic interactions of components in the enzyme complexes. This information allows me the opportunity to concentrate entirely on the major cellulolytic bacterial genera which have been studied as to their potential as agents in biotechnological applications. There are obviously other cellulolytic genera which may fulfill important roles in nature, such as *Cytophaga* which is abundantly widespread in both soil and water. (Brock & Madigan, 1988) and *Erwinia* which is an important plant pathogen (Alexander, 1977). However, these organisms have been relatively neglected as potential sources of cellulases either because their enzymes are not excreted in large quantities from the cells or because their unsupplemented cellulase systems are not capable of degrading crystalline cellulose (Chang & Thayer, 1977; Boyer *et al.*, 1984).

This review of bacterial cellulolytic systems will be divided into aerobic and anaerobic groups. Such a division can be based on more than just oxygen requirements since there are other pertinent points of contrast. Descriptions of each genus will include some morphological and ecological characteristics as well as, physical and chemical properties of their individual cellulases (including those obtained by recombinant DNA techniques).

2 AEROBIC CELLULOLYTIC BACTERIA

2.1 *Acidothermus*

The upper Norris Geyser basin area of Yellowstone National Park in the American west has recently been the source of some interesting cellulolytic thermophiles. The acid springs of that area have temperatures of 45–65°C and a pH range of 4–5·5. Mohagheghi *et al.* (1986) isolated 12 cellulolytic strains from mud and decaying vegetation. No thermophilic actinomycetes or fungi were obtained. The three isolates which had the highest cellulolytic activity were all Gram-variable, non-sporulating and non-pigmented, aerobic rods having temperature and pH optima for growth of 55°C and 5, respectively. Morphology (electron micrograph, Fig. 1) and a variety of other characteristics were similar to those of *Thermus* species isolated earlier from similar environments (Brock & Freeze, 1969). During growth on cellulose, most of the cellulase activity was released during late exponential and early stationary growth. The isolates were classified as *Acidothermus cellulolyticus*.

The cellulases of *A. cellulolyticus* appear to be among the most thermostable of the known cellulases. Although this thermophile has an optimum growth temperature of only 55°C, the optimum for total cellulase activity of crude, cell-free culture fluids was 75°C. These fluids contained both endoglucanase

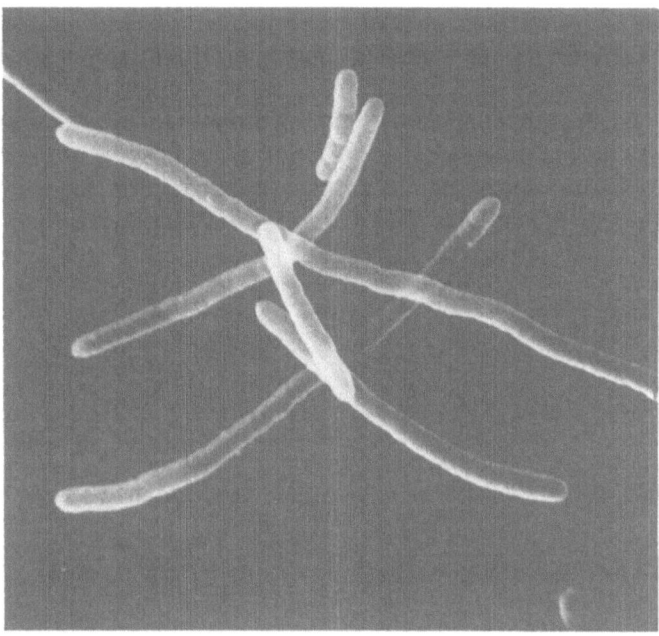

Fig. 1. Scanning electron micrograph of *A. cellulolyticus* (photograph courtesy of Dr Ali Mohagheghi).

and exoglucanase enzymes as well as β-glucosidase, and produced a mixture of glucose and cellobiose from insoluble cellulose. The endoglucanase was the most thermostable (optimum of 83°C and retains more than half its activity at 95°C); the upper temperature limit for cellulolysis was set by the other two enzymes which had optima of 75°C (Seltzer, 1987). Such a high optimum for the β-glucosidase is particularly unusual in comparison with its counterparts in other thermophiles such as *Clostridium* (Ait *et al.*, 1979) and *Thermomonospora* (Rabinovich *et al.*, 1983). In these bacteria, the upper temperature limit for activity of the cytosolic β-glucosidase closely matches (or perhaps even establishes) the optimum for growth on cellulose (about 60°C). Since *A. cellulolyticus* has been under scrutiny only very recently, much remains to be done in the extensive purification and characterization of these thermocellulases for comparison with those of other thermophiles. In any case, these enzymes deserve further study both as to their ecological significance in vegetation decay in hot acidic environments and as to their potential for application in high-temperature biomass conversions.

2.2 Actinomycetes

The actinomycetes are a morphologically, physiologically and ecologically diverse group of bacteria (Goodfellow & Williams, 1983). They establish numerically dominant populations in a variety of soil types; cellulolytic activity is exhibited by one-third to one-half of the actinomycete soil isolates (Isizawa & Araragi, 1976). Another important ecological niche is heating plant biomass such as hay, bagasse, timber wastes and municipal refuse composts (Forsyth & Webley, 1948; Pepys *et al.*, 1963; Fergus, 1964; Stutzenberger *et al.*, 1970; Lacy, 1971; Baecker *et al.*, 1983). Such environments provide initially high concentrations of lignocellulosic materials for growth of the aerobic, thermophilic actinomycetes which secrete amylases, cellulases, hemicellulases, pectinases and lignolytic enzymes. This versatility for biopolymer degradation is responsible in large part for the success of the thermophilic actinomycetes in

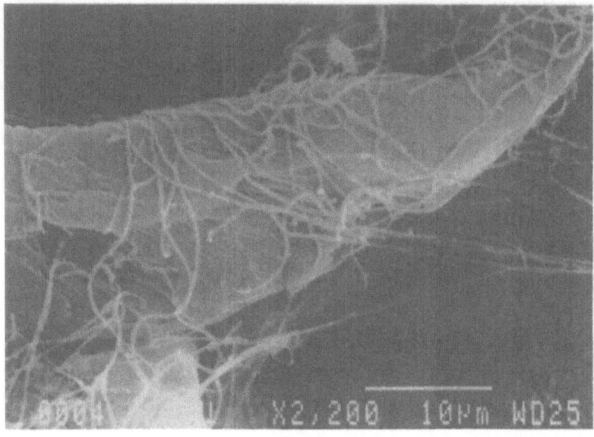

Fig. 2a. Mycelia of the thermophilic actinomycete, *Thermomonospora curvata*, bound to the surface of a native cotton fibre.

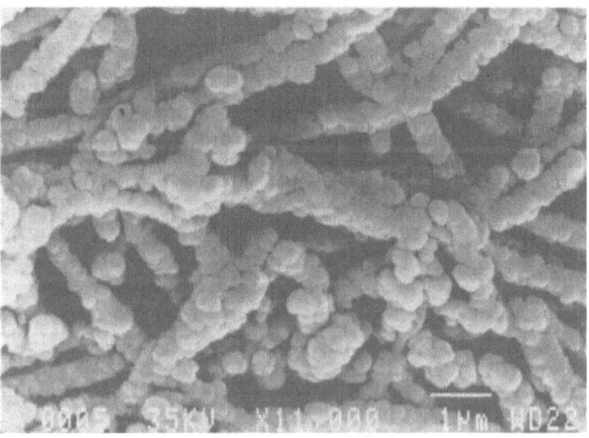

Fig. 2b. Mycellia of *T. curvata* showing profusion of surface-bound cellulases in cellulosomal complexes during growth on cellulose. See section on *Clostridium* for a detailed discussion of cellulosome structure and function.

the composting environment (Catton, 1983). Surface-bound cellulase components enable them to bind to cellulosic fibre (Figs 2a and 2b).

The three actinomycete genera most studied for their cellulolytic abilities have been *Streptomyces, Microbispora* and *Thermomonospora*. The streptomycetes have received more attention than other groups of actinomycetes due to their proclavity for the production of medically-useful antibiotics.

2.2.1 Streptomyces

There are at least 500 recognized *Streptomyces* species (Pridham & Tresner, 1974). Many thermophilic species are found in soil. *S. thermodiastaticus* produces heat-stable endoglucanase levels nearly equal to those of *Thermomonospora fusca*. Maximal hydrolysis of carboxymethyl cellulose (CMC) occurs at pH 6·5, 55°C and releases a mixture of glucose, cellobiose and oligosaccharides (Crawford & McCoy, 1972). *S. thermovulgaris,* during growth on cellulose powder, CMC, and several non-cellulosic substrates (arabinose, pyruvate, malate and citrate) produces at least five cellulolytic components including endoglucanases, an exoglucanase and a cellobiase (Rao & Dhala, 1981). This thermophilic actinomycete is initially present in high numbers in composting animal wastes, which contain about 40% cellulose (Gray *et al.,* 1971), but is superseded as the dominant cellulolytic population by other actinomycetes, such as, *Micromonospora chalcae* and *Pseudonocardia thermophile,* when the temperature of the composting mass reaches 60°C (Godden & Penninchx, 1984). These thermophiles secrete thermostable cellulases having maximal activity at pH 6–7. Some thermophilic strains can also degrade hemicellulose, but lignin degradation appears to be confined mainly to the streptomycetes (McCarthy, 1987).

Other cellulolytic *Streptomyces* species, such as *S. lividans* and *S. flavogriseus,* are thermotolerant, but, the thermostability of their cellulases is poor (endoglucanases have half-lives of 20–30 min at temperatures of 50–60°C, (MacKenzie *et al.,* 1984; Kluepfel *et al.,* 1986). Extracted β-glucosidase is even more labile and has a half-life of 5 min at 40°C (Moldoveanu & Kluepfel, 1983). Therefore, the potential for application of these cellulases as industrial catalysts may be limited. However, the alkalophilic streptomycetes, while not thermophilic in their growth requirements, produce endoglucanases which exhibit greatest activity under rather harsh conditions (pH 9 and 55°C) which, in industrial applications, may be useful in the retardation of contaminant growth (Nakai *et al.,* 1987).

The availability of techniques for the genetic manipulation of streptomyces (Hopwood *et al.,* 1985) have rendered them useful in the study of cellulases both as donors and as recipients of cellulase genes. An endoglucanase gene of a *Streptomyces* species isolated from termite gut has been cloned in *Escherichia coli* (Coppolecchia *et al.,* 1987). Other endoglucanase genes from *Bacillus subtilis* (Horinouchi *et al.,* 1987), *Thermomonospora fusca* (Ghangas & Wilson, 1987*a*), and the alkalophilic *Streptomyces* strain KSM-9 (Nakai *et al.,* 1988) have been cloned in *S. lividans.* The insertion of strong promoters into multicopy and wide host range plasmids, together with the ability to secrete the gene products free of the cells provides an exciting potential for increasing cellulase production in these actinomycetes (Horino-uchi *et al.*, 1987; Shareck *et al.*, 1987).

2.2.2 Microbispora

The reports of Waldron and co-workers (Waldron & Eveleigh, 1986; Waldron *et al.*, 1986) introduced *Microbispora bispora* into the ranks of known cellulose degraders. The most active strain, chosen out of forty aerobic thermophilic actinomycetes isolated from soil samples of thermal areas around the world, produces a 'complete' cellulase complex consisting of extracellular en-doglucanases and cellobiohydrolases and a cell-associated β-glucosidase. Therefore this complex is qualitatively similar to those excreted by fungi such as *Trichoderma reesei*.

The endoglucanases were purified by ion exchange and size exclusion chromatography. Isoelectric focusing and SDS-PAGE revealed four enzymes with pI values of 3·9–4·8 and relative molecular mass (M_r) in the range of $4·4$–$9·5 \times 10^4$. Their action was distinct from the exoglucanases based on their cleavage patterns on CMC or acid-swollen cellulose. The action of the two cellobiohydrolases (pI 4·0–4·1) was determined by release of cellobiose primarily, from the non-reducing ends of reduced oligosaccharides. The synergism between the endo- and exo-enzymes was demonstrated by their inability in isolated form to solubilize cotton fibres or microcrystalline cellulose; when the synergistic enzymes were combined, these recalcitrant substrates were degraded. This degradation rate was stimulated by the addition of the β-glucosidase (Yablonsky *et al.*, 1988). The relatively high resistance of the β-glucosidase to end product repression makes it an excellent candidate for further study as a potential industrial enzyme (Waldron *et al.*, 1986).

Genes for five endoglucanases and two β-glucosidases have been cloned in *E. coli* using the pBR322 plasmid vector. Detailed information on the major endoglucanase gene has recently been published (Yablonsky *et al.*, 1988) and will only be briefly summarized here. By restriction analysis, the major CMCase gene (MBICEL1) was found on a 4·3 kb Pst 1 fragment. When this fragment was subcloned into the Pst 1 site of the pUC13 plasmid (a 10-fold higher copy number vector), the activity of crude cell extracts from *E. coli* had four-fold greater CMCase activity. The level of expression is independent of the gene orientation in pUC13. The coding region of the gene is 1374 nucleotides and encodes a protein with a predicted M_r (Bellamy, 1974; Khan, 1980*a*) consistent with the active enzyme on SDS-PAGE (58 000). The amino acid sequence deduced from MBICEL1 showed extensive homology with the proline- and serine-rich regions of cellulases from other microbes including both *Cellulomonas firmi* and *T. reesei*. Expression in *E. coli* is largely confined to the cytoplasm. When the cloned gene products having β-glucosidase and cellobiohydrolase activities are characterized, a clearer picture of the synergy between bacterial cellulases will be provided.

2.2.3 Thermomonospora

The genus, *Thermomonospora*, originally described by Henssen (1957), currently includes five species, although this taxonomic grouping is tenuous. *Thermomonospora* includes those actinomycetes which form single, heat-sensitive spores on aerial hyphae alone or in combination with substrate hyphae, with cell walls containing *meso*-diaminopimelic acid but no characteristic sugars (Cross & Goodfellow, 1973; McCarthy & Cross, 1984). Studies on their potential for cellulose degradation in large-scale biotechnological applications have apparently included strained misidentifications. The *T. fusca* YX strain used by earlier workers (Hagerdal *et al.*, 1978, 1979*a*; Lee & Humphrey, 1979) was initially identified as a *Thermoactinomyces* species while our laboratory has been working with a strain initially identified as *T. curvata* (Stutzenberger, 1971), but which shares characteristics between that species and *T. fusca*. As pointed out by McCarthy & Cross (1984), further taxonomic studies on *Thermomonospora* and related organisms will be required. In any case, the white *Thermomonospora* deserve extensive study as sources of a wide range of thermostable enzymes potentially useful in biomass conversion processes.

A strain identified as *T. curvata* (reference no. CUB 993, University of Bradford Actinomycete Collection, 413) was the numerically dominant organism during the high-temperature phase of municipal solid waste composting (Stutzenberger *et al.*, 1970; Stutzenberger, 1971). Subsequent studies (Stutzenberger, 1972*a, b*, 1979; Stutzenberger & Kahler, 1986) established its ability during growth in chemically defined medium to secrete enzymes which degrade a variety of cellulosic substrates including the recalcitrant native cotton fibres. CUB 993 secretes three endoglucanases (M_r, 23 000, 46 000 and 146 000) during early exponential growth; this endoglucanase pattern is markedly altered during stationary phase to include endoglucanases with M_r values of 52 000, 106 000 and 114 000 (Lupo & Stutzenberger, 1988). The production of the extracellular endoglucanases is cyclic AMP dependent and therefore subject to catabolite repression by soluble sugar accumulation (Fennington *et al.*, 1983; Wood *et al.*, 1984). A β-glucosidase (M_r, 66 000) is largely intracellular (Bernier & Stutzenberger, 1989*b*) although during growth on legume residues, the enzyme is liberated into the culture fluid (Bernier & Stutzenberger, 1988*a*). The stability of these extracellular enzymes, as well as enzymes active against other biopolymers, is dependent on a variety of factors (Stutzenberger & Lupo, 1986; Bernier & Stutzenberger, 1988*b*, 1989*a*; Bernier *et al.*, 1988).

T. fusca strain YX isolated by Bellamy (1974) has also been the subject of much attention. It produces at least five extracellular endoglucanases (E1–E5) and an intracellular β-glucosidase during growth on cellulose (Hagerdal *et al.*, 1978, 1979*a, b*). These endoglucanases have been purified and characterized (Wilson, 1988). The major components (designated E1 and E2, (Calza *et al.*, 1985)) possess the respective characteristics: M_r, 94 000 and 46 000 (both monomeric glycoproteins), isoelectric points of pH 3·5 and 4·5, temperature optima of 74°C and 58°C, broad pH optima centered at 6·0, and a great degree of sensitivity to Hg^{2+}. Cellulose hydrolysis by E1 releases mainly cellobiose with a small amount of glucose, while E2 produces a mixture of glucose, cellobiose, cellotriose and higher oligomers. Production of E5 appears to be regulated at the transcriptional level by both induction and repression, since there is a good correlation between the level of mRNA and the quantity of enzyme produced under inductive and repressive culture conditions (Lin & Wilson, 1988*a*). This recent finding probably applies to all the *T. fusca* endoglucanase genes since they are coordinately induced (Lin & Wilson, 1987). There appears to be three closely linked promoters in a unique A + T-rich region which are controlled by the binding of an activator protein which is only apparent when the cells are cultured under cellulase-inductive conditions (Lin & Wilson, 1988*b*).

No true exoglucanase activity has yet been detected in culture fluids, although *Thermomonospora* strains produce a 'complete' cellulase complex in that they are able to degrade crystalline cellulose (Hagerdal *et al.*, 1978, 1979*b*; Stutzenberger, 1979*b*). Although the rate of cellulose depletion in thermophilic actinomycete cultures may be greater than that in mesophilic fungal cultures (Su & Paulavicius, 1975), the wild type strains produce less cellulase than do the hyperproducing mutant strains of *Trichoderma reesei* (Aguirre *et al.*, 1986). The potential for hyperproduction by these actinomycetes through mutagenesis has been explored, and some promising, although not spectacular, results have been obtained (Fennington *et al.*, 1982, 1984; Meyer & Humphrey, 1982; Montenecourt *et al.*, 1983). The thermostability of their cellulases is an important factor in their application and confers several advantages over enzymes from mesophiles (Margaritis & Merchant, 1986).

Recent cloning of *Thermomonospora* endoglucanases has offered incentive for further study of these thermophiles at both the molecular and at the applied levels. The endoglucanases, E2 and E5 are the smallest (M_r values 42 000 and 45 000, respectively) cellulases excreted by *T. fusca* YX and are quite distinct from their companion cellulases in terms of amino acid sequence, kinetic properties and antigenic reactivity (Wilson, 1988). The genes encoding these two enzymes were inserted into several shuttle plasmids which could replicate in both *E. coli* and *Streptomyces lividans* (Ghangas & Wilson, 1987*a*, *b*). *S. lividans* protoplast transformants initially expressed and excreted E5, but their ability was unstable; deletion of the gene was apparently caused by the presence of ColE1 replication sequences. Construction of a plasmid containing only inserted *T. fusca* DNA was more stable and resulted in transformants from which one-half of total excreted protein was E5. The E2 activity of *Streptomyces* transformant cultures was about 300-fold that observed in *E. coli* transformants. This high enzyme production did not, however, confer the ability to use cellulose on the *Streptomyces* transformants. *Bacillus subtilis* was also transformed with E5, but there was extensive deletion of plasmid DNA and no evidence of E5 activity.

2.3 *Bacillus*

The genus, *Bacillus,* encompasses about 35 currently recognized species, plus scores of proposed new species. The broad range of DNA (guanosine + cytosine) content (32–51%) of recognized species is an indication of their genetic heterogeneity (Vandemark & Batzing, 1987).

Their morphology ranges from rod-shaped to spherical to mycelial. The major unifying morphological characteristic is endospore formation. This genus exhibits a wide diversity of physiological abilities; although the majority are mesophilic, there are also psychrophilic and thermophilic species. Some are acidophiles while others are alkalophiles. Some are haloduric. Some grow well on a single organic carbon/energy source in inorganic salts media while others require amino acids and vitamins. Soil appears to be their major habitat. The ecology of this diverse group is very poorly defined (Sneath, 1986); conclusions as to specific habitats may be fallacious if based merely on the presence of their endospores (Norris *et al.,* 1981). Their primary ecological role appears to be the degradation of biomass polymers. However, *Bacillus* species, despite their ubiquity and their acknowledged ability to excrete a wide variety of depolymerizing enzymes, have apparently achieved only marginal status as cellulose degraders. Their proteolytic and amylolytic activities were recognized very early (Allen, 1953), and later studies have shown that all of the recognized *Bacillus* species secrete extracellular enzymes (Priest, 1977). Production of *Bacillus* proteases and amylases accounts for over one-half the total commercial enzyme volume (Crueger & Crueger, 1982), but until recently, these organisms have been assigned a secondary role (if any at all) compared to the prominent cellulolytic, anaerobic thermophiles such as *Clostridium thermocellum.* However, Zemek *et al.* (1981), showed that 11 out of 25 *Bacillus* species secrete β-1,4-glucanases. All of the *B. stearother-mophilus* strains, 85% of the *B. coagulans* strains, and 50% of the *B. licheniformis* strains, (three species of which grow well at 50–60°C (Gibson & Gordon, 1974)) produce thermostable cellulase components.

Cellulase production by *Bacillus* is interesting both in regard to the regulation of enzyme biosynthesis and the wide variety of characteristics demonstrated by the cellulases. Fukumori *et al.* (1985) studied CMCase production in an alkalophilic *Bacillus* species. Growth on CMC yielded twice the specific activity obtained on cellobiose. A variety of other carbon sources such as glucose, sucrose and maltose severely repressed production. The CMCase had M_r 92 000, an isoelectric point of 3·1, and a K_m for CMC of 0·48 mg/ml under optimal conditions (pH 9·0 and 40°C). The ability of glucose and other soluble sugars to repress CMCase is in contrast to studies on other *Bacillus* strains. When *B. licheniformis* was grown on a variety of carbon sources, glucose supported the highest production of a CMCase. The enzyme had pH and temperature optima of 6·1 and 55°C, respectively (Dhillon *et al.,* 1985). Endoglucanase production also appeared to be constitutive in a Group I *Bacillus* strain from heat-shocked soil samples (Robson & Chambliss, 1984). A

B. subtilis strain, isolated from cotton waste compost and mutagenized with
N-nitrosoguanidine to obtain a four-fold increase in CMCase production,
yielded more enzyme during growth on raffinose than on the classic inducers,
cellobiose or insoluble celluloses (Can & Au, 1987). This CMCase had a broad
pH optimum of 5·0–8·5 and a relatively high temperature optimum of 65°C.

Particularly interesting are thermostable *Bacillus* cellulases with extreme pH
optima. Kawai *et al.* (1988) recently described an alkaline CMCase from a
strain similar to *B. pumilus*. It had a pH optimum of about 10 and retained
about 80% of maximal activity even at pH 11. It was resistant to a variety of
proteases, heavy metals, chelating agents and detergents. Horikoshi *et al.*
(1984) isolated a soil strain closely resembling *B. pasteurii*. Two endoglucanases active on cellotetraose had pH optima of 10·0. At this extreme
pH, one was stable at 60°C, the other at 80°C. As pointed out by Wood (1985),
for economically feasible application of cellulases in industrial scale biomass
conversions, the hydrolyses must be capable of proceeding rapidly under
non-sterile conditions. Combinations of extreme temperature and pH should
be helpful in preventing significant contamination effects. Whether *Bacillus*
cellulases would be useful in such applications would depend on their ability
to degrade a variety of lignocellulosic substrates. Since these enzymes have
shown little or no activity against crystalline forms of cellulose, it appears
likely that they would be used in conjunction with cellulases from other
sources. Some combinations of *Bacillus* enzymes with other sources already
appears attractive. Vlasenko *et al.* (1985) successfully employed a *B. subtilis*
β-mannanase in conjunction with *Trichoderma reesei* cellulases for the
degradation of plant glucomannans.

Perhaps the most significant role for the *Bacillus* species in cellulose
conversion biotechnology will eventually be an expression system for cellulase
genes cloned from other species. Doi *et al.* (1986) have reviewed the
advantages of *B. subtilis* for production of foreign proteins:

(1) it is non-pathogenic,
(2) it can be manipulated by current genetic engineering techniques,
(3) it lacks both endotoxins (a characteristic important in the production of
    proteins for medical or foodstuff application) and protein modification-
    mechanisms which may create inactive enzyme forms,
(4) it can be grown more easily and has greater rates of protein synthesis
    than many eucaryotic systems, and
(5) its ability to secrete a wide variety of proteins far exceeds that of its
    procaryotic competitor, *E. coli*.

Disadvantages include:

(1) the tendency of *B. subtilis* to produce several proteases able to degrade
    foreign proteins either intracellularly or extracellularly,
(2) the absence of modification systems for some proteins requiring post-
    translational modification for activity and
(3) low export rates for some eucaryotic proteins.

The model system having the highest potential success for such cloning would be transfer of cellulase genes between *Bacillus* species. Kim & Pack (1988) recently cloned a *B. subtilis* endoglucanase into a cellulase-negative *B. megaterium*. The result was a fully excreted enzyme with a uniform M_r, 33 000, isoelectric point of 7·2, K_m of 1·6 mg CMC/ml at the optimal pH of 5·5 and temperature of 60°C. The action of the enzyme on cello-oligosaccharides yielded cellobiose with traces of glucose. As with other *Bacillus* endoglucanases, there was little activity against microcrystalline cellulose. Successful cloning of enzymes highly active against the crystalline substrates may largely depend on the development of multiple protease-negative *Bacillus* host strains such as *B. subtilis* DB104 (Wong *et al.*, 1986*a*).

2.4 *Cellulomonas*

This genus currently is considered, on the basis of DNA homologies, to encompass seven species. They may be isolated from natural sources by plating at 30–35°C under aerobic conditions on a mineral salts medium containing yeast extract and cellulose. Young cultures produce slender irregular rods which stain weakly Gram-positive. Some form filaments which fragment into short rods or spheres (Stackebrandt & Keddie, 1986).

The regulation and properties of the cellulases produced by *Cellulomonas* species appear to be qualitatively similar, although large quantitative differences have been observed in the production of both cellulolytic and xylolytic enzymes (Thayer *et al.*, 1984; Rajoka & Malik, 1986). Due to the earlier confusion as to taxonomic divisions, many of the earlier studies did not designate species. Most of the physiological and genetic studies have concentrated on the species, *C. firmi*, *C. flavigena* and *C. uda*, although Rajoka & Malik (1986) found *C. biazotea* produced slightly higher levels of both extracellular and cellular cellulolytic and hemicellulolytic enzymes than other species when grown on either synthetic or natural cellulosic substrates. *Cellulomonas*, as is the case with many other microbes, produces the highest levels of cellulolytic enzymes when grown on insoluble cellulose; both cellobiose and sophorose are relatively poor inducers (Beguin & Eisen, 1977). Cellulase biosynthesis is regulated by a complex interaction of catabolite repression, induction and enzyme inhibition (Stewart & Leatherwood, 1976). In cultures growing exponentially on cellulose, much of the extracellular enzymes is substrate-bound. A small amount of CMCase remains cell-associated as does the β-glucosidase (Stoppok *et al.*, 1982; Rodriguez & Volfoa, 1984). The surfactant, Tween 80 (0·1%) doubled the free concentration of cellulases in *C. flavigena* cultures, but it is not clear whether this effect was due to increased release from the cells or inhibition of adsorption to the substrate (Sami *et al.*, 1988). The enzymes can be readily removed from the culture fluids by adsorption to filter paper; the adsorbed enzymes release cellobiose from the filters when incubated at pH 6 and 40°C (Antheunisse,

1984). Although the temperature optima for all the *Cellulomonas* cellulases are in the 40–50°C range, their stability at that temperature under assay conditions is poor in the absence of substrate (Rodriguez *et al.*, 1988).

Although some preliminary characterization of cellulolytic enzymes from native *Cellulomonas* strains has been done (Beguin & Eisen, 1978; Stoppok *et al.*, 1982; Wakarchuk *et al.*, 1984; Rodriguez *et al.*, 1988), almost all of the fine structure work in this regard has been done on products from genes cloned in *E. coli*. *C. uda* genes encoding β-glucosidase and CMCase have been cloned and expressed (Nakamura *et al.*, 1986*a*, *b*), but the cellulases of *C. firmi* have received more attention at the molecular level. This facultative anaerobe produces β-1,4-endoglucanases (Beguin & Eisen, 1977; Langsford *et al.*, 1984), at least one β-1,4-exoglucanase (Gilkes *et al.*, 1984*a*; O'Neill *et al.*, 1986*a*) and a beta-glucosidase (Wakarchuk *et al.*, 1984). The cenA and cenB genes (which code for endoglucanases) and the cex gene (which codes for the exoglucanase) have been cloned in *E. coli* (Whittle *et al.*, 1982; Gilkes *et al.*, 1984*b*, Owolabi *et al.*, 1988) or *Brevibacterium lactofermentum* (Paradis *et al.*, 1987), and characterized (O'Neill *et al.*, 1986*a*, *b*, *c*; Wong *et al.*, 1986*b*; Greenberg *et al.*, 1987; Owolabi *et al.*, 1988). An interesting bifunctional exoglucanase–endoglucanase fusion protein has been obtained by the joining of the Cex and cenA genes by expression in *E. coli*. The hybrid enzyme had both activities, but did not bind to microcrystalline cellulose for lack of a substrate-binding region (Warren *et al.*, 1987). These findings, and others related to them, have been recently reviewed (Miller *et al.*, 1988) and will only be briefly summarized here. The Cex and cenA genes, cloned by insertion into the pBR322 plasmid vector, encode non-glycosylated primary sequences (Exg and EngA respectively) which have identical substrate specificities to the corresponding genes expressed in *C. firmi*. Both enzymes contain three regions: (1) a twenty amino acid sequence termed the 'Pro–Thr box' which contains only proline and threonine, (2) a hydroxyamino acid-rich region having four Asn–Xaa–Ser/Thr sites and little charge density or predicted secondary structure, (3) an ordered region with Glu–Xaa$_7$–Asn–Xaa$_6$–Thr sequences of high charge density which probably contains the active site. Although there is nearly complete conservation of the box and at least 50% conservation of irregular regions in EngA and Exg, the active site region appears not to be highly conserved. Since the order of the regions is reversed in the two enzymes, the genes could have evolved by shuffling of the two conserved regions with either one of the two other sequences (Warren *et al.*, 1986).

Production of non-glycosylated forms of Exg and EngA in *E. coli* provides the opportunity for comparison with the corresponding native enzymes produced in *C. firmi*. Although many extracellular microbial enzymes are glycosylated (a condition which may stabilize enzymes against proteolytic degradation or inactivation by extremes of temperature or pH), glycosylation of Exg or EngA is not essential for either binding to the cellulose surface or form hydrolysis of the polymer bonds. However, glycosylation is necessary for effective cellulose digestion in that it protects the enzymes against fragmentation by the *C. firmi* extracellular serine protease at the carboxyl side of Pro-Thr box. It has been suggested (Miller *et al.*, 1988) that the box forms a hinge region which links the catalytic and cellulose-binding domains, since the truncated proteolysis products retain some catalytic activity but lose their ability to bind strongly to cellulose.

2.5 *Cellvibrio*

Members of the genus, *Cellvibrio*, are aerobic, Gram-negative, polar-flagellated cells which may be isolated from soil by enrichment techniques using a mineral salts–microcrystalline cellulose medium incubated at 28°C (Blackall *et al.*, 1985). As a recognized genus, *Cellulomonas* is only recently becoming accepted again after being discredited in the 8th edition of *Bergey's Manual* and the *World Directory of Collections of Cultures of Microorganisms*.

The extracellular cellulases of *C. gilvus* were purified and characterized quite early (Storvick & King, 1960). Four endoglucanases (CMCases) were identified. They adsorbed to alkali-swollen cellulose with cellobiose as the only detectable product. The pH and temperature optima were 7·0 and 37°C respectively (enzymes were inactivated rapidly above 40°C).

Carpenter & Barnett (1967) were the first to publish a purification protocol for the localization and characterization of *C. gilvus* cell-bound cellulases. They showed that although most of the cellulase (CMCase) was present as cell-free enzyme, about one-third of the total activity was bound to cell membrane and ribosomal fragments. The β-glucosidase activity was almost entirely cell-associated as ribosomal fractions. The distribution of cellulases is carbon source dependent (Berg, 1975).

The formation of these enzymes is induced most effectively during growth on cellulose and repressed by the presence of glucose or cellobiose (Berg *et al.*, 1972*a*). CMCase patterns change markedly between mid-exponential and late-stationary phases, probably as a result of glycosylation reactions (Oberkotter & Rosenberg, 1980). The enzymes degrade intact cotton fibres very slowly, but broken ends provide susceptible sites exposing internal fibre structure for cell attachment and enzyme action (Berg *et al.*, 1972*b*). Physical contact between cells and the cellulosic fibre was necessary for maximal cellulase induction (Breuil & Kushner, 1976). Although the specific CMCase activity per cell is about the same whether grown on limiting concentrations of cellobiose or on cellulose fibres, the total culture cellulase is much higher when grown on the cellulose (particularly components degrading crystalline cellulose), indicating that much of the enzyme is bound tightly to the substrate and that cell–fibre contact specifically induces some cellulase components (van Hofsten & Berg, 1972).

Although relatively little has been published on the characterization of *Cellvibrio* cellulases, the cloning of a gene cluster (a 94-kilobase fragment) which encodes cellulase, chitinase, amylase and pectinase) in *C. mixtus* (Wynne & Pemberton, 1986) should provide the opportunity for detailed molecular studies. A comparison of the degradative genes carried and expressed by 50 cosmid clones in *E. coli* suggested the gene sequence as: yellow pigmentation, starch hydrolysis, exculin hydrolysis, cellobiose utilization, chitin hydrolysis, cellulose adherence, CMC hydrolysis and polygalacturonate hydrolysis. Unfortunately, the degradation of microcrystalline cellulose was not detected in any cosmid clone, indicating that additional genes may be necessary to code for a complex active against native substrate.

2.6 *Pseudomonas*

This genus encompasses nearly 100 species of Gram-negative aerobic, motile rods which exhibit a remarkable range and diversity of metabolic activities. In addition to carbohydrates and amino acids as carbon and energy sources, many strains can utilize aromatic compounds such as substituted benzoates and salicylates. Because of their metabolic versatility, widespread occurrence in nature, and their growing medical importance, *Pseudomonas* particularly *P. aeruginosa* species have been subjects for genetic manipulations (Sussman *et al.*, 1988). Their resistance plasmids are particularly attractive vectors for intergeneric transfer (Couturier *et al.*, 1988).

Although not a highly cellulolytic genus rivaling other Gram-negative aerobes such as *Cellulomonas*, the pseudomands have attracted the attention of several research groups over the last two decades. Yamane and colleagues (Yamane *et al.*, 1965, 1970*a*, *b*), in their early studies on *P. fluorescens* var. *cellulosa*, showed that cellulase (CMCase) biosynthesis was controlled by both induction and catabolite repression, that sophorose and isomaltose were better inducers than cellobiose and equal to cellulose. However, with cellulose, most of the cellulase was released from the cells, whereas during growth on sugars, 90% remained cell-associated. The β-glucosidase was entirely cytoplasmic (Ramasamy & Verachtert, 1980). Two extracellular forms of endoglucanase (A & B) and a cell-bound form C, although differing in their localization, demonstrated remarkable similarity in their amino acid and carbohydrate compositions, heat stabilities, and pH-dependence. However, substrate specificity and product patterns were different (Yamane *et al.*, 1970*b*, 1971; Ramasamy & Verachtert, 1980). A subsequent study (Yishikawa *et al.*, 1974) showed that component proportions were subject to cultural influences and suggested that B was converted to A by post-translational modification mechanisms (glycosylation or proteolysis) in the latter stages of growth. The work of Wolff *et al.* (1986) indicated only one CMCase gene per *Pseudomonas* genome; the product was an endoglucanase with an M_r 23 000. However, recent evidence for four distinct CMCase genes having no homology with each other provides another plausible alternative for CMCase multiplicity (Gilbert *et al.*, 1987). The most detailed description of a cloned pseudomonas cellulase gene product has been submitted by Lejeune and co-workers (Lejeune *et al.*,

1986, 1988*a*, *b*). Cloned into *E. coli* via plasmid pRUCL150 (a derivative of pBR322 containing a 11·1 kilobase DNA fragment from *P. fluorescens* subspecies cellulose NCIB10462 inserted at the *Eco*R1 site), the gene encoded an endoglucanase (M_r, 36 000) which cleaved CMC but was inactive on microcrystalline cellulose or cellobiose. The properties of this periplasmic enzyme do not correspond to any of those published previously; possibly this enzyme is the membrane-bound C form, which remains unglycosylated in *E. coli*. This gene has been introduced and expressed in *Zymomonas mobilis* (Lejeune *et al.*, 1988*c*), a homofermentative ethanologen with high alcohol tolerance but limited capacity to ferment carbohydrates.

Characterization of an endoglucanase gene and its product in *P. solanacearum* (a phytopathogenic species causing lethal wilting diseases in over 200 host plants) has been reported by Schell and associates (Roberts *et al*, 1988; Schell, 1987). The gene (egl) was cloned on a cosmid by the insertion of a 2·7 kilobase Xhol-Sall DNA fragment. In *E. coli*, transcription of the gene was controlled by its own promoter; the product was restricted to the cytoplasm. An endoglucanase purified from the native *P. solanacearum* had the following characteristics: M_r 43 000, pH optimum of 7·5, temperature optimum of 50°C, no metal requirement, rapid depolymerizing activity of CMC, but no soluble release from crystalline cellulose.

3 ANAEROBIC CELLULOLYTIC BACTERIA

3.1 *Acetivibrio*

Acetivibrio cellulolyticus has provided a relatively new addition to our knowledge of cellulolytic anaerobic bacteria. This Gram-negative, non-spore-forming anaerobe was originally isolated from a methanogenic enrichment culture from municipal sewage by dilution of the inoculum and plating on mineral salts–vitamin–cellulose agar under 80% N_2–30% CO_2 at 35°C (Khan, 1977; Patel *et al.*, 1980). The initial isolate (strain CD2) produced a yellow pigment which adhered to the cellulose particles in a manner similar to the YAS pigment of *Clostridium thermocellum* (Ljungdahl *et al.*, 1983).

A. cellulolyticus is a rather pleomorphic, often-filamenting straight to slightly curved encapsulated rod, 0·5–0·9 μm wide by 4–10 μm long (Fig. 3). A single flagellum, located about one-third of the distance from the end of the cell, provides an odd tumbling form of motility. This anaerobe has an extremely restrictive carbohydrate fermentation pattern. Of a wide range of hexoses, pentoses and polysaccharides, only cellobiose, salicin or cellulose supported growth and acid production (Patel, 1984). *A. cellulolyticus* slowly adapts to growth on glucose, while the other known species, *A. ethanolgignens*, grows on fructose, galactose, lactose, maltose, mannitol, mannose and pyruvate as well as glucose and pectin but cannot utilize cellobiose nor degrade cellulose (Robinson & Richie, 1981). Growth of *A. cellulolyticus* for eight days on

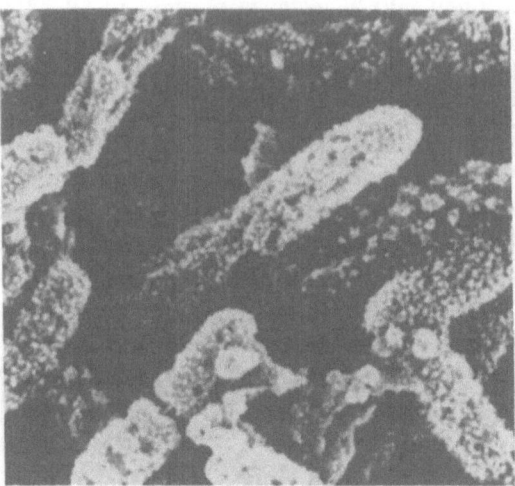

Fig. 3. Cellulosomal arrangement on surfaces of *A. cellulolyticus* (from Lamed *et al.*, 1987).

cellobiose in peptone–yeast extract broth yielded acetic acid, hydrogen, CO_2 and traces of propanol and butanol as fermentation products. Some ethanol is produced during growth on cellulose but the extreme sensitivity of the cells to ethanol (about 50% growth inhibition at 0·4% ethanol) precludes the direct fermentation of cellulose to fuels. Butanol, propanol, succinate and lactate are not produced; of the strains tested to date, acetic acid appears to be the sole or major product of carbohydrate fermentation (Patel, 1984).

The physical and chemical requirements for the growth of *A. cellulolyticus* are rather restrictive; the extreme pH and temperature ranges for growth are 6·5–7·7 and 20–40°C, respectively (Patel *et al.*, 1980). The sulfur requirements for cellulose degradation in a chemically defined medium are in the range of 0·8–1·7 mM in the form of sulphate, thiosulphate, sulphite, sulphide, cysteine or methionine; at higher concentrations, reduction of cellulose to methane is inhibited (Khan & Trotter, 1978). The inhibitory effects of high sulfur levels on methanogenesis in *A. cellulolyticus* are similar to those observed in lake sediment ecosystems (Winfrey & Zeikus, 1977). The distribution and diversity of *Acetivibrio* spp. in nature is not known due to its relatively recent identification (only two species currently identified) and to the paucity of established phylogenetic relationships with other genera (Patel, 1984).

Although *A. cellulolyticus* has been shown to utilize a wide range of cellulosic forms, both crystalline and amorphous (Khan *et al.*, 1979), Avicel microcrystalline cellulose supported the greatest liberation of extracellular protein and the highest specific endoglucanase and exoglucanase cellulase activities (Saddler *et al.*, 1980). When co-cultured with *Methanosarcina barkeri* (a methanogen which prevented the accumulation of acetic acid and H_2), cellulose degradation was increased about three-fold (Khan, 1980*b*). Cellobiose was a relatively good cellulase inducer compared to salicin, while other sugars, including sophorose (which is a good inducer of cellulase biosynthesis in *Trichoderma* (Lowenberg & Chapman, 1977)), were completely ineffective. The effect of adding either glucose or cellobiose (or both) to cultures growing on cellulose was small, relative to the marked effects these soluble sugars have on other bacterial species under similar conditions. Saddler *et al.* (1980) have suggested that since *A. cellulolyticus* attaches itself to the insoluble polymer surface, cell–cellulose contact is necessary for induction of cellulase biosynthesis and that this contact precludes a strong repressive effect when soluble sugars are added to cellulose-grown cultures. Therefore, it would appear that in this anaerobe, induction, rather than catabolite repression, plays the major role in the control of cellulase biosynthesis. Whether this relatively weak potential for control by catabolite repression is due to the limited ability of *A. cellulolyticus* to utilize most sugars has not been determined. For example, although cultures can be adapted to grow on glucose, adaptation is extremely slow and requires several transfers at weekly intervals; moreover, the specific growth rate under continuous culture conditions of glucose-adapted cells is at best only one-sixth that of cellobiose-grown cells, apparently due to the feeble ability for glucose uptake (Patel & MacKenzie, 1982).

The *Acetivibrio* cellulase system has not been studied as extensively as those of the thermophilic anaerobes such as *Clostridium thermocellum*. A preliminary study (Khan, 1980a) showed that this anaerobe produced both cell-associated and cell-free enzymes which could degrade a variety of insoluble celluloses (absorbent cotton fibres, paper products and microcrystalline cellulose). Cellulose solubilization was strongly stimulated by Mg^{2+} and Ca^{2+} (MacKenzie & Bilous, 1982). Filter paper specific activity (IU/mg protein) of *A. cellulolyticus* culture fluids was comparable to that of wild type *T. reesei* and *Aspergillus niger* cultures grown on cellulose (Saddler & Khan, 1980). Saddler & Khan (1981) used polyacrylamide gel electrophoresis (PAGE) to achieve partial purification of the cell-free enzymes of *A. cellulolyticus* culture fluids after six days growth on cellulose. Two endoglucanases, C2 (M_r 33 000) and C3 (M_r 10 400), an exoglucanase (M_r 81 000), released by cell lysis, was apparent. C2 appeared to be a trimeric form of C3; treatment of C2 with sodium dodecyl sulphate and mercaptoethanol caused disassociation to C3-like monomers. C2 and C3 conversion apparently occurred under cultural conditions also, since the C3/C2 ratios increased with culture age and varied depending on the cellulosic form employed as carbon and energy source. These enzymes require thiols for maximal activity and Ca^{2+} for stabilization on prolonged reaction (MacKenzie *et al.*, 1984, 1985, 1987) but detailed studies on requirements and characteristics remain to be done.

3.2 *Bacteroides*

Members of this mesophilic genus are non-spore-forming, Gram-negative anaerobes (Holdeman *et al.*, 1986). Some species are important members of the cellulolytic bovine rumen flora. Of the major cellulolytic rumen bacteria, *Ruminococcus* and *Bacteroides*, the latter were found to hydrolyse native cotton fibres and cellulose powder more effectively (Halliwell & Bryant, 1963; Dehority & Scott, 1967; Stewart *et al.*, 1980). The products of this hydrolysis includes glucose, cellobiose and cellodextrins (Gaudet, 1987). In addition to *B. succinogenes,* the major rumen species, a relatively newly isolated methanogenic, *B. cellulosolvens,* is also receiving attention as to its cellulolytic properties (described later in this section).

Because of its importance to cellulose digestion in commercial ruminents, the regulation and characteristics of the cellulases of *B. succinogenes* have been studied. Groleau & Forsberg (1981) grew cultures on filter paper or microcrystalline cellulose and found that the cells adhered to the insoluble substrate. Cell-free culture supernatants had very little ability to solubilize these substrates, although most of the CMCase activity could be found associated with membranous fragments in the fluids. The cellobiase was produced constituitively and remained tightly associated with the cells. These enzymes are apparently released in a vesicular form by bleb formation from the outer membrane of intact cells adhering to the insoluble cellulose (Forsberg *et al.,* 1981). This observation of a release mechanism other than the conventional model of bacterial protein export (Randall & Hardy, 1984) was important to more than our understanding of plant digestion in the rumen insofar as it led to the development of our contemporary concept of surface-bound cellulase structure and function.

Since that early study, *B. succinogenes* has been shown to produce a variety of cellulolytic enzymes, but purification of the complexes has been difficult because of aggregate formation and the extreme heterogeneity of molecular size and charge. Cellulolytic activity exists in three forms: about 60% is associated with membrane fragments, 10% as non-sedimentable enzyme with an M_r greater than 4×10^6, and 30% as a free form with M_r 45 000. The membranous fragments in culture fluid contain CMCases, a xylanase, and a cellobiase. The CMCases could be detergent-solubilized freeing four distinct endoglucanases (Groleau & Forsberg, 1983*a*; Schellhorn & Forsberg, 1984). The major product was cellobiose with smaller amounts of glucose and oligosaccharides.

The structure/function relationship of vesicular enzymes in the rumen was further elucidated by the report that the membrane-bound CMCase and xylanase were solubilized by bovine pancreatic trypsin (Groleau & Forsberg, 1983*b*). The released CMCase was slightly smaller (M_r, 43 000) than found in the fluids of in-vitro cultures growing on cellulose. This release mechanism was specific to the membranous vesicles, as trypsin treatment of intact cellobiose-grown or cellulose-grown cells yielded little enzyme.

Of the non-sedimentable enzymes in *B. succinogenes* culture fluid, two apparently had exoglucanase activity (Schellhorn & Forsberg, 1984). Two endoglucanases have been recently purified and characterized by McGavin & Forsberg (1988). Endoglucanase EG1 accounted for 32% of the total non-sedimentable endoglucanase, had an M_r 65 000, and a pI of 4·8. Temperature and pH optima were 39°C and 6·4. The K_m for CMC was 3·6 mg/ml. EG1 did not bind to acid-swollen cellulose and the products formed were cellobiose and cellotriose. EG2 (11% of total non-sedimentable endoglucanase activity) existed as two forms, one with M_r 118 000, the other a proteolytic degradation product with M_r 94 000. The larger EG2 had a K_m for CMC of 12·2 mg/ml, a pI of 9·4, and temperature and pH optima of 39°C and 5·8, respectively. EG2 bound specifically to acid-swollen cellulose and yielded cellotetraose. The comparative characteristics of these two enzymes suggested that EG2 initiates the attack on the cellulose polymer and EG1 acts primarily on the products of EG2 activity.

The cellobiosidase in *B. succinogenes* culture fluid has also been recently characterized (Huang *et al.*, 1988). It is a glycoprotein (M_r, 75 000, 12% carbohydrate) with a pI of 6·7 and pH and temperature optima of 6·5 and 39°C. These optima were considerably broadened and catalytic activity stimulated by the presence of chloride ions. Stimulations were also observed with other halides, nitrate and nitrite. The enzyme released cellobiose from cellotriose and cellobiose and cellotriose from long chain oligosaccharides and acid-swollen cellulose. The enzyme had binding affinity for highly crystalline cellulose, but could not hydrolyse it or CMC to any appreciable extent. Although the characterization of this halide-stimulated enzyme indicates that the rumen environment provides nearly optimal conditions for its activity in many respects, its interaction with other enzymes in the complex extracellular synergism of *B. succinogenes* cellulolysis remains to be defined.

The ability of the relatively newly identified *B. cellulosolvens* to degrade cellulose has been extensively studied by Murray and colleagues. Originally isolated from municipal sewage sludge, this mesophile ferments only cellulose and cellobiose to produce acetic acid, CO_2, H_2, ethanol and traces of lactic acid (Murray *et al.*, 1984). Although the growth rate is faster on cellobiose, more biomass and metabolites (ethanol, acetate, and lactate) are produced on cellulose (Murray, 1987a). A variety of substrates, including crystalline cellulose, wood chips, and cotton fibres is consumed; H_2 accumulation slows the rate of growth and shifts the production pattern to ethanol at the expense of acetate (Murray, 1986). Accumulation of cellobiose catabolite represses cellulase biosynthesis and feedback inhibits the enzyme activity; glucose can exert neither effect (Murray, 1987b). The cellulase system remains largely cell surface-bound, giving the cells a distinctive fluffy morphology (Fig. 4) compared to the smooth-surface cellobiose-grown cells; these surface structures probably provide both adherence to, and degradation of, cellulosic substrates in a manner similar to the cellulosomes of other anaerobes (Murray *et al.*, 1986).

3.3 *Caldocellum*

An extremely thermophilic anaerobe, tentatively identified as *Caldocellum saccharolyticum*, has been recently isolated from New Zealand thermal pools by microcrystalline cellulose enrichment culturing at 75°C (Sissons *et al.*,

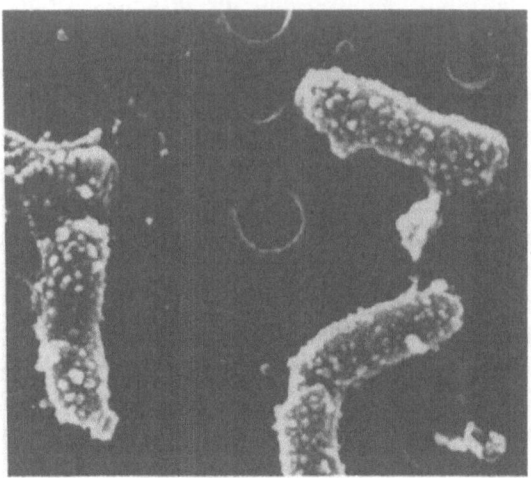

Fig. 4. Cellulosomal arrangement on *B. cellulosolvens* (from Lamed *et al.*, 1987).

1987). Figure 5 illustrates the morphology of this non-motile, non-flagellated Gram-negative rod. Total cellulase production by this thermophile was roughly comparable to, but with greater thermostability than that of *Clostridium thermocellum* (Reynolds *et al.*, 1986). It grew on a wide range of pentoses and hexoses with ethanol and acetate as the major fermentation products. *C.*

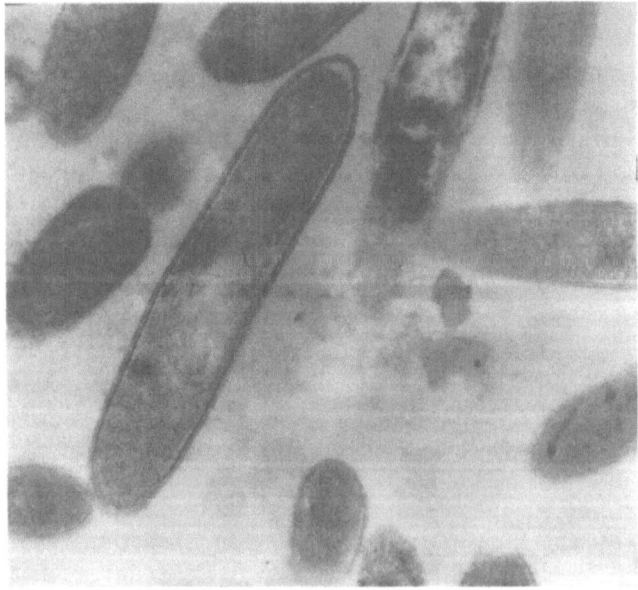

Fig. 5. Thin section of *C. saccharolyticum* TP8 (photograph made by Colin Monk and provided by Dr J. Andrew Hudson, University of Waikato, New Zealand).

saccharolyticum can degrade a variety of insoluble substrates including microcrystalline cellulose, wood pulps, deproteinized alfalfa fibre and hemi-cellulosic materials (Donnison *et al.*, 1989). The cellulolytic system was remarkably resistant to the action of detergents and chelating agents (Sissons *et al.*, 1987).

The initial difficulty in purification of the components in the cellulase complex from *C. saccharolyticum* led to the cloning of their genes into *E. coli* (Bergquist *et al.*, 1987; Love & Streiff, 1987). The thermostability of the cellulases in a thermolabile background allowed the use of heat treatment (one hour at 70°C) for the elimination of most contaminating host proteins (Patchett *et al.*, 1989; Schofield *et al.*, 1989). The cloned β-glucosidase has already been characterized as follows (Plant *et al.*, 1988):

(1) broad specificity for β-D-glucosides, galactosides, fucosides and xylo-sides, with order of linkage preference in β-linked glucose dimers being 1–3 > 1–2 > 1–4 > 1–6;
(2) pH optimum of 6·0–6·5;
(3) thermostable to 80°C with a buffer-dependent temperature optimum around 70°C;
(4) destabilized by a variety of metals;
(5) requirement for free thiol groups for maximal activity.

The thermostability and chaotropic agent resistance of this β-glucosidase (characteristically the most labile enzyme in the cellulolytic systems of other microbes) portends a bright future for the eventual industrial application of *C. saccharolyticum* enzymes.

3.4 *Clostridium*

This genus encompasses at least 85 identified species (Cato *et al.*, 1986). The clostridia are ubiquitous in soil and sewage, droppings of both domesticated and wild animals, marine sediments and decaying vegetation. Their endospores survive for long periods of time under harsh conditions. The majority are mesophiles, although some, such as *C. thermocellum* and *C. thermo-hydrosylfuricum,* grow best at 60–70°C. Most species are obligate anaerobes; some exhibit sufficient oxygen tolerance to grow in the presence of air. Clostridia are, with few exceptions, chemoorganotrophic and produce mixtures of organic acids (such as propionate and butyrate) and alcohols (such as ethanol and butanol) during growth on carbohydrates. Their widespread presence on rotting plant materials in nature is indicative of their ability to produce a variety of extracellular depolymerizing enzymes. However, only about 12% of the known species have been studied closely enough to characterize extracellular enzymes such as amylases, cellulases and pectinases. *C. thermocellum* and *C. stercorarium* are the only two confirmed species of thermophilic, anaerobic, cellulolytic bacteria (Lamed & Bayer, 1990).

C. thermocellum has by far been the more thoroughly characterized. It is the oldest validly-described anaerobic, thermophilic cellulose degrader (Viljoen *et al.*, 1926; McBee, 1948, 1950), but its cellulases have been characterized only relatively recently.

The enzymes exist in three forms whose ratios vary with the age of the culture and nature of the carbon source. In young cultures grown on cellulose, much of the enzyme activity is cell-bound or bound to substrate; as cultures mature and the cellulose is solubilized, more enzyme can be detected in the free fluid (Hon-Nami *et al.*, 1985, 1986). Cultures grown on soluble sugars produce an 'incomplete' cellulose system lacking the cellobiohydrolase activity necessary to degrade crystalline cellulose (Hon-Nami *et al.*, 1987). The cellulases in *C. thermocellum* culture fluids require both Ca^{2+} and a reducing thiol compound for significant cellulose solubilization (Johnson *et al.*, 1982). These cellulases are mainly endoglucanases which are active against CMC, but show little activity against insoluble substrates. They may be generally characterized as follows: M_r values in the range 40 000–150 000, isoelectric points from 4·5–8·7, pH and temperature of 5–6 and 60–70°C, respectively, and major hydrolysis products cellobiose and glucose (Lee & Blackburn, 1975; Ng *et al.*, 1977; Ng & Zeikus, 1981; Hon-Nami *et al.*, 1986).

When grown on insoluble cellulose, *C. thermocellum* cells bind to the substrate and secrete a characteristic yellow pigment (Ng *et al.*, 1977). The yellow pigment attracted greater attention when it was found to be an affinity substance in the cellulolytic system (Ljungdahl *et al.*, 1983). The yellow affinity substance (YAS) attached to the cellulose fibrils during the early culture stages and retained most of the total culture endoglucanase activity. As the cellulose was hydrolysed, the endoglucanase was gradually released into the free culture fluid. However, the free enzyme had only about one-fifth the specific activity against CMC compared to the bound fraction. The insoluble fraction was recalcitrant to washing with buffer or salt solutions, but was easily extracted with distilled water. YAS was water-insoluble, but could be extracted from the cellulose by acetone or ethanol. Ljungdahl *et al.* (1988) have recently described a purification protocol and properties of the YAS: the pigment appears to be carotenoid in nature, based on its Ramen, infrared and nuclear magnetic resonance spectroscopic characteristics. Its elemental composition ($C_{52}H_{94}O_{19}N$) does not include either carbohydrate or amino acids. Although YAS adheres strongly to cellulose and facilitates the binding of the cellulolytic enzyme complex to the polymer, its function in nature is as yet unknown. However, its dual affinity for both cellulose and cellulase has proven useful in the laboratory; YAS–cellulose columns have been used to advantage in the affinity chromatography of the free cellulases of the culture fluid (Hon-Nami *et al.*, 1986).

The adherence of the clostridia to cellulose is dependent on cell surface-bound complexes termed 'cellulosomes' which are formed under cellulase-inductive culture conditions but which are absent when cellulase synthesis is repressed (Lamed *et al.*, 1983a). Although other bacteria, particularly those of rumen origin (Leatherwood, 1973; Groleau & Forsberg, 1981; Wood *et al.*, 1982; Guiliano & Khan, 1984) have been shown to adhere to cellulosic substrates, the art of cellulosome watching has been taken to its highest form in *C. thermocellum*. Scanning electron microscopy has shown the binding of the cells to amorphous cellulose. When cells are grown on cellobiose, the cellulosomes are observed in polycellulosomal organelles termed protuberances (Fig. 6). These protuberances undergo extensive protraction when they come into contact with insoluble cellulosic substrates so that the cellulosomes attach to the substrate surface while an underlying fibrous material connects the cellulosomes to the cell surface. The loss of the ability to form these structures results in adherence-defective cells (Bayer *et al.*, 1983).

The cellulosome is a particulate, multi-subunit structure of relatively uniform size (diameter of 18 nm) which tend to aggregate in the culture fluid when freed from the cell surfaces (Lamed *et al.*, 1983*b*). Over one-half of the cellulosomal peptide composition is accounted for by two proteins (S1, M_r, 210 000, and S8, 75 000). S1 function is as yet incompletely defined, although it is an integral part of cellulosomal structure and has not been observed detached in a free form. It contains a novel O-linked tetrasaccharide composed of galactose, galactitol and 3-*O*-methyl-*N*-acetyl glucosamine (Gerrit *et al.*, 1989). It demonstrates no cellulase enzyme activity, nor does it appear to bind directly to the substrate (Lamed & Bayer, 1988*b*). It may serve to organize the component parts of the cellulosome or serve to attach the complex to the cell surface. Like S1, S8 does not appear in an uncomplexed state, although its contribution to cellulosome composition is markedly diminished during growth on cellobiose compared to that on cellulose (Bayer *et al.*, 1985). It probably serves a function in the degradation of amorphous cellulose in a Ca^{2+} activated, cellobiose-inhibited manner, but has little activity against CMC

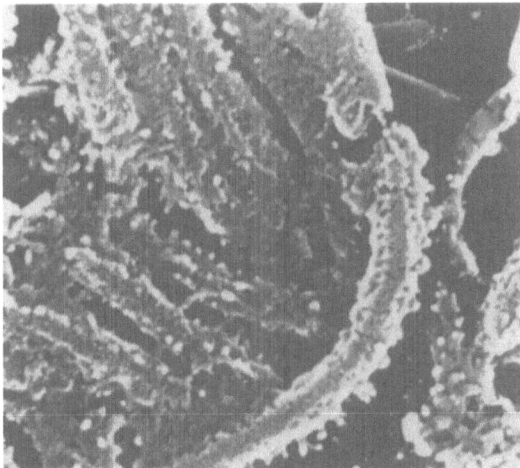

Fig. 6. *Clostridium thermocellum* YS with polycellulosomal protuberances (from Lamed *et al.*, 1987).

(Lamed & Bayer, 1988*a*). About 75% of the extracellular CMCase activity is cellulosome-associated in the form of other subunits (mainly S9, 10 and 13). The S13 subunit is probably the product of the celA genes (Beguin *et al.*, 1985). The rest of the CMCase enzymes are soluble enzymes in the M_r range 30–250 000: those which adsorb to DEAE-cellulose interact synergistically with the cellulosomal cellobiohydrolase in a thiol- and Ca^{2+} independent manner (Lamed & Bayer, 1988). Recent detailed reviews of cellulosome structure and function, include those of Lamed & Bayer, 1985, 1988.

Cellobiose, the major product from cellulosomal degradation of cellulose, is the preferred carbon source for *C. thermocellum* (Ng & Zeikus, 1982). Accumulation of the disaccharide represses cellulase biosynthesis (Johnson *et al.*, 1985) and markedly inhibits the ability of purified cellulosomal preparations to degrade crystalline cellulose (Lamed *et al.*, 1985). *C. thermocellum* utilizes cellobiose through the action of cellobiose phosphorylase and β-glucosidase. The periplasmic β-glucosidase has been purified and characterized (Ait *et al.*, 1979, 1982) it has M_r 50 000, isoelectric point at pH 4.7, maximal activity at pH 6 and 65°C, and exhibits both aryl-β-glucosidase and cellobiase activities. It is probable that this enzyme is the product of a gene recently cloned in *E. coli* via the pBR325 vector; the gene(s) encoding 4-methylumbelliferyl-β-D-glucoside and cellobiose activities are closely associated on the *C. thermocellum* chromosome (Kadam *et al.*, 1988).

Considerable progress has also been made in the cloning and characterization of *C. thermocellum* cellulase genes (Beguin *et al.*, 1983, 1985, 1986; Cornet *et al.*, 1983*a*, *b*; Millet *et al.*, 1985; Grepinet & Beguin, 1986; Ha *et al.*, 1987; Romaniec *et al.*, 1987*a*, *b*; Schwarz *et al.*, 1987, 1988; Soutschek-Bauer & Staudenbauer, 1987; Babykin *et al.*, 1988; Grabnitz & Staudenbauer, 1988; Hazelwood *et al.*, 1988; Mann, 1988; Piruzyan *et al.*, 1988). The most thoroughly characterized cloned gene product to date is the endoglucanase D (EGD). EGD constitutes about 3% of the protein and 30% of the total endoglucanase in the cellulosome (Lamed & Bayer, 1988*a*) *E. coli* cells carrying the pCT600 plasmid with an EcoR1 6·3 kb insert containing the celD gene overproduced the EGD as cytoplasmic crystals. Solubilized EGD was a monomer with an M_r 65 000 an isoelectric point of 5.4, pH and temperature optima of 6·0 and 60°C; the enzyme had high specific activity against CMC but negligible activity against crystalline cellulose (Joliff *et al.*, 1986). Another gene has been obtained from an unclassified thermophilic *Clostridium* species; this gene cloned into *E. coli* encoded an interesting endoglucanase which had a temperature optimum at 70°C and dual pH optima at 7·5 and 10·5 (Ishizaki & Kawauchi, 1988). A variety of cellulase genes has also been cloned from the mesophilic clostridia, including *C. cellulolyticum* (Faure *et al.*, 1988; Perez-Martinez *et al.*, 1988) and *C. acetobutylicum* (Zappe *et al.*, 1986, 1988). The sequencing of these genes has shown DNA homologies useful in uncovering interesting degrees of relatedness between cellulases of the clostridia and other bacterial genera.

3.5 *Ruminococcus*

There are eight recognized species of *Ruminococcus*; these anaerobic, mesophilic Gram-positive cocci are well adapted to their ecological niche in the rumen, rumen-like forestomachs, the caecum and large intestine of a variety of animals (Bryant, 1986). Fermentation of cellulose and other polysaccharides in these complex, controlled environments produces acetate, formate, CO_2 and H_2 as major products (Halliwell & Bryant, 1963). Those species of rumen origin (*R. albus* and *R. flavefaciens*) are active fermenters of both cellulose and xylan, while those of human origin are not (Bryant, 1986). Rumen flora cellulases serve a dual function, providing the ruminant with usable nutrients and stimulating the germination of seeds passing through this ecosystem (Howard & Elliott, 1988). The rate of cellulose utilization by these rumen bacteria is not diminished by accumulations of either glucose or cellobiose (Hiltner & Dehority, 1983). However, the cellulolytic ability of these rumen strains is diminished by phenolic acids derived from lignin in forage and by sutstituted cellulose derivatives (Chesson *et al.*, 1982; Rasmussen *et al.*, 1988).

During growth on cellulose or cellobiose, the cellulases (CMCases) and xylanases of *R. albus* are localized first on the cell surface (Fig. 7) then bind strongly to cellulosic fibres (Smith *et al.*, 1973; Morris & Cole, 1987). Wood *et al.* (1982) found that the surface-bound enzymes were easily released by washing with 0·5 M phosphate buffer at pH 7·5. Under routine culture conditions, most of the cellulase activity, including the cellobiosidase (Ohmiya *et al.*, 1982) was released from the cells in late stationary phase. The solubilized complexes had high M_r values, 2 000 000 for endoglucanase and 200 000 for the cellobiosidase, and could be dissociated by urea or guanidine

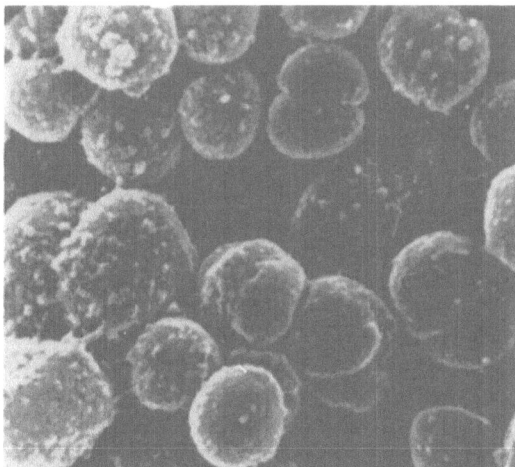

Fig. 7. *Ruminococcus albus* with cell surface-associated cellulase complexes (from Lamed *et al.*, 1987).

into monomers, M_r, 30 000 which had optimal activity under reduced conditions at pH 6·5 and 45°C. A similar enzyme from another *R. albus* strain had activity against CMC, but had little ability to solubilize solid cellulose to detectable products (Ohmiya *et al.*, 1987) The description of one exoglucanase (among seven possible forms) in *R. flavefaciens* (Gardner *et al.*, 1987) indicates the required synergy between the cellulases, whether of bacterial or fungal origin (Kopecny & Williams, 1988). The exoglucanase, a dimer having subunits M_r 118 000 was inactive against CMC, but released cellobiose as the sole product from insoluble celluloses. The enzyme had pH and temperature optima of 5·0 and 45°C, was stimulated by Ca^{2+}, but markedly inhibited by other divalent cations. Similar sensitivities were exhibited by the *R. albus* β-glucosidase (Ohmiya *et al.*, 1985).

The cloning of *Ruminococcus* endoglucanase genes has only recently received attention. A *R. flavefaciens* FD1 gene was cloned on the plasmid vector *pEco*R251 into *E. coli* which constitutively expressed CMCase (but not activity against crystalline cellulose) in the periplasm during late exponential growth (Barros & Thomson, 1987). The *R. albus* endoglucanase, cloned into *E. coli* via the pBR322 vector, was unstable in the periplasm; however, lowering the pH to 6·5 and the NaCl concentration to 80 mM increased the yield of the cloned product about 10-fold (Ohmiya *et al.*, 1988). Endoglucanase, exoglucanase and β-glucosidase activities were all identified in a genomic library of *R. albus* constructed using the *E. coli* lambda phase (Howard & White, 1988). Genetic manipulations of *R. albus* cellulase genes now include intergeneric protoplast fusions between *E. coli* (carrying the CMCase gene on pBR322) and an anaerobic mutant fusant of *Fusobacterium varium* and *Enterococcus faecium* (Chen *et al.*, 1988).

REFERENCES

Aquirre, M. V., Phillips, J. A., Bostwick, L. J. & Montenecourt, B. S. (1986). *Chemical Engineering Communications*, **45**, 93.
Ait, N., Creuzet, N. & Cattaneo, J. (1979). *Biochemical and Biophysical Research Communications*, **90**, 537.
Ait, N., Creuzet, N. & Cattaneo, J. (1982). *Journal of General Microbiology*, **128**, 569.
Alexander, M. (1977). *Introduction to Soil Microbiology*, 2nd ed. John Wiley and Sons, New York, p. 30.
Allen, M. B. (1953). *Bacteriological Reviews*, **17**, 125.
Antheunisse, J. (1984). *Antonie van Leeuwenhoek*, **50**, 7.
Babykin, M. M., Velikodvorskaya, G. A., Mogutov, M. A., Piruzyan, E. S. & Shestakov, S. V. (1988). *Akademia Nauk USSR Seriya Biologii*, **15**, 186.
Baecker, A. A. W., Dyker, R. M. P. & King, B. (1983). In *Biodeterioration*, eds T. A. Oxley & S. Berry. John Wiley & Sons, New York, p. 64.
Barros, M. E. C. & Thomson, J. A. (1987). *Journal of Bacteriology*, **169**, 1760.
Bayer, E. A., Kenig, R. & Lamed, R. (1983). *Journal of Bacteriology*, **156**, 618.
Bayer, E. A., Setter, E. & Lamed, R. (1985). *Journal of Bacteriology*, **163**, 552.
Beguin, P. & Eisen, H. (1977). *Journal of General Microbiology*, **101**, 191.
Beguin, P. & Eisen, H. (1978). *European Journal of Biochemistry*, **87**, 525.

Beguin, P., Cornet, P. & Millet, J. (1983). *Biochemistry*, **65**, 495.

Beguin, P., Cornet, P. & Aubert, J. P. (1985). *Journal of Bacteriology*, **162**, 102.

Beguin, P., Rocancourt, M., Chebrou, M. C. & Aubert, J. P. (1986). *Molecular and General Genetics*, **202**, 251.

Bellamy, W. D. (1974). *Biotechnology and Bioengineering*, **16**, 869.

Berg, B. (1975). *Canadian Journal of Microbiology*, **21**, 51.

Berg, B., van Hofsten, B. & Pattersson, G. (1972a). *Journal of Applied Bacteriology*, **35**, 201.

Berg, B., van Hofsten, B. & Pattersson, G. (1972b). *Journal of Applied Bacteriology*, **35**, 215.

Bergquist, P. L., Love, D. R., Croft, J. E., Streiff, M. G., Daniel, R. M. & Morgan, W. H. (1987). *Biotechnology and Genetic Engineering Reviews*, **5**, 199.

Bernier, R. F. & Stutzenberger, F. J. (1988a). *Letters in Applied Microbiology*, **7**, 103.

Bernier, R. F. & Stutzenberger, F. J. (1988b). *Journal of Biotechnology*, **7**, 293.

Bernier, R. F. & Stutzenberger, F. (1989a). *Letters in Applied Microbiology*, **8**, 9.

Bernier, R. F. & Stutzenberger, F. J. (1989b). *MIRCEN Journal of Applied Microbiology and Biotechnology*, **5**, 15.

Bernier, R., Kopp, M., Trakas, B. & Stutzenberger, F. (1988). *Journal of Applied Bacteriology*, **65**, 411.

Blackall, L. L., Hayward, A. C. & Sly, L. I. (1985). *Journal of Applied Bacteriology*, **49**, 81.

Boyer, M. H., Chambost, J. P., Magnan, M. & Cattaneo, J. (1984). *Journal of Biotechnology*, **1**, 229.

Breuil, C. & Kushner, D. J. (1976). *Canadian Journal of Microbiology*, **22**, 1776.

Brock, T. D. & Freeze, H. (1969). *Journal of Bacteriology*, **98**, 289.

Brock, T. D. & Madigan, M. T. (1988). *Biology of Microorganisms*, 5th edn. Prentice Hall, Englewood Cliffs, NJ, p. 726.

Bryant, M. P. (1986). In *Bergey's Manual of Systematic Bacteriology*, Vol. 2, ed. P. H. A. Sneath. Williams and Wilkins, Baltimore, MD, p. 1093.

Calza, R. E., Irwin, D. C. & Wilson, D. B. (1985). *Biochemistry*, **24**, 7797.

Can, K. Y. & Au, S. (1987). *Antonie van Leeuwenhoek*, **53**, 125.

Carpenter, S. A. & Barnett, L. B. (1967). *Archives of Biochemistry and Biophysics*, **122**, 1.

Cato, E. P., George, W. L. & Finegold, S. M. (1986). In *Bergey's Manual of Systematic Bacteriology*, ed. P. H. A. Sneath. Williams and Wilkins, Baltimore, MD, p. 1141.

Catton, C. (1983). *New Scientist*, **100**, 38.

Chang, W. T. H. & Thayer, D. W. (1977). *Canadian Journal of Microbiology*, **23**, 1285.

Chen, W., Ohmiya, K. & Shimizu, S. (1988). *Applied and Environmental Microbiology*, **54**, 2300.

Chesson, A., Stewart, C. S. & Wallace, R. J. (1982). *Applied and Environmental Microbiology*, **44**, 597.

Coppolecchia, R., Dessi, M. R., Giacomini, A., Lepidi, A., Mastromei, G., Nuti, M. P. & Polsinelli, M. (1987). *Biotechnology Letters*, **9**, 495.

Cornet, P., Millet, J., Beguin, P. & Aubert, J. P. (1983a). *Biotechnology*, **1**, 589.

Cornet, P., Tronik, D., Millet, J. & Aubert, J. P. (1983b). *FEMS Microbiology Letters*, **16**, 137.

Coughlan, M. P. & Ljungdahl, L. G. (1988). In *Biochemistry and Genetics of Cellulose Degradation*, eds J. Aubert, P. Beguin & J. Miller. Academic Press, London, p. 11.

Couturier, M., Bex, F., Bergquist, P. L. & Maas, W. K. (1988). *Microbiological Reviews*, **52**, 375.

Crawford, D. L. & McCoy, E. (1972). *Applied Microbiology*, **24**, 150.

Cross, T. & Goodfellow, M. (1973). In *Actinomycetes. Characteristics and Practical Importance,* eds G. Sykes & F. A. Shinner. Academic Press, London, p. 11.

Crueger, W. & Crueger, A. (1982). *Biotechnology: A Textbook of Industrial Microbiology.* Sinauer Associates, Sunderland, MA, p. 162.

Dehority, B. A. & Scott, H. W. (1967). *Journal of Dairy Science,* **50,** 1136.

Dhillon, N., Chibber, S., Saxena, M., Pajni, S. & Valdehra, D. V. (1985). *Biotechnology Letters,* **7,** 695.

Doi, R. H., Wong, S. L. & Kawamura, F. (1986). *Trends in Biotechnology,* **4,** 232.

Donnison, A. M., Brockelsby, C. M., Morgan, H. W. & Daniel, R. M. (1989). *Biotechnology and Bioengineering,* **33,** 1495.

Faure, E., Bagnara, C. & Belaich, J. P. (1988). *Gene,* **65,** 51.

Fennington, G., Lupo, D. & Stutzenberger, F. (1982). *Biotechnology and Bioengineering,* **24,** 2487.

Fennington, G., Neubauer, D. & Stutzenberger, F. (1983). *Biotechnology and Bioengineering,* **25,** 2271.

Fennington, G., Neubauer, D. & Stutzenberger, F. (1984). *Applied and Environmental Microbiology,* **47,** 201.

Fergus, C. L. (1964). *Mycologia,* **56,** 267.

Forsberg, C. W., Beveridge, T. W. & Hellstrom, A. (1981). *Applied and Environmental Microbiology,* **42,** 886.

Forsyth, W. G. & Webley, D. M. (1948). *Proceedings of the Society for Applied Bacteriology,* **11,** 34.

Fukumori, F., Kudo, T. & Horikoshi, K. (1985). *Journal of General Microbiology,* **131,** 3339.

Gardner, R. M., Doerner, K. C. & White, B. A. (1987). *Journal of Bacteriology,* **169,** 4581.

Gaudet, G. (1987). *Reproduction Nutrition Development,* **27,** 239.

Gerrit, J., deWaard, P., Kamerling, J. P., Viliegenthart, J. F. G., Morgenstern, E., Lamed, R. & Bayer, E. A. (1989). *Journal of Biological Chemistry,* **264,** 1027.

Ghangas, G. S. & Wilson, D. B. (1987a). *Applied and Environmental Microbiology,* **53,** 1470.

Ghangas, G. S. & Wilson, D. B. (1987b). *Applied and Environmental Microbiology,* **54,** 2521.

Gibson, T. & Gordon, R. E. (1974). In *Bergey's Manual of Determinative Bacteriology,* 8th edn. eds R. E. Buchanan & N. E. Gibbons. Williams and Wilkins Co., Baltimore, p. 540.

Gilbert, H. J., Jenkins, G., Sullivan, D. A. & Hall, J. (1987). *Molecular and General Genetics,* **210,** 551.

Gilkes, N. R., Langsford, M. L., Kilburn, D. G., Miller, R. C. & Warren, R. A. J. (1984a). *Journal of Biological Chemistry,* **259,** 10455.

Gilkes, N. R., Kilburn, D. G., Langsford, M. L., Miller, R. C., Wakarchuk, W. W., Warren, R. A. J., Whittle, D. J. & Wong, K. R. (1984b). *Journal of General Microbiology,* **130,** 1377.

Godden, B. & Penninchx, M. J. (1984). *Annales de Microbiologie (Institute Pasteur),* **135B,** 69.

Goodfellow, M. & Williams, S. T. (1983). *Annual Review of Microbiology,* **37,** 189.

Grabnitz, F. & Staudenbauer, W. L. (1988). *Biotechnology Letters,* **10,** 73.

Gray, K., Sherman, K. & Biddlestone, A. J. (1971). *Process Biochemistry,* **6,** 1.

Greenberg, N. M., Warren, R. A. J., Kilburn, D. G. & Miller, R. C. (1987). *Journal of Bacteriology,* **169,** 646.

Grepinet, O. & Beguin, P. (1986). *Nucleic Acids Research,* **14,** 1791.

Groleau, D. & Forsberg, C. W. (1981). *Canadian Journal of Microbiology,* **27,** 517.

Groleau, D. & Forsberg, C. W. (1983a). *Canadian Journal of Microbiology,* **29,** 504.

Groleau, D. & Forsberg, C. W. (1983b). *Canadian Journal of Microbiology,* **29,** 710.

Guiliano, C. & Khan, A. W. (1984). *Applied and Environmental Microbiology*, **48**, 446.

Ha, J. H., Han, S. S., Kim, U. H. & Lee, Y. H. (1987). *Korean Journal of Applied Microbiology and Bioengineering*, **15**, 346.

Hagerdal, B. G. R., Ferchak, J. D. & Pye, E. K. (1978). *Applied and Environmental Microbiology*, **36**, 606.

Hagerdal, B., Harris, H. & Pye, E. K. (1979a). *Biotechnology and Bioengineering*, **21**, 345.

Hagerdal, B., Ferchak, J. & Pye, E. K. (1979b) *Advances in Chemistry Series*, **81**, 331.

Halliwell, G. & Bryant, M. P. (1963). *Journal of General Microbiology*, **32**, 441.

Hazelwood, G. P., Romaniec, M. P. M., Davidson, K., Grepinet, O., Beguin, P., Millet, J., Raynaud, O. & Aubert, J. P. (1988). *FEMS Microbiology Letters*, **51**, 231.

Henssen, A. (1957). *Archives fur Mikrobiologie*, **26**, 373.

Hiltner, P. & Dehority, B. A. (1983). *Applied and Environmental Microbiology*, **46**, 642.

Holdeman, L. V., Kelley, R. W. & Moore, W. E. C. (1986). In *Bergey's Manual of Systematic Bacteriology*, Vol. 1, eds N. R. Kreig & J. G. Holt. Williams and Wilkins, Baltimore, p. 602.

Hon-Nami, H., Coughlan, M. P., Hon-Nami, K. & Ljungdahl, L. G. (1987). *Proceedings Royal Irish Academy*, Section B, Biology, **87**(5), Royal Irish Academy, Dublin, p. 83.

Hon-Nami, K., Coughlan, M. P., Hon-Nami, H., Carreira, L. H. & Ljungdahl, L. G. (1985). *Biotechnology and Bioengineering Symposium*, **15**, 191.

Hon-Nami, K., Coughlan, M., Hon-Nami, H. & Ljungdahl, L. G. (1986). *Archives of Microbiology*, **145**, 13.

Hopwood, D. A., Bibb, M. J., Chater, K. F., Kieser, T., Bruton, C. J., Kieser, H. M., Lydiate, D. J., Smith, C. P., Ward, J. M. & Schrempf, H. (1985). *Genetic Manipulation in Streptomyces: A Laboratory Manual*. The John Innes Foundation, Norwich, UK.

Horikoshi, K., Nakao, M., Kurono, Y. & Sashihara, N. (1984). *Canadian Journal of Microbiology*, **30**, 774.

Horinouchi, S., Nishiyama, M., Nakamuba, A. & Beppu, T. (1987). *Molecular and General Genetics*, **210**, 468.

Howard, G. T. & Elliott, L. P. (1988). *Applied and Environmental Microbiology*, **54**, 4581.

Howard, G. T. & White, B. A. (1988). *Applied and Environmental Microbiology*, **54**, 1752.

Huang, L., Forsberg, C. W. & Thomas, D. Y. (1988). *Journal of Bacteriology*, **170**, 2923.

Ishizaki, A. & Kawauchi, H. (1988). *Agricultural and Biological Chemistry*, **52**, 2937.

Isizawa, S. & Araragi, M. (1976). *Actinomycetes: The Boundary Microorganisms*, ed. T. Arai. Toppan Ltd, Tokyo, p. 97.

Johnson, E. A., Sakajoh, M., Halliwell, G., Madia, S. & Demain, A. L. (1982). *Applied and Environmental Microbiology*, **43**, 1125.

Johnson, E. A., Bouchot, F. & Demain, A. L. (1985). *Journal of General Microbiology*, **131**, 2303.

Joliff, G., Beguin, P., Jugy, M., Millet, J., Ryter, A., Poljak, R. & Aubert, J. P. (1986). *Biotechnology*, **4**, 896.

Kadam, S., Demain, A. L., Millet, J., Beguin, P. & Aubert, J. P. (1988). *Enzyme and Microbial Technology*, **10**, 9.

Kawai, A., Okoshi, H., Ozaki, K., Shikata, S., Katutoski, A. & Ito, S. (1988). *Agricultural and Biological Chemistry*, **52**, 1425.

Khan, A. W. (1977). *Canadian Journal of Microbiology*, **23**, 1700.

Khan, A. W. (1980a). *Journal of General Microbiology*, **121**, 499.

Khan, A. W. (1980b). *FEMS Microbiology Letters*, **9**, 233.

Khan, A. W. & Trotter, T. M. (1978). *Applied and Environmental Microbiology*, **35**, 1027.

Khan, A. W., Trotter, T. M., Patel, G. B. & Margin, S. M. (1979). *Journal of General Microbiology*, **112**, 365.

Kim, H. & Pack, M. Y. (1988). *Enzyme and Microbial Technology*, **10**, 347.

Kluepfel, D., Shareck, F., Mondau, F. & Morosoli, R. (1986). *Applied Microbiology and Biotechnology*, **24**, 230.

Kopecny, J. & Williams, A. G. (1988). *Folia Microbiologica*, **23**, 208.

Lacy, J. (1971). *Journal of General Microbiology*, **66**, 327.

Lamed, R. & Bayer, E. A. (1988a). In *Biochemistry and Genetics of Cellulose Degradation*, eds J. Aubert, P. Beguin & J. Millet. Academic Press, London, p. 101.

Lamed, R. & Bayer, E. A. (1988b). *Advances in Applied Microbiology*, **33**, 1.

Lamed, R. & Bayer, E. A. (1990). In *Biosynthesis and Degradation of Cellulose and Cellulosic Materials*, eds P. Weimer & C. A. Haigler. Marcel Dekker, Inc, New York. (In press.)

Lamed, R., Setter, E. & Bayer, E. A. (1983a) *Journal of Bacteriology*, **156**, 828.

Lamed, R., Setter, E., Kenig, R. & Bayer, E. R. (1983b). *Biotechnology and Bioengineering Symposium*, **13**, 163.

Lamed, R., Kenig, R., Setter, E. & Bayer, E. A. (1985). *Enzyme and Microbial Technology*, **7**, 37.

Lamed, R., Naimark, J., Morgenstern, E. & Bayer, E. A. (1987). Specialized cell surface structures in cellulolytic bacteria. *J. Bacteriol.* **169**, 3792.

Langsford, M. L., Gilkes, N. R., Wakarchuk, W. W., Kilburn, D. G., Miller, R. C. & Warren, R. A. J. (1984). *Journal of General Microbiology*, **130**, 1367.

Leatherwood, J. M. (1973). *Federation Proceedings of the Federation of American Societies for Experimental Biology*, **32**, 1814.

Lee, B. H. & Blackburn, T. H. (1975). *Applied Microbiology*, **30**, 346.

Lee, S. E. & Humphrey, A. E. (1979). *Biotechnology and Bioengineering*, **21**, 1277.

Lejeune, A., Colson, C. & Eveleigh, D. E. (1986). *Journal of Industrial Microbiology*, **1**, 79.

Lejeune, A., Courtois, S. & Colson, C. (1988a). *Applied and Environmental Microbiology*, **54**, 302.

Lejeune, A., Dartois, V. & Colson, C. (1988b). *Biochimica et Biophysica Acta*, **950**, 204.

Lejeune, A., Eveleigh, D. E. & Colson, C. (1988c). *FEMS Microbiology Letters*, **49**, 363.

Lin, E. & Wilson, D. B. (1987). *Applied and Environmental Microbiology*, **53**, 1352.

Lin, E. & Wilson, D. B. (1988a). *Journal of Bacteriology*, **170**, 3838.

Lin, E. & Wilson, D. B. (1988b). *Journal of Bacteriology*, **170**, 3842.

Ljungdahl, L. G., Pettersson, B., Eriksson, K. E. & Wiegel, J. (1983). *Current Microbiology*, **9**, 195.

Ljungdahl, L. G., Coughlan, M. P., Mayer, F., Mori, Y., Hon-Nami, H. & Hon-Nami, K. (1988). In *Methods in Enzymology*, eds W. A. Wood & S. T. Kellogg. Academic Press, San Diego, Vol. 160, Part A, p. 483.

Love, D. R. & Streiff, M. B. (1987). *Biotechnology*, **5**, 384.

Lowenberg, J. R. & Chapman, C. M. (1977). *Archives of Microbiology*, **113**, 61.

Lupo, D. & Stutzenberger, F. (1988). *Applied and Environmental Microbiology*, **54**, 588.

MacKenzie, C. R. (1986). In *Biotechnology and Renewable Energy*, eds M. Moo-Young, S. Hasnain & J. Lamptey. Elsevier Applied Science Publishers, London, p. 76.

MacKenzie, C. R. & Bilous, D. (1982). *Canadian Journal of Microbiology,* **28,** 1158.
MacKenzie, C. R., Bilous, D. & Johnson, K. G. (1984). *Biotechnology and Bioengineering,* **26,** 590.
MacKenzie, C. R., Bilous, D. & Patel, G. B. (1985). *Applied and Environmental Microbiology,* **50,** 243.
MacKenzie, C. R., Patel, G. B. & Bilous, D. (1987). *Applied and Environmental Microbiology,* **53,** 304.
McBee, R. H. (1948). *Journal of Bacteriology,* **56,** 653.
McBee, R. H. (1950). *Bacteriological Reviews,* **14,** 51.
McCarthy, A. J. (1987). *FEMS Microbiology Reviews,* **46,** 145.
McCarthy, A. J. & Cross, T. (1984). *Journal of General Microbiology,* **130,** 5.
McGavin, M. & Forsberg, C. W. (1988). *Journal of Bacteriology,* **170,** 2914.
Mann, S. P. (1988). *Letters in Applied Microbiology,* **7,** 119.
Margaritis, A. & Merchant, R. F. J. (1986). *Chemical Rubber Company Critical Reviews in Biotechnology,* **4,** 327.
Meyer, H. & Humphrey, A. E. (1982). *Biotechnology and Bioengineering,* **24,** 1901.
Miller, R. C., Gilkes, N. R., Greenberg, N. M., Kilburn, D. G., Langsford, M. L. & Warren, R. A. J. (1988). In *Biochemistry and Genetics of Cellulose Degradation,* eds J. P. Aubert, P. Beguin & J. Millet. Academic Press, London, p. 235.
Millet, J., Petre, D., Beguin, P., Raynaud, O. & Aubert, J. P. (1985). *FEMS Microbiology Letters,* **29,** 145.
Mohagheghi, A., Grohmann, K., Himmel, M., Leighton, L. & Updegraff, D. M. (1986). *International Journal of Systematic Bacteriology,* **36,** 435.
Moldoveanu, N. & Kluepfel, D. (1983). *Applied and Environmental Microbiology,* **46,** 17.
Montenecourt, B. S., Carter, J. M., Phillips, J. A. & Blumenthal, H. (1983). In *Wood and Agricultural Residues,* ed. E. J. Soltes. Academic Press, NY, p. 271.
Morris, E. J. & Cole, O. J. (1987). *Journal of General Microbiology,* **133,** 1023.
Murray, W. D. (1986). *Biomass,* **10,** 47.
Murray, W. D. (1987a). *Journal of Industrial Microbiology,* **1,** 393.
Murray, W. D. (1987b). *Biotechnology and Bioengineering,* **29,** 1151.
Murray, W. D., Sowden, L. C. & Colvin, J. R. (1984). *International Journal of Systematic Bacteriology,* **34,** 185.
Murray, W. D., Sowden, L. C. & Colvin, J. R. (1986). *Letters in Applied Microbiology,* **3,** 69.
Nakai, R., Horinouchi, S., Uozumi, T. & Beppu, T. (1987). *Agricultural and Biological Chemistry,* **51,** 3061.
Nakai, R., Aorinouchi, S. & Beppu, T. (1988). *Gene,* **65,** 229.
Nakamura, K., Misawa, N. & Kitamura, K. (1986a). *Journal of Biotechnology,* **3,** 239.
Nakamura, K., Misawa, N. & Kitamura, K. (1986b). *Journal of Biotechnology,* **3,** 247.
Ng, T. K. & Zeikus, J. G. (1981). *Biochemical Journal,* **199,** 341.
Ng, T. K. & Zeikus, J. G. (1982). *Journal of Bacteriology,* **150,** 1391.
Ng, T. K., Weimer, P. J. & Zeikus, J. G. (1977). *Archives of Microbiology,* **114,** 1.
Norris, J. R., Berkeley, R. C. W., Logan, N. A. & O'Donnell, A. G. (1981). In *The Prokaryotes. A Handbook of Habitats, Isolation, and Identification of Bacteria,* ed. M. P. Starr, Springer-Verlag, Berlin, p. 1711.
Oberkotter, L. W. & Rosenberg, F. A. (1980). In *4th International Biodeterioration Symposium,* eds T. A. Oxley, D. Allsopp & G. Becker. Pitman Publishers, London, p. 731.
Ohmiya, K., Shimizu, M., Taya, M. & Shimizu, S. (1982). *Journal of Bacteriology,* **150,** 407.
Ohmiya, K., Shirai, M., Kurachi, Y. & Shimizu, S. (1985). *Journal of Bacteriology,* **161,** 432.
Ohmiya, K., Maeda, K. & Schimizu, S. (1987). *Carbohydrate Research,* **166,** 145.

Ohmiya, K., Nagashima, K., Kajino, T., Goto, E., Tsukada, A. & Shimizu, S., (1988). *Applied and Environmental microbiology*, **54**, 1511.

O'Neill, G., Goh, S. H., Warren, R. A. J., Kilburn, D. G. & Miller, R. C. (1986a). *Gene*, **44**, 325.

O'Neill, G. P., Kilburn, D. G., Warren, R. A. J. & Miller, R. C. (1986b). *Applied and Environmental Microbiology*, **52**, 737.

O'Neill, G. P., Warren, R. A. J., Kilburn, D. G. & Miller, R. C. (1986c). *Gene*, **44**, 331.

Owalabi, J. B., Beguin, P., Kilburn, D. G., Miller, R. C. & Warren, R. A. J. (1988). *Applied and Environmental Microbiology*, **54**, 518.

Paradis, R. W., Warren, R. A. J., Kilburn, D. G. & Miller, R. C. (1987). *Gene*, **61**, 199.

Patchett, M. L., Neal, T. L., Schofield, L., Strange, R. C., Daniel, R. M. & Morgan, H. W. (1989). *Enzyme and Microbial Technology*, **11**, 113.

Patel, G. B. (1984). In *Bergey's Manual of Systematic Bacteriology*, eds N. R. Krieg & J. G. Holt. Williams and Wilkins, Baltimore, London, p. 658.

Patel, G. B. & MacKenzie, C. R. (1982). *European Journal of Applied Microbiology and Biotechnology*, **16**, 212.

Patel, G. B., Khan, A. W., Agnew, B. J. & Colvin, J. R. (1980). *International Journal of Systematic Bacteriology*, **30**, 179.

Pepys, J., Jenkins, P. A., Festenstein, G. N., Lacy, M. E., Gregory, P. H. & Skinner, F. A. (1963). *Lancet* **7308**, 607.

Perez-Martinez, G., Gonzalez-Candelas, L., Polania, J. & Flors, A. (1988). *Journal of Industrial Microbiology*, **3**, 365.

Piruzyan, E. S., Mogutov, M. A., Velikodvorskaya, G. A. & Pushkarkaya, T. A. (1988). *Genetika*, **24**, 204.

Plant, A. R., Oliver, J. E., Patchett, M. L., Daniel, R. M. & Morgan, H. W. (1988). *Archives of Biochemistry and Biophysics*, **262**, 181.

Pridham, T. G. & Tresner, H. D. (1974). In *Bergey's Manual of Determinative Bacteriology*, eds R. E. Buchanan & N. E. Gibbons. Williams and Wilkins, Baltimore, p. 474.

Priest, F. G. (1977). *Bacteriological Reviews*, **41**, 711.

Rabinovich, M. L., Buachidze, T. X., Klesov, A. A. & Kvesitadze, G. I. (1983). *Prikladnaya Biokhimiya i Mikrobiologiya*, **19**, 779.

Rajoka, M. I. & Malik, K. A. (1986). *Biotechnology Letters*, **8**, 753.

Ramasamy, K. & Verachtert, H. (1980). *Journal of General Microbiology*, **117**, 181.

Randall, L. L. & Hardy, S. J. S. (1984). *Microbiological Reviews*, **48**, 290.

Rao, M. R. & Dhala, S. A. (1981). *International Journal of Microbiology*, **21**, 216.

Rasmussen, M. A., Hespell, R. B., White, B. A. & Bothast, R. J. (1988). *Applied and Environmental Microbiology*, **54**, 890.

Reynolds, P. H. S., Sissons, C. H., Daniel, R. M. & Morgan, H. W. (1986). *Applied and Environmental Microbiology*, **51**, 12.

Roberts, D. P., Denny, T. P. & Schell, M. A. (1988). *Journal of Bacteriology*, **170**, 1445.

Robinson, I. M. & Richie, A. E. (1981). *International Journal of Systematic Bacteriology*, **31**, 333.

Robson, L. M. & Chambliss, G. H. (1984). *Applied and Environmental Microbiology*, **47**, 1039.

Rodriguez, H. & Volfoa, O. (1984). *Applied Microbiology in Biotechnology*, **19**, 134.

Rodriguez, H., Volfoa, O. & Klyosov, A. (1988). *Applied Microbiology and Biotechnology*, **28**, 394.

Romaniec, M. P. M., Clarke, N. G. & Hazelwood, G. P. (1987a). *Journal of General Microbiology*, **133**, 1297.

Romaniec, M. P. M., Davidson, K. & Hazelwood, G. P. (1987b). *Enzyme and Microbial Technology*, **9**, 474.

Saddler, J. N. & Khan, A. W. (1980). *Canadian Journal of Microbiology*, **26**, 760.

Saddler, J. N. & Khan, A. W. (1981). *Canadian Journal of Microbiology*, **27**, 288.

Saddler, J. N., Khan, A. W. & Martin, S. M. (1980). *Microbiology*, **28**, 97.

Sami, A. J., Akhtar, M. W., Malik, N. N. & Naz, B. A. (1988). *Enzyme and Microbial Technology*, **10**, 626.

Schell, M. A. (1987). *Applied and Environmental Microbiology*, **53**, 2237.

Schellhorn, H. E. & Forsberg, C. W. (1984). *Canadian Journal of Microbiology*, **30**, 930.

Schofield, L. R., Neal, T. L., Patchett, M. L., Strange, R. C., Daniel, R. M. & Morgan, H. W. (1989). *Annals of the New York Academy of Sciences*, **542**, 240.

Schwarz, W. H., Schimming, S. & Staudenbauer, W. L. (1987). *Applied Microbiology and Biotechnology*, **27**, 50.

Schwarz, W. H., Schimming, S., Rucknagel, K. P., Burgschwaiger, S., Kreil, G. & Staudenbauer, W. L. (1988). *Gene*, **63**, 23.

Seltzer, R. (1987). *Chemical and Engineering News*, **4**, 22.

Shareck, F., Mondou, F., Morosoli, R. & Kluepfel, D. (1987). *Biotechnology Letters*, **9**, 169.

Sissons, C. H., Sharrock, K. R., Daniel, R. M. & Morgan, H. W. (1987). *Applied and Environmental Microbiology*, **53**, 832.

Smith, W. R., Yu, I. & Hungate, R. E. (1973). *Journal of Bacteriology*, **114**, 729.

Sneath, P. H. A. (1986). In *Bergey's Manual of Systematic Bacteriology*, eds P. H. A. Sneath, N. S. Mair, M. E. Sharpe & J. G. Holt. Williams and Wilkins, Baltimore, MD, p. 1104.

Soutschek-bauer, E. & Staudenbauer, W. L. (1987). *Molecular and General Genetics*, **208**, 537.

Stackebrandt, E. & Keddie, R. M. (1986). In *Bergey's Manual of Systematic Bacteriology*, Vol. 2, eds P. H. A. Sneath, N. S. Mair, M. E. Sharpe & J. G. Holt. Williams and Wilkins, Baltimore, MD, p. 1325.

Stewart, B. J. & Leatherwood, J. M. (1976). *Journal of Bacteriolgoy*, **128**, 609.

Stewart, C. S., Dinsdale, D., Cheng, K. J. & Paniagua, C. (1980). In *Straw Decay and its Effects on Disposal and Utilization*, ed. E. Grossbard. John Wiley and Sons, New York, p. 123.

Stoppok, W., Rapp, R. & Wagner, F. (1982). *Applied and Environmental Microbiology*, **44**, 44.

Storvick, W. O. & King, K. W. (1960). *Journal of Biological Chemistry*, **235**, 303.

Stutzenberger, F. J. (1971). *Applied Microbiology*, **22**, 147.

Stutzenberger, F. J. (1972a). *Applied Microbiology*, **24**, 77.

Stutzenberger, F. J. (1972b). *Applied Microbiology*, **24**, 83.

Stutzenberger, F. J. (1979). *Biotechnology and Bioengineering*, **21**, 83.

Stutzenberger, F. J. & Kahler, G. (1986). *Journal of Applied Bacteriology*, **61**, 225.

Stutzenberger, F. & Lupo, D. (1986). *Enzyme and Microbial Technology*, **8**, 205.

Stutzenberger, F. J., Kaufman, A. J. & Lossin, R. D. (1970). *Canadian Journal of Microbiology*, **16**, 553.

Su, T. & Paulavicius, I. (1975). *Proceedings of the 8th Cellulose Conference*, ed. T. E. Timell. *Applied Polymer Symposium*, **28**, 221.

Sussman, M., Collins, C. H., Skinner, F. A. & Stewart-Tull, D. E. (1988). In *Proceedings of the First International Conference on the Release of Genetically-Engineered Microorganisms*. Academic Press, London.

Thayer, D. W., Lowther, S. V. & Phillips, J. G. (1984). *International Journal of Systematic Bacteriology*, **34**, 432.

van Hofsten, B. & Berg, B. (1972). Cellulase formation in *Cellvibrio fulvus. Proc. 4th*

International Fermentation Symposium, Fermentation Technology Today, Uppsala, p. 731.

Vandemark, P. J. & Batzing, B. L. (1987). *The Microbes*. Benjamin/Cummings Co., Menlo Park, CA, p. 153.

Viljoen, J. A., Fred, E. B. & Peterson, W. H. (1926). *Journal of Agricultural Science*, **16**, 1.

Vlasenko, E. Y., Sinitzyn, A. P., Scherbukhin, V. D. & Klesov, A. A. (1985). *Biotekhnologiya*, **6**, 76.

Wakarchuk, W. W., Kilburn, D. G., Miller, R. C. & Warren, R. A. J. (1984). *Journal of General Microbiology*, **130**, 1385.

Waldron, C. R. & Eveleigh, D. E. (1986). *Applied Microbiology and Biotechnology*, **24**, 487.

Waldron, C. R., Becker-Vallone, C. A. & Eveleigh, D. E. (1986). *Applied Microbiology and Biotechnology*, **24**, 477.

Warren, R. A. J., Beck, C. F., Gilkes, N. R., Kilburn, D. G., Langsford, M. L., Miller, R. C. Jr, O'Neill, G. P., Scheufens, M. & Wong, W. K. R. (1986). *Proteins: Structure, Function and Genetics*, **1**, 335.

Warren, R. A. J., Gerhard, B., Gilkes, N. R., Owolabi, J. B., Kilburn, D. G. & Miller, R. C. (1987). *Gene*, **61**, 421.

Whittle, D. J., Kilburn, D. G., Warren, R. A. J. & Miller, R. C. (1982). *Gene*, **17**, 139.

Wilson, D. B. (1988). In: *Methods in Enzymology*, (eds W. A. Wood & S. T. Kellogg). Academic Press, San Diego, Vol. 160, Part A, p. 483.

Winfrey, M. R. & Zeikus, J. G. (1977). *Applied and Environmental Microbiology*, **33**, 275.

Wolff, B. R., Mudry, T. A., Glick, B. R. & Pasternak, J. J. (1986). *Applied and Environmental Microbiology*, **51**, 1367.

Wong, S. L., Kawamura, F. & Doi, R. H. (1986a). *Journal of Bacteriology*, **168**, 1005.

Wong, W. K. R., Gerhanr, B., Gou, Z. M., Kilburn, D. G., Warren, R. A. & Miller, R. C. (1986b). *Gene*, **44**, 315.

Wood, D. A. (1985). *Annals of the Proceedings of the Phytochemical Society of Europe*, **25**, 295.

Wood, T. M., Wilson, C. A. & Stewart, C. S. (1982). *Biochemical Journal*, **205**, 129.

Wood, W. E., Neubauer, D. G. & Stutzenberger, F. J. (1984). *Journal of Bacteriology*, **160**, 1047.

Wynne, E. C. & Pemberton, J. M. (1986). *Applied and Environmental Microbiology*, **52**, 1362.

Yablonsky, M. C., Bartlye, T., Ellison, K. O., Kahars, S. K., Shalita, Z. P. & Eveleigh, D. E. (1988). In *Biochemistry and Genetics of Cellulose Degradation*, eds J. P. Aubert, P. Beguin & J. Millet. Academic Press, London, p. 249.

Yamane, K., Suzuki, H., Hirotani, M., Ozawa, H. & Nisizawa, K. (1965). *Fermentation Technology*, **43**, 721.

Yamane, K., Suzuki, H., Hirotani, M., Ozawa, H. & Nisizawa, K. (1970a). *Journal of Biochemistry*, **67**, 9.

Yamane, K., Suzuki, H. & Nisizawa, K. (1970b). *Journal of Biochemistry*, **67**, 19.

Yamane, K., Yoshikawa, T., Suzuki, H. & Nisizawa, K. (1971). *Journal of Biochemistry*, **69**, 771.

Yishikawa, T., Suzuki, H. & Nisizawa, K. (1974). *Journal of Biochemistry*, **75**, 531.

Zappe, H., Jones, W. A., Jones, D. T. & Woods, D. R. (1988). *Applied and Environmental Microbiology*, **54**, 1289.

Zappe, H., Jones, D. T. & Woods, D. R. (1986). *Journal of General Microbiology*, **132**, 1367.

Zemek, J., Augustin, J., Borriss, R., Kuniak, L., Svabova, M. & Pacova, Z. (1981). *Folia Microbiologica*, **26**, 403.

Chapter 3

RECENT ADVANCES IN MICROBIAL AMYLASES

WILLIAM M. FOGARTY & CATHERINE T. KELLY

*Department of Industrial Microbiology, University College, Belfield,
Dublin 4, Republic of Ireland*

CONTENTS

1 INTRODUCTION

A wide variety of microorganisms produce, and in most cases secrete extracellularly, amylases having different specificities and some rather interesting properties (Fogarty & Kelly, 1980; Fogarty, 1983). Gram-positive bacteria, and particularly the genus *Bacillus,* are prolific producers of amylases, although very few are found among Gram-negative bacteria. A wide range of moulds produce amylases, particularly the genus *Aspergillus.* In recent years interest and awareness in the amylolytic activities of yeasts has generated a number of studies of these systems. It is not the purpose of this review to provide an exhaustive or detailed treatise of microbial amylases. Other publications have dealt with the subject area extensively (Fogarty & Kelly, 1979, 1980; Fogarty, 1983) and this work has as its aim an update of developments which have taken place in the area in recent years.

2 α-AMYLASES—GENERAL PROPERTIES

2.1 Bacterial α-Amylases

Most industrial applications of α-amylase (EC 3.2.1.1, 1.4-α-D-glucan glucanohydrolase, endo-amylase) require their use at high temperatures and this sustains the ongoing search for new enzymes having increasingly better thermostability properties. One approach to this is to screen new sources for thermostable amylases. Alternatively, the thermostability properties of an existing enzyme might be improved by site-directed mutagenesis. The molecular mechanisms of enzyme inactivation must, however, be known, in the first instance, in order to construct rational thermostabilisation programmes (Ahern & Kilbanov, 1985; Zale & Klibanov, 1986). An investigation of irreversible thermoinactivation, at 90°C and at the relevant pH values, of the α-amylases of *Bacillus amyloliquefaciens* and *Bacillus stearothermophilus* was recently undertaken (Tomazic & Klibanov, 1988*a*). These enzymes have 60% homology in their amino acid sequence (Nakajima *et al.*, 1986) and differ significantly in thermostability (Pfueller & Elliott, 1969; Ogasahara *et al.*, 1970; Singleton & Amelunxen, 1973; Yutani *et al.*, 1973). At pH 5·0, 6·5 and 8·0 *B. amyloliquefaciens* α-amylase inactivates irreversibly, by a monomolecular conformational process, resulting in the formation of scrambled structures which undergo aggregation. The presence of the substrate starch suppresses this process at pH 8·0. However, an additional process, deamidation of glutamine and/or asparagine residues, is now a major factor in thermoinactivation. In the case of *B. stearothermophilus* α-amylase, monomolecular conformational scrambling is the major cause of irreversible inactivation at 90°C and pH values of 5·0, 6·5 and 8·0. An additional contributing inactivation process at pH 6·5 is the deamidation of amino acid amide residues and at pH 8·0 oxidation of cysteine. Half-lives of the α-amylases of *B. amyloliquefaciens, B. stearothermophilus* and *B. licheniformis* at 90°C and pH 6·5 were 2, 50 and 270 min, respectively (Tomazic & Klibanov, 1988*b*). Unlike the first two systems which are inactivated by monomolecular conformational scrambling, the mechanism of irreversible thermoinactivation of *B. licheniformis* α-amylase is caused by deamidation of asparagine and/or glutamine residues. The half-life of *B. licheniformis* α-amylase is comparable with the enzyme of *B. stearothermophilus* under similar conditions in the presence of starch, where the deamidation of asparagine and/or glutamine residues was thought to be the thermoinactivation process. The much greater stability against thermoinactivation of a thermophilic α-amylase, compared to a mesophilic enzyme, is essentially due to extra salt bridges caused by two or three specific lysine residues. Thus, with regard to heat-induced unfolding, these electrostatic interactions stabilise the enzyme. The information gained in this work (Tomazic & Klibanov, 1988*a, b*) may be beneficial in attempts to stabilise α-amylases by recombinant DNA technology. Thus, if amino acid residues 88 and 253 in the enzyme molecule were replaced by lysine and/or arginine, the half-life of the α-amylase of *B. stearothermophilus* would be expected to increase. Likewise, the α-amylase of *B. amyloliquefaciens* might be considerably improved if residues 88, 253 and 385 were replaced by lysine and/or arginine residues.

α-Amylases have been reported in a number of new sources (Table 1). David *et al.* (1987) described an extremely novel amylolytic enzyme in *Bacillus megaterium*. Production of the enzyme was increased by a factor of 10 000 by genetic engineering. *B. megaterium* amylase (BMA) does not attack α-1,6-linkages and hydrolyses pullulan to panose. Its most outstanding property, besides hydrolysis, is that it catalyses transfer reactions and these become more pronounced in the presence of suitable acceptors. The latter can be sugars having the same configuration as dextrose in the C-2, C-3 and C-4 positions. In excess of 1% of oligosaccharides remain unhydrolysed in industrial production of dextrose from starch. These oligomers contain 35% α-1,6- linkages and can be degraded through panosyl transfer to dextrose as a result of BMA

Table 1
Some Recently Described α-Amylases

Bacillus acidocaldarius A-2	Kanno (1986)
B. licheniformis	Krishnan & Chandra (1983)
B. megaterium	David et al. (1987)
B. megaterium S-218	Stark et al. (1982)
B. subtilis	Orlando et al. (1983)
Lactobacillus cellobiosus D-39	Sen & Chakrabarty (1986)
Thermoactinomyces sp.	Obi & Odibo (1984)
Schwanniomyces alluvius	Simoes-Mendes (1984)
Candida antarctica	De Mot & Verachtert (1987)
Filobasidium capsuligenum	De Mot & Verachtert (1985)
Trichosporon pullulans	De Mot & Verachtert (1986b)
Thermomyces lanuginosus	Jensen et al. (1988)
Pencillium expansum	Doyle et al. (1989)
Penicillium amagasakiense	Doyle et al. (1988)

synthesising 6^3-α-glucosylmaltotriose which can then be degraded to dextrose by amyloglucosidase. A practical consequence is the increase in yield of dextrose in the saccharification of starch (Hebeda et al., 1988).

Stark et al. (1982) reported the separation of α- and β-amylases, pullulanase and α-glucosidase from the culture liquor of B. megaterium S218 by chromatofocusing.

α-Amylase produced by Bacillus licheniformis CUMC 305 (Krishnan & Chandra, 1983) had maximal activity at 90°C and pH 9·0 (Table 2) and in the

Table 2
Properties of Some α-Amylases

Enzyme source	pH optimum	Temp. optimum (°C)	Relative molecular mass (M_r)	Isoelectric point (pI)	Reference
Bacillus megaterium	5·5	75	55 000	9·5	David et al. (1987)
Bacillus licheniformis cumc 305	9·0	90	28 000	—	Krishnan & Chandra (1983)
Bacillus subtilis	6–7	—	93 000	5·0	Orlando et al. (1983)
Bacillus acidocaldarius A-2	3·5	70	66 000	—	Kanno (1986)
Lactobacillus cellobiosus D-39	7·3	50	22 500	—	Sen & Chakrabarty (1986)
Thermoactinomyces sp.	7·0	80	47 800	—	Obi & Odibo (1984)
Schwanniomyces alluvius	6·3	40	62 000	—	Simoes-Mendes (1984)
Candida antarctica	4·2	62	50 000	10·3	De Mot & Verachtert (1987)
Filobasidium capsuligenum	5·6	50	64 000	—	De Mot & Verachtert (1985)
Thermomyces lanuginosus	4·6	—	54 000–57 000	3·4	Jensen et al. (1988)
Aspergillus kawachii I	4·0–5·0	70	104 000	4·25	Mikami et al. (1987)
II	5·0	70	66 000	4·2	Mikami et al. (1987)
Aspergillus awamori	4·8–5·0	50	54 000	4·2	Bhella & Altosaar (1985)
Paecilomyces sp.	4·0	45	69 000	—	Zenin & Park (1983)
Penicillium expansum	4·5	60	69 000	3·9	Doyle et al. (1989)
Penicillium amagasakiense	4·5	50	63 000	—	Doyle et al. (1988)

Table 3
Properties of α-Amylase in Strains of *Bacillus acidocaldarius*

Strain	M_r	pH optimum	Temp. optimum (°C)	K_m (mg starch/ml)	Activated by Ca^{2+}	Major products from starch	Reference
B. acidocaldarius Agnano 101	68 000	3·5	75	0·8–0·9	+	G_1–G_3	Buonocore *et al.* (1976)
B. acidocaldarius 104-1A	—	4·5	60	—	+	G_2	Boyer *et al.* (1979)
B. acidocaldarius 11-1S	54 000	2·0	70	1·64	−	G_2, G_3	Uchino (1982)
B. acidocaldarius A-2	66 000	3·5	70	1·6	−	G_2, G_3	Kanno (1986)

presence of substrate (starch) was fully stable at 100°C for 4 h. Blakeney & Stone (1985) suggest that a change in substrate affinity in *B. licheniformis* α-amylase is induced at the active centre sub-sites in ethanol–aqueous buffer media. The pattern of oligosaccharides produced from amylose changed with ethanol concentration.

Bacillus acidocaldarius(A2) α-amylase showed greater stability to heat under acidic conditions (Kanno, 1986) compared to other strains of the same organism. On prolonged incubation with starch the major end-products were maltose (18%) and maltotriose (21%). There is a degree of relatedness (Table 3) between this enzyme and those produced by *B. acidocaldarius* Agnano 101 (Buonocore *et al.*, 1976) and *Bacillus* sp. 11-1S (Uchino, 1982), whereas the α-amylase of *B. acidocaldarius* 104-1A is different and less robust in relation to temperature and pH characteristics (Boyer *et al.*, 1979) and, furthermore, maltose is the major product formed on hydrolysis of starch. *Lactobacillus cellobiosus* D-39 isolated from vegetable wastes (Sen & Chakrabarty, 1986) produced an α-amylase maximally active at 50°C in the pH range 6·3–7·9 and released glucose and maltose as the major end-products from starch. A highly stable α-amylase detected in culture filtrates of *Thermoactinomyces* sp. No. 15 (Obi & Odibo, 1984) was purified and had maximum activity at pH 7·0 and 80°C and retained 74% activity after 30 min at 100°C. Although a detailed study of end-products formed on hydrolysis of starch was not carried out it was shown that glucose and maltose were the main sugars released from a number of starches.

2.2 Fungal α-Amylases

2.2.1 Yeast α-Amylases

Interest in amylolytic yeasts has increased in recent years as their potential to utilise starch for the production of single cell protein and ethanol is recognised (Spencer-Martins & van Uden, 1977; Frelot *et al.*, 1982; Touzi *et al.*, 1982; Wilson *et al.*, 1982; Laluce & Mattoon, 1984). The extracellular amylolytic enzymes produced by species of *Lipomyces* (Spencer-Martins, 1982, 1984; Spencer-Martins & van Uden, 1979; Kelly *et al.*, 1985), *Saccharomycopsis* (Ebertova, 1966; Sukhumavasi *et al.*, 1975; Kato *et al.*, 1976) and *Schwanniomyces* (Oteng-Gyang *et al.*, 1981; Sills *et al.*, 1984; Simoes-Mendes, 1984) have been characterised.

Several species of yeasts have been reported to utilise starch as a sole carbon source and secrete amylolytic enzymes. The amylase system of *Schwanniomyces castellii* possesses both α-amylase and amyloglucosidase activities (Clementi *et al.*, 1980; Oteng-Gyang *et al.*, 1981; Wilson & Ingledew, 1982; Sills *et al.*, 1984). Several characteristics of the α-amylase have been reported (Dubreucq *et al.*, 1989) and kinetics of the hydrolysis of various substrates examined. Pasari *et al.* (1988) studied product distributions and starch hydrolysis rates using the same system. *Filobasidium capsuligenum* is capable of extensive starch hydrolysis (De Mot *et al.*, 1984*a*) and consists of α-amylase and two forms of amyloglucosidase (De Mot & Verachtert, 1985).

Using β-cyclodextrin as sole carbon source, De Mot & Verachtert (1986*a*) observed that *Candida antarctica* secreted significantly higher levels of amylolytic activity than many of the currently examined yeast species. Both α-amylase and amyloglucosidase were purified to homogeneity from the culture fluid (De Mot & Verachtert, 1987). Kinetic analyses showed that both enzymes hydrolysed high molecular mass substrates preferentially, including some raw starches. α-Amylase was active on cyclodextrins whereas some debranching activity was associated with amyloglucosidase. *Candida tsukuba-ensis* produces only low levels of α-amylase with amyloglucoside as the major activity (De Mot *et al.*, 1985). The amylases of *Trichosporon pullulans* (De Mot & Verachtert, 1986*b*) were partially purified and separated into an α-amylase and amyloglucosidase.

2.2.2 Mould α-Amylases

The thermophilic fungus, *Thermomyces lanuginosus* (formerly *Humicola lanuginosa*) has been shown to secrete amylolytic enzymes (Barnett & Fergus, 1971; Adams, 1981; Jensen *et al.*, 1987). Jensen *et al.* (1986) showed that highest amylolytic activity was produced in static culture and separated the activity into three components; an α-amylase, an amyloglucosidase and an α-glucosidase (Jensen *et al.*, 1988). *Aspergillus kawachii* α-amylase I and II (Mikami *et al.*, 1987) had relative molecular mass (M_r) values of 104 000 and 66 000, respectively, and the optimum temperatures of both α-amylases were around 70°C at pH 5·0. Maltose and maltotriose were the dominant products detected by HPLC on prolonged hydrolysis of starch. Data presented by Bhella & Altosaar (1985) showed that the α-amylase from *Aspergillus awamori* is similar to α-amylases from other *Aspergillus* spp. in almost all physicochemical properties.

2.3 Amino Acid Sequence and Structure

The amino acid sequences, in whole or in part, of a number of α-amylases have been reported (Sachdev & Friedberg, 1981; Kuhn *et al.*, 1982; Takkinen *et al.*, 1983; Ihara *et al.*, 1985; Yuuki *et al.*, 1985; Nakajima *et al.*, 1986). These have been deduced from nucleotide sequences or by direct amino acid sequencing (Toda *et al.*, 1982; Nagata *et al.*, 1985).

The amino acid sequence of *B. licheniformis* α-amylase had 65·4% and 80·3% homology with *B. stearothermophilus* and *B. amyloliquefaciens* α-amylases, respectively (Yuuki *et al.*, 1985). Comparisons of amino acid sequences have indicated a fair degree of homology (*c.* 60%) between *B. stearothermophilus* and *B. amyloliquefaciens* thermostable α-amylases (Nakajima *et al.*, 1986). The well established differences between saccharifying and liquefying α-amylase is further evident from work to-date on their amino acid sequences (Palva *et al.*, 1981; Ohmura *et al.*, 1983; Yang *et al.*, 1983; Nagata,

1985). Despite the low degree of homology between α-amylases, four highly homologous regions were noted in the sequences of all α-amylases investigated and these may be involved in active sites and/or substrate binding sites. Detailed analysis of amino acid sequence alignments by MacGregor (1988) has indicated further homologies. Although α-amylases from different sources vary widely in amino acid sequence, they are believed to have the same basic secondary structure, i.e. $(\alpha/\beta)_8$—a barrel of eight parallel β-strands surrounded by eight α-helices (MacGregor, 1988; MacGregor & Svensson, 1989).

The α-amylase of *A. oryzae* (Taka amylase A) is the only microbial α-amylase for which the three dimensional structure has been described (Matsuura *et al.*, 1984). Friedberg (1983) suggested the possibility of a family of carbohydrases where certain amino acid residues compose the active site and although the sequence is not necessarily homologous the arrangement of the active site may be. More recently it has again been suggested that, apart from similarities between α-amylases, a wide group of starch and oligosaccharide-degrading enzymes may be structurally related (Svensson, 1988) and may share key structural features in the active centres. By combining computerised predictions of secondary structure along with sequence similarities MacGregor & Svensson (1989) suggest strong correlation between the three dimensional structures of the catalytic areas of *A. oryzae* α-amylase, a yeast α-glucosidase and cyclodextrin glucanotransferases.

2.4 New Enzymes Possessing Both α-Amylase and Pullulanase Activities

Enzymes possessing activity towards both starch and pullulan (Table 4) have been reported in a number of instances and because of their unique properties it is envisaged that they could have special industrial applications. The first

Table 4
Starch–Pullulan Enzyme Degrading Complexes

Source	Product of action on		Reference
	Starch	Pullulan	
Thermoactinomyces			Shimizu *et al.* (1978)
vulgaris	Maltose	Panose	Sakano *et al.* (1982a)
Bacillus			
stearothermophilus KP 1064	Maltose	Panose	Suzuki & Imai (1985)
Bacillus subtilis TU	Maltotriose, maltose	Maltotriose	Takasaki (1987)
Bacillus megaterium	Maltooligomers	Panose	David *et al.* (1987)
Bacillus			
thermoamyloliquefaciens KP 1071	Maltose	Panose	Suzuki *et al.* (1987c)
Thermoanaerobium Tok 6-BI	Maltose, maltotriose, maltotetraose	Maltotriose	Plant *et al.* (1987a, b)
Thermoanaerobium brockii	Maltooligomers	Maltotriose	Coleman *et al.* (1987)
Clostridium			
thermohydrosulfuricum E101-69	Maltooligomers	Maltotriose	Melasniemi (1987a, b)

enzyme reported to possess α-amylase and pullulanase activities was detected in *Thermoactinomyces vulgaris* (Shimizu *et al.*, 1978). Hydrolysis of starch gave maltose (74%) and glucose (12%) whereas degradation of pullulan gave panose (97%). K_m, V_{max} and K_i values supported the view that the hydrolytic action of the enzyme on starch and pullulan was due to a single catalytic site (Sakano *et al.*, 1982*a*). Both pullulan-hydrolysing and starch-hydrolysing activities were inhibited by *p*-chloromercuribenzoate, maltotriitol and microbial α-amylase inhibitors to about the same extent. Various investigations reported the capacity of the system to hydrolyse α-1,4- and α-1,6- linkages (Fukushima *et al.*, 1982; Sakano *et al.*, 1983*a*, 1985*a*). The sub-site structure of the enzyme was estimated from its action and rates of hydrolysis of maltooligosaccharides. Results indicated that this α-amylase had six sub-sites with the catalytic site located between the third and fourth sub-sites (Sano *et al.*, 1985).

Bacillus stearothermophilus KP 1064 (Suzuki & Imai, 1985) produces an enzyme that strongly resembles the system from *T. vulgaris*. Pullulan was hydrolysed quantitatively to panose, and maltose was the dominant product formed from amylose. The system also degraded cyclodextrins and maltotriose. It did not attack glycogen and amylopectin was degraded at less than 10% of the rate achieved with amylose.

Takasaki (1987) reported an amylase–pullulanase complex in a strain of *Bacillus subtilis*. In the initial stages of degradation of starch considerably more G_3 (maltotriose) than G_2 (maltose) was formed. However, when the amount of G_3 formed reached a yield equivalent to 50–55%, G_3 gradually decreased while G_2 gradually increased without appreciable formation of glucose. Pullulan was almost totally degraded to G_3, thereafter G_3 gradually decreased while G_2 was formed without significant amounts of glucose being formed. Laboratory scale saccharification of a starch hydrolysate (30% dissolved solids; dextrose equivalent 7·7) with a combination of amyloglucosidase and this new enzyme gave increased yields of dextrose of 1–3% compared with the yield obtained using amyloglucosidase alone. The enzyme could also be used to produce a syrup containing G_3 (50–55%), G_2 (20–40%) and G (glucose) (2–5%).

As mentioned in Section 2.1, an amylolytic enzyme from *Bacillus megaterium* has been described (David *et al.*, 1987) which hydrolyses pullulan exclusively to panose (yield 98%). Furthermore, it hydrolyses crosslinked Phadebas starch used for determination of α-amylase, and it also catalyses transfer reactions. This new system could be used instead of pullulanase in industrial dextrose production. Amyloglucosidase-resistant structures are degraded by coupling reactions using dextrose as acceptor substrate to maltooligosaccharides which are then suitable substrates for amyloglucosidase.

Bacillus thermoamyloliquefaciens KP 1071 (Suzuki *et al.*, 1987*c*), a facultative thermophile, produces two distinct α-amylases. α-Amylase I is a maltotriogenic endo-acting enzyme whereas α-amylase II is a maltogenic exo-acting enzyme capable of hydrolysing α-1,6- bonds in amylopectin.

Furthermore, α-amylase II produces panose as the major end-product from pullulan and hydrolyses maltotriose to maltose and glucose. This enzyme is markedly similar to the maltogenic α-amylases of *B. stearothermophilus* KP 1064 (Suzuki & Imai, 1985) and *Thermoactinomyces vulgaris* in terms of substrate specificity (Shimizu *et al.*, 1978; Sakano et al., 1982*a*).

Thermoanaerobium Tok6-Bl, an extremely thermophilic bacterium (Plant *et al.*, 1987*a*), secretes an extracellular enzyme having dual specificity towards α-1,4- and α-1,6- glucosidic linkages. Thus, the system was active on α-1,4- linkages in amylose, amylopectin and glycogen, but not pullulan, and on the α-1,6- linkages of pullulan, amylopectin and glycogen (Plant *et al.*, 1987*b*). Evidence supports the view that this dual specificity is located at a single active centre in the enzyme. The dual specificity in *T. vulgaris* (Sakano *et al.*, 1982*a*) towards starch and pullulan is located at a single active site. However, in this case, action on pullulan is towards α-1,4- linkages and panose is the principal product formed, whereas with the *Thermoanaerobium* enzyme only α-1,6- bonds in pullulan are hydrolysed and thus maltotriose is the main product formed. Interestingly, the latter system is active not only on linear α-1,6- linkages in pullulan but also on branch-point α-1,6- linkages in amylopectin and glycogen. A system from *Thermoanaerobium brockii* has been cloned in *Escherichia coli* and *Bacillus subtilis*. The cloned enzyme attacked none of the α-1,4- bonds but all of the α-1,6- glucosidic linkages in pullulan and, in contrast, hydrolysed mainly α-1,4- and very few α-1,6- glucosidic linkages in starch (Coleman *et al.*, 1987). Maltotriose was the only product formed from pullulan whereas starch was degraded to a series of maltooligomers.

Thermostable extracellular α-amylase and pullulanase activities are properties of the same protein in *Clostridium thermohydrosulfuricum* E 101-69 (Melasniemi, 1987*a*). Both activities had pH and temperature optima at pH 5·6

Table 5
Properties of Some Amylase–Pullulanase Complexes

Source	M_r	pH optimum	Temp. optimum (°C)	pI	Sensitive to —SH reagent	Reference
Thermoactinomyces						Shimizu *et al.* (1978)
vulgaris	—	5·0	70	5·2	–	Sakano *et al.* (1982*a*)
Bacillus megaterium	55 000	5·5	75	9·5	–	David *et al.* (1987)
Bacillus						
stearothermophilus KP 1064	115 000	5·8	55	4·4	+	Suzuki & Imai (1985)
Bacillus						
thermoamyloliquefaciens KP 1071						
α-Amylase II	67 000	6·2	63	4·7	+	Suzuki *et al.* (1987*c*)
Bacillus subtillus TU	450 000	6·0–7·0	50–60	—	–	Takasaki (1987)
Thermoanaerobium Tok 6-B1	120 000	5·5	70	—	–	Plant *et al.* (1987*a, b*)
Clostridium						
thermohydrosulfuricum E101-69						
Enzyme I	370 000					Melasniemi
Enzyme II	330 000	5·6	85–90	4·25	–	(1987*a*, 1988)

and 85–90°C, respectively (Table 5). The α-amylase–pullulanase complex was purified as two forms (I and II). Both forms appeared to be dimers of two similar sub-units, having M_r values of 190 000 for enzyme I and 180 000 for enzyme II according to SDS/polyacrylamide gradient gel electrophoresis (Melasniemi, 1988).

The existence of these new enzyme systems possessing both α-1,4- and α-1,6-glucosidic activities opens up new avenues and potential new approaches in starch processing technology.

3 NEW ENZYMES, INCLUDING β-AMYLASES, PRODUCING SPECIFIC MALTOOLIGOSACCHARIDES

3.1 Introduction

Since the original discovery of a maltotetraose-forming amylase in *Pseudomonas stutzeri* (Robyt & Ackerman, 1971), a series of bacterial maltooligosaccharide-forming amylases have been reported (Table 6). At present, the series ranges from maltose to maltohexaose. Specific maltooligosaccharides have a range of potential uses in the food, pharmaceutical and fine chemical industries because of their unique nature and special properties. In the series maltotriose to maltohexaose they have been used as research and clinical reagents for the determination of α-amylase. They are all highly soluble and produce clear viscous solutions which are palatable and are superior nutrient foods for infants and the aged. The price of pure maltooligosaccharides is extremely high because of the difficulties encountered in producing them.

3.2 Maltose-Producing Enzymes

The enzymes currently used commercially for production of maltose syrups, i.e. the mould-saccharifying amylases, yield syrups having a maltose content of *c.* 50–60%, depending on conversion conditions. Large quantities of syrups containing more than 75–80% maltose are not produced in Europe but are used in Japan and to a lesser extent in the USA. World demand for these high maltose syrups is expected to increase. They have a number of properties which distinguish them from dextrose syrups (Table 7).

Four approaches are being considered in the development of suitable enzymes for production of maltose syrups:

1. Development of microbial β-amylases.
2. Development of endo-acting α-amylases of Actinomycetes.
3. Production of new maltogenic α-amylases from Bacilli, e.g. *B. stearothermophilus.*
4. Development of new maltogenic enzymes from moulds.

Table 6
Distribution of Some Novel Bacterial Amylases

Major end-product	Source	Reference
Maltose	Bacillus polymyxa	Fogarty & Griffin (1973, 1975)
	Bacillus sp. IMD 198	Bourke & Fogarty (1975, 1979, 1985)
	Bacillus megaterium	Thomas et al. (1980)
	Bacillus megaterium G-2	Takasaki (1989)
	Bacillus cereus var. mycoides	Takasaki (1976, 1979)
	Bacillus cereus BQ10-S1	Shinke et al. (1979a, b), Namori et al. (1983, 1987)
	Bacillus circulans	Napier (1977), Fogarty & Kelly (1983)
	Clostridium thermosulphurogenes	Hyun & Zeikus (1985a, b)
	Streptomyces praecox	Wako et al. (1978)
	Thermoactinomyces vulgaris	Shimizu et al. (1978), Sakano et al. (1982a)
	Streptomyces hygroscopicus	Hidaka & Adachi (1980)
	Streptomyces limosus	Fairbairn et al. (1985)
	Streptomyces albus	Hyslop & Sleeper (1964), Andrews & Ward (1987)
	Bacillus stearothermophilus	Outtrup & Norman (1984)
	Bacillus stearothermophilus KP 1064	Suzuki & Imai (1985)
	Bacillus stearothermophilus	Brosnan et al. (1989)
	Pencillium expansum	Doyle et al. (1989)
Maltotriose	Streptomyces griseus NA-468	Wako et al. (1979)
	Bacillus sp. 11-1S	Uchino (1982)
	Bacillus subtilis	Takasaki (1985)
	Bacillus thermoamyloliquefaciens KP 1071	Suzuki et al. (1987c)
Maltotetraose	Pseudomonas stutzeri	Robyt & Ackerman (1971), Sakano et al. (1982b, 1983b)
	Pseudomonas saccharophila	Zhou et al. (1989)
	Bacillus circulans	Takasaki (1983)
Maltopentaose	Bacillus licheniformis	Saito (1973)
	Bacillus cereus NY-14	Yoshigi et al. (1985a, 1986)
	Pseudomonas sp. (KO-8940)	Okemoto et al. (1986)
Maltohexaose	Aerobacter aerogenes (Klebsiella pneumoniae)	Kainuma et al. (1972) Nakakuki et al. (1982)
	Bacillus subtilis	Kennedy & White (1979)
	Bacillus circulans G-6	Takasaki (1982)
	Bacillus circulans F-2	Taniguchi et al. (1983)
	Bacillus sp. H-167	Hayashi et al. (1988)
	Bacillus amyloliquefaciens	Welker & Campbell (1967), Norman (1979)

Table 7
Properties of Maltose Syrups

Resistance to crystallisation
Low hygroscopicity
Low viscosity
Low sweetness
Lack of colour
Good heat stability

3.2.1 Microbial β-Amylases

The occurrence of β-amylase (EC 3.2.1.2, 1,4-α-D-glucan maltohydrolase, saccharogenic amylase) as an extracellular enzyme in microorganisms was first established in our laboratories (Fogarty & Griffin, 1973, 1975; Griffin & Fogarty, 1973a, b, c; Bourke & Fogarty, 1975, 1979; W. M. Fogarty & M. T. O'Reilly, unpublished data) and detected subsequently in a number of other bacteria (Higashihara & Okada, 1974; Shinke *et al.*, 1974, 1979a, b; Takasaki, 1976; Napier, 1977; Murao *et al.*, 1979; Thomas *et al.*, 1980).

3.2.1.1 BACILLUS POLYMYXA β-AMYLASE

The amylase system of *Bacillus polymyxa* was studied by Rose (1948) and Robyt & French (1964). It produced maltose in a yield of 92–94% from starch and was thought to possess an endo-mechanism of substrate attack and break or bypass α-1,6- linkages in amylopectin or starch (Robyt & French, 1964). Using high resolution techniques the system was resolved into two components—a β-amylase and a debranching enzyme (Fogarty & Griffin, 1973, 1975; Griffin & Fogarty, 1973a, b, c). The establishment and identification of the presence of two enzymes explained the very high yields of maltose obtained with the system. It also established for the first time the presence of β-amylase in microorganisms. The β-amylase was similar in many of its properties to the corresponding plant enzymes (Fogarty & Griffin, 1975). It was inhibited by Schardinger cyclodextrins, *p*-chloromercuribenzoate (*p*-CMB) and *N*-bromosuccinimide. Inhibition by *p*-CMB could be reversed in the presence of cysteine and mercaptoethanol. The suggestion that this is not a thiol enzyme (Marshall, 1974) is not compatible with these observations. The optimum pH of 6·8 is higher than that detected in similar plant enzymes. The enzyme had an M_r value of 59 000 (Bourke & Fogarty, 1979).

3.2.1.2 OTHER MICROBIAL β-AMYLASES

Following its isolation and characterisation in culture filtrates of *B. polymyxa*, β-amylase has been detected in a number of bacteria and principally in the genus *Bacillus* (Table 8). Recent work has dealt with other β-amylase-producing bacteria. *Bacillus cereus* strain BQ10-SI was isolated following UV-irradiation of wild-type BQ10 (Shinke *et al.*, 1979a) and secreted ten times more β-amylase than its parent. β-Amylase production in strain BQ10-SI was repressed by maltose and ceased at the onset of spore formation. An asporogenous mutant, strain BQ10-SI Spo II, that was rifampin-resistant was selected from BQ10-SI and gave a four-fold increase in β-amylase production (Nanmori *et al.*, 1983). Again enzyme production decreased in the presence of maltose or glucose and was not increased by addition of starch to the culture medium. Following mutagenesis with *N*-methyl-*N'*-nitro-*N*-nitrosoguanidine, a strain designated BQ10-SI Spo III was isolated from strain BQ10-SI Spo II (Nanmori *et al.*, 1987). The overall increase in β-amylase production by this mutant was 220-fold greater than in the original wild-type strain, *B. cereus* BQ10. Enzyme production in this mutant increased in the presence of glucose,

Table 8
Some Properties of β-Amylases

Organism	pH optimum	Temp. optimum (°C)	pI	M_r	Reference
Bacillus cereus	7·0	40	8·3	62 000	Nanmori *et al.* (1983)
Clostridium					
thermosulphurogenes H-12-1	5·5	75	5·1	210 000	Saha *et al.* (1987)
					Shen *et al.* (1988)
Bacillus sp. IMD 198	6·8	55	—	58 000	Bourke & Fogarty (1979)
					Fogarty & Bourke (1983)
Bacillus circulans	7·0–7·5	60	4·5	60 000	Fogarty & Kelly (1983)
Bacillus sp. IMD 273	6·5–7·0	50	5·5	57 000	W. M. Fogarty & N. M. Nash, unpublished data
Bacillus megaterium G-2[a]	7·0	60	—	60 000	Takasaki (1989)

[a] This enzyme is similar to a β-amylase in all respects except that the α-anomeric form of maltose is produced as the end-product on hydrolysis of starch.

maltose or starch. The enzyme was immunologically identical to the wild-type enzyme. An unusual maltose-forming amylase, which solely hydrolyses α-1,4-D-glucosidic bonds in starch and releases α-D-maltose, was found in culture filtrates of *B. megaterium* G-2 (Takasaki, 1989). Its action pattern and mode of action, and its general and enzymatic properties, are essentially similar to those of β-amylase, except for the one property relating to the configuration of the product (α-D-maltose), produced.

Clostridium thermosulfurogenes, an anaerobic bacterium, produced an extracellular β-amylase which was optimally active at 75°C and pH 5·5–6·0 (Hyun & Zeikus, 1985*a*). Like other β-amylases, the enzyme was inhibited by *p*-CMB. It was not competitively inhibited by cyclodextrins. The enzyme was inducible and subject to catabolite repression. It was expressed only at high levels when the bacterium was grown on maltose or carbohydrates possessing maltose units. A hyperproductive mutant, H12-1 gave eight times more β-amylase on starch medium than did the wild-type (Hyun & Zeikus, 1985*b*). Purification and properties of the enzyme produced by mutant H12-1 were reported in a separate publication (Shen *et al.*, 1988). Studies on stability of β-amylase productivity, yield and levels of activity showed that continuous production was possible under thermophilic and anaerobic conditions (Nipkow *et al.*, 1989). Saha *et al.* (1987) reported on raw starch adsorption, digestion and elution behaviour of the enzyme.

Friedberg & Rhodes (1986) cloned the β-amylase gene from *Bacillus polymyxa*, although no sequence data were published. The gene encoding β-amylase was cloned into *Escherichia coli* HB101 from *B. polymyxa* 72 (Kawazu *et al.*, 1987). Multiform β-amylases were produced both by *B. polymyxa* and *E. coli.* Nucleotide sequence analysis of the cloned DNA showed that it contained one open reading frame of 2808 nucleotides without a translational stop codon. The deduced amino acid sequence is 936 amino acids including a signal peptide of 33 or 35 residues at the amino-terminal end. The occurrence of a β-amylase larger than M_r 100 000, predicted on the results of nucleotide sequence analysis of the gene, was confirmed from examination of culture filtrates. Its existence was only transient during cultivation but multiform β-amylases existed for a long time. When a protease inhibitor, such as chymostatin, was added, β-amylase M_r 160 000 existed for much longer thereby suggesting that multiform β-amylases result from proteolytic cleavage.

A gene encoding the β-amylase of *Bacillus circulans* was isolated and sequenced (Siggens, 1987). The gene contained an open reading frame of 1725 base pairs encoding a protein of 575 amino acids. Translation of the open reading frame gives a polypeptide having a predicted M_r 62 830. When the gene was expressed in *Escherichia coli* two active forms of the enzyme were detected. The larger of the two was greater than that secreted by *B. circulans* whereas the smaller was present at a low level in the latter organism and could possibly result from proteolytic hydrolysis. The β-amylase of *B. circulans* shows an 81% degree of homology with that reported for the *B. polymyxa* system (Kawazu *et al.*, 1987). The corresponding DNA sequence showed a 72% degree of homology.

3.2.2 New Maltose-Producing Amylases of Actinomycetes

α-Amylases used commercially produce one of two types of end-product profile, e.g. in the case of thermolabile enzymes (i.e. from *A. oryzae*) it is the di- and tri-saccharides, maltose and maltotriose that are the main end-products, whereas in the case of thermostable enzymes (i.e. from *B. amyloliquefaciens, B. stearothermophilus* and *B. licheniformis*) the degree of polymerisation of the end-products formed is larger, and thus maltohexaose and maltopentaose are the dominant products. Recently, a number of α-amylases have been reported in Actinomycetes, and principally Streptomycetes, which exhibit highly interesting properties (Table 9), including some unusual end-product profiles on hydrolysis of starch. Despite the prevalence of starch-hydrolysing enzymes in Actinomycetes, relatively few have been studied in great detail. Certain exceptions exist and some have unique and special properties. The following discussion relates to some of the systems examined and which produced maltose as the predominant product from starch.

Table 9
Properties of Some Maltogenic Amylases of Actinomycetes

Organism	pH optimum	Temp. optimum (°C)	pI	M_r	Major end-product	Reference
Streptomyces limosus	7·0	35	—		G_2	Fairbairn *et al.* (1985)
Thermoactinomyces sp.	7·0	80	—	47 800	$G_1 + G_2$	Obi & Odibo (1984)
Thermoactinomyces vulgaris	4·5	70	5·2		G_2	Shimizu *et al.* (1978) Sakano *et al.* (1983a)
Streptomyces aureofaciens	4·6–5·6	40	—	40 000	G_2	Hostinova *et al.* (1979)
Streptomyces praecox	6·0	47	—	—	G_2	Wako *et al.* (1978)

An α-amylase from *Streptomyces hygroscopicus* (Hidaka & Adachi, 1980) degraded starch to maltose in 75% yield. Maltotriose and maltotetraose were the major initial products formed. Maltotetraose was then degraded to maltose. Conversion of maltotriose to maltose was achieved by either (1) a condensation–hydrolytic process and/or (2) a synthetic step followed by a hydrolytic reaction. The enzyme is therefore unusual as its mechanism of action involves both synthetic and hydrolytic reaction stages and this results in the production of maltose as the major end-product. Some properties of the system are given in Table 9.

An α-amylase secreted into the culture filtrate of *Streptomyces praecox* NA-273 (Wako *et al.*, 1978) was purified and shown to contain three isoenzymes (Suganuma *et al.*, 1980). It converted starch to maltose in 80% yield. Glucose was not produced during hydrolysis of starch, although maltooligosaccharides G_2 to G_5 are found initially and G_4 accumulates in relatively large amounts. Maltose is the final major end-product detected. The enzyme produced maltose exclusively from maltotriose without formation of glucose. A transformation mechanism involving transglycosylation has been proposed to account for this reaction (Suganuma *et al.*, 1980). Maltotriose is degraded to maltose via maltotetraose as a transfer product. Other properties of the system are given in Table 9.

Streptomyces limosus produces an extracellular amylase which is induced by maltose, malt extract or starch but not by sucrose or lactose (Fairbairn *et al.*, 1985). The endo-acting enzyme degrades amylopectin to maltose in a yield of 68%. The initial products formed are maltotetraose and maltotriose which disappear as the reaction proceeds and maltose accumulates. No further information was provided on the mode of action of this enzyme. The enzyme degrades raw starch granules and shares this property with *S. hygroscopicus* and *S. praecox*.

It is not known whether the amylolytic activity of *Streptomyces albus* is due to two completely separate amylases of M_r 67 000 and 55 000 or to a single protein with an active breakdown product (Andrews & Ward, 1987). The amylolytic activity, which is inducible by maltose, displays a typical α-amylase end-product profile. Hydrolysis of starch results in maltose, 50%; maltotriose, 25–30%; dextrins, 20–25%; and a small amount of glucose (Hyslop & Sleeper, 1964). Maltotetraose is produced transiently and subsequently disappears.

Both extracellular and cell-bound amylases are produced by *Streptomyces aureofaciens*. Both of these enzymes cleaved starch substrates by an endo-mechanism of action (Hostinova *et al.*, 1979). During the initial stages of hydrolysis, quantities of maltooligosaccharides (G_4–G_8) appeared. Subsequently G_1 to G_3 increased while the levels of G_4 to G_8 decreased. Further information on the nature and function of this enzyme is required to elucidate its mode of action.

Thermomonospora curvata extracellular α-amylase has optima for activity at pH 5·5–6·0 and 65°C. It has an M_r value of 62 000 (Glymph & Stutzenberger, 1977). Identification of the products of the α-amylase action on starch revealed only maltotetraose and maltopentaose and in this context is different from other Actinomycetes. By comparison, the amylase of *T. vulgaris* (Allam *et al.*, 1975) produced only glucose and maltose from starch. Comparative analysis of the action of the enzymes of these two species would resolve the question of differences in end-product profiles.

Thermostable α-amylase from *Thermoactinomyces* sp. No. 15 had optima for activity at pH 7·0 and 80°C. It was stable in the pH range 5–10 and retained 74% activity after 30 min at 100°C (Obi & Odibo, 1984). Glucose and maltose were the major products formed from starch but no information was available on mechanism or mode of action. α-Amylase of *Thermoactinomyces vulgaris* (Shimizu *et al.*, 1978; Sakano *et al.*, 1982a) possessed both α-1,4- and α-1,6-hydrolysing activities and gave maltose (74%) and glucose (12%) from starch and produced panose (97%) exclusively from pullulan. This enzyme is dealt with more fully in Section 2.4 dealing with enzymes possessing α-amylase and pullulanase activities.

3.2.3 New Maltogenic α-Amylases from Bacilli

A third approach to the production of high maltose syrups involves the use of another type of maltogenic enzyme from *Bacillus* spp. and, specifically, *Bacillus stearothermophilus*. An isolate of *B. stearothermophilus* (Outtrup & Norman, 1984) produced high levels of maltose from starch but the strain was difficult to cultivate and enzyme yields were low. The maltogenic amylase gene was cloned into *B. subtilis* and this made large scale production of the enzyme feasible. This enzyme was designated an exo-acting maltogenic α-amylase with optima at 60–70°C and pH 5·3 (Table 10). It produces maltose from non-reducing chain ends of the substrate to give a product having the α-configuration. Other important properties which distinguish it from β-amylase are its capacity to hydrolyse maltotriose, its insensitivity to —SH reagents and its lack of inhibition by, and hydrolysis of, Schardinger dextrins. This thermostable and relatively acid stable maltogenic α-amylase can be used to produce syrups containing 60–80% maltose from starch. Maltotriose is hydrolysed to glucose and maltose and therefore a small amount of glucose is present in high maltose syrups formed by this enzyme (Diderichsen & Christiansen, 1986; Outtrup, 1986).

The system reported in *B. stearothermophilus* KP 1064 (Table 10) by Suzuki

Table 10
Properties of Maltogenic Amylases of *Bacillus stearothermophilus*

Source	pH optimum	Temp. optimum (°C)	pI	M_r	Reference
B. stearothermophilus	5·3	60–70	8·5	70 000	Outtrup & Norman (1984)
B. stearothermophilus KP 1064	5·8	55	4·4	115 000	Suzuki & Imai (1985)
B. stearothermophilus J	5·0	80	—	—	Brosnan *et al.* (1989)

& Imai (1985) is similar in that it hydrolyses maltotriose and Schardinger dextrins, but it differs in that it is strongly inhibited by heavy metals and —SH reagents. It did not attack glycogen and amylopectin was degraded at less than 10% of the rate achieved with amylose, while panose was the major product from pullulan—in this latter property it resembles the α-amylase of *T. vulgaris*, a system discussed earlier. This system is claimed to be endo-acting and produces the α- configuration of maltose. Its lack of activity towards glycogen and amylopectin suggests some form of steric hindrance with highly branched substrates.

3.2.4   High Maltose-Producing α-Amylases of Moulds

Another approach to production of high maltose syrups is that being taken in our laboratories dealing with mould-saccharifying α-amylases. Following an initial screening programme, one of the organisms selected for further study was a strain of *Penicillium expansum* (Doyle *et al.*, 1989). The extracellular-saccharifying α-amylase was produced in fed-batch culture and had optima at pH 4·5 and 60°C. It has an M_r value of 69 000 and isoelectric point of 3·9. The most outstanding feature of the *P. expansum* enzyme is its ability to yield 14% more maltose and 17·1% less maltotriose than a currently used commercial enzyme (Table 11). This may be partly explained by the greater affinity of the *P. expansum* enzyme for maltotriose ($K_m = 0·76$ mM) relative to the commercial enzyme, Fungamyl ($K_m = 2·9$ mM). The *P. expansum* enzyme is unique among fungal α-amylases in being able to produce such high levels of maltose.

3.3   Maltotriose-Forming Amylases

Wako *et al.* (1978) described an amylase in *Streptomyces griseus* (Table 12) which hydrolysed starch from the non-reducing chain ends by an exo-mechanism of attack resulting in the formation of maltotriose. Neither glucose nor maltose was formed from soluble starch. An amylase, from a strain of *Bacillus subtilis*, which produced maltotriose as the main product from starch

Table 11

Comparison of Properties of Maltogenic Amylases of *Pencillium expansum* and Fungamyl (*Aspergillus oryzae*)

Property	P. expansum α-amylase	*Fungamyl* (A. oryzai α-amylase)
pH optimum	4·5	4·7
Temperature optimum	60°C	55°C
Thermostability with CaCl$_2$ (250 ppm)	Retained 27·2% activity at 60° for 1 h	—
M_r	68 000	56 000
pI	3·9	3·6
K_m (soluble starch)	0·86 mg/ml	2·62 mg/ml
K_m (maltotriose)	0·76 mM	2·91 mM
End-products from starch	74%, G$_2$; 21%, G$_1$; 3%, G$_3$	60%, G$_2$; 11%, G$_1$; 20%, G$_3$

Table 12

Properties of Some Maltotriose- and Maltotetraose-Forming Amylases

Organism	pH optimum	Temp. optimum (°C)	pI	M_r	Reference
Maltotriose-forming amylases:					
Streptomyces griseus	5·6–6·0	45	—	—	Wako *et al.* (1978)
Bacillus subtilis	6·0–7·0	50	—	25 000	Takasaki (1985)
Bacillus sp. 11-1S	2·0	70	—	54 000	Uchino (1982)
Maltotetraose-forming amylases:					
Pseudomonas stutzeri	8·0	50	—	12 500–125 000	Robyt & Ackerman (1971)
	6·0–6·5	45	5·3	57 000	Sakano *et al.* (1982*b*)
P. stutzeri F1	8·0	45	5·6	55 000 ⎱	Sakano *et al.* (1983*b*)
F2	8·0	45	5·3	55 000 ⎰	
B. circulans	7·0	50	—	45 000	Tanasaki (1983)

was described by Takasaki (1985). Glucose and maltose were also formed. Hydrolysis of starch by this system gave maltotriose (56%) maltose (26%) and glucose (5%). In the presence of pullulanase the yield of maltotriose could be increased to 73% while that of maltose and glucose fell to 10% and 5%, respectively. The enzyme did not degrade either pullulan or cyclodextrins and had optima at pH 6–7 and 50°C. A thermophilic and unusually acidophilic extracellular amylase of *Bacillus* sp. 11-1S (Uchino, 1982) produced maltotriose (39%) and maltose (22%) as the major products in the hydrolysis of starch. The enzyme had optima for activity at pH 2·0 and 70°C (Table 12).

3.4 Maltotetraose-Forming Amylases

A maltotetraose-forming amylase was detected originally in *Pseudomonas stutzeri* (Robyt & Ackerman, 1971). This extracellular, exo-acting amylase attacked the fourth α-1,4- glucosidic bond from the non-reducing chain ends in starch substrates and had an optimum for activity at pH 8·0 and rapidly lost activity above 40°C. In a study using the same strain (*Pseudomonas stutzeri* NRRL B-3389), it was reported that the enzyme had optima at pH 6·0–6·5 and 45°C (Table 12) and it was confirmed that the end-product had the α-anomeric form (Sakano *et al.*, 1982*b*). Two active forms of the enzyme having essentially similar properties were subsequently examined (Sakano *et al.*, 1983*b*).

Kimura *et al.* (1988) examined a continuous production process for maltotetraose using the *P. stutzeri* NRRL B-3389 amylase immobilised on a macroporous hydrophobic resin. The yield of tetramer was greatly influenced by a number of factors including the flow rate of substrate solution, its concentration and the activity of the immobilised enzyme. In an examination of the stability of the immobilised enzyme it was observed that the inactivation process of the enzyme obeyed first-order kinetics and the system became more stable in the presence of 20–30% substrate and calcium ions (Kimura *et al.*, 1989).

The gene (amyP) coding for maltotetraose-forming amylase in *Pseudomonas stutzeri* MO-19 was cloned and the nucleotide sequence determined. An extract of *Escherichia coli* carrying the cloned amyP had amylolytic activity (exo-maltotetraohydrolase) (Fujita *et al.*, 1989). A section in the primary structure of this enzyme had homology with other amylases. The exo-maltotetraohydrolase gene of *Pseudomonas saccharophila* was also cloned in *E. coli* (Zhou *et al.*, 1989). The expressed enzyme activity was four times greater than that in *P. saccharophila*.

An amylase from *Bacillus circulans* (Takasaki, 1983) produced maltotetraose (40%) and maltopentaose (30%) from starch as the major initial products and these were subsequently degraded to a mixture of glucose, maltose and maltotriose.

Maltotetraose is used as a substrate in studying the mode of action of α-amylases and when coupled to a chromogenic compound it forms a highly sensitive substrate for detection of α-amylase. It is also being examined as a food additive to improve texture and moisture retention in foods.

3.5 Maltopentaose-Forming Amylases

Maltopentaose is widely used as a substrate for determination of serum α-amylase. It has been used as a nutrient food for patients having renal failure and those in a condition of calorie-deprivation (Berlyne *et al.*, 1969). It was reported as the major product (33%) formed in the hydrolysis of starch by *Bacillus licheniformis* α-amylase (Saito, 1973). An extracellular α-amylase from a culture medium of *Bacillus cereus* NY-14 (Yoshigi *et al.*, 1985a, 1986) produced maltopentaose in a yield of 44% from soluble starch with smaller amounts of other oligomers. This enzyme had optima at pH 6·0 and 55°C (Table 13) whereas that from *B. licheniformis* has optima in the range pH 5–8 and 76°C (Saito, 1973). Okemoto *et al.* (1986) isolated *Pseudomonas* sp. (KO-8940) from soil and showed it produced an enzyme that formed maltopentaose specifically during the initial stage of degradation of starch. Usui & Murata (1988) described a novel method for the enzymic synthesis of p-nitrophenyl-α-D-maltopentaoside (p-NP-α-D-G$_5$) by utilising the transferase

Table 13
Properties of Maltopentaose-Forming Amylases

Organism	pH optimum	Temp. optimum (°C)	pI	M_r	Reference
Bacillus lichenformis	5·0–8·0	76	—	22 500	Saito (1973)
Bacillus cereus NY-14	6·0	55	6·13	55 000	Yoshigi *et al.* (1985a)

$$G_5 + E \rightleftharpoons G_4 - E + G$$
$$G_4 - E + p\text{-NPG} \rightleftharpoons p\text{-NPG}_5 + E$$
$$G_4 - E + H_2O \rightleftharpoons G_4 + E$$

(Adapted from Usui & Murata, 1988).
Abbreviations: E, A maltotetraose forming amylase of *P. stutzeri*; G_1, glucose; G_4, maltotetraose; G_5, maltopentaose; *p*-NPG, *p*-nitrophenyl-α-D-glucoside.

Fig. 1. Formation of *p*-nitrophenyl-α-D-maltopentaoside.

activity of the maltotetraose-forming amylase of *P. stutzeri* in high methanol concentrations. Maltopentaose serves as the donor and *p*-nitrophenyl-α-D-glucoside as the acceptor in these reactions (Fig. 1). *p*-NP-α-D-G_5 is specifically synthesised by transfer of G_4 residues to *p*-NP-α-D-G which acts as acceptor in a medium having high concentrations of methanol.

3.6 Maltohexaose-Forming Amylases

Aerobacter aerogenes (*Klebsiella pneumoniae*) synthesises a maltohexaose-forming amylase which can be disassociated from the organism using sodium dodecyl sulphate (Kainuma *et al.*, 1972). It produces the hexamer from starch, amylopectin and amylose by an exo-mechanism of attack and is not active on cyclodextrins or pullulan. In the hydrolysis of glycogen and amylopectin limit dextrins are formed in addition to maltohexaose. β-Limit dextrin is degraded to yield branched oligosaccharides. It was suggested that the binding site of the enzyme may be able to accept a substrate chain where one of the α-1,4- bonds is replaced by an α-1,6- linkage. Optima at pH 6·8 and 50°C were reported for the system (Table 14). Nakakuki *et al.* (1982) obtained an extracellular exo-maltohexaohydrolase from an ultraviolet induced mutant of *A. aerogenes* (*K. pneumoniae*) which had maxima for activity at pH 7·0 and 52°C. Maltohexaose was slowly cleaved into maltose and maltotetraose at a rate less than 2% of the activity on starch. Monma *et al.* (1983) showed that this enzyme was capable of forming maltohexaose from maltotetraose by a transferase type of action which was dependent on substrate concentration.

An amylase which produces maltohexaose as the main product from starch was found in the culture filtrates of *Bacillus circulans* G-6 (Takasaki, 1982). The optimum pH and temperature for the enzyme were pH 8·0 and 60°C, respectively. The yield of maltohexaose from soluble starch was about 30%. The presence of pullulanase increased the yield of maltohexaose by 12%. Substrate attack by this enzyme was by an endo-mechanism and the products had an α- configuration. It was concluded that the enzyme was a type of α-amylase.

An extracellular amylase produced by *Bacillus circulans* F-2 (Taniguchi *et al.*, 1983) produced maltohexaose from the non-reducing chain ends of starch and was subsequently hydrolysed to maltotetraose and maltose. The enzyme

Table 14

Properties of Maltohexaose-Forming Amylases

Organism	pH optimum	Temp. optimum (°C)	pI	M_r	Reference
Aerobacter aerogenes	6·8	50			Kainuma *et al.* (1972)
(*Klebsiella pneumoniae*)	7·0	52	7·2–7·8	65 000[a] 48 000[b] }	Nakakuki *et al.* (1982)
Bacillus circulans G-6	8·0	60	—	76 000	Takasaki (1982)
B. circulans F-2	6·0–6·5	60	4·9	93 000	Taniguchi *et al.* (1983)
Bacillus sp. H167					
H-I-1	10·5	60	4·1	73 000 }	
H-I-2	10·5	60	3·5	59 000 }	Hayashi *et al.* (1988)
H-II	10·5	60	4·3	80 000 }	
Bacillus amyloliquefaciens	5·9	70	5·2	49 000	Welker & Campbell (1967), Norman (1979), Fogarty (1983)

[a] SDS-PAGE.
[b] Gel filtration.

had optima at pH 6·0–6·5 and 60°C. In the early stages of hydrolysis maltohexaose was the only product detected. When 10.8% hydrolysis of amylopectin was attained the amount of maltohexaose formed was 21%, with amounts of maltotetraose and maltose being 12% and 7%, respectively. Maltohexaose is hydrolysed by the purified amylase at an initial rate of 56% of that of soluble starch. In contrast, the system from *A. aerogenes* (Nakakuki *et al.*, 1982) hydrolysed maltohexaose at a rate less than 2% of that obtained with soluble starch. It was considered that maltotetraose and maltose were formed only from maltohexaose rather than from amylopectin or starch and that the enzyme attacks substrates by an exo-mechanism. The major difference between this enzyme and that reported by Takasaki (1982) is the ability of the former system to digest raw starch granules.

Bacillus amyloliquefaciens α-amylase produces mainly maltohexaose on hydrolysis of starch (Norman, 1979) and, unlike some of the other enzymes described, appears incapable of hydrolysing this structure any further. Yields of hexamer may be as high as 30–34% (W. M. Fogarty & C. T. Kelly, unpublished data).

Three alkaline maltohexaose-forming amylases were detected in culture filtrates of an alkaline bacterium, *Bacillus* sp. H-167 (Hayashi *et al.*, 1988). They had similar properties, except for M_r values and isoelectric point. Optima reported for activity were pH 10·5 and 60°C. Starch was hydrolysed to produce mainly maltohexaose in the early stages of the reaction; the maximum yield being 25–30%. As the reaction proceeded both maltotetraose and maltose were formed. Pullulan, cyclodextrins or dextran were not degraded. The gene for maltohexaose-forming amylase in *Bacillus* sp. H-167 was cloned into *E. coli* HB101 using pBR329 as a cloning vector (Shirokizawa *et al.*, 1989). The properties of the system produced by *E. coli* HB101 were essentially similar to those of *Bacillus* sp. H-167 and it was also produced in a multiform fashion with M_r values of 90 000, 73 000 and 60 000. Studies using a protease inhibitor demonstrated that multiple expression was caused by intracellular proteolytic hydrolysis of the enzyme.

Kennedy *et al.* (1979) described an enzyme, α-amylase K, from culture filtrates of a strain of *Bacillus subtilis*. No information was provided on the source or properties of the organism or on the method of preparation of the enzyme, although it may have been *Bacillus amyloliquefaciens*. The dominant product on hydrolysis of starch was maltohexaose under most conditions, with varying amounts of maltotriose, maltopentaose and maltoheptaose being formed depending on the time of hydrolysis.

Exo-maltohexaohydrolase from *A. aerogenes* (*K. pneumoniae*) was immobilised by co-polymerisation as a result of irradiation of some synthetic monomers and enzyme (Nakakuki *et al.*, 1983). The pH stability of the enzyme was improved dramatically by immobilisation and when starch was used as substrate the K_m value was three times greater than the native enzyme.

4 RECENTLY DESCRIBED AMYLOGLUCOSIDASES

4.1 Introduction

Amyloglucosidase (EC 3.2.1.3, AG, glucoamylase, 1,4-α-D-glucan glucohydrolase, α-amylase) is an exo-acting enzyme yielding β-D-glucose (Ono *et al.*, 1965) from the non-reducing chain-ends of amylose, amylopectin and glycogen by consecutive bond hydrolysis. Although the enzyme has highest affinity for α-1,4- linkages it will also hydrolyse α-1,6- and α-1,3- bonds (Pazur & Kleppe, 1962) but at a much slower rate (Table 15). Amyloglucosidases occur almost exclusively in fungi (Table 16) and are generally secreted as extracellular enzymes. The enzymes used commercially for conversion of malto-dextrins into D-glucose (Underkofler *et al.*, 1965) originate from strains of *Aspergillus niger* or *Rhizopus* spp. Rates of substrate hydrolysis are affected both by molecular size and structure and also by the next bond in sequence (Abdullah

Table 15

Hydrolysis of disaccharides with different Linkages by amyloglucosidase of *Aspergillus niger*

Disaccharide	Linkages	Relative rate of hydrolysis
Maltose	α-1,4-	100
Isomaltose	α-1,6-	1·0
Nigerose	α-1,3-	0·2

Source: W. M. Fogarty & C. P. Benson, unpublished data.

Table 16
Fungi Producing Single and Multiple Forms of Amyloglucosidase

---

1. *Organisms producing single forms of AG*
   Aspergillus niger (Abe et al., 1982), A. awamori (Yamasaki et al., 1977b), A. oryzae (Ono et al., 1988c), Paecilomyces varioti (Takeda et al., 1985), Rhizopus delemar (Pazur & Odaka, 1967), Thermomyces lanuginosa (Basaveswara et al., 1981), Trichoderma viride (Okada, 1974), Candida antarctica (De Mot & Verachtert, 1987), C. tsukubaensis (De Mot et al., 1985), Saccharomyces diastaticus (Tucker et al., 1984), Schwanniomyces alluvius (Simoes-Mendes, 1984), Schwanniomyces castellii (Sills et al., 1984).

2. *Organisms producing two or more forms of AG*
   Aspergillus niger (Svensson et al., 1982; Fogarty & Benson, 1983; Ono et al., 1988b), A. oryzae (Morita et al., 1968; Miah & Ueda, 1977a, b), A. candidus (Manjunath & Raghavendra Rao, 1979), A. hennebergi (Alazard & Baldensperger, 1982), A. saitoi (Takahashi et al., 1981), Cladosporium resinae (McCleary & Anderson, 1980), Monascus kaoliang (Iizuka & Mineki, 1977), Monascus sp. (Yasuda et al., 1989), Rhizopus sp. (Saha & Ueda, 1983b), Chalara paradoxa (Ishigami et al., 1985), Pencillium oxalicum (Yamasaki et al., 1977a), Mucor rouxianus (Tsuboi et al., 1974), Candida pelliculosa (Kawamura & Sawai, 1968), Filobasidium capsuligenum (De Mot & Verachtert, 1985), Saccharomyces diastaticus (Erratt & Stewart, 1980).

---

et al., 1963; Fleming, 1968). Thus, the rate of hydrolysis of α-1,4- bonds increases linearly with the relative molecular mass of the substrate at least up to maltopentaose (Fukui & Nikuni, 1969; Smiley et al., 1971; Krzechowsk & Urbanck, 1975; Kulp, 1975; Yamasaki et al., 1977a, b). The enzyme from *A. niger* (Fogarty & Benson, 1983) had relative affinities for starch, maltotriose and maltose of 100, 68 and 31, respectively.

AG was shown to hydrolyse the α-1,6- linkage in isomaltose (Pazur & Ando, 1960; Pazur & Kleppe, 1962) and the pH activity curves of the maltose- and isomaltose-hydrolysing activities were the same and could not be separated by paper or gel electrophoresis. The rate of enzymic hydrolysis is affected not only by substrate size but also by structure and by the next bond in sequence. Using a highly purified preparation from *A. niger*, Fogarty & Benson (1983) showed that AG hydrolysed maltose 100 times faster than isomaltose, although panose was hydrolysed at 45% the rate of maltose. Thus, it appears that α-1,6- linkages close to α-1,4- bonds are attacked faster than the same linkage in isolation. Both the α-1,4- linkage of maltose and the α-1,4- and α-1,6- linkages of panose are hydrolysed by a single active centre in the AG from *Rhizopus delemar* (Hiromi et al.,1966). With α-1,6- linkages an increase in chain length causes an increase of the rate of hydrolysis. AG has greater affinity for longer chains, thus the maximum rate increases with length while Michaelis constants decrease. Kusunoki et al. (1982) described a model for the action of AG on starch employing modified Michaelis–Menten kinetics with product inhibitory effects. It was shown that both V_{max} and K_m vary linearly with substrate molecular mass.

4.2 Amyloglucosidases of Moulds

Amyloglucosidase frequently exists in multiple forms (Table 16) which are widely referred to as G I and G II or AG I and AG II, etc. Four forms of AG having small differences in electrophoretic mobilities, sedimentation coefficients and pH stabilities, but possessing identical enzymic and physico-chemical properties, were produced by *Aspergillus oryzae* (Morita *et al.*, 1968; Ohga *et al.*, 1966). Saha & Ueda (1983*b*) detected five active forms of AG in *Rhizopus niveus* and Ishigami *et al.* (1985) reported that the AG of *Chalara paradoxa* possessed six active components. Multiple forms of AG have been reported in *A. niger* (Alazard & Baldensperger, 1982; Ramasesh *et al.*, 1982; Svensson *et al.*, 1982) although some strains may only possess one active form (Abe *et al.*, 1982). Svensson *et al.* (1982) resolved the active components from a commercial preparation into two forms, named G I and G II. Both enzymes were composed of a single glycosylated polypeptide chain. The amino acid content of the two forms were very similar. Sedimentation equilibrium ultracentrifugation gave relative molecular mass values of 52 000 and 46 000 for G I and G II, respectively. Cyanogen fragments and N-terminal amino acid sequence analysis were identical, indicating considerable homology in the primary structure of the two components. Analysis by high pressure gel permeation chromatography of proteolytic digests of G I and G II indicated a high degree of similarity in the respective polypeptide chains. G I, as expected, was shown to produce glucose from raw starch, whereas G II had little activity towards the substrate. G II possessed about 75% of the activity of G I on soluble polymeric substrates, whereas both forms were equally active in hydrolysis of α-1,4- and α-1,6- linkages in low molecular substrates. Ono *et al.* (1988*a*) used an immobilised acarbose column to selectively purify AG free from α-amylase and α-glucosidase in a commercial (*A. niger*) AG preparation. By gel chromatography followed by anion-exchange chromatography, the active component was separated into six sub-fractions. Each sub-fraction liberated β-D-glucose as sole product from soluble starch. However, the six components differed from one another in certain respects. Five forms were fully characterised (Ono *et al.*, 1988*b*) and differed in relative molecular mass, sedimentation constant, chemical composition, isoelectric point, kinetic parameters and other enzymatic properties. Paszczyński *et al.* (1982) also used an affinity chromatography method with amylopectin-AH-Sepharose or amylopectin-porous glass to resolve the AG of *A. niger* C into two components. The efficiency of AG elution from the porous glass column was 32% greater than from the AH-Sepharose column. Yasuda *et al.* (1989) described the purification and properties of two forms of AG from *Monascus* sp. No. 3403 which is used for making Tofuyo. The purified enzymes, GA-I and GA-II, were shown to be homogeneous by disc electrophoresis and ultracentrifugation. Some of the properties of the systems are given in Table 17. The AG from *A. oryzae*, purified by acarbose affinity chromatography (Ono *et al.*, 1988*c*), was shown to be homogeneous by gel filtration, PAGE, SDS-PAGE ultracentrifu-

Table 17
Properties of the Two Forms of AG Produced by *Monascus* sp.
No. 3403 (Yasuda *et al.*, 1989)

	GA-I	*GA-II*
Relative molecular mass (M_r)		
Sephadex gel filtration	52 000	51 000
SDS gel electrophoresis	53 000	54 000
Carbohydrate content	13·5%	9·4%
Maxima for activity:		
pH	4·5	4·8
Temperature	55°C	55°C
K_m values (soluble starch)	2·9 mg/ml	2·5 mg/ml

gation and isoelectric focussing. It was a typical AG with highest activity at 56°C and pH 4·0–4·5, relative molecular mass 68 400, isoelectric point 3·64 and a carbohydrate content of 7·8%. Other AG preparations studied include *Aspergillus candidus* LINK var. *aureus* (Kolhekar *et al.*, 1985) and *Paecilomyces varioti* (Takeda *et al.*, 1985). An AG having an unusually high temperature maximum of 70°C for activity (Table 18) was reported by Fogarty & Benson (1982, 1983) and Benson *et al.* (1982). Another notable feature of this enzyme was that it showed considerable stability to pH over the range pH 2·0–11·0.

AG from *A. niger*, to which additional amino groups had been introduced, was coupled to periodate-oxidised dextran T-70 in the presence of NaBH$_3$CN (Lenders & Crichton, 1988) resulting in enhanced thermostability of the conjugate compared to the native enzyme. Furthermore, operational stability of the conjugate was at least twice that of the native AG at 70°C with 18% w/v maltodextrin. The authors concluded that enzymes possessing a low content of available amino groups require introduction of additional amino functions, prior to crosslinking to an oxidised polysaccharide, in order to enhance thermostability. α-Cyclodextrin (100 mM) exerts a thermostabilising effect on AG at 60°C (Ezure *et al.*, 1988). In contrast, β-and γ-cyclodextrin and other sugars had only a weak effect. Significantly, the stabilising effect of α-

Table 18
Properties of Thermophilic Amyloglucosidase
of *Aspergillus niger* IMDCC No. 1203 (Fogarty
& Benson, 1983)

Maxima for activity:	
pH	4·5
Temperature	70°C
Isoelectric point (pI)	4·0
Relative molecular mass (M_r)	63 000
K_m values	
Maltose	1·42 mM
Starch	0·025%

cyclodextrin was weaker at lower temperatures, e.g. 30°C and 40°C. The mechanism of the stabilising effect of α-cyclodextrin requires further investigation. Hydrolysis of soluble starch, maltotriose and maltose by AG (*A. niger*) was investigated in aqueous solutions of methanol, ethanol, ethylene glycol and 1,4-dioxane at 40°C (Nagamoto *et al.*, 1986). Kinetic analyses showed that the intrinsic rate constant was electrostatically affected by the dielectric constant of the organic solutions. Bhella & Altosaar (1984) described a simple method for large scale purification of AG using chromatography on an anion-exchange polystyrene column. The AG from *Aspergillus awamori* was purified to electrophoretic homogeneity with yields of 80%.

Medium composition and growth conditions were reported to influence production of multiple forms of AG (Hayashida, 1975; Saha *et al.*, 1979; Alazard & Raimbault, 1981). Degradation of native AG by glycosidase or protease activity during cultivation of an organism or during preparation of an extract may also influence formation of multiple forms of an enzyme. Thus, AG II could be formed from AG I in *A. awamori* var. *kawachi* by protease or glycosidases (Yoshino & Hayashida, 1978). Protease was instrumental in converting the AG from *Rhizopus* sp. into three active forms (Takahashi *et al.*, 1982). A protease-negative mutant of *A. awamori* var. *kawachi* was shown to produce but one form of AG (Hayashida & Flor, 1981). Four forms of AG (AG I, AG II, AG III and AG IV) showed different susceptibilities to two proteases (Paszczyński *et al.*, 1985) present in the culture medium of *A. niger* C. The effects of proteases on the different AG fractions during growth of the fungus were established and the results suggest that these may regulate AG activity and heterogeneity, which are functions of the age of the culture in the case of *A. niger* C.

4.3 Amyloglucosidases of Yeasts

Amyloglucosidase has been purified and characterised from a number of yeasts (Table 19). An extensive survey by De Mot *et al.* (1984a) demonstrated amylolytic activity in several yeast species not previously characterised. The

Table 19
Distribution of Recently Described Amyloglucosidases in Some Yeasts

Candida antarctica	De Mot & Verachtert (1987)
Candida tsukubaensis	De Mot *et al.* (1985)
Filobasidium capsuligenum	De Mot & Verachtert (1985)
Saccharomyces diastaticus	Tucker *et al.* (1984)
	Yamashita *et al.* (1986)
Schwanniomyces alluvius	Simoes-Mendes (1984)
Schwanniomyces castellii	Sills *et al.* (1984)
	Pasari *et al.* (1988)

Table 20
Characteristics of Some Purified Amyloglucosidases of Yeasts

	M_r	pH optimum	Temp. optimum (°C)	pI	K_m Maltose	K_m Soluble starch	Reference
Saccharomyces diastaticus	306 000	5·5	60	4·1	18·2 mM	—	Tucker et al. (1984)
Schwanniomyces alluvius	117 000	4·5	50	—	—	22·22 g/litre	Simoes-Mendes (1984)
Candida tsukubaensis	56 000	2·4–4·8	55	—	—	—	De Mot et al. (1985)
Filobasidium capsuligenum							
GI	60 000	5·0–5·6	55	—	—	—	De Mot & Verachtert (1985)
GII	60 000	4·8–5·3	50	—	—	—	Buttner et al. (1987)
Trichosporon adeninovorans	225 000	4·0–4·5	60–70	—	11 mM	1·2 g/litre	De Mot & Verachtert (1987)
Candida antarctica	48 500	4·2	57	10·1	15·8 mM	0·44 g/litre	Pasari et al. (1988)
Schwanniomyces castellii	—	4·2–5·5	40–50	—	0·535 mg/ml	0·3 mg/ml / 10·31 mg/ml	Sills et al. (1984)

AG from *Saccharomyces diastaticus* was purified to homogeneity (Tucker *et al.*, 1984). The enzyme was a glycoprotein having native and sub-unit relative molecular masses of 306 000 and 186 000, respectively, and highest activity at pH 5·5 and 60°C. *Schwanniomyces alluvius*, an alcohol-producing yeast, produces both α-amylase and AG and both systems have been purified (Simoes-Mendes, 1984). The properties of the glucose-producing enzyme are summarised in Table 20. The major starch-degrading activity produced by *Candida tsukubaensis* was isolated and characterised (Table 20) as AG (De Mot *et al.*, 1985) and had highest activity in the range pH 2·4–4·8, whereas most yeast AGs tend to be most active in the range pH 4·5–5·5. The enzyme was strongly inhibited by acarbose—a pseudo-tetrasaccharide produced by Actinomycetes. Acarbose is an extremely powerful inhibitor of microbial amylolytic enzymes and has the most pronounced effect on AG. The concentration required to give 50% inhibition of the *C. tsukubaensis* AG (25 μg/ml) is close to that recorded for the same enzyme from *A. niger* (40 μg/ml) and much lower than that observed for *A. oryzae* α-amylase (180 μg/ml) and a *Saccharomyces* sp. α-glucosidase (70 μg/ml) (Truscheit *et al.*, 1981). Clarke & Svensson (1984*a, b*) claim that two specific tryptophanyl residues are associated with the strong but reversible binding of acarbose to the AGs of *A. niger*. Two forms of AG were produced by *Filobasidium capsuligenum* having identical relative molecular masses (60 000) but differing in pH and temperature maxima for activity, electrophoretic mobility and substrate specificity (De Mot & Verachtert, 1985). Both forms of AG were strongly inhibited by acarbose—71 and 79 μg/ml gave 50% inhibition of G I and G II, respectively. An interesting feature of the AG from *Trichosporon adeninovorans* was that it had identical relative activities on maltose and soluble starch (Büttner *et al.*, 1987). The enzyme from *Candida antarctica* had a relative molecular mass of 48 500, possessed debranching activity and was strongly adsorbed onto raw starches (De Mot & Verachtert, 1987). The specific affinity of acarbose for AG was exploited during purification by coupling the pseduo-tetrasaccharide to AH-Sepharose 4B in an affinity chromatography procedure. Sills *et al.* (1984) isolated and characterised AG from *Schwanniomyces castellii* and Pasari *et al.* (1988) recently studied the α-amylase and AG from this source to characterise the product distributions and starch hydrolysis rates. Interestingly, maltose and isomaltose were hydrolysed almost to the same extent and no carbohydrate was detected in association with AG (Sills *et al.*, 1984).

4.4 Bacterial Amyloglucosidases

Recent reports suggest the occurrence of AG in bacteria. The system produced by *Clostridium thermohydrosulfuricum* (Hyun & Zeikus, 1985*a, b, c*) has unique thermostable and thermoactive characteristics (Table 21). AG was completely stable at 75°C in the presence of 5% w/v starch and retained 50% activity after treatment at 85°C for 1 h. Okada and colleagues (1988*a, b*, 1989)

Table 21
Some Properties of Bacterial Amyloglucosidases

Organism	M_r	pH optimum	Temp. optimum (°C)	pI	K_m soluble starch	Reference
Clostridium thermohydrosulfuricum		4·0–6·0	75	—	0·41 mg/ml	Hyun & Zeikus (1985c)
Arthrobacter globiformis[a]	120 000	6·0	45	4·31	—	Okada *et al.* (1988b) Okada & Unno (1989)

[a] System contains both amyloglucosidase and glucodextranase activities.

described a simple and efficient procedure for the purification of glucodextra-nase from *Arthrobacter globiformis*. In addition to the dextranase activity, it was noted that a highly purified preparation also possessed 'AG-type' activity. Many of the properties of the two systems were identical (Okada & Unno, 1989). Results strongly suggest that a single enzyme molecule was responsible for both activities. Further data on the nature of the active centre(s) of the system would confirm the true nature of the enzyme. Bender (1981a) isolated a cell-associated 'glucoamylase' type enzyme in a *Flavobacterium* sp.

4.5 Carbohydrate and Protein Composition of Amyloglucosidases

Amyloglucosidases are invariably glycoprotein in nature containing from 4–18% carbohydrate, although a carbohydrate content as high as 80% has been reported for the AG from *S. diastaticus* (Erratt & Nasim, 1987; Kleinman *et al.*, 1988). The carbohydrate component generally contains galactose, glucose, glucosamine and mannose. Relative molecular masses of AGs vary between 50 000–250 000 (Flor & Hayashida, 1983; De Mot & Verachtert, 1987). The carbohydrate fractions of AG I from *A. niger* are linked glycosidi-cally to L-threonine and L-serine (Pazur *et al.*, 1980). In this enzyme the carbohydrate fractions are present as 20 individual D-mannose units, 11 disaccharides with the structure 2-*O*-D-mannopyranosyl-D-mannose, 8 trisac-charides and 5 tetrasaccharides composed of various combinations of D-glucose, D-mannose and D-galactose joined by (1→3) and (1→6) glycosidic bonds. Such an arrangement of carbohydrate residues in a glycoprotein is very unusual and it may account for some of the unique properties of AG. The carbohydrate components undoubtedly play an important role in the stability properties of AG, since removal of this moiety reduces both the stability and activity of the enzyme (Pazur *et al.*, 1970). The random distribution of the carbohydrate chains along the polypeptide chain might account for its considerable resistance to heat inactivation and stability on storage at low temperatures (Pazur *et al.*, 1987).

The amino acid composition of a number of AGs has been determined (see Fogarty, 1983). Svensson *et al.* (1982) and Pazur *et al.* (1971) showed that both forms of AG from *A. niger* had almost identical amino acid compositions, whereas identical carbohydrate fractions but different amino acid compositions were reported for AG I and AG II of *A. niger* by Lineback & Aira (1972).

The amino acid sequences in AG I and AG II of *A. niger* have recently been elucidated. The sequence of amino acids in AG II is identical to residues 1–512 of the polypeptide chain of AG I, which is comprised of 616 residues (Boel *et al.*, 1984; Svensson *et al.*, 1985). Whereas the amino acid sequence of *A. niger* AG was determined by direct analysis (Svensson *et al.*, 1983), the sequence of many other AGs has been deduced from DNA sequence data (Boel *et al.*, 1984; Nunberg *et al.*, 1984; Ashikari *et al.*, 1985, 1986; Yamashita *et al.*, 1985; Itoh *et al.*, 1987). It has been proposed that tryptophan residues play important functions in the catalytic action of *A. niger* AG (Clarke & Svensson, 1984*a, b*) and *A. saitoi* AG (Inokuchi *et al.*, 1982). Trp 120 is especially responsible for binding of substrates (Clarke & Svensson, 1984*a, b*). The best-fit alignment of amino acid sequences in a number of AGs from different sources (Itoh *et al.*, 1987) showed five highly homologous segments (S1–S5). S5 was considered non-essential to catalytic activity because of its absence in AGs from *Saccharomyces* spp. Entire sequences of AGs were examined for local hydropathicities and secondary structures. Most parts of the conserved segments (S1–S5) were shown to be conformationally similar.

Static and kinetic methods have been used in detailed studies of the binding of *Rhizopus niveus* AG with different substrates and analogues (Ohnishi & Hiromi, 1978; Kusunoki *et al.*, 1982; Tanaka *et al.*, 1982, 1983*a*) in an attempt to establish sub-site structure (Tanaka *et al.*, 1983*b*).

5 RECENT ADVANCES IN α-GLUCOSIDASES

5.1 Introduction

The object of this report is to collate current data on microbial α-glucosidases with a view to examining those properties which might suggest a logical classification of these enzymes. Microbial α-glucosidases have been reviewed in earlier reports (Fogarty, 1983; Kelly & Fogarty, 1983). α-Glucosidase (EC 3.2.1.20, α-D-glucoside glucohydrolase) catalyses the hydrolysis of terminal, non-reducing α-1,4- and/or α-1,6- linked glucopyranose residues in certain substrates releasing α-D-glucose. They degrade disaccharides and oligosaccharides preferentially, relative to larger structures, and polysaccharides are attacked slowly if at all.

5.2 Occurrence and Properties of Bacterial α-Glucosidases

α-Glucosidases are widely distributed throughout the animal, plant and microbial kingdoms. They occur in a very wide range of microorganisms (Table 22). α-Glucosidases have a number of potential applications and such

Table 22
Occurrence of Some Microbial α-Glucosidases

Organism	Reference
Bacillus amyloliquefaciens	Fogarty & Kelly (1983), Fogarty *et al.* (1985)
Bacillus caldovelox	Giblin *et al.* (1987)
Bacillus cereus	Yoshigi *et al.* (1985*b*)
Bacillus sp. ATCC 21591	Kelly *et al.* (1983)
Bacillus sp. NCIB 11203	Kelly *et al.* (1987)
Bacillus licheniformis	Kelly *et al.* (1986)
Escherichia coli	Olusanya & Olutiola (1986)
Bacillus amylolyticus	Kelly *et al.* (1980)
Mycoplasma mycoides subsp. *capri*	Wadher & Miles (1988)
Lactobacillus acidophilus	Li & Chan (1983)
Streptomyces tosaensis	Behan *et al.* (1988)
Lipomyces starkeyi	Kelly *et al.* (1985)
Saccharomyces cerevisiae	Tabata *et al.* (1984), Krakenaite & Glemzha (1983)
Saccharomyces carlsbergensis	Federoff *et al.* (1982)
Schizosaccharomyces pombe	Chiba & Yamana (1980)
Paecilomyces varioti	O'Mahony *et al.* (1987)
Thermoascus aurantiacus	Bedino *et al.* (1983*a, b*)
Aspergillus flavus	Olutiola (1981)

usefulness is reflected in the current interest being displayed in these enzymes. One highly important property of these enzymes could be a capacity for *in situ* conversion of high maltose to dextrose syrups. A second application could be the conversion of α-limit dextrins in starch hydrolysates. Although these enzymes are collectively referred to as α-glucosidases, during the course of this review they will be discussed under two headings:

(1) α-glucosidases with highest activity towards maltose and referred to as maltases;
(2) α-glucosidases with highest activity towards aryl-α-D-glucosides and referred to as α-glucosidases.

α-Glucosidases exhibit a range of substrate specificities. Of those from *Bacillus* spp., the enzymes from *B. amylolyticus* (Kelly *et al.*, 1980), *B. brevis* (McWethy & Hartman, 1979), *B. cereus* (Yamasaki & Suzuki, 1974), *B. megaterium* (Kelly & Fogarty, 1983), and *B. subtilis* P-11 (Wang & Hartman, 1976) have been termed maltases since they have a high activity towards maltose and little, if any, activity towards p-nitrophenyl-α-D-glucopyranoside. The other group of glucosidases includes those enzymes that hydrolyse p-nitrophenyl-α-D-glycopyranoside more rapidly than maltose. The α-glucosidases of *Bacillus* sp. ATCC 21591 (Kelly *et al.*, 1983) and *B. caldovelox* (Giblin *et al.*, 1987) show preferential affinity for p-nitrophenyl-α-D-glucopyranoside. They also hydrolyse maltose and maltotriose but are inactive on α-1,6- linkages, i.e. isomaltose. Interestingly, hydrolysis of α-1,6- linkages is a feature of the maltases of *B. brevis* (McWethy & Hartman, 1979) and *B. subtilis* P-11 (Wang & Hartman, 1976) which hydrolyse both maltose and isomaltose. Although both α-1,4- and α-1,6- bonds are hydrolysed, it is believed that the α-1,6- activity is an intrinsic property of the maltase in both cases.

Although α-glucosidases occur widely in bacteria and fungi, this review will deal principally with bacterial enzymes and the reader is referred to earlier works (Fogarty, 1983; Kelly & Fogarty, 1983) in regard to the mould and yeast enzymes.

α-Glucosidases differ in their substrate specificities and these differences have been used to classify them. However, much confusion reigns in this area and many anomalies exist. Recent work in our laboratories has thrown new light on the nature, function and characteristics of these enzymes and, as more is learnt about their properties, certain patterns are beginning to emerge. In bacterial systems there would appear to be a number of different types of α-glucosidases rather than structures with very broad substrate specificities, as has been shown with the systems from *Bacillus amyloliquefaciens* (Fogarty & Kelly, 1983; Fogarty *et al.*, 1985) and *B. licheniformis* (Kelly *et al.*, 1986).

α-Glucosidase of *B. amyloliquefaciens* was first examined by Urlaub & Wöber (1978) and was reported as displaying highest activity towards sucrose and *p*-nitrophenyl-α-D-glucoside (Table 23). In addition it hydrolysed maltose, maltotriose, isomaltose and isomaltotriose. No other α-glucosidase has been reported with such high activity towards sucrose. On re-examination this system was resolved into three active components (Table 24) which were identified as a maltase, an α-glucosidase and an invertase (Fogarty & Kelly, 1983; Fogarty *et al.*, 1985).

The α-glucosidase showed activity only towards *p*-nitrophenyl-α-D-glucoside, isomaltose and isomaltotriose whereas the maltase had highest activity towards maltose, sucrose and maltotriose and the invertase was only active on sucrose and raffinose. Differences in physicochemical properties of

Table 23

Substrate Specificity of *Bacillus amyloliquefaciens* α-Glucosidase

Substrate	Relative activity (%)
Maltose	40
Maltotriose	32
Isomaltose	30
Isomaltotriose	18
Sucrose	140
p-Nitrophenyl-α-D-glucoside	100

Urlaub & Wöber (1978).

Table 24
Properties of Three Enzymes of *Bacillus amyloliquefaciens*

	Enzyme		
	α-*Glucosidase*	*Maltase*	*Invertase*
pH optimum	4·9	6·5	6·5
Relative molecular mass	26 000	72 000	30 000
Isoelectric point	4·6	4·7	4·2
Substrate	*Relative activity (%)*		
p-Nitrophenyl-α-D-glucoside	100	0	0
Maltose	0	100	0
Maltotriose	0	32	0
Isomaltose	24	0	0
Isomaltotriose	13	—	0
Sucrose	0	84	100
Raffinose	0	0	22

the systems were also noted (Table 24). Thus, the three enzymes are quite separate and distinct entities.

Similarly, Thirunavukkarasu & Priest (1984) detected a single enzyme (α-glucosidase) in *B. licheniformis* having highest activity towards isomaltose. Subsequently Kelly *et al.* (1986) detected two components in this system. The first of these activities, a maltase, hydrolysed maltose preferentially and had slight activity on isomaltose, p-nitrophenyl-α-D-glucoside and sucrose. The second enzyme was an α-glucosidase and displayed highest activity on p-nitrophenyl-α-D-glucoside followed by isomaltose, sucrose and maltose. Subsequently, Moore & Priest (1987) confirmed these findings.

In the light of these results it is imperative with these enzyme complexes that thorough and rigorous purification protocols are established. Furthermore, systems already quoted in the literature showing broad substrate specificities should be examined closely as the likelihood exists that they are composed of two or more enzymes.

5.3 Bacterial Maltases

The maltases of *B. amylolyticus* (Kelly *et al.*, 1980), *B. amyloliquefaciens* (Fogarty *et al.*, 1985), *B. cereus* (Yoshigi *et al.*, 1985b), *B. brevis* (McWethy & Hartman, 1979) and *B. subtilis* P-11 (Wang & Hartman, 1976) all display virtual total specificity for the α-1,4- glucosidic linkage in maltose and maltotriose (Table 25) with the exception of some minor activity for isomaltose in the case of *B. subtilis* P-11 maltase (Wang & Hartman, 1976). Trace activity towards isomaltose was also detected in the case of maltases from *B. cereus* (Yoshigi *et al.*, 1985b) and *B. brevis* (McWethy & Hartman, 1979). During the purification of the maltases of *B. brevis* and *B. subtilis* P-11 the ratio of maltase to isomaltase activities remained constant and it was claimed that the

Table 25

Substrate Specificities and Other Properties of Some Bacterial Maltases

	Bacillus amylolyticus (Kelly *et al.*, 1980)	*Bacillus cereus* (Yoshigi *et al.*, 1985b)	*Bacillus amyloliquefaciens* (Fogarty *et al.*, 1985)	*Bacillus brevis* (McWethy & Hartman, 1979)	*Bacillus subtilis* P-11 (Wang & Hartman, 1976)	*Bacillus* sp. NCIB 11203 (Kelly *et al.*, 1987)	*Bacillus licheniformis* (Kelly *et al.*, 1986)	*Bacillus* sp. RKll (W. M. Fogarty & C. T. Kelly, unpublished data)	*Bacillus* sp. NCIB 11204 (W. M. Fogarty, C. T. Kelly & M. Melvin, unpublished data)	*Bacillus* sp. ATCC 31075 (C. T. Kelly, W. M. Fogarty & M. P. Brosnan, unpublished data)
						Relative activity				
Maltose	100	100	100	100	100	100	100	100	100	100
Maltotriose	61	28	36	22	16	64	40	31	12	ND
p-NPG	0	<5	0	0	0	17	20	8	31	72
Isomaltose	0	0	0	5	15	0	21	0	0	0
Isomaltotriose	0	0	0	0	0	0	40	0	0	0
Sucrose	0	0	90	0	0	0	7·4	0	0	0
Starch	0	0	0	4	4	0	0	0	0	0
pH optimum	7·0	7·0	6·5	6·5	6·0	7·0	6·0	8·0	7·5	7·5
Temp. optimum (°C)	40	24	30	48–50	45	30	40	ND	40	60
pI	ND	5·74	4·7	5·2	6·0	ND	4·5	ND	ND	ND
M_r	ND	57 000	64 000	52 000	33 000	ND	160 000	ND	ND	ND
Location[a]	E	I	I	I	E	E	E	I	E	E

[a] Abbreviations: E, extracellular; I, intracellular; ND, not determined; *p*-NPG, *p*-nitrophenyl-α-D-glucoside.

α-1,6- activity was an inherent property of the maltase enzyme. Notwithstanding this intrinsic, but minor, property the outstanding features (Table 25) of the five systems referred to above are:

(1) highest activity is displayed on α-1,4- glucosidic linkages in each case, and
(2) none of the systems is capable of degrading p-NPG.

The maltase of *B. amyloliquefaciens* (Fogarty *et al.*, 1985) was different from the other four in its ability to degrade sucrose.

Five other maltases (Table 25) which included *Bacillus* sp. NCIB 11203 (Kelly *et al.*, 1987), *Bacillus licheniformis* (Kelly *et al.*, 1986), *Bacillus* sp. RK11 (W. M. Fogarty & C. T. Kelly, unpublished data), *Bacillus* sp. NCIB 11204 (W. M. Fogarty, C. T. Kelly & M. Melvin, unpublished data) and *Bacillus* sp. ATCC 31075 (C. T. Kelly, W. M. Fogarty & M. P. Brosnan, unpublished data), also showed highest affinity for α-1,4- glucosidic linkages and surprisingly possessed distinct activity towards p-nitrophenyl-α-D-glucoside. This capacity to degrade an aryl-glucoside clearly distinguishes these enzymes from the five discussed earlier. The capacity to hydrolyse the α-1,6-linkage in isomaltose is an inherent property of the maltase of *B. licheniformis* (Kelly *et al.*, 1986).

5.4 Bacterial α-Glucosidases

The α-glucosidases of *Bacillus* sp. ATCC 21591 (Kelly *et al.*, 1983), *B. caldovelox* (Giblin *et al.*, 1987) and *B. megaterium* No. 7196 (Fogarty, 1983) displayed highest activity for p-NPG (Table 26) and lesser activity for the α-1,4- glucosidic linkages in maltose and maltotriose. The enzymes were inactive on isomaltose and isomaltotriose and neither sucrose nor starch were degraded.

B. amyloliquefaciens (Fogarty *et al.*, 1985) and *Bacillus* sp. NCIB 11203 (Kelly *et al.*, 1987) produced α-glucosidases having highest affinity for p-NPG and lesser activity towards α-1,6- glucosidic linkages in isomaltose and isomaltotriose but were inactive on other naturally-occurring substrates including maltose and maltotriose.

B. licheniformis (Kelly *et al.*, 1986) and *Streptomyces tosaensis* (Behan *et al.*, 1988) showed highest affinity for p-NPG but also possessed the ability to degrade both α-1,4- and α-1,6- glucosidic linkages, e.g. in maltose and isomaltose, respectively, and in this property were quite distinctive from the two previous groups.

An important feature of these α-glucosidases is their inability to degrade starch. The enzyme from *B. caldovelox* (Giblin *et al.*, 1987) showed highest activity in the range pH 5·5–6·0 and was the most thermostable of those examined, displaying maximal activity in the range 50–60°C.

Table 26

Substrate Specificities and Other Properties of Some Bacterial α-Glucosidases

	Bacillus sp. ATCC 21591 (Kelly *et al.*, 1983)	*Bacillus caldovelox* (Giblin *et al.*, 1987)	*Bacillus megaterium* No. 7196 (Fogarty, 1983)	*Bacillus amyloliquefaciens* (Fogarty *et al.*, 1985)	*Bacillus* sp. NCIB 11203 (Kelly *et al.*, 1987)	*Bacillus licheniformis* (Kelly *et al.*, 1986)	*Streptomyces tosaensis* (Behan *et al.*, 1988)
	Relative activity						
p-PNG	100	100	100	100	100	100	100
Maltose	49	49	60	0	0	10	12·5
Maltotriose	16	34	13	0	0	84	ND
Isomaltose	0	0	0	18	0·84	91	49
Isomaltotriose	0	0	0	10	2·4	84	ND
Sucrose	0	0	0	0	0	28	16
Starch	0	0	0	0	0	0	0
pH optimum	7·0	5·5–6·0	7·0	5·3	7·0	6·0	7·0
Temp. optimum (°C)	40	50–60	30	40	40	40	40
pI	ND	5·0	ND	4·6	ND	4·5	5·2
M_r	ND	30 000	81 000	27 000	ND	66 000	79 400
Location[a]	E	I	I	I	E	E	E

[a] Abbreviations: E, extracellular; I, intracellular; ND, not determined; *p*-NPG, *p*-nitrophenyl-α-D-glucoside.

5.5 Classification of α-Glucosidases

Data emanating from this and other laboratories regarding the nature of these enzymes would suggest a more detailed classification than a straightforward sub-division into maltases and α-glucosidases. In the light of evidence to date, it would seem logical to base this classification on affinity towards the two primary substrates, maltose and *p*-NPG. Using this as a basis, an examination of the properties of systems described here would suggest a classification containing a number of sub-groups (Table 27). In the case of maltases (Table 25), two sub-divisions may be made:

(1) Enzymes possessing α-1,4-glucosidase activity only;
(2) enzymes possessing α-1,4-glucosidase as the primary activity but also having activity on *p*-NPG.

Bacterial α-glucosidases (Table 26) may be divided into three sub-divisions:

(1) Enzymes possessing highest affinity for *p*-NPG and lesser activity towards maltose (i.e. α-1,4-glucosidase activity);

Table 27
Proposed Classification of Bacterial Maltases and α-Glucosidases

Bacterial maltases

	Sub-division I	Sub-division II	
Maltose	+ +	+ +	
p-NPG	0	+	
Isomaltose	0	0	

Bacterial α-glucosidases

	Sub-division I	Sub-division II	Sub-division III
p-NPG	+ +	+ +	+ +
Maltose	+	0	+
Isomaltose	0	+	+

Symbols: + +, primary activity; +, secondary (lesser) activity; 0, no activity.

(2) enzymes possessing highest affinity for p-NPG and lesser activity towards isomaltose (i.e. α-1,6-glucosidase activity);
(3) enzymes possessing highest affinity for p-NPG and lesser activity towards maltose and isomaltose (i.e. α-1,4- and α-1,6-glucosidase activities).

The classification outlined in Table 27 conforms readily with the data presented in Tables 25 and 26, with the exception of α-1,6-glucosidase activity associated with the maltases of *B. subtilis* P-11 and *B. licheniformis*. Should this property merit a further sub-group it will become evident with characterisation of further maltases.

5.6 Oligo-glucosidases

A group of related enzymes have been described as oligo-α-1,4- and oligo-α-1,6-glucosidases. Only one example of an oligo-α-1,4-glucosidase has been described in a bacterium. It was detected in *Bacillus stearothermophilus* (Suzuki *et al.*, 1984). The enzyme degrades aryl-glucosides and hydrolyses the terminal α-1,4- linkages in a successive manner from the non-reducing ends of oligomaltosaccharides (degree of polymerisation (DP), 2–6), α-dextrins and α-glucans releasing α-D-glucose. The relative molecular mass, isoelectric point, maximum pH and temperature for activity were 47 000, 5·2, 6·4 and 70°C, respectively. No activity was detected towards isomaltosaccharides (DP 2–6), α- and β-cyclodextrin, pullulan and dextrin.

Oligo-α-1,6-glucosidase (EC 3.2.1.10, dextrin 6-α-D-glucanohydrolase) catalyses the exo-hydrolysis of α-1,6- linked D-glucose units from the non-reducing ends of isomaltooligosaccharides and aryl-glucosides, but show no activity towards α-1,4- linked D-glucose units in maltooligosaccharides. Oligo-α-1,6-glucosidases have been characterised in the following sources: *B. cereus* (Suzuki *et al.*, 1982); *B. coagulans* (Suzuki & Tomura, 1986); *B. thermo-*

glucosidius (Suzuki *et al.*, 1979); *B. flavocaldarius* (Suzuki *et al.*, 1987*a*); *Bacillus* sp. KP1071 (Suzuki *et al.*, 1987*b*) and *Thermoanaerobium* Tok 6-B1 (Plant *et al.*, 1988).

6 RECENT DEVELOPMENTS IN DEBRANCHING ENZYMES

6.1 Introduction

Debranching enzymes catalyse the hydrolysis of α-1,6- glucosidic bonds in amylopectin and/or glycogen and related polymers. Their affinity for the α-1,6- bond distinguishes these enzymes from other amylases which have primary affinity for α-1,4- glucosidic linkages. Debranching enzymes are classified into two groups, i.e. direct and indirect. The former hydrolyse α-1,6- glucosidic bonds of unmodified amylopectin and glycogen. Direct debranching enzymes may be classified into pullulanases or isoamylases on the basis of substrate specificity. Pullulanases degrade pullulan, and isoamylases have little if any affinity for this substrate (Lee & Whelan, 1971). Indirect debranching enzymes require a modification of the substrate by another enzyme or enzymes prior to their debranching action.

6.2 Direct Debranching Enzymes—Pullulanases and Isoamylases

Microbial pullulanase (EC 3.2.1.41, α-dextrin 6-glucanohydrolase) hydrolyses amylopectin and its β-limit dextrins but cannot significantly degrade native glycogen although it can attack the partially degraded polymer. Pullulanase requires that each of the two chains linked by an α-1,6- glucosidic bond contains at least two α-1,4- linked glucose units. Thus, the smallest substrate for pullulanase is the tetrasaccharide 6^2-α-maltosylmaltose (Kainuma *et al.*, 1978). Bacterial pullulanases hydrolyse α-1,6- linkages in a random fashion in pullulan, releasing maltotriose and a series of maltotriosyl oligosaccharides which are ultimately broken down to the trimer.

Microbial isoamylase (EC 3.2.1.68, glycogen 6-glucanohydrolase) is distinguished from pullulanase on the basis of substrate specificity. It is the only known enzyme that will totally debranch glycogen and is virtually incapable of degrading pullulan. Another property of isoamylase is its inability to hydrolyse 2- and 3-glucose unit side chains in β-limit and α-limit dextrins. Thus, simultaneous action of β-amylase and isoamylase cannot quantitatively convert amylopectin to maltose.

A pullulanase from *Bacillus acidopullulyticus* (Jensen & Norman, 1984; Schülein & Højer-Pedersen, 1984) was both thermostable and acidophilic, having optima at 60°C and pH 4·5–5·0. These properties made the enzyme very compatible with *A. niger* amyloglucosidase and suitable for increasing the efficiency of starch conversion to glucose. Two active forms (F1 and F2) of this enzyme were purified and they had essentially the same physicochemical

Table 28
Properties of Some Pullulanases

Source	pH optimum	Temp. optimum (°C)	pI	M_r	Reference
Bacillus stearothermophilus	6·0	65	ND	83 000	Kuriki *et al.* (1988*b*)
Bacillus acidopullulyticus	5·0	60–65	5·0	115 000	Kusano *et al.* (1988)
Thermus aquaticus	6·4	95[a]	ND	83 000	Plant *et al.* (1986)
Clostridium sp.	5·0	60–70	ND	ND	Madi *et al.* (1987)

[a] Half-life ($t_{1/2}$) 4·5 h.
ND, Not determined.

properties (Kusano *et al.*, 1988). A cell-associated pullulanase from *Thermus aquaticus* (Plant *et al.*, 1986) had a relative molecular mass of 83 000, pH optimum of 6·4 and retained over 80% of its activity after 10 h at 85°C (Table 28). The thermostability of this enzyme is significantly higher than that of other reported pullulanases, including the pullulanase activity in *B. stearothermophilus* KP1064 (Suzuki & Chishiro, 1983) and *Clostridium thermohydrosulfuricum* (Hyun & Zeikus, 1985*a*). The mechanism for the regulation of the synthesis of the cell-bound pullulanase of *C. thermohydrosulfuricum* was studied and a hyperproductive mutant, Z 21-109, developed (Hyun & Zeikus, 1985*b*). Activity was selectively removed from cells by treatment with detergent and lipase and the solubilised pullulanase was purified and characterised (Saha *et al.*, 1988). The enzyme was optimally active at pH 5·5 and 90°C and was inhibited by cyclodextrins. Acarbose, a strong inhibitor of amyloglucosidase, had virtually no effect on the activity of this pullulanase. Thermostable pullulanase activities have been identified in other sources including *Thermoanaerobium* Tok 6-Bl (Plant et al., 1987*a*), *Thermoanaerobium brockii* (Coleman *et al.*, 1987), *Clostridium* sp. (Madi *et al.*, 1987; Madi & Antranikian, 1989), and *C. thermohydrosulfuricum* E101-69 (Melasniemi, 1987*a*, 1988). In the case of *Clostridium* sp. (Madi *et al.*, 1987), which was identified as *Clostridium thermosulfurogenes* (Madi & Antranikian, 1989), highest pullulanase activity was induced in media containing maltose as carbon source, whereas maltotriose induced highest levels of α-amylase and little more than one-third of the pullulanase activity obtained with maltose. It would appear that these two activities are related to separate proteins, whereas in the cases of *Thermoanaerobium* Tok 6-B1, *T. brockii* and *C. thermohydrosulfuricum*, both activities appear to be associated with a single thermostable enzyme. A thermostable pullulanase from a newly isolated strain of *B. stearothermophilus* (TRS 128) had optima at pH 6·0 and 65°C (Kuriki *et al.*, 1988*b*). The extracellular constitutive pullulanase of *Sclerotium rolfsii* (Kelkar *et al.*, 1988) had optima at pH 4·2 and 50°C and a K_m of 8·33 mg/ml, with pullulan as substrate. The system was not purified and other physico-chemical properties were not described.

In addition to pullulanase, two new related types of enzyme have been described: (a) isopullulanase (EC 3.2.1.57) from *A. niger* which hydrolyses the α-1,4- bonds in pullulan to produce isopanose (Sakano *et al.*, 1971); (b) neopullulanase, which hydrolyses α-1,4- bonds in pullulan to produce panose (Kuriki *et al.*, 1988*a*, *b*). The latter is produced by *B. stearothermophilus* (Imanaka & Kuriki, 1989) and hydrolyses not only α-1-,4- glucosidic bonds but also specific α-1,6- glucosidic bonds of several branched oligosaccharides. With pullulan as substrate, panose, maltose and glucose were produced in the ratio 3:1:1. *B. stearothermophilus* neopullulanase hydrolysed starch also, but less efficiently (13%) than it did pullulan (45%). It had a pH optimum of 6·0, relative molecular mass of 62 000 and 90% activity was retained after 60 min at 60°C (Kuriki *et al.*, 1988*a*, *b*). *B. stearothermophilus* KP1064 pullulan-hydrolysing enzyme also produced panose from pullulan (Suzuki & Chishiro, 1983; Suzuki & Imai, 1985). It had a very broad substrate specificity and could hydrolyse the α-1,4- linkages in amylose, soluble starch and related polymers. In this case amylose was a far better substrate than pullulan and the major end-product was α-D-maltose. It had M_r 115 000, was made up of two identical sub-units and had maximum activity at 55°C and pH 5·8. Isopullulanase activity was detected with an isomaltodextranase in *Arthrobacter globiformis* T6 (Okada *et al.*, 1988*a*). It was suggested that both activities were part of the one enzyme. The isopullulanase had maximum activity at pH 4·8 and 60°C, an endo-mode of action, and produced isopanose from pullulan.

Considerable interest has been shown in starch-degrading yeasts in recent times. De Mot *et al.* (1984*b*) investigated the production of debranching activity in a number of starch-degrading yeasts. Highest values of extracellular pullulanase activities were detected in the cases of *Filobasidium capsuligenum* (CCY 71-1-1), *Lipomyces kononenkoae* (IGC 4052), *Leucosporidium cap-suligenum* (CCY 64-2-4), *Schwanniomyces castellii* (R 91) and *Trichosporon pullulans* (IGC 3488). Extracellular isoamylase is produced by *Lipomyces kononenkoae* (Spencer-Martins, 1982). Optimum pH and temperature for activity of the enzyme were 5·6 and 30°C. It had M_r 65 000 and pI 4·7–4·8. As reported for other isoamylases (Marshall, 1973*b*; Kitagawa *et al.*, 1975), this enzyme was not inhibited by β-cyclodextrin (β-CD).

7 CYCLODEXTRIN-PRODUCING ENZYMES

7.1 Introduction

Cyclodextrins (cycloamyloses, Schardinger dextrins, CDs) are cyclic, non-reducing maltooligosaccharides produced from starch by cyclodextrin glycosyl-transferase (EC 2.4.1.19, CGT, 1,4-α-D-glucan 4-α-D-(1,4-α-D-glucano)-transferase (cyclising)). To-date, these enzymes have been detected exclusively in bacteria and predominantly in *Bacillus* spp. CGTs detach short chain lengths

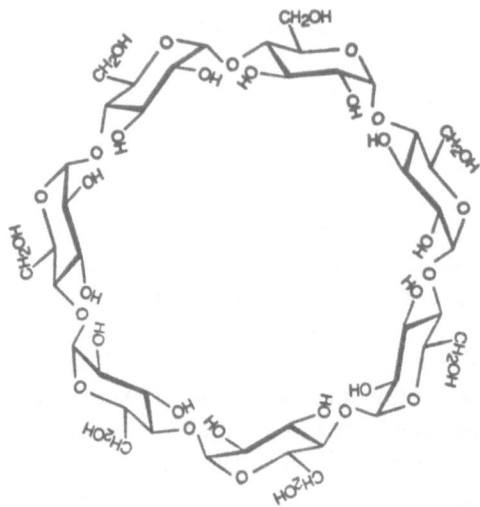

Fig. 2. Chemical structure of β-cyclodextrin (cycloheptaamylose).

from starch and then link the two ends of each fragment to form a cyclic molecule. Since the enzymes do not detach specific chain lengths, the resulting cyclical dextrins may contain 6–12 glucopyranose residues per ring structure. The principal fractions formed, however, possess six, seven or eight α-1,4-D linked glucopyranose units and are designated α-cyclodextrin (cyclohexa-amylose, cyclomaltohexaose, α-CD), β-cyclodextrin (cycloheptaamylose, cyclo-maltoheptaose, β-CD), γ-cyclodextrin (cyclootaamylose, cyclomaltooctaose, γ-CD), respectively. The relative quantities of the three cyclodextrins formed depend on the type and source of enzyme used and may be effected by the presence of organic compounds (French, 1957; Mifune & Shima, 1977; Bender, 1986).

The ring-shaped structure of cyclodextrins consists of a hydrophilic or polar exterior and a hydrophobic central cavity (Figs 2 and 3). The most outstanding

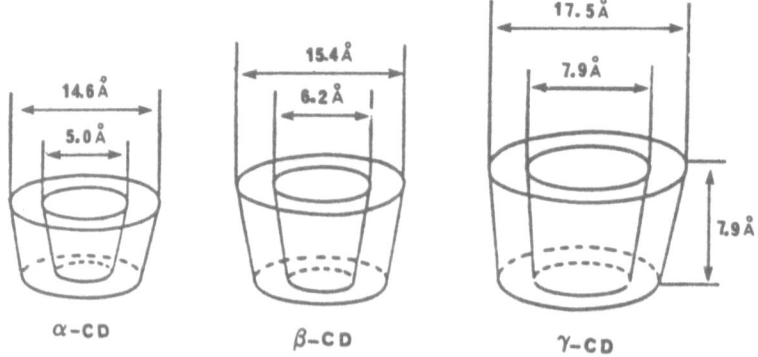

Fig. 3. Dimensions of α-, β- and γ-cyclodextrins.

Table 29
Physical Properties of cyclodextrins

Solubility at 30°C in water (g/100 ml)	14·5	1·85	23·2
Relative molecular mass	972	1135	1297
Cavity diameter (Å)	5·0	6·2	7·9
Height of torus (Å)	7·9	7·9	7·9
Specific rotation—$[\alpha]_D^{25}$	+150·5	+162·5	+177·4

property of CDs is their capacity to form inclusion complexes in aqueous solution in which 'guest' molecules of suitable dimensions are included within the cavity of the 'host' CD without any covalent bonds being formed. All types of 'guest' molecules are accommodated, ranging from non-polar aliphatic and aromatic hydrocarbons to polar compounds such as acids and amines. Cyclodextrin inclusion compounds are seen to have considerable potential in the food, pharmaceutical, fermentation and fine chemical industries in the areas of preparation, separation and purification, as well as the protection of products such as fragrances, flavours, various pharmaceuticals, e.g. steroids, etc. Controlled release of biologically active compounds, e.g. medicines, drugs, and pesticides, from inclusion complexes is of particular interest. CDs also form stereospecific complexes and may be used in the separation of enantiomers. Like enzymes, CDs possess specific catalytic properties and have formed the basis of an intensive study as enzyme models. Reviews include those of Saenger (1984), Bender (1986) and Szejtli (1987).

Solubilities of α-, β- and γ-CDs at 20°C are 14·5, 1·85 and 23·2 g/100 ml water, respectively (Table 29). The lack of solubility of β-CD for many practical applications is the subject of considerable interest, specifically with a view to forming derivatives having increased solubility properties.

7.2 Cyclodextrin Glycosyltransferases

CGTs catalyse two reversible α-1,4-D-glucopyranosyltransfer reactions:

$$G_n \underset{\text{coupling}}{\overset{\text{cyclisation}}{\rightleftharpoons}} G_{(n-x)} + \text{CDs}$$

$$G_n + G_o \overset{\text{disproportionation}}{\rightleftharpoons} G_{(n-x)} + G_{(o+x)}$$

where G_n and G_o are α-1,4-D-glucopyranosyl chains having n and o D-glucopyranosyl units, x is a part of the α-1,4-D-glucopyranosyl chain. Because CGTs catalyse two-substrate reactions, coupling and disproportionation, they are considered to possess two binding sites (Bender, 1980, 1985).

A number of bacteria produce extracellular CGTs (Table 30). Maximal activity is generally in the pH range 5·0–7·0, with the exception of *Bacillus* No. 38-2, which produces three enzymes having maximal activity at pH values of 4·6, 7·0 and 9·5 (Table 31). The systems have essentially similar M_r values.

Table 30
Bacteria Producing Cyclodextrin Glycosyltransferases

Organism	Reference
Alkalophilic *Bacillus* No. 38-2	Nakamura & Horikoshi (1976*a, b*), Mäkelä *et al.* (1988)
Alkalophilic *Bacillus* sp. strain 1011	Kimura *et al.* (1987*a, b*)
Bacillus circulans	Kitahata & Okada (1982*b*)
Bacillus macerans	French (1957), Kobayashi *et al.* (1978), Miskolci-Török *et al.* (1980)
Bacillus megaterium	Kitahata *et al.* (1974)
Bacillus ohbensis	Yagi & Iguchi (1974)
Bacillus stearothermophilus	Kitahata & Okada (1982*a, b*), Kubota *et al.* (1988)
Klebsiella pneumoniae M5 al	Bender (1977*a, b*)
Micrococcus sp.	Yagi *et al.* (1980)

Although all CGTs produce α-, β- and γ-CDs simultaneously, they can nevertheless be classified into two types—the one producing α-CD as the predominant end-product and the other producing β-CD as the main product. CGTs producing γ-CD preferentially are unknown. Methods of preparing individual CDs have been summarised (Bender, 1986).

CDs are known to be competitive inhibitors of β-amylases (Thoma & Koshland, 1960; Marshall, 1973*a*; Fogarty 1983; Fogarty & Kelly, 1983; Bourke & Fogarty, 1985), amyloglucosidases (W. M. Fogarty, unpublished data), potato phosphorylase (French & Wild, 1953) and pullulanase of

Table 31
Properties of Some Cycloglycosyltransferases

Source	Maximum pH for activity	M_r	Reference
CGTs producing α-CD			
Bacillus macerans	5.0–6.0	79 000	László *et al.* (1981)
Bacillus stearothermophilus	6.0	68 000	Kitahata & Okada (1982*a*)
Klebsiella pneumoniae M5 al	6.0–7.2	68 000	Bender (1977*a, b*)
CGTs producing β-CD			
Alkalophilic *Bacillus* No. 38-2	4·6	88 000	Nakamura & Horikoshi (1976*b, c*)
	7·0		Nakamura & Horikoshi (1976*b, c*)
	9·5		Nakamura & Horikoshi (1976*b, c*)
	7·0	70 500	Mäkelä *et al.* (1988)
Bacillus megaterium	5·0–5·7	—	Kitahata *et al.* (1974)
Micrococcus sp.	5·0–8·0	88 000	Yagi *et al.* (1980)

Klebsiella pneumoniae (Marshall, 1973*b*). The α-amylase of *B. subtilis* was inhibited by β-CD (Ohnishi, 1971), whereas both α- and β-CD inhibited sweet potato β-amylase (Marshall, 1973*a*). CD-substituted epoxy-activated Sepharose-6B was used to purify sweet potato β-amylase (Vretblad, 1974), *B. polymyxa* β-amylase (Bourke & Fogarty, 1975; Fogarty & Kelly, 1983) and cereal α-amylase (Hoschke *et al.*, 1976; Silvanovich & Hill, 1976; Weselake & Hill, 1982) by affinity chromatography. Evidence indicated that β-CD bound to a non-catalytic site of the cereal α-amylase molecule (Weselake & Hill, 1982). Because of the strong competitive inhibition of β-amylase by β-CD, it is plausible that, in this instance, it may be the active site of the enzyme that binds to the cyclodextrin. Hydrolysis of starch granules by α-amylase is also inhibited by CDs (Weselake & Hill, 1983). α-CD is known to have a stabilising effect on amyloglucosidase (Ezure *et al.*, 1988)—see Section 4.2.

The CGT from *Bacillus* sp. No. 38-2 was cloned in *E. coli* using pBR322 and the nucleotide and amino acid sequences of this gene and its enzyme product, respectively, were determined (Kaneko *et al.*, 1988). Similarly, the β-CGT of Bacillus sp. No. 1011 was cloned in *E. coli* using pBR322 and both nucleotide and amino acid sequences of the gene and enzyme, respectively, were determined (Kimura *et al.*, 1987*a, b*). The amino acid sequence at the NH_2-terminal side of the *Bacillus* sp. No. 1011 CGT had a considerable homology with sequences identified in the active centres of α-amylases.

One of the major difficulties in quantitative determination of CGTs is the lack of specific assay methods. Enzymes forming α-CD as the primary initial product may be measured by detecting the 2·5-fold increase in absorption at 290 nm when iodine complexes with the CD (Bender, 1981*b*). Cyclisation activity may also be measured by following the decrease in absorption at 546 nm of methylorange when complexing with α-CD or β-CD (Landert *et al.*, 1982). Initial cyclisation activity may also be determined by quantitative HPLC (Bender, 1985; Sato *et al.*, 1985). New-specific assay methods are based on reduction of viscosity or iodine colour of starch solutions (Bender, 1981*b*). Recently, Mäkelä & Korpela (1988) described a method using methylorange and maltotriose for measurement of CGTase activity. The method is based on measurement of CDs formed by CGTase using maltotriose as substrate.

7.3 Branched Cyclodextrins

The reverse action of hydrolytic enzymes is significant because it allows synthesis of some interesting materials, including branched CDs. Branched cyclodextrins contain glucose (G), maltose (G_2), maltotriose (G_3), etc. bound to CDs through α-1,6-D- glucosidic linkages. They have assumed special significance because of their increased solubility compared with that of the parent compound and particularly so in the case of β-CD. Maltosyl-α-CD (G_2-α-CD) was prepared from maltose and α-CD by the reverse action of *Bacillus pullulyticus* pullulanase (Sakano *et al.*, 1985*b*). Yield of G_2-α-CD was increased twenty-fold by replacing α-maltose with α-maltosylfluoride (α-G_2F)

(Kitahata *et al.*, 1987). Three branched cyclodextrins isolated by HPLC from the mother liquors in large scale synthesis of CDs by *Bacillus ohbensis* CGT were identified as G-α-CD, G-β-CD and di-G-β-CD (Koizumi & Utamura, 1986). Three debranching enzymes gave branched CDs in different yields (Yoshimura *et al.*, 1987). It would appear that the enzymes have different reactivities towards α-G$_2$F and different acceptor specificities towards CDs. Thus, the yields of G$_2$-α-CD produced by *Aerobacter aerogenes* pullulanase and *Pseudomonas amyloderamosa* isoamylase were about three times greater than that obtained with *Bacillus pullulyticus* pullulanase. Several branched CDs were synthesised through the reverse action of *Pseudomonas* isoamylase (Abe *et al.*, 1986, 1988*a*). The reaction rate was greater with maltotriose than maltose and with increasing size of CD, i.e. α-CD < β-CD < γ-CD. Whereas maltotriose was effective as both a side-chain donor and acceptor substrate, maltose was only effective as a side-chain donor. Hizukuri *et al.* (1989) exploited the reverse action of *Klebsiella aerogenes* pullulanase to produce the branched cyclomaltooactaoses, G$_2$-γ-CD and di-G$_2$-γ-CD. It was observed that CDs were better substrates in the order γ-CD > α-CD > β-CD with this enzyme system. Maltose and maltotriose were equally effective with β-CD and γ-CD but the dimer was less effective than maltotriose with α-CD. Both the formation of α-CD and branched CDs is enhanced when using CGT of *B. macerans* in the presence of sodium dodecyl sulphate (Kobayashi *et al.*, 1983, 1984).

7.4 Enzymic Degradation of Cyclodextrins

7.4.1 *Hydrolysis of Cyclodextrins by Endo-acting Amylases*
CDs are relatively stable molecules in comparison to the corresponding linear maltooligosaccharide (French, 1957; Kondo *et al.*, 1981). A feature of both α-CD and β-CD is their resistance to hydrolysis by most amylases. The α-amylases of *B. subtilis* (Keay, 1970; Moseley & Keay, 1970; Ohnishi, 1971), *A. oryzae* (Svetsugu *et al.*, 1974; Jodál *et al.*, 1984) and *Pencillium africanum* (Ben-Gersham and Leibowitz, 1958) degrade CDs but the rate of hydrolysis is much slower than that of starch and is regarded as a lesser or minor property. A maltogenic α-amylase of *B. stearothermophilus* KP1064 showed relatively high activity towards α- and β-CD (Suzuki & Imai, 1985). Antenucci & Palmer (1984) showed that 24 out of 30 strains of *Bacteroides* isolated from the human colon were capable of degrading CDs. γ-CD is degraded by a series of α-amylases and has been employed for assaying their isoenzymes (Yokobayashi, 1979; Marshall & Miwa, 1981; Miwa & Marshall, 1982). *B. subtilis* 65 α-amylase (Teramoto *et al.*, 1989) digested raw potato starch and hydrolysed γ-CD but not α- or β-CD. Bender (1981*a*) described a *Flavobacterium* sp. which produced an inducible, cell-bound cyclodextrin-degrading glucoamylase. The final end-product with α-, β- and γ-CD and all starch-like substrates was glucose. It should be noted that the K_m for α-CD was 0·142 mM and the enzyme had higher relative activity on CDs than linear substrates. Although

Table 32
Properties of Some Cyclodextrinases

Source	Location	pH optimum	Temp. optimum (°C)	M_r	Reference
Bacillus macerans	Intracellular	6·2–6·4	50–55	—	De Pinto & Campbell (1968*a*, *b*)
Pseudomonas sp. MSI	Intracellular	5·5	50	96 000	Kato *et al.* (1975)
Bacillus coagulans	Intracellular	6·2	50	62 000	Kitahata *et al.* (1983), Kitahata & Okada (1985)
Lipomyces kononenkoae	Extracellular	5·0	60	345 000	Spencer-Martins (1984)

CDs may be slowly hydrolysed by some endo-acting amylases (α-amylases), they are resistant, however, to degradation by exo-acting amylases (amyloglucosidase and β-amylase).

7.4.2 Hydrolysis of Cyclodextrins by Cyclodextrinase

Enzymes that hydrolyse CDs faster than starch (EC 3.2.1.54, cyclomaltodextrinase, cyclodextrinase, cyclomaltodextrin dextrinhydrolase (decyclising)) have so far only been reported in a small number of bacteria. The first enzyme described was that of *B. macerans* (Table 32) (De Pinto & Campbell, 1968*a*, *b*). Kato *et al.* (1975) described an intracellular cyclodextrinase from *Pseudomonas* MSI having high activity on CDs and maltooligosaccharides. It hydrolysed CDs rapidly and, although it degraded maltotriose and maltotetraose faster than amylose, it did not hydrolyse maltose. The K_m values for α-CD, amylose and amylopectin were 3·5, 3·3 and 13 mg/ml, respectively. The products formed initially from amylose were saccharides in the range G_1–G_5 having the α- configuration and, thus, the enzyme appears to have many characteristics of an α-amylase. These saccharides were subsequently broken down to a mixture of glucose and maltose. Kitahata *et al.* (1983) detected an intracellular enzyme in *B. coagulans* which had K_m values 10, 2·8, 0·47, 4·5, 4·0, 2·3, 1·5 and 1·5 mM for α-CD, β-CD, γ-CD, maltotriose, maltotetraose, maltopentaose, maltohexaose, and amylose (DP 18), respectively. Two other cyclodextrinases, from *B. macerans* (De Pinto & Campbell, 1968*a*, *b*) and *Pseudomonas* sp. MSI (Kato *et al.*, 1975) have been reported to degrade CDs faster than linear maltooligosaccharides as has the 'glucoamylase' of *Flavobacterium* sp. (Bender, 1981*a*). Degradation of starch and related polymers by the *B. coagulans* enzyme was less than 1% of the value for β-CD and the enzyme failed to hydrolyse maltose. Linear maltooligosaccharides in the series G_4–G_6 were hydrolysed much faster than CDs, thus indicating that the opening of the CD ring structure is the rate limiting step in the overall reaction. The *B. coagulans* cyclodextrinase (Kitahata & Okada, 1985) hydrolyses maltose units from the non-reducing ends of linear maltooligosaccharides. It hydrolysed α-1,4, α-1,2, α-1,3 and α-1,6 D-glucosidic linkages as well as phenyl-α-D-maltoside.

An extracellular cyclodextrinase of the yeast *Lipomyces kononenkoae* purified to homogeneity (Spencer-Martins, 1984) had a relative molecular mass, isoelectric point and pH and temperature maxima for activity of 345 000, 5·2–5·4, 5·0 and 60°C, respectively. The enzyme produces glucose from a range of substrates including maltose. K_m values for α-CD, β-CD, γ-CD, maltose, p-NPG and soluble starch were 0·12 mM, 0·16 mM, 0·16 mM, 1·3 mM, 2·3 mM and 19·7 g/litre. Glucose competitively inhibited hydrolysis of p-NPG.

De Mot & Verachtert (1986*a*) detected significant levels of cyclodextrinase activities (1–6 U/ml) in the yeasts *Candida antarctica*, *Cryptococcus flavus*, *Filobasidium capsuligenum* and *Lipomyces starkeyi*, although these levels were outstandingly less than the level detected in *L. kononenkoae* (40 U/ml).

8 AMYLOLYTIC ENZYMES AND HYDROLYSIS OF NATIVE STARCH

8.1 Fungal Enzymes and Degradation of Raw Starch

In the native state, starch exists as granules in a polycrystalline state. These granules are insoluble in water and resistant to many chemicals and enzymes. Gelatinisation by heating in water enhances chemical reactivity towards liquefying and subsequently saccharifying enzymes, and this has now become the accepted method of starch hydrolysis. In order to reduce costs associated with the high temperatures required in gelatinisation, there has been much interest recently in enzymes capable of digesting raw starch granules (Saha & Ueda, 1983*b*, *c*, 1984*a*; Tadasa & Takeda, 1986; Tani et *al.*, 1986; Mikuni *et al.*, 1987; Hayashida *et al.*, 1989*a*, *b*; Itkor *et al*, 1989). *Aspergillus* sp. K-27, a soil isolate from a screening programme aimed at selection of microorganisms producing raw starch-degrading enzymes (Abe *et al.*, 1988*b*, *c*, Bergmann *et al.*, 1988), produced an extracellular α-amylase and amyloglucosidase. The latter had a specific starch binding site and the enzymes acted synergistically on raw corn and raw potato starch. The synergistic affect is not only specific to the *Aspergillus* sp. K-27 amylase system but was also observed with other systems and is more evident with α-amylases which have lower activities for raw starch. One theory explaining the synergism is that the action of α-amylase on the surface of the granules supplies new non-reducing end-groups for the amyloglucosidase. The amyloglucosidase acting on these new end-groups strips the molecule from the surface and thereby allows access of the α-amylase to the next layer of the starch granule (Fujii *et al.*, 1988).

An inhibitory factor (IF) from *A. niger* (Saha & Ueda, 1984*b*; Towprayoon *et al.*, 1988) caused inhibition of raw starch digestion by amyloglucosidases, α-amylases and a combination of both enzymes. It caused greater inhibition of raw starch hydrolysis by the latter than by individual enzymes. It was completely adsorbed onto raw starch, was a glycoprotein and had a relative molecular mass of about 10 500. The presence of this inhibitory factor initially became evident with the inhibition of raw starch digestion with increasing amyloglucosidase concentration (Ueda & Saha, 1980; Saha & Ueda, 1984*b*).

This inhibitory factor was isolated from a commercial enzyme preparation. It contained about 25% carbohydrate and had a relative molecular mass of 12 000 (Saha & Ueda, 1983a).

Amyloglucosidase of *Rhizopus niveus* was resolved into five forms, all of which had debranching activity. They were each adsorbed onto, and digested, raw starch (Saha & Ueda, 1983b). This is quite different from the amylogluco- sidase systems of other fungi, e.g. *A. awamori* var. *kawachi* (Hayashida, 1975; Hayashida *et al.*, 1982) and *A. oryzae* (Miah & Ueda, 1977a, b), the multiple forms of which varied in their capacity to adsorb onto and digest raw starch. Amyloglucosidase I of an *Aspergillus* sp. had debranching activity, strong raw starch adsorption and digestion ability, whereas amyloglucosidase II did not have debranching activity, could not be adsorbed onto raw starch and had weak raw starch digestion (Medda *et al.*, 1982a; Saha & Ueda, 1984a). Amyloglucosidase I was maximally adsorbed onto starch at pH 3·4, which is the isoelectric point of the enzyme.

The raw starch-digesting amyloglucosidase I of *A. awamori* var. *kawachi* was shown to be degraded with proteases and glycosidases to the non-raw starch-digesting amyloglucosidases I′ and II (Hayashida *et al.*, 1976; Hayashida & Yoshino, 1978; Yoshino & Hayashida, 1978). A protease-less mutant of *A. awamori* var. *kawachi* was obtained (Hayashida & Flor, 1981) which produced a large amount of raw starch-adsorbable and raw starch-digesting enzyme, similar to amyloglucosidase I. It varied in relative molecular mass and carbohydrate content in comparison with the latter. Amyloglucosidase I had a relative molecular mass of 90 000 and *c.* 7% carbohydrate, while the new enzyme had a relative molecular mass of 250 000 and a carbohydrate content of 24·3% (Flor & Hayashida, 1983). Evidence also suggested the presence of a raw starch affinity site different from the active site.

Proteolysis of amyloglucosidase I of *A. awamori* var. *kawachi* with subtilisin produced a glycopeptide (G_p-I) from this raw starch-adsorbable and raw starch-digesting enzyme. This structure contained the raw starch affinity site essential for raw starch digestion but was enzymatically inactive (Hayashida *et al.* 1989b). It contained 45 amino acid residues, the sequence of which was determined, and 56 mannose residues (Hayashida *et al.*, 1989a). In a comparison of G_p-I with three amyloglucosidases, significantly homologous regions were noted. The workers suggested that the homologous region could be functionally constrained and essential for raw starch-digesting amyloglucosidase.

A protease-negative mutant from *Aspergillus ficum* IFO 4320 produced a raw starch-adsorbable α-amylase which completely solubilised raw corn starch granules within six days (Hayashida & Teramoto, 1986). It had a higher relative molecular mass (88 000), compared to other fungal α-amylases. The raw starch-digesting enzymes of *Corticium rolfii* completely hydrolysed a 20% (w/v) suspension of raw corn starch at pH 4·0 and 40°C in 48 h (Sasaki *et al.*, 1986).

A variety of raw starches were also hydrolysed by the extracellular amyloglucosidase of *Schizophyllum commune* (Shimazaki *et al.*, 1984). The raw starch-digesting amyloglucosidase of *Endomycopsis fibuligera* was completely adsorbed onto raw starch at all the pH values tested (Ueda & Saha, 1983). Adsorption onto raw starch granules may in some cases be dependent on pH (Medda *et al.*, 1982*a*). The pH optimum of *E. fibuligera* amyloglucosidase varies for soluble starch (pH 5·5) and for raw starch (pH 4·5), a phenomenon not unknown among amyloglucosidases (Medda *et al.*, 1982*b*).

Attempts have been made to produce ethanol from raw starch in a single-step process combining liquefaction, saccharification and fermentation by using the enzymes from *A. niger* (Ueda & Koba, 1980; Ueda *et al.*, 1981), *Rhizopus* sp. (Ueda, 1982), *E. fibuligera* (Saha & Ueda, 1983*c* and *Chalara paradoxa* (Mikuni *et al.*, 1987).

8.2 Bacterial Enzymes and Degradation of Raw Starch

Whereas there has almost always been a correlation between raw starch digestion and raw starch adsorption in fungal enzymes, this was not evident with some bacterial α-amylases (Teramoto *et al.*, 1989). *Bacillus subtilis* 65 produced a raw starch-digesting α-amylase which could digest raw potato starch almost as readily as it could corn starch, but was not adsorbed onto any type of raw starch at any pH (Hayashida *et al.*, 1988). However, on treatment of *B. subtilis* 65 α-amylase with a protease (Pronase), the enzyme lost its ability to hydrolyse raw starch (Teramoto *et al.*, 1989). The native enzyme had a relative molecular mass of 68 000 which was modified to 58 000 on digestion with Pronase. The modified enzyme lost the capacity to digest raw starch, indicating that the specific region removed by Pronase is essential for hydrolysis of raw starch. Liquefying α-amylase had a very limited capacity to digest raw starch.

Bacillus circulans F-2 produced an amylase with a relative molecular mass of 93 000 and pH and temperature optima of 6·0–6·5 and 60°C, respectively (Taniguchi *et al.*, 1983). The enzyme appeared to remove G_6 units successively from the non-reducing ends of polysaccharides and subsequently degraded these to G_4 and G_2. The *B. circulans* F-2 enzyme also had the capacity to adsorb onto and digest raw starch and, whereas it had considerable capacity to degrade raw corn starch, it also showed significant activity on raw potato starch.

Raw starch digestion was also observed with an α-amylase from a non-sulphur purple photosynthetic bacterium (Buranakarl *et al.*, 1988). This enzyme adsorbed onto granules of raw starch in the culture medium which facilitated recovery of the enzyme. The starch-binding α-amylase of *Clostridium butyricum* T-7 (Tanaka *et al.*, 1987) could not be liberated from raw starch granules when the intact structure of the granule was lost by the action of the enzyme. The enzyme had a relative molecular mass of 89 000 and displayed maximum activity at pH 5·0 and 60°C. Wheat and corn starch granules were easily digested by the enzyme while potato starch was quite resistant to hydrolysis.

Another strain of *Clostridium butyricum* (McCarthy *et al.*, 1988) produced a raw starch-digesting, raw-starch-adsorbing, α-amylase with maximum activity at pH 5·0 and 50°C and a relative molecular mass of 81 000. The enzyme appeared to initiate its attack on raw starch granules by making small pits and depressions on the peripheral region and these penetrated deeper into the granule on subsequent attacks. The type and the extent of morphological changes in starch granules on hydrolysis by amylolytic enzymes has been well described (Evers *et al.*, 1971; Smith & Lineback, 1976; Ramadas Bhat *et al.*, 1983; Colonna *et al.*, 1988).

REFERENCES

Abdullah, M., Fleming, T. D., Taylor, P. M. & Whelan, W. J. (1963). *Biochemical Journal,* **89,** 35.

Abe, J., Takeda, Y. & Hizukuri, S. (1982). *Biochimica et Biophysica Acta,* **703,** 26.

Abe, J., Mizowaki, N., Hizukuri, S., Koizumi, K. & Utamura, T. (1986). *Carbohydrate Research,* **154,** 81.

Abe, J., Hizukuri, S., Koizumi, K. & Utamura, T. (1988*a*). *Carbohydrate Research,* **176,** 87.

Abe, J.-I., Bergmann, F. W., Obata, K. & Hizukuri, S. (1988*b*). *Applied Microbiology and Biotechnology,* **27,** 447.

Abe, J.-I., Nakajima, K., Nazano, H. & Hizukuri, S. (1988*c*). *Carbohydrate Research,* **175,** 85.

Adams, P. R. (1981). *Mycopathology,* **76,** 97.

Ahern, T. J. & Klibanov, A. M. (1985). *Science,* **228,** 1280.

Alazard, D. & Baldensperger, J. F. (1982). *Carbohydrate Research,* **107,** 231.

Alazard, D. & Raimbault, M. (1981). *European Journal of Applied Microbiology and Biotechnology,* **12,** 113.

Allam, A. M., Hussein, A. M. & Razab, A. M. (1975). *Zeitschrift für Allgemeine Mikrobiologie,* **15,** 393.

Andrews, L. & Ward, J. (1987). *Biochemical Society Transactions,* **15**(3), 522.

Antenucci, R. N. & Palmer, J. K. (1984). *Journal of Agricultural and Food Chemistry,* **32,** 1316.

Ashikari, T., Nakamura, N., Tanaka, Y., Kiuchi, N., Shibano, Y., Tanaka, T., Amachi, T. & Yoshizumi, H. (1985). *Agricultural and Biological Chemistry,* **49,** 2521.

Ashikari, T., Nakamura, N., Taneka, Y., Kiuchi, N., Shibano, Y., Tanaka, T., Amachi, T. & Yoshizumi, Y. (1986). *Agricultural and Biological Chemistry,* **50,** 957.

Barnett, E. A. & Fergus, C. L. (1971). *Mycopathology Mycology Applied,* **44,** 131.

Basaveswara, R., Sastri, N. V. S. & Subra Rao, P. V. (1981). *Biochemical Journal,* **193,** 379.

Bedino, S., Testore, G. & Obert, F. (1983*a*). *Italian Journal of Biochemistry,* **32,** 371.

Bedino, S., Testore, G. & Obert, F. (1983*b*). *Italian Journal of Biochemistry,* **32,** 408.

Behan, J., Kelly, C. T. & Fogarty, W. M. (1988). *Biochemical Society Transactions,* **16,** 180.

Bender, H. (1977*a*). *Archives of Microbiology,* **111,** 271.

Bender, H. (1977*b*). *Archives of Microbiology,* **113,** 49.

Bender, H. (1980). *Carbohydrate Research,* **78,** 133.

Bender, H. (1981*a*). *European Journal of Biochemistry,* **115,** 287.

Bender, H. (1981*b*). *Analytical Biochemistry,* **144,** 158.

Bender, H. (1985). *Carbohydrate Research*, **135**, 291.
Bender, H. (1986). *Advances in Biotechnology Processes*, **6**, 31.
Ben-Gersham, E. & Leibowitz, J. (1958). *Enzymologia*, **20**, 133.
Benson, C. P., Kelly, C. T. & Fogarty, W. M. (1982). *Journal of Chemical Technology and Biotechnology*, **32**, 790.
Bergmann, F. W., Abe, J.-I. & Hizukuri, S. (1988). *Applied Microbiology and Biotechnology*, **27**, 443.
Berlyne, G. M., Booth, E. M., Brewis, R. A. L., Mallick, N. P. & Simons, P. J. (1969). *Lancet*, **1**, 689.
Bhella, R. S. & Altosaar, I. (1984). *Analytical Biochemistry*, **140**, 200.
Bhella, R. S. & Altosaar, I. (1985). *Canadian Journal of Microbiology*, **31**, 149.
Blakeney, A. B. & Stone, B. A. (1985). *FEBS Letters*, **186**, 229.
Boel, E., Hjort, I., Svensson, B., Norris, F., Norris, K. E. & Fill, N. P. (1984). *EMBO Journal*, **3**, 1097.
Bourke, E. J. & Fogarty, W. M. (1975). *Proceedings of the Society of General Microbiology*, **2**, 81.
Bourke, E. J. & Fogarty, W. M. (1979). *The Society for General Microbiology Quarterly*, **6**, 153.
Bourke, E. J. & Fogarty, W. M. (1985). *Biochemical Society Transactions*, **13**, 455.
Boyer, E. W., Ingle, M. B. & Mercer, G. D. (1979). *Starch*, **31**, 166.
Brosnan, M. P., Kelly, C. T. & Fogarty, W. M. (1989). FEBS 19th Meeting Rome. Abstract TU 487.
Buonocore, V., Caporale, C., De Rosa, M. & Gambacorta, A. (1976). *Journal of Bacteriology*, **128**, 515.
Buranakarl, L., Ito, K., Izaki, K. & Takahashi, H. (1988). *Enzyme and Microbial Technology*, **10**, 173.
Büttner, R., Bode, R. & Birnbaum, D. (1987). *Journal of Basic Microbiology*, **27**, 297.
Chiba, S. & Yamana, O. (1980). *Agricultural and Biological Chemistry*, **44**, 549.
Clarke, A. J. & Svensson, B. (1984a). *Carlsberg Research Communications*, **49**, 111.
Clarke, A. J. & Svensson, B. (1984b). *Carlsberg Research Communications*, **49**, 559.
Clementi, F., Rossi, J., Costamagna, L. & Rossi, J. (1980). *Antonie van Leeuwenhoek*, **46**, 399.
Coleman, R. D., Yang, S.-S. & McAllister, M. P. (1987). *Journal of Bacteriology*, **169**, 4302.
Colonna, P., Buleon, A. & Lemarié, F. (1988). *Biotechnology and Bioengineering*, **31**, 895.
David, M.-H., Günther, H. & Roper, H. (1987). *Starch*, **39**, 436.
De Mot, R. & Verachtert, H. (1985). *Applied and Environmental Microbiology*, **50**, 1474.
De Mot, R. & Verachtert, H. (1986a). *Applied Microbiology and Biotechnology*, **24**, 459.
Dc Mot, R. & Verachtert, H. (1986b). *Canadian Journal of Microbiology*, **32**, 47.
De Mot, R. & Verachtert, H. (1987). *European Journal of Biochemistry*, **164**, 643.
De Mot, R., Van Oudendijck, E., Hougaerts, S. & Verachtert, H. (1984a). *FEMS Microbiology Letters*, **25**, 169.
De Mot, R., Van Oudendijck, E. & Verachtert, H. (1984b). *Biotechnology Letters*, **6**, 581.
De Mot, R., Van Oudendijck, E. & Verachtert, H. (1985). *Antonie van Leeuwenhoek*, **51**, 275.
De Pinto, J. A. & Campbell, L. L. (1968a). *Biochemistry*, **7**, 121.
De Pinto, J. A. & Campbell, L. L. (1968b). *Archives of Biochemistry and Biophysics*, **125**, 253.
Diderichsen, B. K. & Christiansen, L. (1986). US Patent No. 4, 598, 048.
Doyle, E. M., Kelly, C. T. & Fogarty, W. M. (1988). *Biochemical Society Transactions*, **16**, 181.
Doyle, E. M., Kelly, C. T. & Fogarty, W. M. (1989). *Applied Microbiology and Biotechnology*, **30**, 492.

Dubreucq, E., Boze, H., Nicol, D., Moulin, G. & Galzy, P. (1989). *Biotechnology and Bioengineering*, **33**, 369.

Ebertova, H. (1966). *Folia Microbiologica*, **11**, 422.

Erratt, J. A. & Nasim, A. (1987). *CRC Critical Reviews in Biotechnology*, **5**, 95.

Erratt, J. A. & Stewart, G. G. (1980). In *Current Developments in Yeast Research*, eds G. G. Stewart & I. Russel. Pergamon Press, Toronto, p. 177.

Evers, A. D., Gough, B. M. & Pybus, J. N. (1971). *Starch*, **23**, 16.

Ezure, Y., Maruo, S., Kojima, M., Yamashita, H. & Sugiyama, M. (1988). *Agricultural and Biological Chemistry*, **52**, 1073.

Fairbairn, D. A., Priest, F. G. & Stark, J. R. (1985). *Enzyme and Microbial Technology*, **8**, 89.

Federoff, H. J., Cohen, J. D., Eccleshall, T. R., Needleman, R. B., Buchferer, B. A., Giacalone, J. & Marmur, J. (1982). *Journal of Bacteriology*, **149**, 1064.

Fleming, I. D. (1968). In *Starch and its Derivatives*, ed. J. A. Radley. Chapman and Hall Ltd, London, p. 498.

Flor, P. Q. & Hayashida, S. (1983). *Applied and Environmental Microbiology*, **45**, 905.

Fogarty, W. M. (1983). In *Microbial Enzymes and Biotechnology*, ed. W. M. Fogarty. Applied Science Publishers, London, p. 1.

Fogarty, W. M. & Benson, C. P. (1982). *Biotechnology Letters*, **4**, 61.

Fogarty, W. M. & Benson, C. P. (1983). *European Journal of Applied Microbiology and Biotechnology*, **18**, 271.

Fogarty, W. M. & Bourke, E. J. (1983). *Journal of Chemical Technology and Biotechnology*, **33B**, 145.

Fogarty, W. M. & Griffin, P. J. (1973). *Journal of Applied Chemistry and Biotechnology*, **23**, 166.

Fogarty, W. M. & Griffin, P. J. (1975). *Journal of Applied Chemistry and Biotechnology*, **25**, 229.

Fogarty, W. M. & Kelly, C. T. (1979). In *Progress in Industrial Microbiology*, Vol. 15 ed. A. M. Bull. Elsevier Publishing Company, Amsterdam, p. 87.

Fogarty, W. M. & Kelly, C. T. (1980). In *Economic Microbiology, Vol. 5: Microbial Enzymes and Bioconversions*, ed. A. H. Rose. Academic Press, London, p. 115.

Fogarty, W. M. & Kelly, C. T. (1983). In *Enzyme Technology*, ed. R. M. Lafferty. Springer-Verlag, Heidelberg, p. 149.

Fogarty, W. M., Kelly, C. T. & Kadam, S. K. (1985). *Canadian Journal of Microbiology*, **31**, 670.

Frelot, D., Moulin, G. & Galzy, P. (1982). *Biotechnology Letters*, **4**, 705.

French, D. (1957). *Advances in Carbohydrate Chemistry*, **12**, 189.

French, D. & Wild, C. M. (1953). *Journal of the American Chemical Society*, **75**, 4490.

Friedberg, F. (1983). *FEBS Letters*, **152**(2), 139.

Friedberg, F. & Rhodes, C. (1986). *Journal of Bacteriology*, **165**, 819.

Fujii, M., Homma, T. & Taniguchi, M. (1988). *Biotechnology and Bioengineering*, **32**, 910.

Fujita, M., Torigoe, K., Nakada, T., Tsusaki, K., Kubota, M., Sakai, S. & Tsujisaka, Y. (1989). *Journal of Bacteriology*, **171**, 1333.

Fukui, T. & Nikuni, Z. (1969). *Agricultural and Biological Chemistry*, **33**, 884.

Fukushima, J., Sakano, Y., Iwai, H., Itoh, Y., Tamura, M. & Kobayashi, T. (1982). *Agricultural and Biological Chemistry*, **46**, 1423.

Giblin, M., Kelly, C. T. & Fogarty, W. M. (1987). *Canadian Journal of Microbiology*, **33**, 614.

Glymph, J. L. & Stutzenberger, F. J. (1977). *Applied and Environmental Microbiology*, **34**, 391.

Griffin, P. J. & Fogarty, W. M. (1973a). *Journal of Applied Chemistry and Biotechnology*, **23**, 301.

Griffin, P. J. & Fogarty, W. M. (1973b). *Biochemical Society Transactions*, **1**, 397.

Griffin, P. J. & Fogarty, W. M. (1973c). *Biochemical Society Transactions*, **1**, 1097.
Hayashi, T., Akiba, T. & Horikoshi, K. (1988). *Applied Microbiology and Biotechnology*, **28**, 281.
Hayashida, S. (1975). *Agricultural and Biological Chemistry*, **39**, 2093.
Hayashida, S. & Flor, P. Q. (1981). *Agricultural and Biological Chemistry*, **45**, 2675.
Hayashida, S. & Teramoto, Y. (1986). *Applied and Environmental Microbiology*, **52**, 1068.
Hayashida, S. & Yoshino, E. (1978). *Agricultural and Biological Chemistry*, **42**, 927.
Hayashida, S., Nomura, T., Yoshino, E. & Hongo, M. (1976). *Agricultural and Biological Chemistry*, **40**, 141.
Hayashida, S., Kunisaki, S.-I., Nakao, M. & Flor, P. Q. (1982). *Agricultural and Biological Chemistry*, **46**, 83.
Hayashida, S., Teramoto, Y. & Inoue, T. (1988). *Applied and Environmental Microbiology*, **54**, 516.
Hayashida, S., Nakahava, K., Kuroda, K., Miyata, T. & Iwanaga, S. (1989a). *Agricultural and Biological Chemistry*, **53**, 135.
Hayashida, S., Nakahava, K., Kanlayakrit, W., Hava, T. & Teramoto, Y. (1989b). *Agricultural and Biological Chemistry*, **53**, 143.
Hebeda, R. E., Styrlund, C. R. & Teaque, M. (1988). *Starch*, **40**, 33.
Hidaka, H. & Adachi, T. (1980). In *Mechanisms of Saccharide Polymerization and Depolymerization*, ed J. J. Marshall. Academic Press, New York, p. 101.
Higashihara, M. & Okada, S. (1974). *Agricultural and Biological Chemistry*, **38**, 1023.
Hiromi, K., Takahashi, K., Hamauzu, Z. I. & Ono, S. (1966). *Journal of Biochemistry (Tokyo)*, **59**, 469.
Hizukuri, S., Kawano, S., Abe, J., Koizumi, K. & Tanimoto, T. (1989). *Biotechnology and Applied Biochemistry*, **11**, 60.
Hoschke, A., László, E. & Holló, J. (1976). *Starch*, **28**, 426.
Hostinova, E., Bacova, M., Polivka, L., Gasperik, J. & Zelinka, J. (1979). *Biologia (Bratislava)*, **34**(12), 939.
Hyslop, P. & Sleeper, B. P. (1964). *Bacteriological Proceedings*, p. 89.
Hyun, H. H. & Zeikus, J. G. (1985a). *Applied and Environmental Microbiology*, **49**, 1162.
Hyun, H. H. & Zeikus, J. G. (1985b). *Journal of Bacteriology*, **164**, 1146.
Hyun, H. H. & Zeikus, J. G. (1985c). *Applied and Environmental Microbiology*, **49**, 1168.
Ihara, H., Sasaki, T., Tsuboi, A., Yamagata, H., Tsukagoshi, N. & Udaka, S. (1985). *Journal of Biochemistry*, **98**, 95.
Iizuka, H. & Mineki, S. (1977). *Journal of General Microbiology*, **23**, 217.
Imanaka, T. & Kuriki, T. (1989). *Journal of Bacteriology*, **171**, 369.
Inokuchi, N., Takahashi, T., Yoshimoto, A. & Irie, M. (1982). *Journal of Biochemistry (Tokyo)*, **91**, 1661.
Ishigami, H., Hashimoto, H. & Kainuma, K. (1985). *Journal of the Japanese Society for Starch Science*, **32**, 197.
Itkor, P., Shida, O., Tsukagoshi, N. & Udaka, S. (1989). *Agricultural and Biological Chemistry*, **53**, 53.
Itoh, T., Ohtsuki, I., Yamashita, I. & Fukui, S. (1987). *Journal of Bacteriology*, **169**, 4171.
Jensen, B. F. & Norman, B. E. (1984). *Process Biochemistry*, **19**, 129.
Jensen, B., Olsen, J. & Allermann, K. (1986). *Abstracts XIV International Congress of Microbiology*, Manchester, UK, p. 165.
Jensen, B., Olsen, J. & Allermann, K. (1987). *Biotechnology Letters*, **9**, 313.
Jensen, B., Olsen, J. & Allermann, K. (1988). *Canadian Journal of Microbiology*, **34**, 218.

Jodál, I., Kandra, L., Harangi, J., Náyási, P., Debrecen, J. & Szejtli, J. (1984). *Starch*, **36**, 140.

Kainuma, K., Kobayashi, S., Ito, T. & Suzuki, S. (1972). *FEBS Letters*, **26**, 281.

Kainuma, K., Kobayashi, S. & Harada, T. (1978). *Carbohydrate Research*, **61**, 345.

Kaneko, T., Hamamoto, T. & Horikoshi, K. (1988). *Journal of General Microbiology*, **134**, 97.

Kanno, M. (1986). *Agricultural and Biological Chemistry*, **50**, 23.

Kato, K., Sugimoto, T., Amemura, A. & Harada, T. (1975). *Biochimica et Biophysica Acta*, **391**, 96.

Kato, K., Kuswanto, K., Banno, I. & Harada, T. (1976). *Journal of Fermentation Technology*, **54**, 831.

Kawamura, S. & Sawai, T. (1968). *Agricultural and Biological Chemistry*, **32**, 114.

Kawazu, T., Nakanishi, Y., Uozumi, N., Sasaki, T., Yamagata, H., Tsukagoshi, N. & Udaka, S. (1987). *Journal of Bacteriology*, **169**, 1564.

Keay, L. (1970). *Starch*, **22**, 153.

Kelkar, H., Lachka, A. H. & Deshpande, M. V. (1988). *Canadian Journal of Microbiology*, **34**, 82.

Kelly, C. T. & Fogarty, W. M. (1983). *Process Biochemistry*, **18**, 6.

Kelly, C. T., Hefferman, M. E. & Fogarty, W. M. (1980). *Biotechnology Letters*, **2**, 351.

Kelly, C. T., O'Reilly, F. & Fogarty, W. M. (1983). *FEMS Microbiology Letters*, **20**, 55.

Kelly, C. T., Moriarty, M. E. & Fogarty, W. M. (1985). *Applied Microbiology and Biotechnology*, **22**, 352.

Kelly, C. T., Giblin, M. & Fogarty, W. M. (1986). *Canadian Journal of Microbiology*, **32**, 342.

Kelly, C. T., Brennan, P. A. & Fogarty, W. M. (1987). *Biotechnology Letters*, **9**, 125.

Kennedy, J. F. & White, C. A. (1979). *Starch*, **31**, 93.

Kennedy, J. F., White, C. A. & Riddiford, C. L. (1979). *Starch*, **31**, 235.

Kimura, K., Takamo, T. & Yamane, K. (1987a). *Applied Microbiology and Biotechnology*, **26**, 149.

Kimura, K., Kataoka, S., Ishii, Y., Takano, T. & Yanane, K. (1987b). *Journal of Bacteriology*, **169**, 4399.

Kimura, T., Ogata, M., Yoshida, M. & Nakakuki, T. (1988). *Biotechnology and Bioengineering*, **32**, 669.

Kimura, T., Ogata, M., Yoshida, M. & Nakakuki, T. (1989). *Biotechnology and Bioengineering*, **33**, 845.

Kitagawa, H., Amemura, A. & Harada, T. (1975). *Agricultural and Biological Chemistry*, **39**, 989.

Kitahata, S. & Okada, S. (1982a). *Journal of Japanese Society for Starch Science*, **29**, 9.

Kitahata, S. & Okada, S. (1982b). *Journal of Japanese Society for Starch Science*, **29**, 13.

Kitahata, S. & Okada, S. (1985). *Carbohydrate Research*, **137**, 317.

Kitahata, S., Tsuyama, N. & Okada, S. (1974). *Agricultural and Biological Chemistry*, **38**, 387.

Kitahata, S., Taniguchi, M., Beltran, S. D., Sugimoto, T. & Okada, S. (1983). *Agricultural and Biological Chemistry*, **47**, 1441.

Kitahata, S., Yoshimura, Y. & Okada, S. (1987). *Carbohydrate Research*, **159**, 303.

Kleinman, M. J., Wilkinson, A. E., Wright, I. P., Evans, I. H. & Bevan, E. A. (1988). *Biochemical Journal*, **249**, 163.

Kobayashi, S., Kainuma, K. & Suzuki, S. (1978). *Carbohydrate Research*, **61**, 229.

Kobayashi, S., Kainuma, K. & French, D. (1983). *Dempun Kagaku*, **30**, 62.

Kobayashi, S., Shibuya, N., Yeung, B. M. & French, D. (1984). *Carbohydrate Research*, **126**, 215.

Koizumi, K. & Utamura, T. (1986). *Carbohydrate Research*, **153**, 55.

Kolhekar, S. R., Mahajan, P. B., Ambedkar, S. S. & Borker, P. S. (1985). *Applied Microbiology and Biotechnology*, **22**, 181.

Kondo, H., Nakatani, M. & Hiromi, K. (1981). *Agricultural and Biological Chemistry*, **45**, 2369.

Krakenaite, R. P. & Glemzha, A. A. (1983). *Biokhimiya*, **48**, 62.

Krishnan, T. & Chandra, J. K. (1983). *Applied and Environmental Microbiology*, **46**, 430.

Krzechowsk, M. & Urbanek, H. (1975). *Applied Microbiology*, **30**, 163.

Kubota, M., Mikami, B., Tsujisaka, Y. & Morita, Y. (1988). *Journal of Biochemistry*, **104**, 12.

Kuhn, H., Fietzek, P. P. & Lampen, J. O. (1982). *Journal of Bacteriology*, **149**(1), 372.

Kulp, K. (1975). In *Enzymes in Food Processing*, 2nd edn, ed. G. Reed. Academic Press, New York, p. 54.

Kuriki, T., Okada, S. & Imanaka, T. (1988a). *Journal of Bacteriology*, **170**, 1554.

Kuriki, T., Park, J. H., Okada, S. & Imanaka, T. (1988b). *Applied and Environmental Microbiology*, **54**, 2881.

Kusano, S., Nagahata, N., Takahashi, S., Fujimoto, D. & Sakano, Y. (1988). *Agricultural and Biological Chemistry*, **52**, 2293.

Kusunoki, K., Kawakami, K., Sharaishi, F., Kato, K. & Kaim, M. (1982). *Biotechnology and Bioengineering*, **24**, 347.

Laluce, C. & Mattoon, J. R. (1984). *Applied and Environmental Microbiology*, **48**, 17.

Landert, J. P., Flaschel, E. & Renken, A. (1982). In *Proceedings of the First International Symposium on Cyclodextrins*, ed. J. Szejtli. Akademiai Kiado, Budapest, p. 89.

László, E., Banky, B., Seres, G. & Szejtli, J. (1981). *Starch*, **33**, 281.

Lee, E. Y. C. & Whelan, W. J. (1971). In *The Enzymes*, Vol. 5, ed. P. D. Boyer. Academic Press, New York, p. 191.

Lenders, J.-P. & Crichton, R. R. (1988). *Biotechnology and Bioengineering*, **31**, 267.

Li, K.-B. & Chan, K.-Y. (1983). *Applied and Environmental Microbiology*, **46**, 1380.

Lineback, D. R. & Aira, L. A. (1972). *Cereal Chemistry*, **49**, 283.

MacGregor, A. (1988). *Journal of Protein Chemistry*, **7**(4), 399.

MacGregor, E. A. & Svensson, B. (1989). *Biochemical Journal*, **259**, 145.

Madi, E. & Antranikian, G. (1989). *Applied Microbiology and Biotechnology*, **30**, 422.

Madi, E., Antranikian, G., Ohmiya, K. & Gottschalk, G. (1987). *Applied and Environmental Microbiology*, **53**, 1661.

Mäkelä, M. J. & Korpela, T. K. (1988). *Journal of Biochemical and Biophysical Methods*, **15**, 307.

Mäkelä, M., Mattsson, P., Schinina, M. E. & Korpela, T. (1988). *Biotechnology of Applied Biochemistry*, **10**, 414.

Manjunath, P. & Raghavendra Rao, M. R. (1979). *Journal of Biosciences*, **1**, 409.

Marshall, J. J. (1973a). *European Journal of Biochemistry*, **33**, 494.

Marshall, J. J. (1973b). *FEBS Letters*, **37**, 269.

Marshall, J. J. (1974). *FEBS Letters*, **46**, 1.

Marshall, J. J. & Miwa, I. (1981). *Biochimica et Biophysica Acta*, **661**, 142.

Matsuura, Y., Kusunoki, M., Havada, W. & Kakudo, M. (1984). *Journal of Biochemistry*, **95**, 697.

McCarthy, J., Kelly, C. T. & Fogarty, W. M. (1988). *Biochemical Society Transactions*, **16**, 184.

McCleary, B. V. & Anderson, M. A. (1980). *Carbohydrate Research*, **86**, 77.

McWethy, S. J. & Hartman, P. A. (1979). *Applied and Environmental Microbiology*, **37**, 1096.

Medda, S., Saha, B. C. & Ueda, S. (1982a). *Journal of Fermentation Technology*, **60**, 261.

Medda, S., Saha, B. C. & Ueda, S. (1982b). *Journal of the Faculty of Agriculture Kyushu University*, **26**, 139.

Melasniemi, H. (1987a). *Biochemical Journal*, **246**, 193.

Melasniemi, H. (1987b). *Journal of General Microbiology*, **133**, 883.

Melasniemi, H. (1988). *Biochemical Journal*, **250**, 813.

Miah, M. N. N. & Ueda, S. (1977a). *Starch*, **29**, 191.

Miah, M. N. N. & Ueda, S. (1977b). *Starch*, **29**, 235.

Mifune, A. & Shima, A. (1977). *Journal of Synthetic Organic Chemistry (Japan)*, **35**, 116.

Mikami, S., Iwano, K., Shinoki, S. & Shimada, T. (1987). *Agricultural and Biological Chemistry*, **51**, 2495.

Mikuni, K., Monma, M. & Kainuma, K. (1987). *Biotechnology and Bioengineering*, **29**, 729.

Miskolci-Török, M., Seress, L., Vakalin, H., Szejtli, J. & Jarai, M. (1980). Hungarian Patent No. 17926.

Miwa, I. & Marshall, J. J. (1982). *Chemical and Pharmaceutical Bulletin*, **30**, 362.

Monma, M., Nakakulei, T. & Kainuma, K. (1983). *Agricultural and Biological Chemistry*, **47**, 1769.

Moore, J. & Priest, F. G. (1987). *Letters in Applied Microbiology*, **4**, 1.

Morita, Y., Ogha, M. & Shimizu, K. (1968). *Memoirs of the Research Institute of Food Science, Kyoto University*, **29**, 191.

Moseley, M. H. & Keay, L. (1970). *Biotechnology and Bioengineering*, **12**, 251.

Murao, S., Ohyama, K. & Arai, M. (1979). *Agricultural and Biological Chemistry*, **43**, 719.

Nagamoto, H., Yasuda, T. & Inove, H. (1986). *Biotechnology and Bioengineering*, **28**, 1172.

Nagata, Y. (1985). *Agricultural and Biological Chemistry*, **49**(7), 1933.

Nagata, Y., Suga, S., Ohkawa, S. & Maruo, B. (1985). *Agricultural and Biological Chemistry*, **49**, 1923.

Nakajima, R., Imanaka, T. & Aiba, S. (1986). *Applied Microbiology and Biotechnology*, **23**, 355.

Nakakuki, T., Azuma, K., Monma, M. & Kainuma, K. (1982). *Journal of the Japanese Society for Starch Science*, **29**, 188.

Nakakuki, T., Hayashi, T., Monma, M., Kawashima, K. & Kainuma, K. (1983). *Biotechnology and Bioengineering*, **XXV**, 1095.

Nakamura, N. & Horikoshi, K. (1976a). *Agricultural and Biological Chemistry*, **40**, 753.

Nakamura, N. & Horikoshi, K. (1976b). *Agricultural and Biological Chemistry*, **40**, 935.

Nakamura, N. & Horikoshi, K. (1976c). *Agricultural and Biological Chemistry*, **40**, 1785.

Nanmori, T., Shinke, R., Aoki, K. & Nishira, H. (1983). *Agricultural and Biological Chemistry*, **47**, 609.

Nanmori, T., Numata, Y. & Shinke, R. (1987). *Applied and Environmental Microbiology*, **53**, 768.

Napier, E. J. (1977). U.S. Patent No. 4, 011, 136.

Nipkow, A., Shen, G.-J. & Zeikus, J. G. (1989). *Applied and Environmental Microbiology*, **55**, 689.

Norman, B. (1979). In *Microbial Polysaccharides and Polysaccharases*, eds R. C. W. Berkeley, G. W. Gooday & D. C. Ellwood. Academic Press, London, p. 339.

Nunberg, J. H., Meade, J. H., Cole, G., Lawyer, F. C., McCabe, P., Schweickart, V., Tal, R., Wittman, V. P., Flatgaard, J. E., & Innis, M. A. (1984). *Molecular and Cell Biology*, **4**, 2306.

Obi, S. K. C. & Odibo, F. J. C. (1984). *Canadian Journal of Microbiology*, **30**, 780.

Ogasahara, K., Imanishi, A. & Isemura, T. (1970). *Journal of Biochemistry (Tokyo)*, **67**, 77.

Ohga, M., Shimizu, K. & Morita, Y. (1966). *Agricultural and Biological Chemistry*, **30**, 967.

Ohmura, K., Yamazaki, H., Takeichi, Y., Nakayama, A., Otozai, K., Yamare, K., Yamasaki, M. & Tamura, G. (1983). *Biochemical and Biophysical Research Communications*, **112**, 678.

Ohnishi, M. (1971). *Journal of Biochemistry (Tokyo)*, **69**, 181.

Ohnishi, M. & Hiromi, K. (1978). *Carbohydrate Research*, **61**, 335.

Okada, G. (1974). *Journal of the Japanese Society for Starch Science*, **21**, 283.

Okada, G. & Unno, T. (1989). *Agricultural and Biological Chemistry*, **53**, 223.

Okada, G., Takayanagi, T., Miyahava, S. & Sawai, T. (1988a). *Agricultural and Biological Chemistry*, **53**, 829.

Okada, G., Unno, T. & Sawai, T. (1988b). *Agricultural and Biological Chemistry*, **52**, 2169.

Okemoto, H., Kobayashi, S., Momma, H., Hashimoto, H., Hara, K. & Kainuma, K. (1986). *Applied Microbiology and Biotechnology*, **25**, 137.

Olusanya, O. & Olutiola, P. O. (1986). *FEBS Microbiology Letters*, **36**, 239.

Olutiola, P. O. (1981). *Mycologia*, **73**, 1130.

O'Mahony, M. R., Kelly, C. T. & Fogarty, W. M. (1987). *Biotechnology Letters*, **9**, 317.

Ono, K., Shintani, K., Shigeta, S. & Oka, S. (1988a). *Agricultural and Biological Chemistry*, **52**, 1689.

Ono, K., Shintani, K., Shigeta, S. & Oka, S. (1988b). *Agricultural and Biological Chemistry*, **52**, 1699.

Ono, K., Shigeta, S. & Oka, S. (1988c). *Agricultural and Biological Chemistry*, **52**, 1707.

Ono, S., Hiromi, K. & Hamauzu, Z.-I. (1965). *Journal of Biochemistry (Tokyo)*, **57**, 34.

Orlando, A. R., Ade, P., Di Maggio, D., Fanelli, C. & Vittozzi, L. (1983). *Biochemical Journal*, **209**, 561.

Oteng-Gyang, K., Moulin, G. & Galzy, P. (1981). *Zeitschrift für Allgemeine Mikrobiologie*, **21**, 537.

Outtrup, H. (1986). US Patent No. 4, 604, 355.

Outtrup, H. & Norman, B. E. (1984). *Starch*, **36**, 405.

Palva, I., Pettersson, R. F., Kalkkinen, N., Lehtovaara, P., Sarvas, M., Söderlund, H., Takkinen, T. & Kääriäinen, L. (1981). *Gene*, **15**, 43.

Pasari, A. B., Korus, R. A. & Heimsch, R. C. (1988). *Enzyme and Microbial Technology*, **10**, 156.

Paszczyński, A., Miedziak, I., Lobarzewski, J., Kochmańska, J. & Trojanowski, J. (1982). *FEBS Letters*, **149**, 63.

Paszczyński, A., Fiedurek, J., Ilczuk, Z. & Ginalska, G. (1985). *Applied Microbiology and Biotechnology*, **22**, 434.

Pazur, J. H. & Ando, T. (1960). *Journal of Biological Chemistry*, **235**, 297.

Pazur, J. H. & Kleppe, K. (1962). *Journal of Biological Chemistry*, **237**, 1002.

Pazur, J. H. & Okada, S. (1967). *Carbohydrate Research*, **4**, 371.

Pazur, J. H., Knull, H. R. & Simpson, D. L. (1970). *Biochemical and Biophysical Research Communications*, **40**, 110.

Pazur, J. H., Knull, H. R. & Cepure, A. (1971). *Carbohydrate Research*, **20**, 83.

Pazur, J. H., Tominga, Y., Forsberg, L. S. & Simpson, D. L. (1980). *Carbohydrate Research*, **84**, 103.

Pazur, J. H., Liv, B., Tyke, S. & Baumrucker, C. R. (1987). *Journal of Protein Chemistry*, **6**, 517.

Pfeuller, S. L. & Elliott, W. H. (1969). *Journal of Biological Chemistry*, **244**, 48.

Plant, A. R., Morgan, H. W. & Daniel, R. M. (1986). *Enzyme and Microbial Technology*, **8**, 668.

Plant, A. R., Clemens, R. M., Daniel, R. M. & Morgan, H. W. (1987*a*). *Applied Microbiology and Biotechnology*, **26**, 427.

Plant, A. R., Clemens, R. M., Morgan, H. W. & Daniel, R. M. (1987*b*). *Biochemical Journal*, **246**, 537.

Plant, A. R., Parratt, S., Daniel, R. M. & Morgan, H. W. (1988). *Biochemical Journal*, **255**, 865.

Ramadas Bhat, U., Paramahans, S. V. & Tharanathan, R. N. (1983). *Starch*, **35**, 261.

Ramasesh, N., Sreekantish, K. R. & Murthy, V. S. (1982). *Starch*, **34**, 346.

Robyt, J. R. & Ackerman, R. J. (1971). *Archives of Biochemistry and Biophysics*, **145**, 105.

Robyt, J. & French, D. (1964). *Archives of Biochemistry and Biophysics*, **104**, 338.

Rose, D. (1948). *Archives of Biochemistry and Biophysics*, **16**, 349.

Sachdev, O. & Friedberg, F. (1981). *International Journal of Peptide and Protein Research*, **18**, 228.

Saenger, W. (1984). In *Inclusion Compounds 2*, eds J. L. Attwood, J. E. D. Davies & D. D. MacNicol. Academic Press, London, p. 231.

Saha, B. C. & Ueda, S. (1983*a*). *Agricultural and Biological Chemistry*, **47**, 2773.

Saha, B. C. & Ueda, S. (1983*b*). *Journal of Fermentation Technology*, **61**, 67.

Saha, B. C. & Ueda, S. (1983*c*). *Biotechnology and Bioengineering*, **25**, 1181.

Saha, B. C. & Ueda, S. (1984*a*). *Journal of the Japanese Society of Starch Science*, **31**, 8.

Saha, B. C. & Ueda, S. (1984*b*). *Applied Microbiology and Biotechnology*, **19**, 341.

Saha, B. C., Mitsue, T. & Ueda, S. (1979). *Starch*, **31**, 307.

Saha, B. C., Shen, G.-J. & Zeikus, J. G. (1987). *Enzyme and Microbial Technology*, **9**, 598.

Saha, B. C., Mathupala, S. P. & Zeikus, J. G. (1988). *Biochemical Journal*, **252**, 343.

Saito, N. (1973). *Archives of Biochemistry and Biophysics*, **155**, 290.

Sakano, Y., Masuda, N. & Kobayashi, T. (1971). *Agricultural and Biological Chemistry*, **35**, 971.

Sakano, Y., Hiraiwa, S., Fukushima, J. & Kobayashi, T. (1982*a*). *Agricultural and Biological Chemistry*, **46**, 1121.

Sakano, Y., Kashiwagi, Y. & Kobayashi, T. (1982*b*). *Agricultural and Biological Chemistry*, **46**, 639.

Sakano, Y., Fukushima, J. & Kobayashi, T. (1983*a*). *Agricultural and Biological Chemistry*, **47**, 2211.

Sakano, Y., Kashiyama, E. & Kobayashi, T. (1983*b*). *Agricultural and Biological Chemistry*, **47**, 1761.

Sakano, Y., Sano, M. & Kobayashi, T. (1985*a*). *Agricultural and Biological Chemistry*, **49**, 3041.

Sakano, Y., Sano, M. & Kobayashi, T. (1985*b*). *Agricultural and Biological Chemistry*, **49**, 3391.

Sano, M., Sakano, Y. & Kobayashi, T. (1985). *Agricultural and Biological Chemistry*, **49**, 2843.

Sasaki, H., Kurosawa, K. & Takao, S. (1986). *Agricultural and Biological Chemistry*, **50**, 1661.

Sato, M., Yagi, Y., Nagano, H. & Ishikara, T. (1985). *Agricultural and Biological Chemistry*, **49**, 1189.

Schülein, M. & Højer-Pedersen, B. (1984). *Annals of the New York Academy of Sciences*, **434**, 271.

Sen, S. & Chakrabarty, S. L. (1986). *Journal of Applied Bacteriology*, **60**, 419.

Shen, G.-J., Saha, B. C., Lee, Y.-E., Bhatnager, L. & Zeikus, J. G. (1988). *Biochemical Journal*, **254**, 835.

Shimazaki, T., Hara, S. & Sato, M. (1984). *Journal of Fermentation Technology,* **62,** 165.

Shimizu, M., Kanno, M., Tamura, M. & Suekane, M. (1978). *Agricultural and Biological Chemistry,* **42,** 1681.

Shinke, R., Nishira, H. & Mugibayashi, N. (1974). *Agricultural and Biological Chemistry,* **38,** 665.

Shinke, R., Aoki, K., Nishira, H. & Yuki, S. (1979a). *Journal of Fermentation Technology,* **57,** 53.

Shinke, R., Kunimi, Y., Aoki, K. & Nishira, H. (1979b). *Journal of Fermentation Technology,* **55,** 103.

Shirokizawa, O., Akiba, T. & Horikoshi, K. (1989). *Agricultural and Biological Chemistry,* **53,** 491.

Siggens, K. W. (1987). *Molecular Microbiology,* **1,** 86.

Sills, A. M., Sauder, M. E. & Stewart, G. G. (1984). *Journal of the Institute of Brewing,* **90,** 311.

Silvanovich, M. P. & Hill, R. D. (1976). *Analytical Biochemistry,* **73,** 430.

Simoes-Mendes, B. (1984). *Canadian Journal of Microbiology,* **30,** 1163.

Singleton, R., Jr & Amelunxen, R. E. (1973). *Bacteriological Reviews,* **37,** 320.

Smiley, K. L., Hensley, D. E., Smiley, M. J. & Gasdorf, H. J. (1971). *Archives of Biochemistry and Biophysics,* **144,** 694.

Smith, J. S. & Lineback, D. R. (1976). *Starch,* **28,** 243.

Spencer-Martins, I. (1982). *Applied and Environmental Microbiology,* **44,** 1253.

Spencer-Martins, I. (1984). *International Journal of Microbiology,* **2,** 31.

Spencer-Martins, I. & Van Uden, N. (1977). *European Journal of Applied Microbiology,* **4,** 29.

Spencer-Martins, I. & Van Uden, N. (1979). *European Journal of Applied Microbiology and Biotechnology,* **6,** 241.

Stark, J. R., Stewart, T. B. & Priest, F. G. (1982). *FEMS Microbiology Letters,* **15,** 295.

Suganuma, T., Mizukami, T., Moori, K., Ohnishi, M. & Hiromi, K. (1980). *Journal of Biochemistry (Tokyo),* **88,** 131.

Sukhumavasi, J., Kato, K. & Harada, T. (1975). *Journal of Fermentation Technology,* **53,** 559.

Suzuki, Y. & Chishiro, M. (1983). *Applied Microbiology and Biotechnology,* **17,** 24.

Suzuki, Y. & Imai, T. (1985). *Applied Microbiology and Biotechnology,* **21,** 20.

Suzuki, Y. & Tomura, Y. (1986). *European Journal of Biochemistry,* **158,** 77.

Suzuki, Y., Ueda, Y., Nakamura, N. & Abe, S. (1979). *Biochimica et Biophysica Acta,* **566,** 62.

Suzuki, Y., Aoki, R. & Hayashi, H. (1982). *Biochimica et Biophysica Acta,* **704,** 476.

Suzuki, Y., Shinji, M. & Eto, N. (1984). *Biochimica et Biophysica Acta,* **787,** 281.

Suzuki, Y., Fuju, H., Uemura, H. & Suzuki, M. (1987a). *Starch,* **39,** 17.

Suzuki, Y., Oishi, K., Nahano, H. & Nagayama, T. (1987b). *Applied Microbiology and Biotechnology,* **26,** 546.

Suzuki, Y., Nagayama, T., Nakano, H. & Oishi, K. (1987c). *Starch,* **39,** 211.

Svensson, B. (1988). *FEBS Letters,* **230,** 72.

Svensson, B. T., Pedersen, T. G., Svendsen, I., Sakai, T. & Ottesen, M. (1982). *Carlsberg Research Communications,* **47,** 55.

Svensson, B., Larsen, K., Svendsen, I. & Boel, E. (1983). *Carlsberg Research Communications,* **48,** 529.

Svensson, B., Larsen, K. & Gunnarsson, A. (1985). *European Journal of Biochemistry,* **154,** 497.

Svetsugu, N., Koyama, S., Takeo, K. & Kuge, T. (1974). *Journal of Biochemistry (Tokyo),* **76,** 57.

Szejtli, J. (1987). *Chimica Oggi (Maggio),* 49.

Tabata, S., Ide, T., Umemura, Y. & Torü, K. (1984). *Biochemica et Biophysica Acta,* **797,** 231.

Tadasa, K. & Takeda, K. (1986). *Journal of Fermentation Technology,* **64,** 81.

Takahashi, T., Inokuchi, N. & Irie, M. (1981). *Journal of Biochemistry (Tokyo),* **89,** 125.

Takahashi, T., Tsuchida, Y. & Irie, M. (1982). *Journal of Biochemistry (Tokyo),* **92,** 1623.

Takasaki, Y. (1976). *Agricultural and Biological Chemistry,* **40,** 1523.

Takasaki, Y. (1979). *Agricultural and Biological Chemistry,* **43,** 1599.

Takasaki, Y. (1982). *Agricultural and Biological Chemistry,* **46,** 1539.

Takasaki, Y. (1983). *Agricultural and Biological Chemistry,* **47,** 2193.

Takasaki, Y. (1985). *Agricultural and Biological Chemistry,* **49,** 1091.

Takasaki, Y. (1987). *Agricultural and Biological Chemistry,* **51,** 9.

Takasaki, Y. (1989). *Agricultural and Biological Chemistry,* **53,** 341.

Takeda, Y., Matsui, H., Tanida, M., Takao, S. & Chiba, S. (1985). *Agricultural and Biological Chemistry,* **49,** 1633.

Takkinen, K., Pettersson, R. F., Kalkkinen, N., Palva, I., Söderlund, H. & Kääriäinen, L. (1983). *Journal of Biological Chemistry,* **258,** 1007.

Tanaka, A., Ohnishi, M., Hiromi, K., Miyata, S. & Murao, S. (1982). *Journal of Biochemistry (Tokyo),* **91,** 1.

Tanaka, A., Yamashita, T., Ohnishi, M. & Hiromi, K. (1983a). *Journal of Biochemistry (Tokyo),* **93,** 1037.

Tanaka, A., Fukuchi, M., Ohnishi, M., Hiromi, K., Aibaba, S. & Morita, Y. (1983b). *Agricultural and Biological Chemistry,* **47,** 573.

Tanaka, T., Ishimoto, E., Shimomura, Y., Taniguchi, M. & Oi, S. (1987). *Agricultural and Biological Chemistry,* **51,** 399.

Tani, Y., Vongsuvanlert, V. & Kumuanta, J. (1986). *Journal of Fermentation Technology,* **64,** 405.

Taniguchi, M., Chung, M. J., Yoshigi, N. & Maruyama, Y. (1983). *Agricultural and Biological Chemistry,* **47,** 511.

Teramoto, Y., Kira, I. & Hayashida, S. (1989). *Agricultural and Biological Chemistry,* **53,** 601.

Thirunavukkarasu, M. & Priest, F. G. (1984). *Journal of General Microbiology,* **130,** 3135.

Thoma, J. A. & Koshland, D. E. (1960). *Journal of the American Chemical Society,* **82,** 3329.

Thomas, M., Priest, F. G. & Stark, J. R. (1980). *Journal of General Microbiology,* **118,** 67.

Toda, H., Kondo, K. & Narita, K. (1982). *Proceedings of the Japanese Academy,* **58,** 208.

Tomazic, S. J. & Klibanov, (1988a). *Journal of Biological Chemistry,* **263,** 3086.

Tomazic, S. J. & Klibanov, A. M. (1988b). *Journal of Biological Chemistry,* **263,** 3092.

Touzi, A., Prebois, J. P., Moulin, G., Deschamps, F. & Galzy, P. (1982). *European Journal of Applied Microbiology and Biotechnology,* **15,** 232.

Towprayoon, S., Saha, B. C., Fujio, Y. & Ueda, S. (1988). *Applied Microbiology and Biotechnology,* **29,** 289.

Truscheit, E., Frommer, W., Junge, B., Müller, L., Schmidt, D. D. & Wingender, W. (1981). *Angewandte Chemie,* **20,** 744.

Tsuboi, A., Yamasaki, Y. & Suzuki, Y. (1974). *Agricultural and Biological Chemistry,* **38,** 543.

Tucker, M., Grohmann, K. & Himmel, M. (1984). *Biotechnology and Bioengineering,* **14,** 279.

Uchino, F. (1982). *Agricultural and Biological Chemistry,* **46,** 7.

Ueda, S. (1982). *Journal of the Japanese Society of Starch Science,* **29,** 123.

Ueda, S. & Koba, Y. (1980). *Journal of Fermentation Technology*, **58**, 237.

Ueda, S. & Saha, B. C. (1980). *Starch*, **32**, 420.

Ueda, S. & Saha, B. C. (1983). *Enzyme and Microbial Technology*, **5**, 196.

Ueda, S., Zenin, C. T., Monteiro, D. A. & Park, Y. K. (1981). *Biotechnology and Bioengineering*, **23**, 291.

Underkofler, L. Z., Denault, L. J. & Hou, E. F. (1965). *Starch*, **17**, 179.

Urlaub, H. & Wöber, G. (1978). *Biochimica et Biophysica Acta*, **522**, 161.

Usui, T. & Murata, T. (1988). *Journal of Biochemistry (Tokyo)*, **103**, 969.

Vretblad, P. (1974). *FEBS Letters*, **47**, 86.

Wadher, B. J. & Miles, R. J. (1988). *FEMS Microbiology Letters*, **49**, 459.

Wako, K., Takahashi, C., Hashimoto, S. & Kaneda, J. (1978). *Journal of the Japanese Society of Starch Science*, **25**, 155.

Wang, L.-H. & Hartman, P. A. (1976). *Applied and Environmental Microbiology*, **31**, 108.

Welker, N. E. & Campbell, L. L. (1967). *Biochemistry*, **6**, 3681.

Weselake, R. J. & Hill, R. D. (1982). *Carbohydrate Research*, **108**, 153.

Weselake, R. J. & Hill, R. D. (1983). *Cereal Chemistry*, **60**, 98.

Wilson, J. J. & Ingledew, W. M. (1982). *Applied and Environmental Microbiology*, **44**, 301.

Wilson, J. J., Khachatourians, G. G. & Ingledew, W. M. (1982). *Biotechnology Letters*, **4**, 333.

Yagi, Y. & Iguchi, H. (1974). Japanese Patent No. 74, 124, 285.

Yagi, Y., Kouno, K. & Juni, T. (1980). European Patent No. 0, 017, 242.

Yamasaki, Y. & Suzuki, Y. (1974). *Agricultural and Biological Chemistry*, **38**, 443.

Yamasaki, Y., Suzuki, Y. & Ozawa, J. (1977a). *Agricultural and Biological Chemistry*, **41**, 1443.

Yamasaki, Y., Suzuki, Y. & Ozawa, J. (1977b). *Agricultural and Biological Chemistry*, **41**, 2149.

Yamashita, I., Suzuki, K. & Fukui, S. (1985). *Journal of Bacteriology*, **161**, 567.

Yamashita, I., Suzuki, K. & Fukui, S. (1986). *Agricultural and Biological Chemistry*, **50**, 475.

Yang, M., Galizzi, A. & Henner, D. (1983). *Nucleic Acids Research*, **11**, 237.

Yasuda, M., Kuwar, M. & Matsushita, H. (1989). *Agricultural and Biological Chemistry*, **53**, 247.

Yokobayashi, Y. (1979). Japanese Patent No. 79, 96, 093.

Yoshigi, N., Chikano, T. & Kamimura, M. (1985a). *Agricultural and Biological Chemistry*, **49**, 3369.

Yoshigi, N., Chikano, T. & Kamimura, M. (1985b). *Journal of the Japanese Society of Starch Science*, **32**, 273.

Yoshigi, N., Chikano, T. & Mori, Y. (1986). US Patent No. 4, 591, 561.

Yoshimura, Y., Kitahata, S. & Okada, S. (1987). *Carbohydrate Research*, **168**, 285.

Yoshino, E. & Hayashida, S. (1978). *Journal of Fermentation Technology*, **56**, 289.

Yutani, K., Sasaki, I. & Ogasahara, K. (1973). *Journal of Biochemistry*, **74**, 573.

Yuuki, T., Nomura, T., Tezuka, H., Tsuboi, A., Yamagata, H., Tsukagoshi, N. & Udaka, S. (1985). *Journal of Biochemistry*, **98**, 1147.

Zale, S. E. & Klibanov, A. M. (1986). *Biochemistry*, **25**, 5432.

Zenin, C. T. & Park, Y. K. (1983). *Journal of Fermentation Technology*, **61**, 109.

Zhou, J., Takano, T. & Kobayashi, S. (1989). *Agricultural and Biological Chemistry*, **53**, 301.

Chapter 4

MICROBIAL PECTOLYTIC ENZYMES

JOHN R. WHITAKER

Department of Food Science and Technology, University of California, Davis, California 95616, USA

CONTENTS

1 INTRODUCTION

Pectolytic enzymes are widely distributed in higher plants and microorganisms. They are not found in higher animals but are found in some protozoa, nematodes and insects. In higher plants, they are important during growth, permitting cell elongation. They are also important for softening of some plant tissues during maturation and storage and in the decomposition and recycling of plant materials. Microbial pectolytic enzymes are known to play key roles in plant pathogenicity and in much of fruit and vegetable spoilage involving rotting. Pectolytic enzymes are commercially important in a number of industrial processes including retting of flax and other vegetable fibers, extraction, clarification and depectinization of fruit juices, extraction of vegetable oils and maceration of fruits and vegetables to give unicellular foods. Pectin methylesterase is important in the firming of plant tissue, in the presence of Ca^{2+}.

The reader is referred to excellent recent reviews on use of pectic enzymes in the food industry (Rombouts & Pilnik, 1986; Voragen & Pilnik, 1989). Previous reviews of pectic enzymes include those of Macmillan & Sheiman (1974), Rexová-Benková & Markovič (1976), Rombouts & Pilnik (1980) and Fogarty & Kelly (1983).

2 THE PECTIC SUBSTANCES

An understanding of enzymes and their effects begins with an understanding of the compounds they act upon, the substrates. In the case of the pectic substances, the substrates are as complex as the enzymes; however, they have received more attention than have the enzymes due to extensive research by Aspinall (1982) and many others. We shall use the generic name *pectic substances* for this heterogeneous mixture of compounds. An additional complexity is that polysaccharides, to which the pectic substances belong, have no definite molecular mass size, in contrast to proteins, lipids and nucleic acids. For example, reported relative molecular masses of pectic substances range from 23 000–71 000 from citrus (Luh & Phaff, 1951; Porwal & Chakravarti, 1970), 25 000–35 000 from apple, plum and pear (Luh & Phaff, 1951) to 200 000–360 000 from apple and lemon (Newbold & Joslyn, 1952). Extraction and storage conditions are critical in determining the size (BeMiller, 1986; Fishman *et al.*, 1986).

Higher plants contain a number of polysaccharides. These include starch, cellulose, hemicelluloses, pectic substances and other polysaccharides. Groups of polysaccharides found in cell walls of higher plants are shown in Table 1. Polysaccharides from cell walls of ripe pears were reported to contain 11·5% pectic substances, 16·1% lignin, 21·4% glucosan, 3·5% galactan, 1·1% mannan, 21% xylan and 10% arabinan (Jermyn & Isherwood, 1956). The pectic substances include the galacturonans and rhamnogalacturonans in which the C-6 carbon of galactose is oxidized to a carboxyl group, the arabinans and the galactans and arabinogalactans I (Table 1). The pectolytic enzymes are generally classified as those enzymes that act upon the galacturonans and rhamnogalacturonans. That definition will be used in this chapter.

The pectic substances consist of a number of compounds, depending on the degree and type of enzymatic action. The parent compound in the intact immature tissue is *protopectin,* an insoluble substance located primarily in the

Table 1
Polysaccharides of Cell Walls in Higher Plants[a]

General category	Structural classification
Cellulose	β-D-Glucan (4-linked)
Pectic substances	Galacturonans and rhamnogalacturonans Arabinans Galactans and arabinogalactans I[b]
Hemicelluloses	Xylans (including arabinoxylans and (4-O-methyl)glucuronoxylans) Galactomannans and glucomannans β-D-Glucans (3- and 4-linked) β-D-Glucan-callose (3-linked) Xyloglucans (4-linked β-D-glucans with attached side chains)
Other polysaccharides	Arabinogalactans II[b] Glucuronomannans

[a] Aspinall (1969, 1980).
[b] Arabinogalactans of type I are essentially linear and contain 4-linked β-D-galactan chains, whereas those of type II contain branched 3- and 6-linked β-galactan chains. The polysaccharides may occur in part as proteoglycans or polysaccharide–protein conjugates.

middle lamella that serves as the glue to hold cells together and in the cell walls. Protopectin may be insoluble because of polymer size or due to cross-linking with divalent cations, especially Ca^{2+} and/or other polysaccharides. *Pectin* (polymethylgalacturonate) is the soluble polymeric material in which at least 75% of the carboxyl groups of the galacturonate units are esterified with methanol. *Pectic acid* (polygalacturonic acid) is the soluble polymeric material in which all the methoxyl groups are removed from the galacturonate units. *Pectinic acids* contain >0 and $<75\%$ methylated galacturonate units. *Oligogalacturonates* have lower relative molecular masses and are smaller polymers with two or more galacturonate units, while *oligomethylgalacturonates* are similar polymers with two or more galacturonate units which are partially or completely methylated on C-6.

The rhamnogalacturonans (Fig. 1) are the major constituents of the pectic substances. The primary chains consist primarily of α-D-galacturonate units linked $(1 \rightarrow 4)$, with 2–4% of L-rhamnose units (about one every 25 galacturonate units) linked β-$(1 \rightarrow 2)$ and β-$(1 \rightarrow 4)$ to the D-galacturonate units. The side chains of the rhamnogalacturonans are of various compositions and lengths (Fig. 1). The extended side chains are usually homogeneous polymers of either D-galacturonic acid units (giving the galacturonans), or of L-arabinose units. Voragen & Pilnik (1989) have presented evidence that indicate the neutral sugars are concentrated in blocks of more highly substituted rhamnogalacturonate regions ('hairy') separated by extended unsubstituted ('smooth') regions containing almost exclusively D-galacturonate units. In

Rhamnogalacturonans
Main chain in pectins

-[→4)-α-D-GalpA-(1-]$_n$→2)-L-Rhap-(1→4)-α-D-GalpA-(1→2)-L-Rhap-1-[→4)-α-D-GalpA-(1-]$_n$-

Short side chains in pectins   Extended side chains in pectins

β-D-Xylp-(1→3)-

β-D-Galp-(1→2)-D-Xylp-(1-

α-L-Fucp(1→2)-D-Xylp-(1-

L-Araf-(1→3)-

D-Apif-(1→3)-D-Apif-(1-

→4)-β-D-Galp-(1→4)-β-D-Galp-(1-

→5)-α-L-Araf-(1→5)-α-L-Araf-(1-

$\overset{\text{3}}{\uparrow}$

$\overset{\text{1}}{\underset{}{}}$

α-L-Araf

Fig. 1. Primary structures of rhamnogalacturonans.

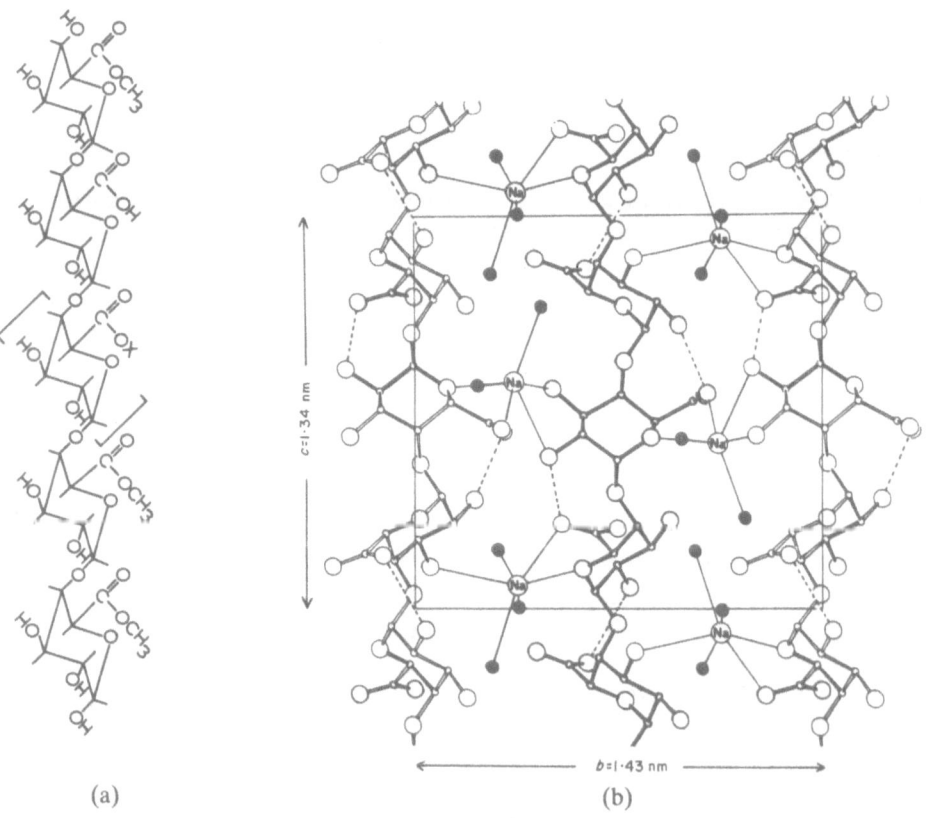

(a)   (b)

Fig. 2. (a) Chemical structure of a segment of pectin. (b) A view of the projection along *a*-axis of the sodium pectate crystal structure. Intramolecular hydrogen bonds are drawn as broken lines. The coordination shells for Na are also shown. o, carbons; O, oxygen; ●, H$_2$O molecules. From Walkinshaw & Arnott (1981*a*) with permission.

apple pectin fractions, the galacturonate units in the hairy regions are almost completely methylated while in the smooth regions about 70% of the galacturonate units are methylated (DeVries *et al.*, 1983). In most plant tissues, about 75% of the D-galacturonate units are methylated. The free carboxyl groups tend to occur in clusters along the chain. The rhamnogalacturonans are negatively charged polymers at pH ≥ 5 (pK_a of 65% DE (degree of esterification) pectin = 3.55; pK_a of 0% DE pectic acid = 4·10; Plaschina *et al.*, 1978).

An extended stretch of D-galacturonate units, in pyranose ring form, along the main chain of a rhamnogalacturonan is shown in Fig. 2(a). As noted

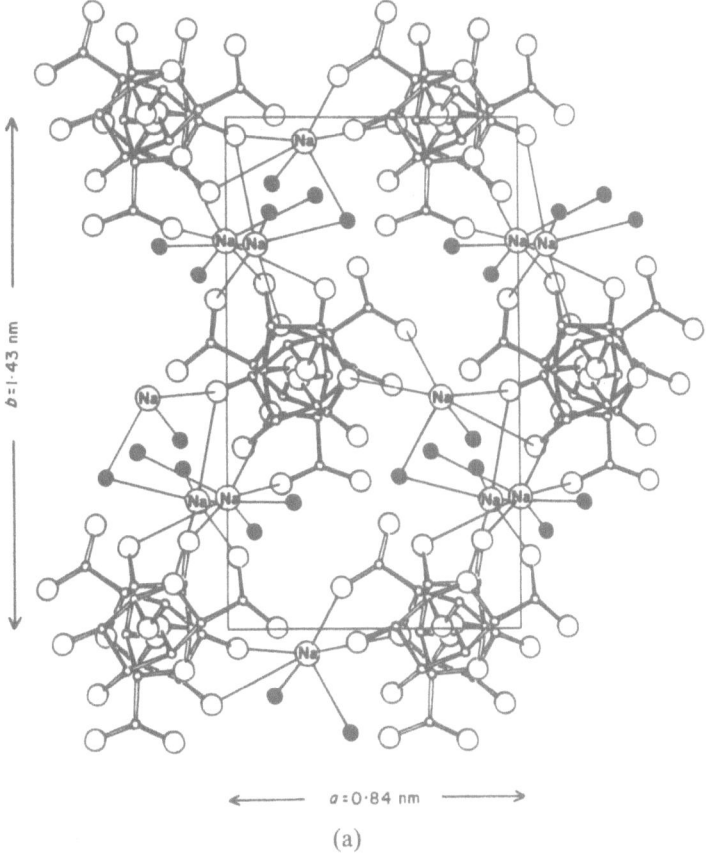

(a)

Fig. 3. (a) The sodium pectate crystal viewed parallel to *c*-axis (looking down the helix). All sodium ions (Na) and H_2O molecules (●) are included. Attractive Na$^+$– – –O interactions are drawn as thin, unbroken lines. o, carbons; O, oxygens. From Walkinshaw & Arnott (1981*a*) with permission. (b) A view of the pectin structure (100% methylated) along the (001) direction (looking down the helix). Parallel chains are packed in a hexagonal lattice. Methyl hydrogens are denoted as ● to emphasize the columnar stacking of methoxy groups. o, carbons; O, oxygens. From Walkinshaw & Arnott (1981*b*) with permission.

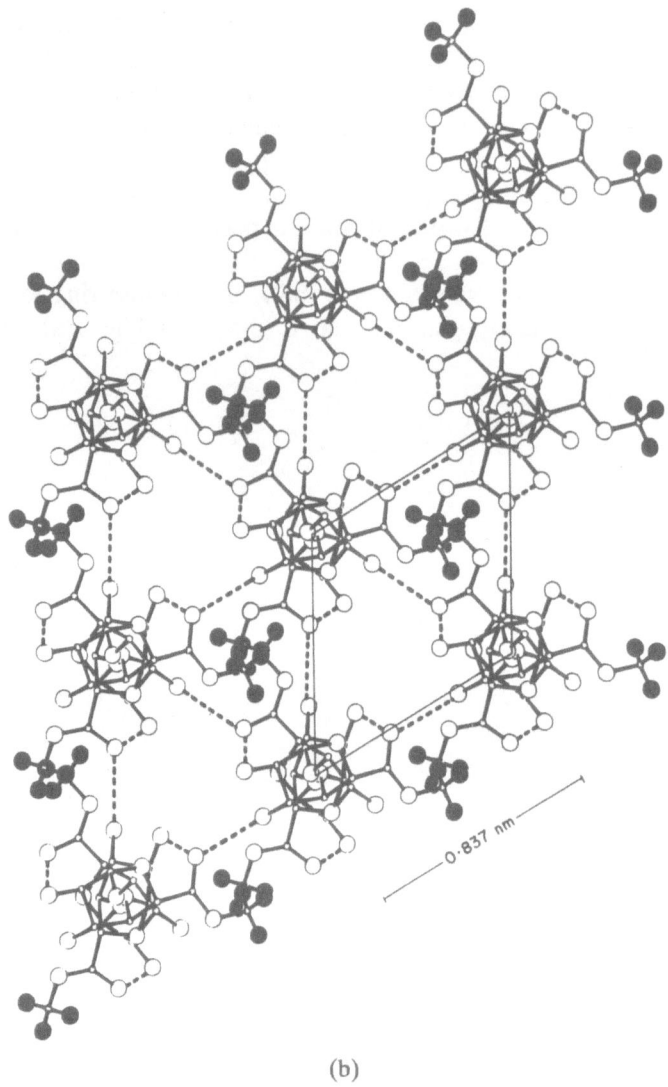

(b)

Fig. 3—(*continued*)

above, the D-galacturonate units along the primary chain, in boat conformation, are linked α-(1→4) via glycosidic bonds. Some of the C-6-carboxyl
groups are methylated; others are in —COO⁻ or —COOH forms depending
on the pH (see above).

The rhamnogalacturonans have secondary structures that differ between
pectic acids and pectins. Polygalacturonates exist in solution in secondary
structures that resemble a threefold screw axis (Figs 2(b), 3(a); Walkinshaw &
Arnott, 1981*a*) while pectins have twofold screw symmetry (Fig. 3(b);

Walkinshaw & Arnott, 1981*b*) because of the ester groups on adjacent galacturonide units which project out on opposite sides of the main chains. These analyses are based on crystallographic studies. Pectin chains, unlike pectic acid chains, tend to associate with each other to give high viscosity solutions (Merrill & Weeks, 1945; Walkinshaw & Arnott, 1981*b*). Note the columnar stacking of methoxy groups in the crystal structure. Both pectates and pectins occur in helical structures with trisaccharide repeat units 1·34 nm in length. Pectic substances are most stable at pH 3–4. At lower and higher pH values demethylation occurs. At lower pH values the galacturonosidic bonds are broken by hydrolysis while in alkaline solutions pectins (but not pectic acids) rapidly undergo *trans* elimination to give a $\Delta4:5$ double bond.

3 THE PECTOLYTIC ENZYMES

The pectolytic enzymes include one esterase (pectin methylesterase), six polygalacturonases (generic name; Whitaker, 1972) and four lyases (Table 2). α-L-Arabinofuranosidase and endoarabanase, which act on arabinans (Table

Table 2
Classification of the Pectolytic Enzymes[a]

Name	EC No.	Primary substrate	Products	Mechanism
Esterase				
Pectin methylesterases (pectinesterases)	3.1.1.11	Pectin	Pectic acid + methanol	Hydrolysis
Polygalacturonases				
Protopectinases		Protopectin	Pectin	Hydrolysis
Endopolygalacturonases	3.2.1.15	Pectic acid	Oligogalacturonates	Hydrolysis
Exopolygalacturonases	3.2.1.82	Pectic acid	Monogalacturonate	Hydrolysis
Oligogalacturonate hydrolases		Trigalacturonate	Monogalacturonate	Hydrolysis
$\Delta4:5$ Unsaturated oligogalacturonate hydrolases		$\Delta4:5$ (Galacturonate)$_n$	Unsaturated mono-galacturonate and saturated $(n-1)$	Hydrolysis
Endopolymethyl-galacturonases		Pectin	Methyl oligo-galacturonates	Hydrolysis
Lyases				
Endopolygalacturonate lyases (endopectate lyases)	4.2.2.2	Pectic acid	Unsaturated oligo-galacturonates	*Trans* elimination
Exopolygalacturonate lyases (exopectate lyases)	4.2.2.9	Pectic acid	Unsaturated diga-lacturonate	*Trans* elimination
Oligogalacturonate lyases	4.2.2.6	Unsaturated digalacturonate	Unsaturated monogalacturonate	*Trans* elimination
Endopolymethyl-galacturonate lyases (endopectin lyases)	4.2.2.10	Pectin	Unsaturated methyl oligo-galacturonates	*Trans* elimination

[a] Adapted from the classification of Demain & Phaff (1957), Deuel & Stutz (1958) and Bateman & Millar (1966).

1), are sometimes included among the pectolytic enzymes (Whitaker, 1984; Voragen et al., 1986). Most enzymologists, microbiologists and food scientists are not familiar with these two enzymes. There is no mention of these enzymes in the some 180 volumes of *Methods in Enzymology* or in the 18 volumes of the third edition of *The Enzymes*. They will not be treated in this chapter.

3.1 Pectin Methylesterases

Pectin methylesterase (pectin pectylhydrolase, EC 3.1.1.11) is synonymous with pectinesterase, pectase, pectin methoxylase, pectin demethoxylase and pectolipase. Pectin methylesterase activity was first reported in 1840 (Frémy, 1848). Pectin methylesterase acts upon pectin to remove the methoxyl groups from the 6-carboxyl group of the galacturonate unit by hydrolysis as shown in eqn (1). The enzyme is a carboxylic acid esterase, belonging to the hydrolase group of enzymes.

Pectin methylesterase

$$\text{Polymethylgalacturonate (pectin)} + H_2O \xrightarrow{\text{pectin methylesterase}} \text{polygalacturonate} + \text{methanol} + H^+ \quad (1)$$

The products of the reaction are: (1) a further deesterified pectin which eventually becomes a pectinic acid and finally a pectic acid; (2) methanol; and (3) a H^+ from the ionization of the newly formed carboxyl group.

Pectin methylesterase activity is best determined continuously in a pH-stat between pH 4·5–7·5 (determined by the pH optimum of the enzyme and pH where the carboxyl group is ionized). As indicated by eqn (1), H^+ are released from the newly formed carboxyl groups. The $[H^+]$ is equivalent to the number of carboxyl groups formed above pH 5·5. At lower pH values, the total carboxyl groups formed can be calculated from the measured $[H^+]$ released using the Henderson–Hasselbalch equation (eqn (2)):

$$\log\frac{[-COO^-]}{[-COOH]} = pH - pK_a \quad (2)$$

where the pK_a of the newly formed carboxyl group is about 3·6 and the pH is that of the reaction medium. For example, at pH 3·6 and 4·6, 50% and 91% of the carboxyl groups would be ionized. Therefore, the pH-stat method determines 50% and 91% of the newly formed carboxyl groups at pH 4·5 and 5·5, respectively. Knowing the percentage of ionization, one can calculate the true number of carboxyl groups formed. In the absence of a pH-stat, the reaction can be followed precisely by use of a pH meter either by titrating manually with a standard NaOH solution to maintain the pH constant or by observing the initial rate of decrease in pH from a fixed value. The reaction may also be followed readily in a spectrophotometer in the presence of a pH indicator with a pK_a within the pH range used. The precision of the titration method depends on pH of the reaction. At pH 5·0 one can readily determine 1×10^{-6} M newly formed $[H^+]$, while at pH 7·0, the determinability is 1×10^{-8} M.

There are several other ways of determining pectin methylesterase activity (Rexová-Benková & Markovič, 1976). The methanol can be determined on a discontinuous basis by gas chromatography (GC) or by high performance liquid chromatography (HPLC). There are a number of chemical methods for measuring methanol, none of which are very useful in the presence of the substrate and other products of the reaction. Enzymatic oxidation of methanol to formaldehyde and measurement of the aldehyde by hydroxylamine/Fe^{3+} reactions are quite sensitive but tedious.

Pectin methylesterases have been purified from fruits [banana (Brady, 1976); orange (Versteeg *et al.*, 1978); tomato (Lee & Macmillan, 1970*a*); Markovič, 1974]; bacteria (*Clostridium multifermentans*; Miller & Macmillan, 1971; Sheiman *et al.*, 1976) and fungi [*Botrytis cinerea* (Schejter & Marcus, 1988); *Corticium rolfsii* (Yoshihara *et al.*, 1977); *Fusarium oxysporum* (Miller & Macmillan, 1971); *Phytophthora infestans* (Förster, 1988)], and have been detected in other sources (Table 3).

Pectin methylesterases are almost absolutely specific for polygalacturonide methyl esters. Pectin methylesterase from orange albedo acted on pectin at least 1000 times faster than on non-polygalacturonide esters (MacDonnell *et al.*, 1950). In comparative studies using pectin methylesterases from alfalfa, tomato, orange and a fungal source, the initial rates of deesterification were (relative basis) 100 for pectin, ~50 for the fully methylated polygalacturonate (except ~80 for the fungal enzyme) and 6–16 for the 50% ethylated polygalacturonate (Kertesz, 1937; MacDonnell *et al.*, 1950).

The degree of polymerization and/or size of the substrate is also important. Alfalfa pectin methylesterase demethylated methyl-D-galacturonate and methyl-D-galacturonate methyl glycoside at ≤1% the initial rate for pectin (Lineweaver & Ballou, 1943).

The distribution of methoxyl groups along the pectin chain was also important for orange pectin methylesterase (Solms & Deuel, 1955). While the enzyme attacked completely methylated pectin slowly, the rate of deesterification increased to a maximum at 50% methylation. Pectin, partially deesterified previously with pectin methylesterase, resulted in a slower rate than the equivalent degree of deesterification by alkali treatment (Schultz *et al.*, 1945;

John R. Whitaker

Table 3
Distribution of Pectolytic Enzymes in Some Microorganisms

Organism	PME[a]	PG	PGL	PMG	PMGL	Reference
Moulds						
Alternaria brassicae			+	+	+	Suri & Mandahar (1982)
brassicola		+[b,c]		+	+	Mandahar & Suri (1981)
sesami		+	+		+	Rajpurohit & Prased (1982)
Aspergillus alliaceus				+	+	Lobanok *et al.* (1985)
awamori	+[d]	+[d]			+[d]	Blieva & Rodionova (1987)
carbonarius				+	+	Lobanok *et al.* (1985)
flavus				+	+	Adisa (1985)
foetidus		+[b,c]				Astapovich *et al.* (1984)
flumigatus		+		+	+	Adisa (1985)
niger	+	+	+		+	Fiedurek & Ilczuk (1983a, b)
		+		+		Chopra & Mehta (1985)
niger 71		+	+		+	Fiedurek (1983)
Athelia (Sclerotium) *rolfsii*		+[b,c]				Scala & Zoina (1983a)
Aureobasidium pullulans						
LV10					+	Parini *et al.* (1988)
		+[d]			+[d]	Manachini *et al.* (1988)
Botrytis cinerea	+[d]	+[d]				Leone & Van Den Heuvel (1987)
		+[d]				Di Lenna & Fielding (1983)
	+			+[b,c]	+	Martinez *et al.* (1988)
Fusarium oxysporum *ricini*	+	+	+			Piplani *et al.* (1981)
tricinctum	+	+	+			Wick & Schroeder (1982)
Heterobasidian annosum (Fr.) Bref.		+[b,c]				Johanssen (1988)
Helminthosporium *sacchari*	+	+	+			Dube & Bordia (1982a, b)
Lachnospira multiparus	+	0	+			Silley (1985)
Myrothecium roridum Tode ex Fr[e]	+	+	+		+	Reddy & Reddy (1983)
Nectia distissima Tu1		+[b,c]				Perrin (1984)
Penicillium crustosum Thom[e]	+	+	+			Awasthi & Mishra (1982)
digitatum				+[b,c]	+	Mikhailova *et al.* (1981); Lobanok *et al.* (1985)
digitatum 24P		+	+			Mikhailova *et al.* (1982)
fellutanum	+					Aizenberg (1983)
implicatus	+					Aizenberg (1983)
multicolor	+					Aizenberg (1983)
thomii	+					Aizenberg (1983)
340 strains[f]						Aizenberg (1983)
Sclerotium rolfsii[g]		+[b]		+[b]		Punja *et al.* (1985)
Yeasts						
Candida kefyr		+[b,d]				Call *et al.* (1985)
macedoniensis		+[d]				Call *et al.* (1985)
pseudotropicalis[h]		+[d]				Call *et al.* (1985)
Kluyveromyces marxianus		+[d]				Call *et al.* (1985)
Saccharomyces vini	+	+				Gogeliya & Machavariani (1983)
Saccharomycopis *fibuliger*	0	+	0			Fellows & Worgan (1984)

Table 3—contd.

Organism	PME[a]	PG	PGL	PML	PMGL	Reference
Bacteria						
Clostridium pectino-						
fermentans 15	+	+[c]			+[c]	Bravova & Kalunyants (1984)
Corynebacterium						
michiganense						
(E. F. Smith) Jensen			+		+	Scala & Zoina (1983b)
Erwinia (545 strains)[f]		+[j]	+[k]			Lobanok et al. (1981)
chrysanthemi		+	+	+	+	Prasad (1987)
chrysanthemi		+[b,c]			+[d]	Ried & Collmer (1986)
carotovora						
subsp. carotovora		+[b,c]			+[d]	Ried & Collmer (1986)
subsp. atroseptica		+[b,c]			+[d]	Ried & Collmer (1986)
dissolvens					+	Khan et al. (1987)
Pseudomonas marginalis					+	Sone et al. (1988)
Streptomyces fradiae			+[b]			Sato & Kaji (1981)
massasporeus			+[c]			Sato & Kaji (1981)
nitrosporeus			+[c]			Sato & Kaji (1981)
Xanthomonas campestris		+	+	+		Pandey & Prasad (1983); Prasad (1987)

[a] PME, pectin methylesterase; PG, polygalacturonase; PGL, polygalacturonate lyase; PMG, polymethylgalac-turonase; PMGL, polymethylgalacturonate lyase. List includes only reports since 1981. See Fogarty & Kelly (1983) for earlier listing.
[b] Endo-splitting.
[c] Exo-splitting.
[d] Multiple forms.
[e] Protopectinase also present.
[f] Represents 38 species; 12% of 340 strains produced variable amounts of pectin methylesterase (PME), 28% produced traces of PME and 60% had no PME.
[g] Forty-three single basidiospore strains and five parental field isolates studied.
[h] Seventy-two strains of *Candida* examined. Only *C. kefya*, *C. macedoniensis* and four strains of *C. pseudotropicalis* were pectolytically active.
[i] All 545 *Erwinia* strains hydrolyzed pectin and had high levels of lyase activity (200–360 units/mg) and low polygalacturonase activity (0·15–1·5 units/mg).
[j] Low activity.
[k] High activity.
+ , Activity detected; 0, no detectable activity.

Speiser *et al.*, 1947). The interpretation is that the pectin methylesterase produces blocks of deesterified galacturonate units while alkali treatment gives random deesterification thereby leading to the hypothesis that the enzyme requires a free carboxyl group adjacent to a methylated group in order to act. Pectin partially reduced by $NaBH_4$ treatment, converting methylgalacturonate residues to galactose residues, was only slowly deesterified (Solms & Deuel, 1955).

Does the pectin methylesterase attack the methylgalacturonate residues randomly or from one of the terminal ends? This question was addressed in a clever manner by Miller & Macmillan (1970). They used the exopolygalac-turonate lyase from *Clostridium multifermentans* and polymethylgalacturonate methyl glycoside. The exolyase degrades polygalacturonic acid in a linear manner, removing $\Delta4:5$ unsaturated digalacturonic acid units from the reducing end of the chain. The heat-treated enzyme (to remove pectin methylesterase activity) did not degrade glycosidic linkages in which the pectin was highly esterified with methanol. In the experiment, pectin methylesterase and lyase activities were monitored automatically in a single reaction containing highly esterified pectin as substrate. Results with partially purified *Fusarium oxysporum* pectin methylesterase indicated 57% of deesterification occurred from the reducing end of the substrate.

Miller & Macmillan (1970) reported that the pectin methylesterase and exopolygalacturonate lyase activities of *Clostridium multifermentans* reside on the same protein (400 000 daltons), since they could not be separated by Sephadex G-200 chromatography, DEAE-cellulose chromatography or rate zonal centrifugation (Miller & Macmillan, 1971). When the pectin methylesterase and exolyase activities were determined simultaneously in the same reaction using highly esterified pectin, the esterase rate was twice that for the lyase rate (based on the $[H^+]$ and ΔAbs at 235 nm). The 2:1 ratio is exactly that expected if the pectin methylesterase deesterifies the substrate only from the reducing end, since two methylgalacturonate residues must be deesterified for each glycosidic bond broken to give $\Delta 4:5$ unsaturated digalacturonate.

Polygalacturonates are often inhibitors of pectic enzymes. As a result, higher salt (NaCl, KCl) concentrations often result in higher specific activities. Sodium polygalacturonate was reported to be a competitive inhibitor with K_i of 2·8 and 0·11 mg/ml, for *Botrytis cinerea* Pers. pectin methylesterases I and II, respectively (Schejter & Marcus, 1988).

Some properties of pectin methylesterases from plants, fungi and bacteria are shown in Table 4. With the exception of *Clostridium multifermantans* pectin methylesterase, the relative molecular masses are in the range of 26 000–45 000 for both higher plant and microbial enzymes. The isoelectric points (pI) range from 6·8–11. The specific activities range from 25–2200 units/mg of protein (leaving out the value for *P. infestans* enzyme); the differences reflect differences in degree of purity as well as intrinsic differences in catalytic efficiency. With the exception of the enzymes from *C. diplodiella* and *C. rolfsii* with lower pH optima, the plant and microbial enzymes have similar pH optima. The K_m values range from a low of 0·004 and 0·0046 mg/ml for the *B. cinerea* Pers. enzyme and orange isoenzyme II to 6·0 mg/ml of pectin for the banana enzyme.

3.2 Protopectinases

Early on in the study of pectolytic enzymes, the existence of a protopectinase was taken for granted (Dore *et al.*, 1927). However, from the 1950s to 1978, researchers assumed the solubilization of protopectin to pectin was due to the action of pectin methylesterase and endopolygalacturonase and/or pectin lyase. Beginning in 1978, Sakai and his colleagues (Sakai & Okushima, 1978, 1980, 1982; Sakai *et al.*, 1982, 1984; Sakai & Yoshitake, 1984; Sakai & Takaoka, 1985) published a series of papers on this enzyme. Much of this research is summarized in Sakai (1988).

Table 4
Properties of Some Plant and Microbial Pectin Methylesterases[a]

Source of enzyme	M_r	pI	Specific activity (U/mg protein)	pH optimum	K_m for pectin (mg/ml)	Reference
Fruits and vegetables						
Banana I[b]	30 000	8·9	457		6·0 ⎫	Brady (1976)
II[b]	30 000	9·4	529	6·0	⎬	
Orange (*Citrus natsudaidai*)			2 200	8·0	2·3 ⎭	Manabe (1973)
Orange (*Citrus sinensis*) I[b]	36 200	10·0	694	7·6	0·083 ⎫	Versteeg *et al.* (1978)
II[b]	36 200	11	762	8·0	0·0046 ⎭	
Plums (*Prunis salicina*)			25	7·5	0·1	Theron *et al.* (1977)
Potato (*Solanum tuberosum*)	25 000			7·5	0·9	Puri *et al.* (1982)
Tomato[b]	27 500		1 150	6–9	0·74	Lee & Macmillan (1968, 1970a, b)
Tomato[b]	26 300	8·4				Delincée & Radola (1970); Delincée (1976)
Tomato			724	8·0	2·40	Nakagawa *et al.* (1970a, b)
Tomato[b]	27 800					Markovič (1974)
Fungi						
Acrocylindrium sp.				7·5		Kimura *et al.* (1973)
Coniothyrium diplodiella I[b]				4·8 ⎫		Endo (1964a, b)
II[b]				4·8 ⎭		
Corticium rolfsii[c]	37 000			3·5		Yoshihara *et al.* (1977)
Fusarium oxysporum	35 000		203	7·0		Miller & Macmillan (1971)
Bacteria						
Botrytis cinierea Pers. I[b]	28 400[d] 28 100[e]	6·8	608	7·0	0·06 ⎫	Schejter & Marcus (1988)
II[b]	27 800[d] 27 600[e]	6·8	710	6·5	0·004 ⎭	
Clostridium multifermentans[g]	400 000		48	9·0	0·74	Miller & Macmillan (1970); Sheiman *et al.* (1976)
Phytophthora infestans[h] I[b]	45 000– 48 000[d]; 45 000[e]		12 100[f]	7		Förster (1988)
II[b]	40 000[d]; 35 000– 37 000[e]		15 350[f]	7		Förster (1988)

[a] From Rombouts & Pilnik (1980) with permission, except for *Botrytis cinerea* Pers. and *Phytophthora infestans*.
[b] (One of) multiple molecular forms or isoenzymes.
[c] This enzyme is active at low pH values; stable in the pH range 1–11.
[d] Relative molecular mass estimated by SDS-PAGE electrophoresis.
[e] Relative molecular mass estimated by gel filtration.
[f] One unit of enzyme defined as amount of enzyme that causes a decrease in pH of the reaction mixture of 0·1 in 30 min.
[g] This enzyme is complexed with exopectate lyase.
[h] The enzymes are glycoproteins.

The reaction catalyzed by protopectinases is shown in eqn (3):

$$\text{Protopectin} + H_2O \xrightarrow{\text{protopectinase}} \text{pectin} \tag{3}$$

(insoluble)   (soluble)

The protopectinase activity was determined on protopectin prepared from the orange albedo. The protopectinase activity was assayed by measuring the

Table 5
Some Physical and Chemical Properties of Protopectinases[a]

Property	F	L	S
Relative molecular mass			
SDS-PAGE electrophoresis	40 000	40 000	40 000
Gel filtration on Sephadex G-75	33 000	30 000	30 000
Sedimentation equilibrium	32 800	29 200	29 300
$E^{1\%}_{280}$	10·0	11·9	9·20
pI	5·0	8·4–8·5	7·6–7·8
N-terminal amino acid	—	Glycine	Glycine
Carbohydrate content (%)	5·9[b]	5·1[c]	1·7[b]
Specific activity (U/mg)	5 900	3 957	5 769

[a] Adapted from Sakai (1988).
[b] Determined as mannose.
[c] Determined as rhamnose.

amount of pectic substance solubilized from protopectin, as determined by the carbazole–sulfuric acid method (Furutani & Osajima, 1965).

Protopectinase was purified and crystallized from *Kluyveromyces fragilis* (protopectinase F), *Galactomyces reessii* (protopectinase L) and *Trichosporon penicillatum* (protopectinase S). Some of the physical and chemical properties of protopectinases F, L and S are shown in Table 5. The relative molecular masses are very similar within a single method of determination and are around 30 000–40 000. The $E^{1\%}_{280nm}$ values are similar but not identical. This difference is reflected in differences in aromatic amino acid composition (Trp-5, 9, and 5 residues/30 000; Tyr-4, 5 and 4 residues/30 000 and Phe-7, 8 and 10 residues/30 000 for protopectinases F, L and S, respectively). The

Table 6
Some Kinetic Properties of Protopectinases[a]

Property	Protopectinase		
	F	L	S
pH optimum	5·0	5·0	5·0
Temperature optimum (°C)	60	55	50
Thermostability (°C)	Up to 40	Up to 50	Up to 55
pH stability	2–8	3–7	3–7
Activity (U/mg)			
Protopectin	556	3 940	5 770
Polygalacturonate	2 050	16 200	21 100
K_m (mg/ml)			
Protopectin	90	50	30
Polygalacturonate[b]	6·6	7·7	9·0

[a] Adapted from Sakai (1988).
[b] Polygalacturonate used for determining K_m had a mean DP of 130.

Table 7
Action Mode of Protopectinases on Galacturonic Acid Oligomers[a]

Enzyme	Substrates	Reaction products	K_m (mM)	V_{max} (μM/U/min)	V_{max}/K_m (μM/U/min mM)
Protopectinase F	*(hexagon diagrams)*	*(hexagon diagrams)*	$>10^2$	$<10^{-10}$	10^{-12}
			1·82	$1·90 \times 10^{-2}$	$1·05 \times 10^{-2}$
			$5·95 \times 10^{-1}$	$2·95 \times 10^{-1}$	0·496
Protopectinase L	*(hexagon diagrams)*	*(hexagon diagrams)*	3·98	$1·96 \times 10^{-3}$	$4·92 \times 10^{-4}$
			2·77	$2·37 \times 10^{-2}$	$8·5 \times 10^{-3}$
			$7·09 \times 10^{-1}$	$9·79 \times 10^{-2}$	0·138
Protopectinase S	*(hexagon diagrams)*	*(hexagon diagrams)*	4·26	$4·79 \times 10^{-4}$	$1·12 \times 10^{-4}$
			2·20	$5·20 \times 10^{-2}$	$2·36 \times 10^{-2}$
			$8·69 \times 10^{-1}$	$2·85 \times 10^{-1}$	0·328

[a] Reactions were done under the optimum conditions for each enzyme. Hexagons express D-galacturonic acid molecules. Adapted from Sakai (1988).

isoelectric points (pI) are different with protopectinase F being an acidic protein and L and S basic proteins. Glycine is the N-terminal amino acid for both protopectinases L and S and all are glycoproteins, but differ in the amount of carbohydrate. Protopectinases L and S are identical immunochemically while F did not react with rabbit antiserum to protopectinase S.

Some kinetic properties of the three protopectinases are shown in Table 6. pH optimum values are the same and the optimum temperatures are similar, as is the range of pH stability. Thermostability differs among the three enzymes as do the specific activities and K_m values. Polygalacturonate (DP (degree of polymerization) 130) was a better substrate than protopectin for all three enzymes, although the assay methods and conditions were not comparable. Protopectin is an insoluble substrate.

As noted in Table 6, the protopectinases have high specific activity on polygalacturonate. By using tri- tetra- and pentagalacturonates, Sakai and colleagues (Sakai, 1988) determined the action patterns and kinetic constants, V_{max} and K_m, for the three enzymes (Table 7). All three enzymes removed one galacturonate unit from the reducing end of trigalacturonate; all three enzymes removed one galacturonate unit or one digalacturonate unit from the reducing end of tetragalacturonate. Protopectinase F and S removed one galacturonate unit or one digalacturonate unit from the reducing end of pentagalacturonate but protopectinase L removed only one galacturonate unit initially from pentagalacturonate. Based on V_{max}/K_m values (specificity coefficient; Whitaker, 1972), specificity for the substrates increased in the order tri-, tetra- and pentagalacturonates for all three enzymes. Sakai and colleagues (Sakai, 1988) reported that the fragmentation patterns differed also depending on the degree of methylation of polygalacturonate (DP = 33). With no methylation, the average DP of the products was 2·3 for all three enzymes. At 30% methylation the average DP of the products was 4–5 and at 70% methylation it was 9–15. Based on these results, Sakai and colleagues (Sakai, 1988) suggested that the protopectinases react with protopectin at sites having three or more non-methylated galacturonate units and cleave the glycosidic bonds by hydrolysis.

3.3 Endopolygalacturonases

Endopolygalacturonases [poly(1,4-α-D-galacturonide)glycanhydrolase; EC 3.2.1.15; endopectate hydrolase] hydrolyze internal glycosidic bonds of polygalacturonates in a more or less random fashion giving a series of intermediate size polygalacturonates (eqn (4)), as determined by the specific enzyme.

The endopolygalacturonases have no activity on highly methylated pectin in the absence of pectin methylesterase. Pectinic acids give ambiguous results since some blocks of deesterified galacturonate units may be large enough to permit endopolygalacturonate activity.

Endopolygalacturonase activity can be followed readily by measuring the rate of formation of reducing groups by either the 3,5-dinitrosalicylate method (Bernfeld, 1955) or the arsenomolybdate–copper reagent (Somogyi, 1952).

Polygalacturonases

$$\text{Polygalacturonate} + n\,\text{H}_2\text{O} \xrightarrow{\text{endopolygalacturonase}} \text{oligogalacturonates} + \text{galacturonate} \quad (4)$$

The results can be correlated readily with the number of glycosidic bonds hydrolyzed. Endopolygalacturonase activity can also be determined by viscosity reduction. While more sensitive in the early stages of hydrolysis of polygalacturonate, there is no direct correlation between viscosity reduction and number of glycosidic bonds hydrolyzed as viscosity reduction depends on the location of the glycosidic bond hydrolyzed within the polygalacturonate chain. The viscosity method cannot be used at lower pH values (below 3) where polygalacturonates form gels.

Neither the reducing group method nor the viscosity reduction method alone distinguishes between an endo- and exopolygalacturonase. In combination, the two generally will permit one to determine between an endo- and exopolygalacturonase. By standardizing two enzyme preparations against the reducing group method, an endopolygalacturonase will initially reduce the viscosity of the reaction much more rapidly than will an expolygalacturonase (Rombouts & Pilnik, 1972). Determination of the products formed early in the reaction is a better way of distinguishing between an endopolygalacturonase, which produces a series of oligosaccharides, but little mono- and digalacturonates and an exopolygalacturonase, which produces a relatively high concentration of mono- and/or digalacturonates.

Endopolygalacturonases do not form products in a truly random pattern, as can be seen by comparing the products formed early in the reaction by a number of different endopolygalacturonases (Koller & Neukom, 1969; Rexová-Benková, 1973; Rexová-Benková & Markovič, 1976). The nature of the products formed are a function of the size of the binding locus of the active sites which affects K_m and V_{max}, and the K_m and V_{max} for the intermediate products. In those enzymes where size is relatively unimportant in binding, few intermediate products may accumulate (often called inappropriately single chain attack) and those enzymes where there is much better binding of large polygalacturonates than smaller polygalacturonates, high concentrations of intermediate products accumulate (incorrectly referred to as multichain attack). As an example, with a purified polygalacturonase from *Kluyveromyces fragilis*, hydrolysis of polygalacturonate occurred in three stages (Demain &

Phaff, 1954). A rapid, initial stage to ~25% hydrolysis produced tetra-, tri-
and digalacturonates; a second stage that occurred at about 2% the initial rate
increased hydrolysis to 50% in which the tetragalacturonate was hydrolyzed to
tri- and monogalacturonates. The final very slow third phase to ~70%
hydrolysis resulted in conversion of the trigalacturonate to di- and monogalacturonates. This is a typical pattern where the specificity of the enzyme
decreases markedly for the smaller oligogalacturonates (Table 8).

Rexová-Benková & Markovič (1976) proposed three models to explain the
various products formed by three groups of endopolygalacturonases. They
proposed that the binding sites differed among the three groups and that
different size substrates could be accommodated differently. The relative
affinity of the subsites for the reducing group end in relation to the catalytic
groups determines where the oligogalacturonate is hydrolyzed.

Endopolygalacturonases have been purified from a number of plant and
microbial sources as listed in Table 9. The endopolygalacturonases often occur
in multiple molecular forms. The relative molecular masses range from
30 000–85 000 with isoelectric points (pI) of 3·8–7·6. The specific activity, a
function of both purity and specificity coefficient, range from 44–7092 units/mg
of protein. All of the endopolygalacturonases have pH optima in the acid
region between pH 2·5 and 6·5 (most are between pH 4 and 5). The K_m values
are remarkably constant, ranging from 0·14–2·7 mg/ml for pectate.

Table 8
Effect of Size of Substrate on Relative Rates for Some Polygalacturonases and Pectate Lyases

Substrate	Enzymes					
N^g	A^a	B^b	C^c	D^d	E^e	F^f
>300	100	100	100	100	100	100
12	52					
8			103	38	36	17
7			78	35	15	13
6	12		41	27	11	7
5		12	32	15	3	2
4	2	2	20	7	0·6	0·6
3	2	0·2	2	1	0·02	0·008
2	0	0				

[a] Polygalacturonase of *Erwinia carotovora* (Nasuno & Starr, 1966).
[b] Polygalacturonase of *Trichosporon penicillatum* SNO-3 (Sakai *et al.*,
1982).
[c] Pectate lyase-1 of *Bacillus polymyxa* (Nagel & Wilson, 1970).
[d] Pectate lyase-2 of *Bacillus polymyxa* (Nagel & Wilson, 1970).
[e] Pectate lyase-3 of *Bacillus polymyxa* (Nagel & Wilson, 1970).
[f] Pectate lyase-4 of *Bacillus polmyxa* (Nagel & Wilson, 1970).
[g] N = number of galacturonate units.

Table 9
Properties of Some Plant and Microbial Endopolygalacturonases[a]

Source of enzyme	M_r	pI	Specific activity (U/mg protein)	pH optimum	K_m for pectin (mg/ml)	Reference
Fruits						
Tomato	52 000		47	4·5	2·7	Takehana *et al.* (1977)
Tomato I[b]	84 000			4·5 ⎫		
II[b]	44 000			5·0 ⎭		Pressey & Avants (1973)
Moulds						
Aspergillus niger I[b]		3·8		4·0		Koller (1966)
II[b]				4·5		
III[b]		4·5		5·5	1·7 ⎫	Koller & Neukom (1969)
Aspergillus niger I[b]	35 000		81	4·1 ⎫		
II[b]	85 000		44	3·8 ⎭		Cooke *et al.* (1976)
Aspergillus niger	46 000		75	5·0	0·54	Heinrichová & Rexová-Benková (1977)
Aspergillus japonicus	35 500		1 362	4·5		Ishii & Yokotsuka
Botrytis cinerea	69 000		2 049	4·0	1·2	(1972b)
						Urbanek & Zalewska-Sobczak (1975)
Botrytis cinerea Pers.						
I[b]	34 600[c]	7·3	2 180	4·5	0·14 ⎫	
	34 100[d]					Schejter & Marcus (1988)
II[b]	56 800[c]	7·6	7 092	4·0	0·21 ⎭	
	56 200[d]					
Corticium rolfsii			126	2·5		Tagawa & Kaji (1988)
Fusarium oxysporum I[e]	37 000	7·0	194	5·0		
II[e]	37 000	7·0	148	5·0	0·54	Strand *et al.* (1976)
Rhizoctonia fragariae I[e]	36 000	6·8	1 866	5·0	0·80	Cervone *et al.* (1977, 1978)
II[e]	36 000	7·1	1 845	5·0	0·75	
Rhizopus arrhizus	30 300		92	5·0	0·54	Liu & Luh (1978)
Trichoderma koningii I[e]	32 000	6·41		5·0	0·80	
II[e]	32 000	6·57		5·0	0·85	Fanelli *et al.* (1978)
Verticillium albo-atrum	30 000		2 075	6·5	1·5	Wang & Keen (1970)
Yeasts						
Kluyveromyces fragilis			168	4·4		Phaff (1966)
I	46 000	6·1		4–5 ⎫		
II	50 000	6·1		4–5 ⎬		Lim *et al.* (1980)
III	30 000	5·8		4–5 ⎭		
Bacteria						
Erwinia carotovora			362	5·3		Nasuno & Starr (1966)
Pseudomonas cepacia			125	4·5		Ulrich (1975)

[a] From Rombouts & Pilnik (1980) with permission, except for *Corticium rolfsii*, *Botrytis cinerea* Pers. and second set of data for *Kluyveromyces fragilis* isoenzymes.
[b] (One of) multiple molecular forms or isoenzymes.
[c] Molecular weight estimated by SDS-PAGE electrophoresis.
[d] Molecular weight estimated by gel filtration.
[e] These are isoenzymes and glycoproteins.

3.4   Exopolygalacturonases

Exopolygalacturonases   [poly(1,4-α-D-galacturonide)galacturohydrolase; EC 3.2.1.67; galacturan 1,4-α-galacturonidase; exopectate hydrolase] act on polygalacturonates (pectate), rather than on polymethylgalacturonates (pectin). Determination of specificity is best using pectate and fully methylated pectin, rather than pectin or pectinic acids. Preparations must be free of pectin methylesterase. The reaction catalyzed by the exopolygalacturonases is shown in eqn (5).

$$\text{Polygalacturonate}_n + H_2O \xrightarrow{\text{exopolygalacturonase}} \text{polygalacturonate}_{n-1} + \text{galacturonate} \quad (5)$$

The usual end product is galacturonate, except for *Erwinia aroideae* and *Pseudomonas* sp. exopolygalacturonase. These two enzymes hydrolyze a digalacturonate unit from the substrate (Hatanaka & Ozawa, 1969a; Hatanaka & Imamura, 1974) and the preferred name is poly(1,4-α-D-galacturonide)digalacturonohydrolase. In all cases, hydrolysis occurs from the non-reducing end (Hatanaka & Ozawa, 1964, 1969a, 1971). Evidence for attack at the non-reducing end includes failure to hydrolyze substrates with an added Δ4:5 unsaturated galacturonate unit or a deoxy sugar at the non-reducing end of the substrate. Carrot exopolygalacturonase gave variable extent of hydrolysis depending on the source of pectate used (Hatanaka & Ozawa, 1964), presumably because of the presence of neutral sugars along the main chain or as side chain constituents. Peach, citrus and partially degraded citrus pectates were hydrolyzed to 7, 50 and 79%, respectively (Ozawa, 1955).

Activity is best determined by measuring the rate of formation of reducing groups by the 3,5-dinitrosalicylate method (Bernfeld, 1955) or the arsenomolybdate–copper method (Somogyi, 1952) or by specific enzymatic measurement of the galacturonate formed (Nagel & Hasegawa, 1967a; Bateman et al., 1970; Wagner & Hollmann, 1976). Reduction in viscosity is a less sensitive method than for endopolygalacturonases since the decrease in viscosity is relatively slow. However, the viscosity reduction method, in combination with the reducing method is a useful method in determining that the enzyme is an exopolygalacturonase (Rombouts & Pilnik, 1972). The other useful methods are to determine the products in the early stages of the reaction by HPLC or by thin-layer chromatography (TLC), or by enzymatic determination of the galacturonate (Nagel & Hasegawa, 1967a; Bateman et al., 1970; Wagner & Hollmann, 1976).

Exopolygalacturonases have been found in a number of plants, including carrots (Pressey & Avants, 1975a; Heinrichová, 1977), peaches (Pressey & Avants, 1975a), citrus (Riov, 1975), cucumbers (Pressey & Avants, 1975b), pears (Pressey & Avants, 1976), apples (Bartley, 1978), oat seedlings (Pressey & Avants, 1977) and insects (Courtois et al., 1968; Foglietti et al., 1971). In contrast to other exopolygalacturonases, the insect enzymes are reported to attack the reducing end of the substrate (Courtois *et al.*, 1968; Foglietti *et al.*, 1971).

Exopolygalacturonases have been reported in fungi (*Aspergillus niger*; Mill 1966*a*, *b*; *Coniothyrium diplodiella*; Endo, 1964*a*; Hatanaka & Ozawa, 1966; *Rhizopus tritici*; McClendon & Kreisher, 1963) and bacteria (*Erwinia aroideae*; Hatanaka & Ozawa, 1969*a*) (See Table 3).

Mill (1966*a*, *b*) isolated two exopolygalacturonases from *Aspergillus niger*. Exopolygalacturonase I had a pH optimum of 4·5 and was activated by Hg^{2+}. One μM Hg^{2+} increased the activity 9-fold. Removal of Hg^{2+} by 2,3-dimercaptopropanol or by EDTA eliminated all activity. K_m for Hg^{2+} was $6·2 \times 10^{-8}$ M. Hatanaka & Ozawa (1969*c*) also reported a marked stimulatory effect of 1 μM Hg^{2+} on an exopolygalacturonase from *Aspergillus niger* that had high activity on digalacturonate. Exopolygalacturonase II had a pH optimum of 5·1 and was not affected by Hg^{2+} or EDTA. The exopolygalacturonase from *Coniothyrium diplodiella* was not affected by Hg^{2+} or EDTA (Hatanaka & Ozawa, 1969*b*).

As described above, the exopolygalacturonases from *Erwinia aroideae* and a *Pseudomonas* sp. hydrolyze digalacturonate units from the non-reducing end of pectate (Hatanaka & Ozawa, 1969*a*) and should be named poly(1,4-α-D-galacturonide)digalacturonohydrolases.

There is little data available on the chemical and physical properties of the microbial exopolygalacturonases. As a basis for further work on these enzymes, Konno (1988*b*) has recently summarized data for carrot (*Daucus carota*) and liverwort (*Marchantia polymorpha*) purified exopolygalacturonases (Konno *et al.*, 1981, 1983). The carrot exopolygalacturonase had a relative molecular mass of 48 000 by gel filtration, pI of 4·75, and K_m and V_{max} of $4·35 \times 10^{-2}$ mM and 97·1 U/mg protein, respectively, for a polygalacturonate with a DP = 52. The pH optimum was 4·8. The relative rates, measured during a 1 h reaction were: polygalacturonate, 100; pectate, 56·2; and pectin, 6·2. The liverwort exopolygalacturonase had a relative molecular mass of 76 000 by gel filtration and a pI of 5·20. The K_m and V_{max} values were $3·23 \times 10^{-2}$ mM and 31·6 U/mg protein, respectively, for polygalacturonate of DP = 52. The pH optimum was 3·6–3·8. The relative rates, measured during a 1 h reaction, were: polygalacturonate, 100; pectate, 55·4; and pectin, 18·2.

Clearly, much more research is needed on this group of enzymes. The only thing that distinguishes them from the oligogalacturonate hydrolases (below) is preference for high molecular weight polygalacturonates.

3.5 Oligogalacturonate Hydrolases

The oligogalacturonate hydrolases differ in activity from the exopolygalacturonases (above) only in that they have higher specific activities on low molecular mass oligogalacturonates than on high molecular mass polygalacturonates and pectate. The rate of hydrolysis decreases rapidly with increasing degree of polymerization. The reaction catalyzed is similar to that shown by eqn (5). However, the substrate would be oligogalacturonates and the products would be smaller oligogalacturonates plus galacturonate.

There are two types of oligogalacturonate hydrolases. One type acts on saturated oligogalacturonates and the other acts on Δ4:5 unsaturated oligogalacturonates (Table 10). The oligogalacturonate hydrolases isolated from a *Bacillus* sp. attacked their substrates from the non-reducing end with the final product being monogalacturonate (Hasegawa & Nagel, 1968). The enzyme was most active on trigalacturonate and the activity rate decreased with increasing degree of polymerization (Table 10). The Δ4:5 unsaturated oligogalacturonate hydrolase from a *Bacillus* sp. is specific only for unsaturated oligogalacturonates and removes the unsaturated Δ4:5 galacturonate from the non reducing end of the chain, leaving the other glycosidic bonds intact (Nagel & Hasegawa,

Table 10

Properties of Some Microbial Oligogalacturonases[a]

Micro-organism	Substrates[b]	Products from		Optimum pH value	Reference
		Dimer[c]	u-Dimer		
Oligogalacturonate hydrolases; attack of substrate from non-reducing end					
Bacillus sp.	Trimer > tetramer > pentamer > dimer > acid-soluble pectic acid	Monomer	None	6·5	Hasegawa & Nagel (1968)
Bacillus sp[d]	u-Dimer > u-trimer > u-tetramer > u-pentamer	None	u-Monomer Monomer	6·5	Nagel & Hasegawa (1968)
Aspergillus niger	Dimer > pectic acid > u-dimer	Monomer	u-Monomer Monomer	—	Hatanaka & Ozawa (1969a, b)
Oligogalacturonate lyases; attack of substrate from reducing end					
Erwinia carotovora	u-Dimer > dimer > trimer > tetramer, pectate	u-Monomer Monomer	u-Monomer	7·2	Moran *et al.* (1968b)
Erwinia aroideae	u-Dimer > u-trimer > dimer > trimer > tetramer > acid-soluble pectic acid	u-Monomer[e] Monomer	u-Monomer[e]	7·0	Hatanaka & Ozawa (1970)
Pseudomonas sp[f]	Tetramer > trimer U-trimer > dimer U-dimer > acid-soluble pectic acid > pectate	u-Monomer Monomer	u-Monomer	7·0	Hatanaka & Ozawa (1971)

[a] From Rombouts & Pilnik (1980) with permission.
[b] These are listed in order of rate of degradation.
[c] Terminology and abbreviations: monomer, D-galacturonic acid; dimer, digalacturonic acid; u-oligomers (unsaturated oligomers), oligogalacturonates of which the terminal unit at the non-reducing end is a 4,5-dehydrogalacturonate; u-monomer; 4-deoxy-5-oxo-D-glucuronic acid (4-deoxy-L-*threo*-5-hexoseulose uronic acid); the degree of polymerization of acid-soluble pectic acid is 15–20.
[d] This enzyme, which attacks the bond adjacent to a terminal unsaturated residue only, is called an unsaturated oligogalacturonate hydrolase.
[e] The product formed in the reaction mixture was identified as 4-deoxy-5-oxo-D-fructuronic acid.
[f] This oligogalacturonate lyase, unlike those of *Erwinia* spp., is stimulated by calcium ions.

1968). The enzyme was most active on the unsaturated digalacturonate and the activity decreased with higher degrees of polymerization. Hatanaka & Ozawa (1969*b*, *c*) prepared a protein from *Aspergillus niger* that had both types of activity, although the activity on Δ4:5 unsaturated digalacturonate was about 1/60 that on digalacturonate (saturated). The two exopolygalacturonases reported by Mill (1966*a*, *b*) from *A. niger* (see previous section) should now be classified as oligogalacturonate hydrolases since they had much higher specific activity on di- and trigalacturonates than on pectate, and produced only monogalacturonate. The oligogalacturonate hydrolases described by Hasegawa & Nagel (1968) are intracellular enzymes, rather than extracellular as is the case for the exopolygalacturonases. Measuring the activity of oligogalacturonate hydrolases involves quantifying hydrolysis of the glycosidic bond via reducing sugar methods (Somogyi, 1952; Bernfeld, 1955).

3.6 Endopolymethylgalacturonases

Endopolymethylgalacturonases (endopectin hydrolase) are reported in a number of organisms by many researchers (Table 3). The presumed reaction is random hydrolysis of polymethylgalacturonates to oligomethylgalacturonates (Albersheim & Killias, 1962; Finkelman & Zajic, 1978) as shown in eqn (6).

$$\text{Polymethylgalacturonate}_n + n\,\text{H}_2\text{O} \xrightarrow{\text{endopectin hydrolase}} \text{oligomethylgalacturonates} \quad (6)$$

Activity can be followed by measuring the rate of increase in reducing groups formed by hydrolysis of the glycosidic bond, using the Bernfeld (1955) or Somogyi (1952) methods. Viscosity reduction could also be used. Some prominent researchers in this field question the existence of this group of enzymes (Rombouts & Pilnik, 1972; Macmillan & Sheiman, 1974), since they have never been able to demonstrate the presence of these enzymes when 100% methylated polymethylgalacturonate is used as substrate in the absence of pectin methylesterase and endopolymethylgalacturonate lyase. Researchers need to be more careful in reporting the presence of endopolymethylgalacturonase activity (see Table 3 for recent examples).

3.7 Endopolygalacturonate Lyases

Endopolygalacturonate lyases (poly-1,4-α-D-galacturonide lyase, EC 4.2.2.2; endopectate lyase) are produced by a number of bacteria and some pathogenic fungi. The reaction catalyzed is shown in eqn (7).

$$\text{Polygalacturonate} \xrightarrow[\text{lyase}]{\text{endopectate}} \Delta 4:5 \text{ unsaturated oligogalacturonates} \quad (7)$$

The endopolygalacturonate lyases differ from the endopolygalacturonases in four important respects. First, they are found only in microorganisms. Secondly, they all have pH optima in the range 8–10—much higher than for other pectolytic enzymes. Thirdly, they have an absolute requirement for Ca^{2+} and fourthly they split the glycosidic bond via a *trans* elimination mechanism, in contrast to hydrolysis, to give a product with a double bond between C-4 and C-5 of the galacturonate. In conjugation with the carbonyl constituent on C-5, there is a maximum absorbance peak at 235 nm due to the double bond in the product (see eqn (8)).

The most convenient method of following the activity of endopoly-galacturonase lyase is with a recording spectrophotometer at 235 nm, where the ε_m is $4 \cdot 6 \times 10^3 \, M^{-1} cm^{-1}$ (Macmillan & Phaff, 1966). The unsaturated product also gives a red color with thiobarbituric acid reagent (Albersheim *et al.*, 1960). Activity can also be measured by reducing sugar methods (Somogyi, 1952; Bernfeld, 1955) or by viscosity reduction. The ratio of reducing groups formed to double bonds formed should be $1 \cdot 0$; if not, the investigator should look for contamination with other pectolytic enzymes. Viscosity reduction, in conjunction with a reducing group method, or along with intermediate product analysis by HPLC or GC, is useful in distinguishing between an endo-polygalacturonate lyase and an exopolygalacturonate lyase.

Endopolygalacturonate lyases are found primarily in bacteria and a few fungi associated with food spoilage and soft rot (Table 11). Filamentous fungi more frequently produce endopolymethylgalacturonate lyases. There is a report of a polygalacturonate lyase in a protozoa, *Ophryoscolex purkynei* (Mah & Hungate, 1965).

Albersheim *et al.* (1960) were the first to report a pectolytic enzyme (of fungal origin) that splits the 1,4-glycosidic bonds of pectin by a *trans* elimination mechanism, rather than by hydrolysis. However, earlier Wood (1951) had reported a bacterial enzyme that degraded polygalacturonate which required Ca^{2+} and had a pH optimum of $8 \cdot 5$–$9 \cdot 0$. Some properties of endopolygalacturonate lyases are shown in Table 11. The relative molecular masses are in the range of 30 000–40 000 and have pH optima between 8 and 10. Most are basic proteins as indicated by the pI values. There is a wide range of specific activities and K_m values for pectate.

Pectates are generally good substrates for the endopolygalacturonate lyases. However, for enzymes from two *Arthrobacter* strains and from *Bacillus polymyxa*, the best substrates are pectinic acids with degrees of esterification of 21, 44 and 26%, respectively (Rombouts, 1972; Pilnik *et al.*, 1973). The 1,4-α-D-galacturan structure is necessary for activity. However, the hydroxyl groups at C-2 and C-3 of the galacturan are thought not to be essential since the Vi antigen from an *Escherichia* sp., containing $(1 \rightarrow 4)$-α-D-linked 2-deoxy,2-acetamido and 3-O-acetyl groups, are degraded by enzymes from *Bacillus sphaericus* and *B. polymyxa* (McNicol & Baker, 1970).

Endopolygalacturonate lyase activity decreases with decreasing chain length of oligogalacturonate substrates. This is primarily due to a decrease in K_m since V_{max} is fairly constant regardless of chain length (Nagel & Hasegawa, 1967b; Nagel & Wilson, 1970; Atallah & Nagel, 1977). Most of the

Table 11
Properties of Some Endopolygalacturonate Lyases[a]

Source of enzyme	M_r	pI	Specific activity (U/mg protein)	K_m for pH optimum	Pectate (mg/ml)	References
Bacteria						
Bacillus circulans I	70 000			10·0		Joyce & Fogarty (1974, 1975)
II	18 000			9·5		
Bacillus polymyxa				8·3–9·6	0·056–0·0065	Nagel & Wilson (1970)
Bacillus subtilis[d]	33 000	9·85		8·5		Chesson & Codner (1978)
	49 000			8·0		Joyce & Fogarty (1974, 1975)
Bacillus stearothermophilus	24 000			9·0		Karbassi & Vaughn (1980) Kelly & Fogarty (1978)
Bacillus sp. RK9				10·0		Kamimiya *et al.* (1977)
Erwinia aroideae	37 000			9·1		
Erwinia aroideae	67 000[b]		31·9[c]	9·3	0·118 mM	Konno (1988a)
Erwinia carotovora			90	8·5		Moran *et al.* (1968a)
Erwinia chrysanthemi[d]	30 000	9·4		9·8		Garibaldi & Bateman (1971); Pupillo *et al.* (1976)
	36 000	4·6		8·2		
Erwinia chrysanthemi		9·4	320	9·0		Basham (1974)
Erwinia rubrifaciens	41 000	6·25	450	9·5	5·0	Gardner & Kado (1976)
Pseudomonas fluorescens	42 300	10·3	956	9·4	0·10	Rombouts *et al.* (1978)
Streptomyces fradiae			176	9·1		Sato & Kaji (1975)
nitrosporeus	41 000	4·6		10·0		Sato & Kaji (1980)
Xanthomonas campestris			1 050	9·5		Nasuno & Starr (1967)
Fungi						
Cephalosporium sp.			364	9·9	0·018	Atallah & Nagel (1977)
Hypomyces solani[d]	32 400	10·2		8·5		Hancock (1976)
	42 000	10·5				

[a] From Rombouts & Pilnik (1980) with permission except for second set of data for *Erwinia aroideae*.
[b] Molecular weight determined by gel filtration.
[c] Determined as polygalacturonase with DP of 52.
[d] These enzymes have multiple molecular forms.

endopolygalacturonate lyases slowly degrade trigalacturonate and unsaturated trigalacturonate, but the endopolygalacturonate lyases of a *Bacillus* sp. (Hasegawa & Nagel, 1966; Nagel & Hasegawa, 1967b) and of *Bacillus pumilus* (Davé & Vaughn, 1971) do not attack these substrates. On this basis, Rexová-Benková & Markovič (1976) proposed two different action patterns for endopolygalacturonate lyases.

Because of the requirement for Ca^{2+}, chelating agents such as EDTA are inhibitors of endopolygalacturonate lyases. With tetragalacturonate as substrate, Atallah & Nagel (1977) showed that the true substrate for endopolygalacturonate lyases is calcium tetragalacturonate and that the free tetragalacturonate anion is a competitive inhibitor.

3.8 Exopolygalacturonate Lyases

Exopolygalacturonate lyases (poly-1,4-α-D-galacturonide lyase, EC 4.2.2.9; exopectate lyase) have been found only in a few bacteria [*Clostridium multifermentans* (Macmillan & Vaughn, 1964; Macmillan *et al.*, 1964; Macmillan & Phaff, 1966; Lee & Macmillan 1970*b*; Miller & Macmillan, 1970); *Erwinia aroideae* (Okamoto *et al.*, 1963, 1964); *Erwinia dissolvens* (Castelien & Pilnik, 1976); *Erwinia* sp. (Hatanaka & Ozawa, 1972, 1973); *Fusarium culmorum* (Urbanek *et al.*, 1976); and *Streptomyces nitrosporeus* (Sato & Kaji, 1977*a, b*)]. (See Table 3.) The preferred substrate is polygalacturonate (eqn 8).

Pectate lyases

$$\text{Polygalacturonate} \xrightarrow[\text{lyase}]{\text{exopectate}} \Delta 4{:}5 \text{ unsaturated galacturonates} \qquad (8)$$

Polymethylgalacturonate-methyl glycoside is not a substrate. The products are generally unsaturated $\Delta 4{:}5$ digalacturonates split from the reducing end of the substrate (the enzyme from *Streptomyces nitrosporeus* produces unsaturated trigalacturonates; Sato & Kaji, 1977*b*). The pH optima are between 8·0 and 9·5. Calcium ions are an absolute requirement for all reported enzymes except for the *Erwinia* spp. enzymes. The enzymes do not have a preference for size since polygalacturonates and oligogalacturonates, including trigalacturonate, are split at about the same rate. The exopolygalacturonate lyase from *C. multifermentans* is complexed (single protein of 400 000 daltons) with a pectin methylesterase. They work in concert to degrade pectins (Miller & Macmillan, 1970).

Activity is assayed as described for the endopolygalacturonate lyases. Viscosity reduction is not as sensitive as for endopolygalacturonate lyases; however, it may be useful in distinguishing between the two. The *C. multifermentans* exopolygalacturonate lyase converted 22% of a 0·5% polygalacturonate substrate solution to unsaturated dimer before the viscosity was reduced to 50% (Macmillan & Vaughn, 1964; Macmillan & Phaff, 1966).

3.9 Oligogalacturonate Lyases

The oligogalacturonate lyases (oligogalacturonide lyases, EC 4.2.2.6) act on oligogalacturonates and unsaturated oligogalacturonates, by *trans* elimination to remove unsaturated monomers from the reducing end of the substrate. The enzymes from *Erwinia carotovora* and *Er. aroideae* have preference for the unsaturated dimer as substrate, while an enzyme from a *Pseudomonas* sp. has preference for tetragalacturonate (Table 10). The pH optima are near 7.

Activity of these enzymes can best be followed by change in absorbance at 235 nm due to formation of a $\Delta 4:5$ unsaturated bond in the product. The reaction can also be followed by measuring the reducing groups formed (Somogyi, 1952; Bernfeld, 1955). The ratio of reducing groups formed to double bonds formed should be 1·0. Otherwise, there is reason to suspect contaminating enzymes that can use these substrates.

3.10 Endopolymethylgalacturonate Lyases

Endopolymethylgalacturonate lyases [poly(methoxygalacturonide)lyase, EC 4.2.2.10; endopectin lyase] have maximum activity on pectins versus pectates. They degrade pectin in a random fashion to the end product tetramethylgalacturonate [trimethylgalacturonate in case of the enzymes from *Aspergillus fonsecaeus* (Edstrom & Phaff, 1964*b*) and *Aspergillus niger* (Voragen, 1972; Van Houdenhoven, 1975)].

The reaction catalyzed by endopolymethylgalacturonate lyases is shown in eqn (9).

$$\text{Polymethylgalacturonate} \xrightarrow[\text{galacturonate lyase}]{\text{endopolymethyl-}} \text{unsaturated methyloligogalacturonates} \quad (9)$$

The preferred method of assay is to measure the increase in absorbance at 235 nm due to formation of the $\Delta 4:5$ double bonds produced at the non-reducing ends of the unsaturated oligomethylgalacturonate products. The ε_m is $5·5 \times 10^3$ M^{-1} cm^{-1} for unsaturated oligomethylgalacturonates (Edstrom & Phaff, 1964*a*). Reducing group methods (Somogyi, 1952; Bernfeld 1955) are also useful methods especially to test for purity of the enzyme, since the ratio of reducing groups formed to double bonds formed should be 1·0. Viscosity reduction may also be a useful method to determine whether the enzyme is an endo- or exo-splitting enzyme. Determination of the products formed early in the reaction is most useful. At 27·5% degradation of 95% esterified pectin, the highly purified endopolymethylgalacturonate lyase from *Aspergillus fonsecaeus* produced a series of unsaturated oligomethylgalacturonates with chain length of 2–8 residues (Edstrom & Phaff, 1964*a*, *b*).

Highly esterified pectins are the best substrates; pectate, pectic acid amide and the glycylester of pectin were not degraded (Voragen *et al.*, 1971; Voragen, 1972; Pilnik *et al.*, 1973, 1974). The pectate lyases do not have an absolute requirement for cations, except for the *Fusarium solani* endopolymethylgalacturonate lyase which requires calcium (Bateman, 1966). However, most of the enzymes are stimulated by Ca^{2+} and other cations, probably as a result of masking the negative charges in incompletely

Table 12
Properties of Some Endopolymethylgalacturonate Lyases[a]

Source of enzyme	M_r	pI	Specific activity (U/mg protein)	pH optimum	K_m for pectin (mg/ml)	Reference
Fungi						
Alternaria mali I	28 000		176	8·7		Hasui *et al.* (1976)
II	31 000		577	8·2		
Aspergillus fonsecaeus			19	5·2		Edstrom & Phaff (1964a, b)
Aspergillus japonicus	32 000	7·7	355	6·0		Ishii & Yokotsuka (1975)
Aspergillus niger		3·5	24	5·2		Albersheim & Killias (1962)
Aspergillus niger		3·5		5·9	2·2	Amado (1970)
Aspergillus niger I[b]	35 400	3·65	17	6·0	5·0	Van Houdenhoven (1975)
II[b]	33 100	3·75	44	6·0	0·9	
Aspergillus sojae	32 000		77	5·5		Ishii & Yokotsuka (1972a)
Dothidea ribesia	31 200	8·9		8·4	3·2	Knobel & Neukom (1974)
Phoma medicaginis var. pinodella	29 500[c]		669	7·5		Pitt (1988)
Bacteria						
Erwinia aroideae	30 000		400	8·1		Kamimiya *et al.* (1972, 1974)

[a] From Rombouts & Pilnik (1980) with permission, except for *Phoma medicaginis* var. *pinodella*.
[b] These enzymes are glycoproteins, with single peptides.
[c] By gel filtration. An 118 000 dalton component was also observed, which is presumed to be a polymer.

methylated pectins (Edstrom & Phaff, 1964a; Voragen, 1972; Ishii & Yokotsuka, 1972a, 1975).

The endopolymethylgalacturonate lyases are produced almost exclusively by fungi; the only known exceptions are a soft-rot pseudomonad (Ohuchi & Tominaga, 1974), a strain of *Er. carotovora* (Almengor-Hecht & Bull, 1978) and an enzyme from *Er. aroideae* (Kamimiya *et al.*, 1972, 1974).

Properties of some endopolymethylgalacturonate lyases are shown in Table 12. The enzymes all have relative molecular masses around 30 000 with isoelectric points (pI) ranging from 3·5–8·9. The majority of the enzymes have pH optima in the acid range, but four are above pH 8·0. The specific activities ranged from 17–669 units/mg of protein while the K_m for pectin ranged from 0·9–5·0 mg/ml.

4 TYPES OF PECTOLYTIC ENZYMES IN MICROORGANISMS

Despite much information on the several types of pectolytic enzymes that may be present in microorganisms and excellent methods for distinguishing among the enzymes, the majority of reports in the literature are not conclusive and/or definitive (See Table 3). For example, since 1981, there are 11 reports of polymethylgalacturonase (Table 3), despite the reservations of two of the major research groups (Rombouts & Pilnik, 1972; Macmillan & Sheiman,

1974) that this enzyme really exists. They pointed out the difficulty of determining whether polymethylgalacturonase is present in the same extract with pectin methylesterase and polymethylgalacturonate lyase. The second issue is determining whether an enzyme is an endo- or exo-splitting enzyme and the third is whether both hydrolase and lyase activities are present. Clearly, some purification of an extract is necessary to permit definitive answers. Also, proper substrate selection is essential. Isoenzymes of most of the enzymes make determination of the type of enzyme even more difficult.

4.1 Partial Purification Methods

At least two methods can be used to give rapid partial separation of the pectolytic enzymes in extracts. These are isoelectric focusing and chromatographic methods.

Cruickshank & Wade (1980) developed a method for simultaneously detecting the isozymes of pectin methylesterase, polygalacturonase and pectin lyase in pectin–acrylamide gels. The electrophoresis system consists of citrus pectin incorporated into a polyacrylamide gel and use of a discontinuous buffer system. Ampholytes may also be used. On completion, the pH is changed

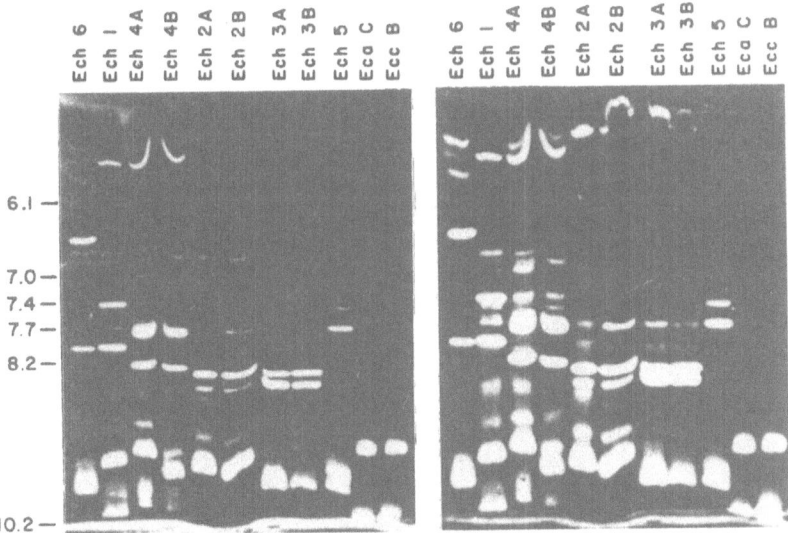

Fig. 4. Isoelectric focusing (IEF) profiles of extracellular PL isozymes produced by strains of *Er. chrysanthemi* (Ech strains), *Er. carotovora* subsp. *atroseptica* (Eca strain), and *Er. carotovora* subsp. *carotovora* (Ecc strain). PL isozymes in concentrated culture supernatant samples were activity stained with a pectate agarose overlay following IEF. The IEF gel was incubated for 8 min with one overlay (left panel) and then for 20 min with another overlay (right panel). The positions of the pI markers (carbonic anhydrase, horse myoglobin minor and major bands, whale myoglobin minor and major bands, and cytochrome *c*; FMC Corp.) are shown. From Ried & Collmer (1986) with permission.

gradually from 8·7 to 3·0, allowing sufficient time for the enzymes to act. After washing, the gel is stained with ruthenium red. Pectin methylesterases show up as red bands, polygalacturonases as white bands and lyases as yellow bands. The utility of the method has been amply demonstrated in studies on the *Botrytis cinerea* enzymes (Di Lenna & Fielding, 1983; Ried & Collmer, 1986; Leone & Van Den Heuvel, 1987). Some typical results from the research of Ried and Collmer (1986) are shown in Fig. 4.

Numerous variations of the method of Cruickshank & Wade (1980) are possible. Calcium ions can be incorporated to prevent the activity of polygalacturonases on polygalacturonate or fully methylated pectin can be used in place of citrus pectin. Nitrocellulose blotting of the gel after pI separation can be used, permitting multiple blotting and analyses. The isoelectric focused gel can be overlaid with an agarose–pectate overlay (or other substrate; Ried & Collmer, 1986), sliced, eluted and enzyme activity determined quantitatively (Di Lenna & Fielding, 1983) or regular polyacrylamide gel electrophoresis used (Martinez *et al.*, 1988).

Partial separation of the enzymes by gel filtration or ion-exchange chromatography before determination of the activities can also be used (Martinez *et al.*, 1988). By use of HPLC or Fast Protein Liquid Chromatography (FPLC), separation can be achieved in a few minutes (Rexová-Benková *et al.*, 1982; Mikeš & Rexová-Benková, 1988).

4.2   Endo- Versus Exo-splitting Enzymes

Voragen & Pilnik (1969) suggested that rate of viscosity reduction, coupled with rate of increase in reducing groups (cleaved glycosidic bonds), could be

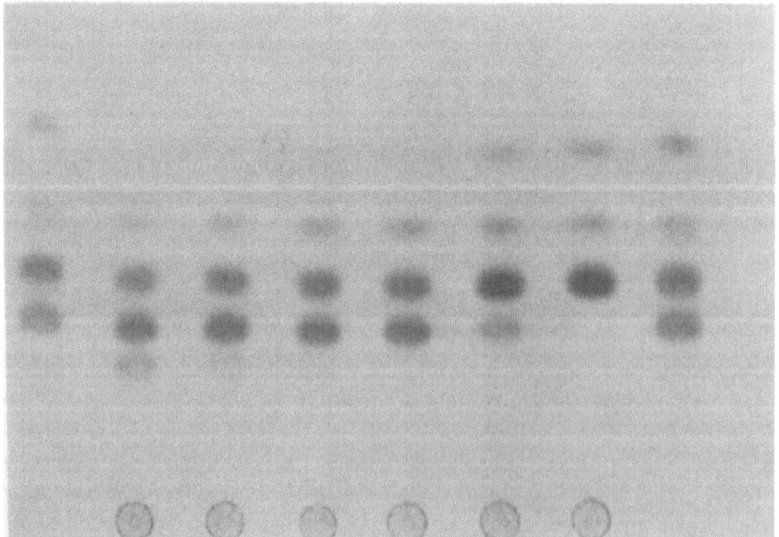

Fig. 5. Thin-layer chromatogram of oligogalacturonates formed by *Candida kefyr* endopolygalacturonase as a function of increasing time (left to right). Times of incubation were: 15, 30, 45, 60 min, 6 h, 30 h. From Call *et al.* (1985) with permission.

used to distinguish between endo- and exo-splitting enzymes. It was noted that a 50% reduction in viscosity accompanied by only 2–3% bond hydrolysis indicated an endo-enzyme while 10% or more hydrolysis indicated an exo-enzyme. Further studies indicated this must be interpreted cautiously (Rombouts & Pilnik, 1980). Other methods must also be used to confirm this. Thin layer chromatography may be used (Call *et al.*, 1985) or HPLC using a gel filtration column is also useful (Carunchio *et al.*, 1988). In these methods, samples are taken at various times near the beginning of the reaction and the products determined. Only end products from starting substrate is indicative of exo-enzymes while initial intermediate products formation with a slower formation of final products is indicative of endo-enzymes (Fig. 5). Obviously, determination of endo- versus exo-splitting cannot be done when both types of enzymes are present.

4.3 Pectate- Versus Pectin-hydrolyzing Enzymes

In determining whether a pectolytic enzyme is a polygalacturonate-hydrolyzing enzyme or a polymethylgalacturonase-hydrolyzing enzyme, fully demethylated polygalacturonates and fully methylated polygalacturonates should be used as substrates. Naturally-occurring pectins (with blocks of deesterified galacturon-

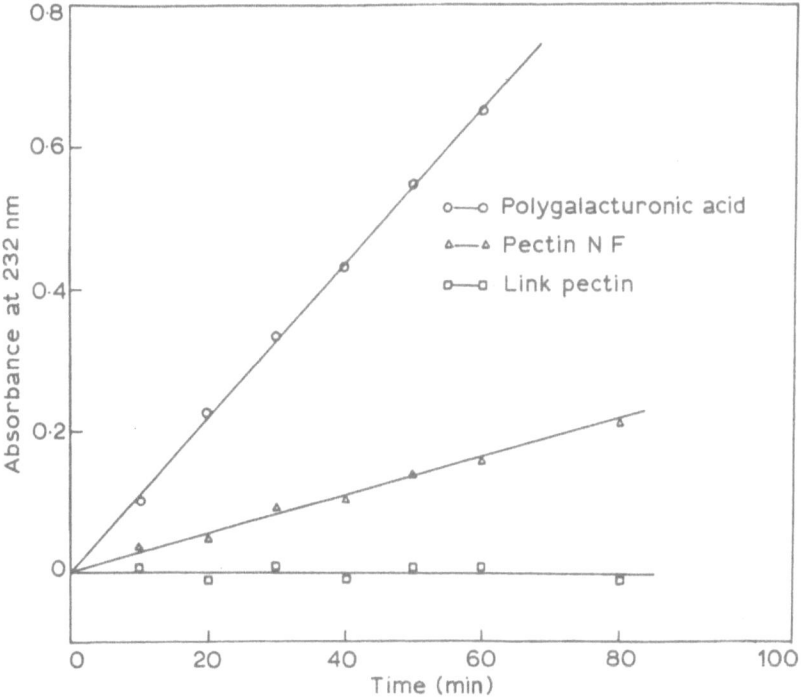

Fig. 6. Substrate specificity of pectate lyase from a thermophilic *Bacillus* sp. Substrates were: O, polygalacturonate; △, pectin NF; □, link pectin (95·5% methylated). From Karbassi & Luh (1979) with permission.

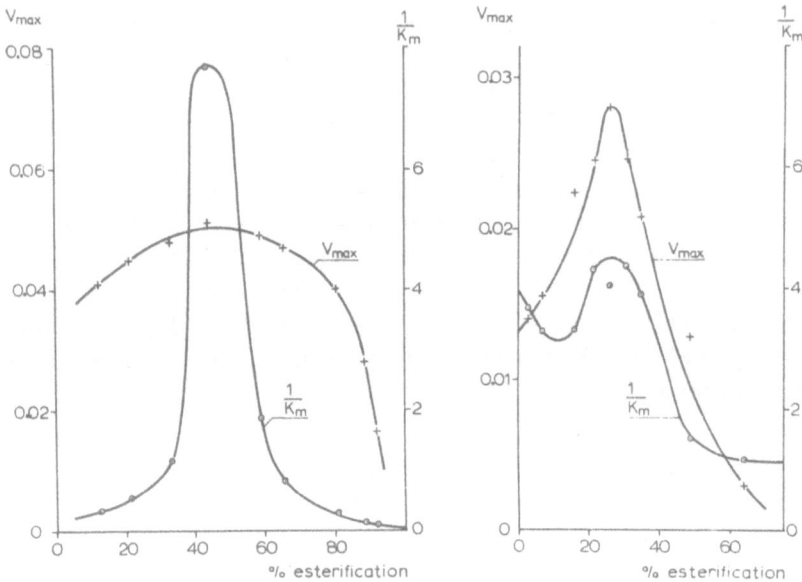

Fig. 7. Influence of degree of esterification of pectin substrate on V_{max} and $1/K_m$ of pectate lyase of *Arthrobacter* 370 (*left*) and *Bacillus polymyxa* (*right*). From Rombouts (1972) with permission.

ate units or pectinic acids) are not suitable. This is illustrated in Figs 6 and 7. Karbassi & Luh (1979) showed that pectin NF could not distinguish whether a *Bacillus* sp. enzyme was a polygalacturonate lyase or a polymethylgalacturonate lyase (Fig. 6). However, use of polygalacturonate and polymethylgalacturonate methyl glycoside (Link pectin, 95·5% esterified) clearly indicated that the enzyme was a polygalacturonate lyase. The difficulty of using naturally-occurring pectins is further illustrated by the data in Fig. 7. Rombouts (1972) showed the polygalacturonate lyases (pectate lyases) of *Arthrobacter* 370 and 547 and *Bacillus polymyxa* have activity on partially esterified substrates, up to ~90% esterification.

4.4 Hydrolase Versus Lyase Activities on Pectate and Pectin

The presence of lyase activity can be readily determined by observing rate of increase of absorbance at 235 nm as discussed above. However, the presence or absence of a hydrolase cannot be determined uniquely in microbial extracts by measuring the rate of increase in reducing groups only. Lyases also produce one reducing group per double bond formed. Therefore, the relative importance of hydrolases and lyases in glycosidic bond splitting can be determined from the ratio of molar concentration of reducing groups formed to molar concentration of double bonds formed during the initial reaction (preferably). If the ratio is 1·0, only lyases are present. If the ratio is 2·0, the hydrolase and lyase activities are equal.

5 NATURALLY-OCCURRING INHIBITORS OF PECTOLYTIC ENZYMES

Fruits and vegetables have been shown to contain proteins which are inhibitors of polygalacturonases. These protein inhibitors should be distinguished from nonspecific phenolic inhibition (Cole & Wood, 1961; Byrde & Archer, 1977). Albersheim & Anderson (1971) purified a protein of 50 000 daltons from red kidney bean (*Phaseolus vulgaris*) that inhibited polygalacturonases secreted by plant pathogens. Chemical and physical properties of the protein were similar to lectins of the bean.

Subsequently, the protein was shown to inhibit the endopolygalacturonases secreted by *Colletotrichum lindemuthianum* and *Aspergillus niger* (Fisher *et al.*, 1973). More recently, Fielding (1981) reported that several plant pathogens elicit production of inhibitors of fungal polygalacturonases. Presence of the inhibitor in peach and plum tissue infected with *Monilinia fructigena* or *M. laxa* was detected because of noticeable increase in polygalacturonase activity on fractionation of extracts where little enzyme activity existed before. The inhibitor is a protein with M_r of 15 000 and is completely inactivated by heating at 90–100°C for 20 min. It was not inhibitory of polygalacturonase preparations from *Rhizopus sexualis*, *R. stolonifer*, *Mucor mucedo*, *Phytophthora infestans* and *Penicillium* sp. Apple tissue treated with *Monilinia fructigena*, *M. laxa*, *Nectria galligena* and *Botrytis cinerea* elicited inhibitors but *Pezicula malicorticis* and *Penicillium* sp. did not elicit inhibitors in apple tissue. *M. fructigena* and *M. laxa* also elicited inhibitors in pear and peach tissue.

Whether the inhibitor is produced by the fruit or by the microorganisms has not been established, although Fielding (1981) favored a host origin. The presence of naturally-occurring inhibitors should not be overlooked in analyzing crude extracts of pectolytic enzymes.

6 MOLECULAR BIOLOGY OF PECTOLYTIC ENZYMES

At least eight research groups, based on publications, are actively working on the molecular biology of the pectolytic enzymes; three patents have been filed (Heim *et al.*, 1988; Takahashi *et al.*, 1988; Tsuyunaski *et al.*, 1988). The degradative pathway of pectin and polygalacturonate, including the enzymes and the genes which code for some of the enzymes are well known in *Erwinia chrysanthemi*. This pathway is shown in Fig. 8 (Hugouvieux-Cotte-Pattat & Robert-Baudouy, 1985).

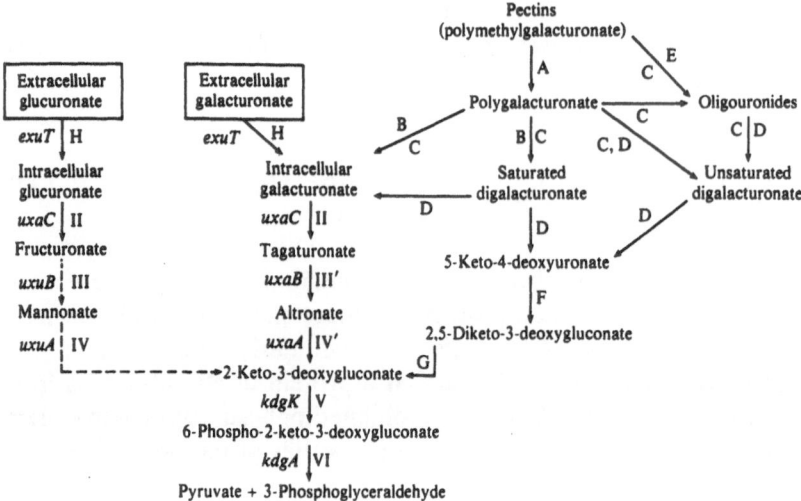

Fig. 8. Degradative pathway of pectin and polygalacturonate in *Erwinia chrysanthemi*. The different steps are catalyzed by the following enzymes: A, pectin methylesterase (EC 3.1.1.11); B, polygalacturonase (EC 3.2.1.15); C and E, polygalacturonate and pectin transeliminase (EC 4.2.2.2); D, oligouronide transeliminase (EC 4.2.2.6); F, 5-keto-4-deoxyuronate isomerase (EC 5.3.1.17); G, 2-keto-3-deoxygluconate oxidore-ductase (EC 1.1.1.127); H, hexuronate transport system; II, uronate isomerase (EC 5.3.1.12); III, mannonate oxidoreductase (EC 1.1.1.57); IV, mannonate hydrolyase (EC 4.2.1.8); III', altronate oxidoreductase (EC 1.1.1.58); IV', altronate hydrolyase (EC 4.2.1.7); V, 2-keto-3-deoxygluconate kinase (EC 2.7.1.45); VI, 2-keto-3-deoxy-6-phosphogluconate aldolase (EC 4.1.2.14). The gene symbols to the left of the arrows denote the corresponding structural genes. From Hugouvieux-Cotte-Pattat & Robert-Baudouy (1985) with permission.

The Robert-Baudouy group has chosen to work on modifying the genetic makeup of *Erwinia chrysanthemi* (Reverchon *et al.*, 1985; Condemine & Robert-Baudouy, 1987). Mutants of *Er. chrysanthemi* containing genetic fusions of the *kdgK* and *kdgA* genes (see Fig. 8) to the *lac Z* gene of *E. coli* were isolated by infection of a *lac Z* mutant of *Er. chrysanthemi* with the phage Mu d(Ap *lac*). In these fusion strains, absence of growth on galacturonate, glucuronate and polygalacturonate, and also β-galacturonase expression, are caused by a single Mu d(Ap *lac*) insertion (Hugouvieux-Cotte-Pattet & Robert-Baudouy, 1985). Reverchon *et al.* (1985) constructed a genetic library from *Sau*3A-digested *Er. chrysanthemi* B374 DNA cloned in the *Bam*HI site of the broad-host-range cosmid pMMB33 grown in *E. coli*. Genes coding for three pectate lyases were localized in a 2·7-kb fragment (pPL03). The pectate lyases were expressed at various levels. The enzymes accumulated in the periplasmic space of the *E. coli* host, rather than being excreted into the culture medium as *Er. chrysanthemi* does. Condemine & Robert-Baudouy (1987) inserted Tn5 into *Er. chrysanthemi* with the transposon vector pM0194. The Tn5 was inserted into the regulatory gene, *kdgR*, resulting in constitutive expression of the genes of the polygalacturonate degradative pathway.

Some other researchers have chosen to use *E. carotovora* as the organism for molecular biology studies. Handa *et al.* (1987) studied the genes affecting virulence of *Er. carotovora* by transposon mutagenesis. They selected mutants based on the inability to cause soft rot on the cut surface of potato tubers. Class 1 mutants were unable to secrete polygalacturonase and pectate lyase. Class 2 mutants were still able to synthesize and secrete pectolytic enzymes. Class 3 mutants were blocked in biosynthesis of pyrimidines and purines. The Class 4 mutants were not able to utilize various carbon sources and were defective in synthesis of UDP-glucose pyrophosphorylase. Takahashi *et al.* (1988) cloned a DNA sequence composed of a promoter, a ribosome-binding region and a signal sequence region from the pectate lyase operon of *Er. carotovora*. *Er. carotovora* transformed with plasmid pNN101 comprising the pectate lyase operon secreted 20-fold more pectate lyase into the medium than did the wild type. Hu *et al.* (1987) also produced a clone containing the gene encoding a pectolytic enzyme of *Er. carotovora* subsp. *carotovora*. The gene was located on a 3·2-kb DNA fragment flanked by a *Bg1II* restriction enzyme site and a *HindII* restriction enzyme site. Via mini-Mud*lac* mutagenesis, a promoter site was located within the gene between the *Bg1II* site and the *EcoRI* site. Inserted into *E. coli*, the system produced a 78,000 Mv protein with pectolytic activity. The enzyme(s) did not macerate potato slices and biosynthesis was not regulated by sodium polypectate.

Heim *et al.* (1988) cloned and sequenced the pectin lyase I gene of *A. niger*. The gene expression system included *A. niger pyrA* mutant, a plasmid containing the pectin lyase I promoter, signal sequence, a terminator (optional) and a plasmid containing the *pyrA* gene. A desulfatohiruden gene was inserted as a marker. The *A. niger* mutant (An 8) was transformed with pLHL5 and pCG5907 and the transformants grown in pectin-containing medium. Desulfatohiruden was excreted into the medium.

Yu *et al.* (1988) cloned the pectate lyase gene from alkali-tolerant *Bacillus* sp. YA-14 into *E. coli* MB1000 by inserting *HindIII*-generated DNA fragments into the *HindIII* site of pBR322 and then screening recombinant transformants for the ability to hydrolyze sodium polypectate on an agar plate. The recombinant plasmid, pYPC29, was isolated and the size of the cloned *HindIII* fragment was determined to be 1·6-kb. The pectate lyase gene was stably maintained and expressed efficiently in *E. coli*. The pectate lyase accumulated largely in the periplasmic space of the *E. coli* clones.

7 METABOLIC REGULATION OF PECTOLYTIC ENZYME BIOSYNTHESIS

Researchers continue to look for ways of changing the culture media in order to give higher yields or higher specificity to pectolytic enzyme products. Much of the effort is focused on the effect of carbon sources; if the enzyme is constitutive, the cheapest source of carbon could be used. If on the other hand, pectin or pectate gives preferential enhancement this may be an advantage. Leon & Van Den Heuvel (1987) found that the carbon source and concentration had a major effect on the total activity and the types of enzymes detected, as well as on the timing at which the enzymes were produced by *Botrytis cinerea.* Martinez *et al.* (1988) found that pectin did not influence maximum pectolytic enzyme activity in autolyzed cultures of *B. cinerea* but maximal activity was reached earlier in cultures with pectin. Carbon source, including pectin or pectate, were reported to have no effect on *Aspergillus foetidus* (Astapovich *et al.*, 1984), *Corynebacterium michiganense* (E. F. Smith) Jensen (Scala & Zoina, 1983*b*), *Erwinia* (545 strains) (Lobanok *et al.*, 1981) and *Penicillium digitatum* (Mikhailova *et al.*, 1981). Therefore, the researchers considered the enzymes to be constitutive.

On the other hand, there are a number of examples where the carbon source was quite important. *Athelia* (*Sclerotium*) *rolfii* produced large amounts of polygalacturonase in liquid culture containing pectin or autoclaved bean hypocotyls as a carbon source (Scala & Zoina, 1983*a*). Glucose gave variable levels of enzyme. *Alternaria brassicae* produced maximum polygalacturonase with glucose as carbon source and less with pectate and pectin. Pectin methylgalacturonase and polygalacturonase lyase production were maximum with pectate as substrate, while pectin methylgalacturonase lyase production was maximum when either pectate or pectin was used. Glucose was less effective for the last three enzymes (Suri & Mandahar, 1982). Pectate was the best carbon source for polygalacturonase and polymethylgalacturonase production by *Aspergillus niger*; pectin as carbon source completely inhibited production of these two enzymes (Chopra & Mehta, 1985).

Dube & Bordia (1982*a*, *b*) reported marked effects of carbon source on pectolytic enzyme production by *Helminthosporium sacchari*. The organism produced pectate lyase in glucose and pectate media and polygalacturonase and pectin methylesterase in pectin-containing media. Galactose repressed pectate lyase formation and induced polygalacturonase. With fructose, sorbose xylose and cellobiose pectin lyase production was favored, while with lactose pectate lyase was favored. With *Myrothecium roridum,* pectin methylesterase and polygalacturonase production was maximum on pectin-containing media. Pectin lyase and pectate lyase were produced only in pectin-containing media of older cultures following 12–16 day incubations (Reddy & Reddy, 1983).

Alternaria brassicola produced endo- and exopolygalacturonases, pectin methylgalacturonase, pectin lyase and pectin methylesterase with 20% glucose as carbon source (Mandahar & Suri, 1981). When pectin or pectate was used as sole carbon source, only pectate lyase was induced; exopolygalacturonase production was inhibited by pectin but stimulated by pectate. Both pectin and pectate inhibited pectin methylgalacturonase production but stimulated pectin lyase production. Pectin methylesterase production was stimulated by pectin but inhibited by pectate. With *Xanthomonas campestris*, pectin induced polymethylgalacturonase and pectate induced polygalacturonase and polygalacturonase lyase (Pandey & Prasad, 1983). Beech bark was a better inducer of pectolytic enzymes in *Nectria ditissima* Tu1 than were other carbon sources (Perrin, 1984).

Pectolytic enzyme synthesis by *Sclerotina sclerotiorum* was stimulated by addition of histidine, glutamate, alanine, asparagine, aspartate, glutamine, arginine and proline to the media. Methionine, isoleucine, tyrosine, phenylala-nine, leucine and serine had no effect while valine and cysteine were inhibitory (Astapovich & Rozhkova, 1983). The highest stimulatory effect was with 0·02% histidine. Chopra & Mehta (1985) reported that cystine inhibited polygalacturonase and polymethylgalacturonase production completely in *Aspergillus niger*. *Penicillium crustosum* Thom produced protopectinase, polygalacturonase, 'pectin depolymerase' and pectin methylesterase when 0·2% asparagine was added to the 1% pectin-containing media (Awasthi & Mishra, 1982).

Maximum growth of *Penicillium digitatum* occurred at 20–25°C; however, maximum enzyme synthesis was at 15°C (Mikhailova *et al.*, 1981). Optimal enzyme production occurred at 28°C for *Saccharomycopis fibuliger* (Fellows & Worgan, 1984). Mitomycin C added at 0·5 μg/ml greatly increased the level (to 190 units/ml) of pectin lyase produced by *Pseudomonas marginalis* (Sone *et al.*, 1988). Increased buffer capacity (60 g/litre KH_2PO_4 versus 5 g/litre normally) and longer incubation time (144–169 h versus 72–96 h) increased the yield of polygalacturonase from 200–300 units to 600–700 units/ml for *Aspergillus niger* production (Friese *et al.*, 1987). Under submerged culture production of pectolytic enzymes by *Aspergillus niger* 71, phosphate buffer were best (Fiedurek, 1983). Polygalacturonase and pectin methylesterase production was best at pH 4·0–4·5 and best conditions for pectin lyase and pectase lyase were at pH 7·0 and 8·0, respectively.

REFERENCES

Adisa, V. A. (1985). *Mycopathologia*, **91**, 101.
Aizenberg, V. L. (1983). *Mikrobiologichnii Zhurnal* (*Kiev*), **45**, 99. *Chemical Abstracts*, **99**, 136625j (1983).
Albersheim, P. & Anderson, A. J. (1971). *Proceedings of the National Academy of Sciences, USA*, **68**, 1815.
Albersheim, P. & Killias, U. (1962). *Archives of Biochemistry and Biophysics*, **97**, 107.
Albersheim, P., Neukom, H. & Deuel, H. (1960). *Archives of Biochemistry and Biophysics*, **90**, 46.
Almengor-Hecht, M. L. & Bull, A. T. (1978). *Archives of Microbiology*, **119**, 163.
Amadò, R. (1970). Dissertation No. 4536; Eidgenoessische Technische Hochschule, Zürich.
Aspinall, G. O. (1969). *Advances in Carbohydrate Chemistry and Biochemistry*, **24**, 333.
Aspinall, G. O. (1980). In *The Biochemistry of Plant Carbohydrates*: *Structure and Function*, Vol. 3 (ed. J. Preiss). Academic Press, New York, p. 473.
Aspinall, G. O. (1982). In *Food Carbohydrates*, eds D. R. Lineback & G. E. Inglett. Avi Publ. Co., Connecticut, p. 356.

Astapovich, N. I. & Rozhkova, Z. A. (1983). *Vestsi Akademii Navuk BSSR, Seryya Biyalagichnykh Navuk,* 49. *Chemical Abstracts,* **98,** 122594r (1983).

Astopovich, N. I., Kondratéva, L. V., Rayabaya, N. E. & Grel, M. V. (1984). *Vestsi Akademii Navuk BSSR, Seryya Biyalagichnykh Navuk,* 64. *Chemical Abstracts,* **100,** 117643a (1984).

Atallah, M. T. & Nagel, C. W. (1977). *Journal of Food Biochemistry,* **1,** 185.

Awasthi, P. B. & Mishra, U. S. (1982). *Science and Culture,* **48,** 114.

Bartley, I. M. (1978). *Phytochemistry,* **17,** 213.

Basham, H. G., III (1974). Injury of plant cells by pectic enzymes. PhD Thesis, Cornell Univ.

Bateman, D. F. (1966). *Phytopathology,* **56,** 238.

Bateman, D. F. & Millar, R. L. (1966). *Annual Review of Phytopathology,* **4,** 119.

Bateman, D. F., Kosuge, T. & Kilgore, W. W. (1970). *Archives of Biochemistry and Biophysics,* **136,** 97.

BeMiller, J. N. (1986). In *Chemistry and Function of Pectins,* ed. M. L. Fishman & J. J. Jen. *American Chemical Society Symposium Series,* **310,** 2.

Bernfeld, P. (1955). *Methods in Enzymology,* **1,** 149.

Blieva, R. K. & Rodionova, N. A. (1987). *Prikladnaya Biokhimiya i Mikrobiologiya,* **23,** 561.

Brady, C. J. (1976). *Australian Journal of Plant Physiology,* **3,** 163.

Bravova, G. B. & Kalunyants, K. A. (1984). *Prikladnaya Biokhimiya i Mikrobiologiya,* **20,** 69. *Chemical Abstracts,* **100,** 116977m (1984).

Byrde, R. J. W. & Archer, S. A. (1977). In *Cell Wall Biochemistry* eds B. Solheim & J. Rae. Tromsø Univ., Norway, p. 213.

Call, H. P., Harding, M. & Emeis, C. C. (1985). *Journal of Food Biochemistry,* **9,** 193.

Carunchio, V., Girelli, A. M., Sinibaldi, M. & Tarola, A. M. (1988). *Chromatographia,* **25,** 870.

Castelein, J. M. & Pilnik, W. (1976). *Lebensmittel-Wissenschaft und Technologie,* **9,** 277.

Cervone, F., Scala, A., Foresti, M., Cacace, M. G. & Noviello, C. (1977). *Biochimica et Biophysica Acta,* **482,** 329.

Cervone, F., Scala, A. & Scala, F. (1978). *Plant Pathology,* **12,** 19.

Chesson, A. & Codner, R. C. (1978). *Journal of Applied Bacteriology,* **44,** 347.

Chopra, S. & Mehta, P. (1985). *Folia Microbiologica* (Prague), **30,** 117. *Chemical Abstracts,* **103,** 344d (1985).

Cole, M. & Wood, R. K. S. (1961). *Annals of Botany,* **25,** 435.

Condemine, G. & Robert-Baudouy, J. (1987). *Federation of European Microbiological Societies Microbiological Letters,* **42,** 39.

Cooke, R. D., Ferber, E. M. & Kanagasabapathy, H. (1976). *Biochimica et Biophysica Acta,* **452,** 440.

Courtois, J.-E., Percheron, F. & Foglietti, M.-J. (1968). *Comptes Rendus de l'Academie des Sciences Paris, Serie D,* **266,** 164.

Cruickshank, R. H. & Wade, G. C. (1980). *Analytical Biochemistry,* **107,** 177.

Davé, B. A. & Vaughn, R. H. (1971). *Journal of Bacteriology,* **108,** 166.

Delincée, H. (1976). *Phytochemistry,* **15,** 903.

Delincée, H. & Radola, B. J. (1970). *Biochimica et Biophysica Acta,* **214,** 178.

Demain, A. L. & Phaff, H. J. (1954). *Journal of Biological Chemistry,* **210,** 381.

Demain, A. L. & Phaff, H. J. (1957). *Wallerstein Laboratory Communications,* **20,** 119.

Deuel, H. & Stutz, E. (1958). *Advances in Enzymology,* **20,** 341.

De Vries, J. A., Rombouts, F. M., Voragen, A. G. J. & Pilnik, W. (1983). *Carbohydrate Polymers,* **3,** 245.

Di Lenna, P. & Fielding, A. H. (1983). *Journal of General Microbiology,* **129,** 3015.

Dore, W. H., Brinton, C. S., Wichmann, H. J., Willaman, J. J. & Wilson, C. P. (1927). *Journal of the American Chemical Society,* **49,** 37.

Dube, H. C. & Bordia, S. (1982a). *Indian Phytopathology*, **35**, 115.
Dube, H. C. & Bordia, S. (1982b). *Indian Phytopathology*, **35**, 434.
Edstrom, R. D. & Phaff, H. J. (1964a). *Journal of Biological Chemistry*, **239**, 2403.
Edstrom, R. D. & Phaff, H. J. (1964b). *Journal of Biological Chemistry*, **239**, 2409.
Endo, A. (1964a). *Agricultural and Biological Chemistry*, **28**, 639.
Endo, A. (1964b). *Agricultural and Biological Chemistry*, **28**, 757.
Fanelli, C., Cacace, M. G. & Cervone, F. (1978). *Journal of General Microbiology*, **104**, 305.
Fellows, P. J. & Worgan, J. T. (1984). *Enzyme and Microbial Technology*, **6**, 405.
Fiedurek, J. (1983). *Annals Universitatis Mariae Curie-Sklodowska, Section C*. 1981. **36**, 37. *Chemical Abstracts*, **102**, 22767g (1985).
Fiedurek, J. & Ilczuk, Z. (1983a). *Acta Alimentaria Polonica* **9**, 89. *Chemical Abstracts*, **101**, 166843f (1984).
Fiedurek, J. & Ilczuk, Z. (1983b). *Acta Alimentaria Polonica* **9**, 101. *Chemical Abstracts*, **101**, 69038f (1984).
Fielding, A. H. (1981). *Journal of General Microbiology* **123**, 377.
Finkelman, M. J. & Zajic, J. E. (1978). *Developments in Industrial Microbiology*, **19**, 459.
Fisher, M. L., Anderson, A. J. & Albersheim, P. (1973). *Plant Physiology*, **51**, 489.
Fishman, M. L., Pepper, L., Damert, W. C., Phillips, J. G. & Barford, R. A. (1986). In *Chemistry and Function of Pectins*, eds M. L. Fishman & J. J. Jen. *American Chemical Society Symposium Series*, **310**, 22.
Fogarty, W. M. & Kelly, C. T. (1983). In *Microbial Enzymes and Biotechnology*, ed. W. M. Fogarty. Applied Science Publishers, London, p. 131.
Foglietti, M.-J., Levaditou, V., Courtois, J.-E. & Chararas, C. (1971). *Comptes Rendus des Seances de la Societe de Biologie et de Ses Filiales*, **165**, 1019.
Förster, H. (1988). *Methods in Enzymology*, **161**, 355.
Frémy, E. (1848). *Annales de Chimie et de Physique*, **24**, 5.
Friese, E., Leuchtenberger, A., Marek, E. & Van Lupin, I. (1987). German (East) DD 246,317 (June 3, 1987). *Chemical Abstracts*, **108**, 93113a (1988).
Furutani, S. & Osajima, Y. (1965). *Science Bulletin of the Faculty of Agriculture, Kyushu Univ.*, **22**, 35.
Gardner, J. M. & Kado, C. I. (1976). *Journal of Bacteriology*, **127**, 451.
Garibaldi, A. & Bateman, D. F. (1971). *Physiological Plant Pathology*, **1**, 25.
Gogeliya, I. E. & Machavariani, F. D. (1983). *Izvestiya Akademii Nauk Gruzinskoi SSR, Seriya Biologicheskaya*, **9**, 414. *Chemical Abstracts*, **100**, 119286w (1984).
Hancock, J. G. (1976). *Phytopathology*, **66**, 40.
Handa, A. K., Bressan, R. A., Lee, L., Charles, D. J., Jayaswal, R. K., Chiu, J. & Bennetzen, J. L. (1987). *Molecular Genetics of Plant-Microbe Interaction, Proceedings of International Symposium, 3rd 1986*, eds D. P. S. Verma & N. Brisson, p. 67.
Hasegawa, S. & Nagel, C. W. (1966). *Journal of Food Science*, **31**, 838.
Hasegawa, S. & Nagel, C. W. (1968). *Archives of Biochemistry and Biophysics*, **124**, 513.
Hasui, Y., Miyairi, K., Okuno, T., Sawai, K. & Sawamura, K. (1976). *Annals of the Phytopathological Society of Japan*, **42**, 228.
Hatanaka, C. & Imamura, T. (1974). *Agricultural and Biological Chemistry*, **38**, 2267.
Hatanaka, C. & Ozawa, J. (1964). *Agricultural and Biological Chemistry*, **28**, 627.
Hatanaka, C. & Ozawa, J. (1966). *Berichte des Ohara Institute fur Landwirtschaftliche Biologie, Okayama University*, **13**, 175.
Hatanaka, C. & Ozawa, J. (1969a). *Agricultural and Biological Chemistry*, **33**, 116.
Hatanaka, C. & Ozawa, J. (1969b). *Nippon Nogei Kagaku Kaishi*, **43**, 85. *Chemical Abstracts*, **70**, 103168k (1969).
Hatanaka, C. & Ozawa, J. (1969c). *Nippon Nogei Kagaku Kaishi*, **43**, 77. *Chemical Abstracts*, **70**, 103167j (1969).
Hatanaka, C. & Ozawa, J. (1970). *Agricultural and Biological Chemistry*, **34**, 1618.

Hatanaka, C. & Ozawa, J. (1971). *Agricultural and Biological Chemistry*, **35**, 1617.
Hatanaka, C. & Ozawa, J. (1972). *Agricultural and Biological Chemistry*, **36**, 2307.
Hatanaka, C. & Ozawa, J. (1973). *Agricultural and Biological Chemistry*, **37**, 593.
Heim, J., Gysler, C., Visser, J. & Kester, H. C. M. (1988). European Patent Applied EP278,355 17 August 1988. *Chemical Abstracts*, **110**, 70492a (1989).
Heinrichová, K. (1977). *Collection of Czechoslovak Chemical Communications*, **42**, 3214.
Heinrichová, K. & Rexová-Benková, L. (1977). *Collection of Czechoslovak Chemical Communications*, **42**, 2569.
Hu, N. T., Wang, Y. M., Lee, W. Y., Yang, S. M. & Tseng, Y. H. (1987). *MGG, Molecular and General Genetics*, **210**, 294.
Hugouvieux-Cotte-Pattat, N. & Robert-Baudouy, J. (1985). *Journal of General Microbiology*, **131**, 1205.
Ishii, S. & Yokotsuka, T. (1972*a*). *Agricultural and Biological Chemistry*, **36**, 146.
Ishii, S. & Yokotsuka, T. (1972*b*). *Agricultural and Biological Chemistry*, **36**, 1885.
Ishii, S. & Yokotsuka, T. (1975). *Agricultural and Biological Chemistry*, **39**, 313.
Jermyn, M. A. & Isherwood, F. A. (1956). *Biochemical Journal*, **64**, 123.
Johanssen, M. (1988). *Physiological and Molecular Plant Pathology*, **33**, 333.
Joyce, A. M. & Fogarty, W. M. (1974). *Proceedings of the Society for General Microbiology*, **2**, 10.
Joyce, A. M. & Fogarty, W. M. (1975). *Proceedings of the Society for General Microbiology*, **2**, 79.
Kamimiya, S., Izaki, K. & Takahashi, H. (1972). *Agricultural and Biological Chemistry*, **36**, 2367.
Kamimiya, S., Nishiya, T., Izaki, K. & Takahashi, H. (1974). *Agricultural and Biological Chemistry*, **38**, 1071.
Kamimiya, S., Itoh, Y., Izaki, K. & Takehashi, H. (1977). *Agricultural and Biological Chemistry*, **41**, 975.
Karbassi, A. & Luh, B. S. (1979). *Journal of Food Science*, **44**, 1156.
Karbassi, A. & Vaughn, R. H. (1980). *Canadian Journal of Microbiology*, **26**, 377.
Kelly, C. T. & Fogarty, W. M. (1978). *Journal of Microbiology*, **24**, 1164.
Kertesz, Z. (1937). *Journal of Biological Chemistry*, **121**, 589.
Khan, S. S., Hall, A. N. & Khan, M. R. (1987). *Pakistan Journal of Biochemistry*, **20**, 23. *Chemical Abstracts*, **110**, 53514S (1989).
Kimura, H., Uchino, F. & Mizushima, S. (1973). *Agricultural and Biological Chemistry*, **37**, 1209.
Knobel, H. R. & Neukom, H. (1974). *Phytopathologische Zeitschrift*, **80**, 244.
Koller, A. (1966). Dissertation No. 3774. Eidgennoessische Technische Hochschule, Zurich.
Koller, A. & Neukom, H. (1969). *European Journal of Biochemistry*, **7**, 485.
Konno, H. (1988*a*). *Methods in Enzymology*, **161**, 381.
Konno, H. (1988*b*). *Methods in Enzymology*, **161**, 373.
Konno, H., Katoh, K. & Yamasaki, Y. (1981). *Plant Cell Physiology*, **22**, 899.
Konno, H., Yamasaki, Y. & Katoh, K. (1983). *Plant Physiology*, **73**, 216.
Lee, M. & Macmillan, J. D. (1968). *Biochemistry*, **7**, 4005.
Lee, M. & Macmillan, J. D. (1970*a*). *Biochemistry*, **9**, 1930.
Lee, M. & Macmillan, J. D. (1970*b*). *Journal of Bacteriology*, **103**, 595.
Leone, G. & Van Den Heuvel, J. (1987). *Canadian Journal of Botany*, **65**, 2133.
Lim, J., Yamasaki, Y., Suzuki, Y. & Ozawa, J. (1980). *Agricultural and Biological Chemistry*, **44**, 473.
Lineweaver, H. & Ballou, G. S. (1943). *Federation Proceedings*, **2**, 66.
Liu, Y. K. & Luh, B. S. (1978). *Journal of Food Science*, **43**, 721.
Lobanok, A. G., Makmiene, B. & Kapitonova, L. S. (1981). *Mikrobiologiya Proizvodstvo*, 69. *Chemical Abstracts*, **97**, 35763f (1982).

Lobanok, A. G., Mikhailova, R. V., Oparina, A. A., Voskresenskaya, L. G., Nikulina, E. A., Selivanova, N. Yu & Pinchuk, Ya. I. (1985). *Prikladnaya Biokhimiya i Mikrobiologiya*, **21**, 219. *Chemical Abstracts*, **102**, 202544r (1985).

Luh, B. S. & Phaff, H. J. (1951). *Archives of Biochemistry and Biophysics*, **33**, 212.

MacDonnell, L. R., Jang, R., Jensen, E. F. & Lineweaver, H. (1950). *Archives of Biochemistry*, **28**, 260.

Macmillan, J. D. & Phaff, H. J. (1966). *Methods in Enzymology*, **8**, 632.

Macmillan, J. D. & Sheiman, M. I. (1974). In *Food Related Enzymes*, ed. J. R. Whitaker. *Advances in Chemistry Series*, **136**, 101.

Macmillan, J. D. & Vaughn, R. H. (1964). *Biochemistry*, **3**, 564.

Macmillan, J. D., Phaff, H. J. & Vaughn, R. H. (1964). *Biochemistry*, **3**, 572.

Mah, R. A. & Hungate, R. E. (1965). *The Journal of Protozoology*, **12**, 131.

Manabe, M. (1973). *Agricultural and Biological Chemistry*, **37**, 1487.

Manachini, P. L., Parini, C. & Fortina, M. G. (1988). *Enzyme and Microbial Technology*, **10**, 682.

Mandahar, C. L. & Suri, R. A. (1981). *Indian Journal of Physical and Natural Sciences*, **1**, 13.

Markovič, O. (1974). *Collection of Czechoslovak Chemical Communications*, **39**, 908.

Martinez, M. J., Martinez, R. & Reyes, F. (1988). *Mycopathologia*, **102**, 37.

McClendon, J. H. & Kreisher, J. H. (1963). *Analytical Biochemistry*, **5**, 295.

McNicol, L. N. & Baker, E. E. (1970). *Biochemistry*, **9**, 1017.

Merrill, R. & Weeks, M. (1945). *Journal of the American Chemical Society*, **67**, 2244.

Mikeš, O. & Rexová-Benková, L. (1988). *Methods in Enzymology*, **161**, 385.

Mikhailova, R. V., Lobanok, A. G. & Sapunova, L. I. (1981). *Mikologiya Fitopatologiya*, **15**, 491. *Chemical Abstracts*, **96**, 82459a (1982).

Mikhailova, R. V., Lobanok, A. G., Sapunova, L. I. & Oparina, A. A. (1982). *Vesti Akademii Navuk BSSR, Seryya Biyalagichnykh Navuk*, 57. *Chemical Abstracts*, **96**, 141137u (1982).

Mill, P. J. (1966*a*). *Biochemical Journal*, **99**, 557.

Mill, P. J. (1966*b*). *Biochemical Journal*, **99**, 562.

Miller, L. & Macmillan, J. D. (1970). *Journal of Bacteriology*, **102**, 72.

Miller, L. & Macmillan, J. D. (1971). *Biochemistry*, **10**, 570.

Moran, F., Nasuno, S. & Starr, M. P. (1968*a*). *Archives of Biochemistry and Biophysics*, **123**, 298.

Moran, F., Nasuno, S. & Starr, M. P. (1968*b*). *Archives of Biochemistry and Biophysics*, **125**, 734.

Nagel, C. W. & Hasegawa, S. (1967*a*). *Analytical Biochemistry*, **21**, 411.

Nagel, C. W. & Hasegawa, S. (1967*b*). *Archives of Biochemistry and Biophysics*, **118**, 590.

Nagel, C. W. & Hasegawa, S. (1968). *Journal of Food Science*, **33**, 378.

Nagel, C. W. & Wilson, T. M. (1970). *Applied Microbiology*, **20**, 374.

Nakagawa, H., Yamagawa, Y. & Takehana, H. (1970*a*). *Agricultural and Biological Chemistry*, **34**, 991.

Nakagawa, H., Yamagawa, Y. & Takehana, H. (1970*b*). *Agricultural and Biological Chemistry*, **34**, 998.

Nasuno, S. & Starr, M. P. (1966). *Journal of Biological Chemistry*, **241**, 5298.

Nasuno, S. & Starr, M. P. (1967). *Biochemical Journal*, **104**, 178.

Newbold, R. & Joslyn, M. A. (1952). *Journal of the Association of Official Agricultural Chemists*, **35**, 872.

Ohuchi, H. & Tominaga, T. (1974). *Annals of the Phytopathological Society of Japan*, **40**, 22.

Okamoto, K., Hatanaka, C. & Ozawa, J. (1963). *Agricultural and Biological Chemistry*, **27**, 596.

Okamoto, K., Hatanaka, C. & Ozawa, J. (1964). *Berichte des Ohara Institute fur Landwirtschaftliche Biologie, Okayama University* **12,** 115.
Ozawa, J. (1955). *Nogaku Kenkyo,* **42,** 157. *Chemical Abstracts,* **49,** 10405e (1955).
Pandey, P. K. & Prasad, M. (1983). *Zentralblatt fuer Mikrobiologie,* **138,** 71.
Parini, C., Fortina, M. G. & Manachi, P. L. (1988). *Journal of Applied Bacteriology,* **65,** 477.
Perrin, R. (1984). *European Journal of Forestry Pathology,* **14,** 219.
Phaff, H. J. (1966). *Methods in Enzymology,* **8,** 636.
Pilnik, W., Rombouts, F. M. & Voragen, A. G. J. (1973). *Chemie, Mikrobiologie Technologie der Lebensmittel,* **2,** 122.
Pilnik, W., Voragen, A. G. J. & Rombouts, F. M. (1974). *Lebensmittel-Wissenschaft und Technologie,* **7,** 353.
Piplani, S., Gemawat, P. D. & Prasad, N. (1981). *Indian Journal of Mycology and Plant Pathology,* **11,** 225.
Pitt, D. (1988). *Methods in Enzymology,* **161,** 350.
Plaschina, I. G., Braudo, E. E. & Tolstoguzov, V. B. (1978). *Carbohydrate Research,* **60,** 1.
Porwal, S. & Chakravarti, B. P. (1970). *Acta Phytopathologica Academic Scientiarum Hungaricae,* **5,** 327.
Prasad, M. (1987). *Plant Pathogenic Bacteria,* Proceedings 6th International Conference, 1985, p. 862.
Pressey, R. & Avants, J. K. (1973). *Biochimica et Biophysica Acta,* **309,** 363.
Pressey, R. & Avants, J. K. (1975a). *Phytochemistry,* **14,** 957.
Pressey, R. & Avants, J. K. (1975b). *Journal of Food Science,* **40,** 937.
Pressey, R. & Avants, J. K. (1976). *Phytochemistry,* **15,** 1349.
Pressey, R. & Avants, J. K. (1977). *Plant Physiology,* **60,** 548.
Pupillo, P., Mazzuchi, U. & Pierini, G. (1976). *Physiological Plant Pathology,* **9,** 113.
Punja, Z. K., Huang, J. S. & Jenkins, S. K. (1985). *Canadian Journal of Plant Pathology,* **7,** 109.
Puri, A., Solomos, T. & Kramer, A. (1982). *Food Chemistry,* **8,** 203.
Rajpurohit, T. S. & Prased, N. (1982). *Indian Journal of Mycology and Plant Pathology,* **12,** 220.
Reddy, S. R. & Reddy, S. M. (1983). *Comparative Physiology and Ecology,* **8,** 69.
Reverchon, S., Hugouvieux-Cotte-Pattat, N. & Robert-Baudouy, J. (1985). *Gene,* **35,** 121.
Rexová-Benková, L. (1973). *European Journal of Biochemistry,* **39,** 109.
Rexová-Benková, L. & Markovič, O. (1976). *Advances in Carbohydrate Chemistry and Biochemistry,* **33,** 323.
Rexová-Benková, L., Omelková, J., Mikeš, O. & Sedláčková, J. (1982). *Journal of Chromatography,* **238,** 183.
Ried, J. L. & Collmer, A. (1986). *Applied and Environmental Microbiology,* **52,** 305.
Riov, J. (1975). *Journal of Food Science,* **40,** 201.
Rombouts, F. M. (1972). Occurrence and Properties of Bacterial Pectate Lyases. PhD Thesis, Agricultural University, Wageningen.
Rombouts, F. M. & Pilnik, W. (1972). *Chemical Rubber Company Critical Reviews of Food Technology,* **3,** 1.
Rombouts, F. M. & Pilnik, W. (1980). *Economic Microbiology,* **5,** 227.
Rombouts, F. M. & Pilnik, W. (1986). *Symbiosis,* **21,** 79.
Rombouts, F. M., Spaansen, C. H., Visser, J. & Pilnik, W. (1978). *Journal of Food Biochemistry,* **2,** 1.
Sakai, T. (1988). *Methods in Enzymology,* **161,** 335.
Sakai, T. & Okushima, M. (1978). *Agricultural and Biological Chemistry,* **42,** 2427.
Sakai, T. & Okushima, M. (1980). *Applied and Environmental Microbiology,* **39,** 908.

Sakai, T. & Okushima, M. (1982). *Agricultural and Biological Chemistry*, **46**, 667.
Sakai, T. & Takaoka, A. (1985). *Agricultural and Biological Chemistry*, **49**, 449.
Sakai, T. & Yoshitake, S. (1984). *Agricultural and Biological Chemistry*, **48**, 1941.
Sakai, T., Okushima, M. & Sawada, M. (1982). *Agricultural and Biological Chemistry*, **46**, 2223.
Sakai, T., Okushima, M. & Yoshitake, S. (1984). *Agricultural and Biological Chemistry*, **48**, 1951.
Sato, M. & Kaji, A. (1975). *Agricultural and Biological Chemistry*, **39**, 819.
Sato, M. & Kaji, A. (1977a). *Agricultural and Biological Chemistry*, **41**, 2193.
Sato, M. & Kaji, A. (1977b). *Agricultural and Biological Chemistry*, **41**, 2199.
Sato, M. & Kaji, A. (1980). *Agricultural and Biological Chemistry*, **44**, 1345.
Sato, M. & Kaji, A. (1981). *Kagawa Daigaku Nogakubul Gakujutso Hokoku*. 1980–81. **32**, 121. *Chemical Abstracts*, **98**, 140188r (1983).
Scala, F. & Zoina, A. (1983a). *Annali della Facolta di Scienze Agrarie della Universita degli Studi di Napoli, Portici*. **17**, 122. *Chemical Abstracts*, **101**, 87494j (1984).
Scala, F. & Zoina, A. (1983b). *Annali della Facolta di Scienze Agrarie della Universita degli Studi di Napoli, Portici*. **17**, 68. *Chemical Abstracts*, **101**, 87093h (1984).
Schejter, A. & Marcus, L. (1988). *Methods in Enzymology*, **161**, 366.
Schultz, T. H., Lotzkar, H., Owens, H. S. & Maclay, W. D. (1945). *Journal of Physical Chemistry*, **49**, 554.
Scheiman, M. I., Macmillan, J. D., Miller, L. & Chase, T. (1976). *European Journal of Biochemistry*, **64**, 565.
Silley, P. (1985). *Journal of Applied Bacteriology*, **58**, 145.
Solms, J. & Deuel, H. (1955). *Helvetica Chimica Acta*, **38**, 321.
Somogyi, M. (1952). *Journal of Biological Chemistry*, **195**, 19.
Sone, H., Sugiura, J., Itoh, Y., Izaki, K. & Takahashi, H. (1988). *Agricultural and Biological Chemistry*, **52**, 3205.
Speiser, R., Copley, M. J. & Nutting, G. C. (1947). *Journal of Physical and Colloidal Chemistry*, **51**, 117.
Strand, L. L., Corden, M. E. & MacDonald, D. L. (1976). *Biochimica et Biophysica Acta*, **429**, 870.
Suri, R. A. & Mandahar, C. L. (1982). *Indian Journal of Mycology and Plant Pathology* **12**, 351.
Tagawa, K. & Kaji, A. (1988). *Methods in Enzymology*, **161**, 361.
Takahashi, H., Isaki, K., Ito, K. & Nikaido, A. (1988). Japan Kodai Tokkyo Koho JP 63 68,087 [88,68087] 26 March 1988. *Chemical Abstracts*, **110**, 52329y (1989).
Takehana, H., Shibuya, T., Nakagawa, H. & Ogura, N. (1977). *Technical Bulletin of the Faculty of Horticulture, Chiba University*, **25**, 29. *Chemical Abstracts*, **89**, 142462p (1978).
Theron, T., DeVilliers, O. T. & Schmidt, A. A. (1977). *Agrochemophysica*, **9**, 7.
Tsuyunashi, S., Kawamura, D., Yagi, J. & Miyamoto, Y. (1988). Japan Kodai Tokkyo Koho JP 63,146,790 [88,146,790] 18 June 1988. *Chemical Abstracts*, **110**, 113224j (1989).
Urbanek, H. & Zalewska-Sobczak, J. (1975). *Biochimica et Biophysica Acta*, **377**, 402.
Urbanek, H., Zalewska-Sobczak, J. & Krzechowsaka, M. (1976). *Bulletin de Academie Polonaise des Sciences, Serie des Sciences Biologiques*, **24**, 635.
Ulrich, J. M. (1975). *Physiological Plant Pathology*, **5**, 37.
Van Houdenhoven, F. E. A. (1975). Pectin Lyase. PhD Thesis, Agricultural University, Wageningen.
Versteeg, C., Rombouts, F. M. & Pilnik, W. (1978). *Lebensmittel-Wissenschaft und Technologie*, **11**, 267.
Voragen, A. G. J. (1972). Characterization of Pectin Lyases on Pectins and Methyl Oligogalacturonates. PhD Thesis, Agricultural University, Wageningen.

Voragen, A. G. J. & Pilnik, W. (1969). Pektindepolymerasen. Mitt. Laborat. Lebensmittelchemie und Lebensmittelmikrobiologie Landwirtschaft. Hochschule Wageningen.

Voragen, A. G. J. & Pilnik, W. (1989). In *Biocatalysis in Agricultural Biotechnology*, eds J. R. Whitaker & P. E. Sonnet. *American Chemical Society Symposium Series*, **389**, 93.

Voragen, A. G. J., Rombouts, R. M. & Pilnik, W. (1971). *Lebensmittel-Wissenshaft und Technologie*, **4**, 126.

Voragen, A. G. J., Schols, H. A., Siliha, H. A. I. & Pilnik, W. (1986). In *Chemistry and Function of Pectins*, eds M. L. Fishman & J. J. Jen. *American Chemical Society Symposium Series*, **310**, 230.

Wagner, G. & Hollmann, S. (1976). *European Journal of Biochemistry*, **61**, 589.

Walkinshaw, M. D. & Arnott, S. (1981a). *Journal of Molecular Biology*, **153**, 1055.

Walkinshaw, M. D. & Arnott, S. (1981b). *Journal of Molecular Biology*, **153**, 1075.

Wang, M. C. & Keen, N. T. (1970). *Archives of Biochemistry and Biophysics*, **141**, 749.

Wood, R. K. S. (1951). *Nature*, **167**, 771.

Whitaker, J. R. (1972). *Principles of Enzymology for the Food Sciences*. Marcel Dekker, Inc., New York.

Whitaker, J. R. (1984). *Enzyme and Microbial Technology*, **6**, 341.

Wick, R. L. & Schroeder, D. B. (1982). *Mycologia*, **24**, 460.

Yoshihara, O., Matsuo, T. & Kaji, A. (1977). *Agricultural and Biological Chemistry*, **41**, 2335.

Yu, Y. H., Park, Y. S., Kim, J. M., Kong, I. S. & Chung, Y. J. (1988). *Sanop Misaengmul Hakhoechi*, **16**, 316. *Chemical Abstracts*, **110**, 2135m (1989).

Chapter 5

GLUCOSE TRANSFORMING ENZYMES

ANNELIESE CRUEGER

Verfahrensentwicklung Biochemie, Bayer AG, Postfach, Wuppertal 1, FRG

& WULF CRUEGER

WV Umweltschutz, 5090 Leverkusen, Bayer AG, FRG

CONTENTS

1   INTRODUCTION

The metabolism of glucose is an important part of the primary metabolism in microorganisms. Therefore, the attack of microorganisms on glucose results in a broad range of reactions. A complete utilization may be observed, resulting in the production of CO_2, H_2O, and cell mass. In other cases only slight alterations of the molecule may occur. These latter reactions, such as incomplete oxidations, aldose–ketose isomerization or glycosyl-transfer may also be used for commercial glucose conversion. This review is chiefly concerned with some of these reactions and will not deal with the many intermediary steps of primary metabolism. The most important glucose transforming enzyme from a commercial viewpoint is glucose isomerase. Other well-known enzymes are the glucose oxidases; the preferred position for the oxidation of glucose is C-1, yielding gluconic acid. Further oxidations of gluconic acid yield after several steps 2-keto-L-gulonic acid (KGA). KGA is an intermediate in commercial production of ascorbic acid, where either a combination of microbial conversion steps via 2-ketogluconic acid and 2,5-diketo-D-gluconic acid results in KGA, or where gluconic acid is oxidized enzymatically to 5-keto-D-gluconate, which after catalytic hydrogenation to calcium L-gulonate, is transformed to KGA by several microorganisms. These pathways are reviewed by Crueger & Crueger (1984). Many efforts were made to optimize these alternative pathways to KGA, but as yet none of these techniques can compete with Reichstein's procedure for production of L-ascorbic acid.

The production of kojic acid (5-hydroxy-2-hydroxymethyl-4-pyrone (Fig. 1)) by oxidation at several C- positions in glucose has only been of minor commercial interest and will not be considered here (for details see Crueger & Crueger, 1984).

Other glucose transforming enzymes include the glucokinases. The ATP:D-glucose 6-phosphotransferase (EC 2.7.1.2) catalyses the reaction (Porter & Chassy, 1982):

$$ATP + \text{D-glucose} \rightarrow ADP + \text{D-glucose 6-phosphate} + H^+$$

This enzyme may be used as a glucose or an ATP sensor. A further glucokinase (Polyphosphate:D-glucose 6-phosphotransferase, EC 2.7.1.63)

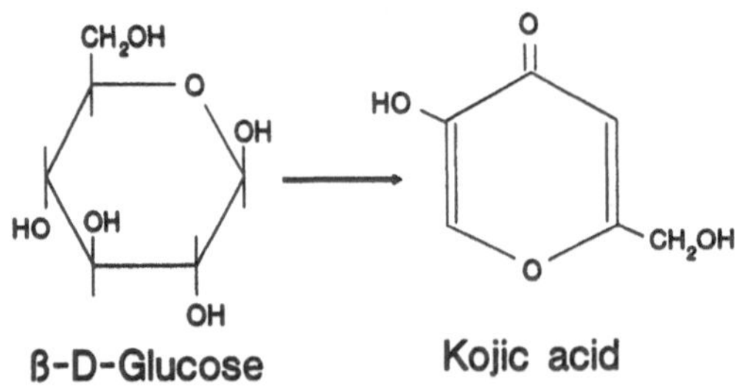

Fig. 1. Kojic acid production by bioconversion of glucose with *Aspergillus ozyzae*.

catalysing the reaction (Szymona & Ostrowski, 1964)

$$(\text{phosphate})_n + \text{D-glucose} \rightarrow (\text{phosphate})_{n-1} + \text{D-glucose 6-phosphate}$$

is only of scientific interest, as is the phosphoglucomutase (D-glucose 1-phosphate: D-glucose 6-phosphotransferase, EC 2.7.5.5; Fujimoto *et al.*, 1965).

Mutarotase (Aldose 1-epimerase, EC 5.1.3.3) catalyses the conversion of α-D-glucose to β-D-glucose (Bentley & Bhate, 1960*a*, *b*) and is added to analytical determinations in cases where the enzyme has a higher specificity for the β-D-anomer. Mutarotase is also used in multi-enzyme biosensors.

The glycosyl transfer, where a glucose is transferred from one molecule to water or to another molecule involves numerous reactions within microbial metabolism (Spencer & Gorin, 1965). The dextran process (Jeanes, 1978) and production of palatinose (Nakajima, 1988) are carried out commercially. A detailed description of all these reactions would exceed the scope of this chapter.

2 GLUCOSE OXIDASES

2.1 Introduction

Several microbial enzymes capable of oxidizing glucose are known. Of these β-D-glucose: oxygen-oxidoreductase (EC 1.1.3.4) is of commercial interest, but in addition there exists several types of glucose dehydrogenases, for example, D-glucose: (acceptor) 1-oxidoreductase (EC 1.1.99.10), D-glucose: NAD(P)$^+$ 1-oxidoreductases (EC 1.1.1.118, EC 1.1.1.119), β-D-glucose: NAD(P)$^+$ 1-oxidoreductase (EC 1.1.1.47), and D-glucose: (pyrrolo-quinoline-quinone) 1-oxidoreductase (EC 1.1.99.17), which is also commercially used. Each of these enzymes produces D-glucono-δ-lactone and a reduced acceptor. In addition, the pyranose: oxygen 2-oxidoreductase (EC 1.1.3.10) oxidizes the C-2-position of D-glucose producing D-arabino-2-hexosulose (glucosone). This enzyme may be of considerable commercial value. Two further enzymes capable of oxidizing the C-3 position of glucose (EC 1.1.99.13) and the C-5 position (EC 1.1.99.11) are at present without any interest and will not be considered further. At the moment it is unlikely that the glucose oxidases will find new applications.

The fungal process for the commercial production of D-glucono-δ-lactone and its free acid form, gluconic acid is catalysed by glucose oxidase (GOD). GOD is commercially available as a by-product of the gluconic acid process. The enzyme has been known since at least 1904 when the glucose-oxidizing capability of extracts of *Aspergillus niger* was demonstrated. Müller (1928) was the first to isolate the enzyme from mycelia of *A. niger* and *Penicillium glaucum*. Gluconic acid as a product of fungal metabolism was first reported by Molliard (1922). GOD which catalyses the conversion of glucose to δ-gluconolactone (Fig. 2) is highly specific for β-D-glucose, the rate of oxidation

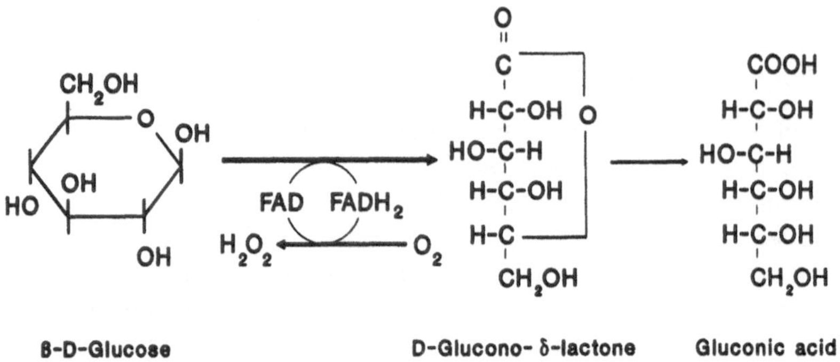

CH₂OH ... O ... OH / HO / OH / OH — **β-D-Glucose**

FAD   FADH₂

H₂O₂ ← O₂

```
         O
         ‖
         C ――┐
      H-C-OH  O
      HO-C-H
      H-C-OH
      H-C ―――
        CH₂OH
```
D-Glucono-δ-lactone

```
         COOH
      H-C-OH
      HO-C-H
      H-C-OH
      H-C-OH
        CH₂OH
```
Gluconic acid

Fig. 2. Oxidation of glucose to gluconic acid by glucose-1-oxidase.

of β-glucose is about 157 times faster than that of the α-D-glucose. GOD (β-D-glucose:oxygen 1-oxidoreductase, glucose oxyhydrase, notatin) was first described as an antibiotic because of its inhibitory effect due to the formation of H_2O_2.

GOD is a flavoprotein, it removes two hydrogens from glucose and is itself reduced. The reduced form is then re-oxidized by molecular oxygen. The developing H_2O_2 is decomposed by catalase to H_2O giving the net reaction

$$\text{D-glucose} + 0.5\ O_2 \rightarrow \text{D-glucono-δ-lactone} + H_2O$$

The δ-lactone hydrolyses spontaneously or by help of the enzyme glucono-δ-lactonase to gluconic acid. The technical process is reviewed by several authors (Ward, 1967; Lockwood, 1975, 1979; Miall, 1978; Milsom & Meers, 1985). GOD is widespread among the fungi. It has been detected in cultures of *Penicillium amagasakiense, P. brunneum, P. diversum, P. egyptiacum, P. funiculosum, P. glaucum, P. notatum, P. paxilli, P. purpurogenum,* and *P. vitale.* Gluconic acid is also the major product from the growth of the penicillin-producing cultures of *P. chrysogenum* when grown in broth containing 5% or more glucose. In addition GOD was found in the genera *Gliocladium, Gonatobotyris,* and *Scopulariopsis. Talaromyces stipitatus* and the mycelial yeast *Endomycopsis fibuliger* were also observed to produce GOD. Many strains of *A. niger* have been examined and nearly all produce a mixture of gluconic, citric and oxalic acid. *A. niger* NRRL 3 (ATCC 9029) appears to be satisfactory in that only glucono-δ-lactone and gluconic acid are produced and it is the most widely used strain for industrial-scale production. Another production strain is *P. amagasakiense,* which produces 2500–3500 units/ml of GOD in submerged culture in one day. A newer strain, *P. purpurogenum* No. 778, produces about 10-times more enzyme (Nakamatsu *et al.,* 1975). Recent reports have also pointed out that the hydrogen peroxide produced by GOD plays an important role in lignin degradation by *Phanerochaete chrysosporium,* a white-rot basidiomycete (Kelley *et al.,* 1986). This enzyme is a flavoprotein with an apparent native M_r 180 000 but does not appear to be a glycoprotein. Van Dijken & Veenhuis (1980) have established that in *A. niger* both GOD and catalase are located in peroxisomes thus claiming an intracellular origin for gluconic acid. Contrary to this Mischak *et al.* (1985) found in a citric acid-producing *A. niger* that GOD was excreted immediately after its synthesis.

2.2 Enzyme Properties

Crystalline preparations of GOD were first obtained from culture filtrates of *P. amagasakiense* by Kusai *et al.* (1960). Data on enzyme structure and properties have been reviewed by Bucke (1983) in the first edition of this book. In summary, the native enzyme is a glycoprotein with a molecular mass of 160 000 daltons (O'Malley & Weaver, 1972), values given by other authors are in the range of 150 000–192 000. GOD contains 2 moles of FAD cofactor per mole of enzyme and 16% carbohydrate (mannose, galactose, and glucosamine in the *N*-acetyl form), the carbohydrate chains are not directly involved in catalysis. The correct structure of the carbohydrate chain is not yet known. Hayashi & Nakamura (1981) separated, from a purified enzyme sample, six components with isoelectric points between 3·9 and 4·3 with identical protein composition but differing in the carbohydrate moieties. GOD can be dissociated into two subunits by cleaving the two disulphide bridges by mercaptoethanol and SDS-treatment (Tominaga & Kelly, 1984), each polypeptide subunit contains one atom of Fe^{2+}. The reaction of the native enzyme with 5,5'dithiobis(2-nitrobenzoate) was very slow, thus indicating that the SH—groups are not easily approachable from the surface. GOD of *A. niger* exhibits a maximal activity at pH 5·6 and at a temperature of 35–40°C. The pH optima of both *A. niger* and *Penicillium* enzymes are broad, activity being nearly constant between pH 4·5 and 7·0. The influence of different additives (polyhydric alcohols, polyethylene glycol, and salts) on the thermostability of GOD at 60°C has been studied by Ye *et al.* (1988). There was a stabilizing effect in the presence of polyhydric alcohols (ethylene glycol, glycerol, erythritol, xylitol, and sorbitol) and for most of the polyethylene glycol used. The GOD half-life was increased by a factor of nearly 25 in the presence of 4 M xylitol. The influence of different salts on GOD stability was also studied ($BaCl_2$, $CaCl_2$, $MgCl_2$, LiCl, NaCl, and KCl) at different concentrations. The influence of the monovalent ions on GOD thermostability can be correlated to the lyotropic series of Hofmeister ($K^+ > Na^+ > Li^+ > Mg^{2+} > Ca^{2+} > Ba^{2+}$). The enzyme obeys Michaelis–Menten kinetics with a V_{max} of 16·95 mmol/min per mg^{-1} protein (Gibson *et al.*, 1974). The K_m of the *P. amagasakiense* enzyme was 11 mM for glucose whilst that of the *Aspergillus* enzyme was 30 mM. The specificity of GOD is very high, the β-form of glucose is oxidized 157 times more rapidly than the α-form and of other substrates examined only 2-deoxy-D-glucose and 6-deoxy-D-glucose were oxidized at rates greater than 10% of that of D-glucose.

The amino acid composition of GOD from *A. niger* was determined by Hayashi & Nakamura (1981) (Table 1).

Table 1
Amino Acid Composition of Glucose
Oxidase Purified from *Aspergillus niger*
(Hayashi & Nakamura, 1981)

Amino acids/FAD	
Lysine	35·9
Histidine	31·3
Arginine	47·5
Aspartic acid	149·5
Threonine	84·0
Serine	78·5
Glutamic acid	108·0
Proline	57·9
Glycine	118·5
Alanine	134·2
Cysteine	4·6
Valine	92·5
Methionine	26·0
Isoleucine	52·1
Leucine	103·0
Tyrosine	53·0
Phenylalanine	37·0
Tryptophan	15·8

Rosenberg *et al.* (1989) have obtained cDNA and genomic clones of *A. niger* enzyme. GOD consists of 583 amino acids. This sequence is preceded by a 22 amino acid N-terminal extension, which must finally be processed. The mature N-terminus shows homology with other AMP-binding proteins.

GOD has been assayed manometrically, polarographically, by differential conductivity, and by a coupled colorimetric assay where 1 U catalyses the oxidation of 1 μmol glucose to gluconic acid per minute at 25°C, pH 7, coupled with peroxidase and *o*-dianisidine (Tsuga & Mitsuda, 1973).

2.3 Genetics

Only a few reports deal with methods of strain improvement such as mutation and selection as a means of increasing GOD activity in production strains (Zamfirescu *et al.*, 1983; Doeppner & Hartmeier, 1984). Fiedurek *et al.* (1986) used UV irradiation and the chemical mutagen *N*-methyl-*N'*-nitro-*N*-nitroso-guanidine for mutagenesis with *A. niger* NRRL 3. Among 960 strains analysed only 12 isolates showed a moderate increase of 1·5–18% compared to the parental strain. Markwell *et al.* (1989) presented a method for the mutagenesis and primary selection of *A. niger* strains with increased GOD activity. A suspension of *A. niger* NRRL 3 spores in sodium acetate buffer was treated with sodium nitrite solution which generated the mutagenic agent, nitrous acid, and plated on diagnostic agar containing 2% (w/v) glycerol, 0·1 M D-glucose, and 0·001% (w/v) methyl red. After 5 days incubation at 25°C, the plates were examined for colonies surrounded by pink zones in the medium due to acidification. The mycelia of 26 presumptive mutants were homogenized and measured for GOD activity. Two strains contained a marked increase in enzyme activity relative to the parent strain at all three D-glucose concentrations (at 0·01 M D-glucose 2·9 and 5·8-fold, at 0·1 M 3- and 20-fold and at 0·5 M 9·1 and 11·3-fold increase).

Rosenberg *et al.* (1989) succeeded in cloning GOD of *A. niger*. Both cDNA and genomic clones were obtained. A full length GOD cDNA was engineered for expression in *S. cerevisiae* using either the GOD or *S. cerevisiae* α-factor pre-prosequences for secretion. Transformants of yeast containing high copy number plasmids secreted more than 100 µg/ml of active GOD into the yeast media. Analysis of the proteins produced, showed more extensive N-linked glycosylation of the yeast-derived enzymes than that from *A. niger*.

2.4 Production of Glucose Oxidase

GOD is a by-product of the fungal process for production of gluconic acid and its salts, the enzyme is recovered from the mycelium at the end of fermentation. The development of this process from surface culture with *P. purpurogenum* or *A. niger*, submerged culture, with a 80–87% conversion of 200 g glucose per litre with *P. purpurogenum*, the rotating-drum technique with a selected strain of *A. niger*, up to modern processes is described in detail by Bucke (1983) in the first edition of this book and by Miall (1978). At present the manufacture is mainly achieved by batch fermentation processes using fungi of the genera *Aspergillus* and *Penicillium* (e.g. *P. amagasakiense*, *P. notatum*, *P. purpurogenum*) with *A. niger* as the most widely used strain for industrial-scale production. The process is a simple fermentation with glucose and oxygen as substrates (Lockwood, 1975, 1979; Miall, 1978; Milsom & Meers, 1985). GOD production is directly coupled to the use of glucose-containing media and the production of D-gluconate (Mischak *et al.*, 1985; Rogalski *et al.*, 1988). Therefore the carbohydrate source for gluconate and GOD production is glucose as glucose monohydrate crystals or dextrose syrup. During fermentation the gluconic acid is quickly neutralized on excretion by the addition of $CaCO_3$ or NaOH and the fermentation broth contains either calcium or sodium gluconate. The glucose concentration that can be handled depends on whether the calcium or the sodium salt is the final product. The solubility of calcium gluconate at 30°C is very low. With an initial glucose content of 100 g/litre, a calcium gluconate fermentation medium at harvest is strongly supersaturated and at higher initial glucose concentrations the risk of sudden crystallization is considerable. Because of the increased solubility of sodium gluconate (at 30°C: 39·6%) it is possible to use initial glucose concentrations as high as 28–30%. In addition to glucose, a nitrogen source (corn steep liquor, ammonium salts, urea), phosphate, potassium, and magnesium must also be provided for the growth of *A. niger*. Too much nitrogen results in an excessive growth and decreased gluconate production. If ammonium salts or urea are used as sole nitrogen source, traces of iron, copper and zinc must be added to the broth. A typical medium for production scale consists of (g/litre): glucose (anhydrous basis), 250–350; corn steep liquor, 3·7; $MgSO_4 \cdot 7H_2O$, 3·5–4·0; KH_2PO_4, 0·2–0·3; $(NH_4)_2HPO_4$, 0·4; urea, 0·1, and H_2SO_4 to pH 4·5 (Lockwood 1979). The seed culture medium has a lower glucose concentration c. 100 g/litre and sources of nitrogen and salts are increased up to two-fold. To increase growth corn steep powder is added The initial pH is adjusted to about 6·5 by NaOH.

Inoculum for production scale may be a suspension of conidia but usually spores which have germinated in seed medium in a special seed tank are used. Spore suspensions are prepared from cultures grown for 5–10 days on a solid sporulation medium (i.e. bran, or beet cossets), or by liquid sporulation cultures. For the germination of spores a water suspension is used to inoculate the seed tank to a final concentration of about 2×10^5 conidia /ml (Rogalski *et al.*, 1988). The culture is vigorously aerated and agitated for about 24 h. The inoculum volume is 3–10% of the main culture volume. The correct time for inoculum may be judged by the amount of mycelial growth or by the rate of production of GOD (Lockwood, 1975). The use of germinated spores reduces the operating cycle in the production stage but one must consider the cost of installation and operation of the seed tanks. Thus the choice of inoculum type depends on plant size and capacity.

The operating conditions for the process are—temperature 28–32°C, aeration 1·0–1·5 vvm (vol. air/vol. nutrient × min) and vigorous agitation for maximum oxygenation efficiency is maintained by using turbomixers. An over pressure on the fermenter of up to 2 bar or the use of oxygen-enriched air are suitable methods to increase the oxygen pressure in the fermentation broth. *A. niger* is well known for its ability to accumulate various organic acids as citric, oxalic and gluconic acids. The type of acid produced is mainly controlled by the pH of the nutrient broth. At a pH below 2, citric acid is produced almost exclusively, whereas at a pH above 5, gluconic acid is synthesized (Mischak *et al.*, 1985). Therefore the pH of the fermentation is maintained at about 5·5–6·0 by the addition of $CaCO_3$ when calcium gluconate is produced. Calcium carbonate is sterilized as an aqueous slurry. Due to the formation of $Ca(OH)_2$ by loss of CO_2 during sterilization the pH of the solution may reach 12·5. To avoid alkalinity of the broth, $CaCO_3$ slurry is added incrementally. In sodium gluconate production the pH is maintained at about 6·8–7·2 by NaOH additions by means of an automatic pH controller.

Depending on the initial concentration of glucose and the type of inoculum, the fermentation usually is finished between 20–36 h but may take up to 100 h. In an optimal fermentation, yields of gluconic acid exceed 90–95%. The theoretical yield of gluconic acid from anhydrous glucose, assuming no production of mycelium or CO_2, is 109%. The maximal specific productivity was determined as 20–30 mmol gluconic acid per g dry cell wt/h per litre.

As a variant of the process the mycelium may be reused up to two further occasions. In the fed-batch process of Ziffer *et al.* (1971) gluconic acid concentrations of about 600 g/litre were achieved. Due to the low solubility of sodium gluconate, glucose is added in stages. Initial glucose concentrations are about 270 g/litre and subsequent additions of glucose (300, 80, 90 g/litre) are made. About 97–99% yields of product expressed as gluconic acid are obtained in 60 to 70 h. In this process neutralization of the broth is carried out only until optimum cell growth and GOD content are achieved. Subsequently pH control is stopped and it drops to values of 3·2–3·5. Under these conditions a mixture of sodium gluconate and free gluconic acid is formed. The solubility of such a mixture is much higher than that of the pure sodium salt. An alternative method of avoiding premature product crystallization is the addition of sodium borate. However, the resulting borogluconate was found to be deleterious to blood vessels of animals and therefore was withdrawn from the market. The fermentation processes described above are very efficient so there is little incentive to develop an immobilized process.

2.5 Purification

For the commercial production of GOD the mycelium at the end of the calcium or sodium gluconate fermentation is used for the recovery of the enzyme (Lockwood, 1975). The mycelium is disrupted by grinding or sonication. For a higher recovery rate the ground cell mass is permitted to autolyse in phosphate buffer (pH 5·0–9·0). The autolysate is purified from suspended solids by filtration. The enzyme may be precipitated using cold ethanol or acetone. Both liquid and dry forms of GOD are on the market. Freeze-dried enzyme usually is stabilized by addition of synthetic polymers such as polyethylene oxide which increases recovery by almost a factor of two.

Further purification of GOD may be achieved by chromatography on DEAE-cellulose or diethylaminohydroxypropyl-cellulose with elution by 1 M NaCl (Kucera *et al.*, 1983). Another technical-scale process is described by Sojka *et al.* (1984), where the crude enzyme solution at pH 3·5 is mixed with polyacrylonitrile powder. The adsorbed GOD is eluted with phosphate buffer (0·2 M, pH 7·5). After precipitation of GOD by an aqueous tannin solution the acetone washed precipitate is extracted with distilled water at 4°C. The extract obtained is again precipitated with cold acetone. The final product has a specific GOD activity of 108 units/mg protein and the yield is 54%.

The commercial preparations, depending on purification procedure, contain various amounts of contaminating enzymes, including catalase, mutarotase, invertase, gluconolactonase and glucoamylase. In many applications GOD is used to generate hydrogen peroxide to quantify the glucose substrate. When catalase is present hydrogen peroxide is destroyed thus making the analysis less accurate or even ineffective. Using conventional gel chromatography alone GOD is very inefficiently separated from catalase. It was found by Keyes (1980) that modifying the carbohydrate moiety of GOD results in a much better separation from the contaminating enzyme. Modification may be achieved by partial hydrolysis with dextranase or other hydrolytic enzymes such as amylase, glucoamylase or cellulase. Following separation on a Sepharose 4 B column, fractions which are about 95% pure GOD with a reduced catalyse content are obtained.

A very effective purification is achieved by the process described by Stärk (1976). Prepurified GOD (15 units/mg of protein and catalase 1 unit/mg of protein) is precipitated with water-soluble salts of acridine bases, preferably diaminoethoxyacridine lactate, where 1 mg of the acridine base is used per 10 mg of protein for 15 h at 4°C. The diaminoethoxyacridine lactate–protein complex containing the GOD is isolated by centrifugation followed by release of GOD by NaCl solution. After dialysis the resulting GOD solution is subjected to chromatography on sulphethyl (SE)-Sephadex C-50. GOD was eluted using a linear gradient of 0–0·3 M NaCl in 0·008 M sodium acetate buffer pH 4. This is followed by two further chromatography steps using DEAE-Sephadex A-50. The GOD fraction is then concentrated by ultrafiltration up to a protein content of 2%. A final product with a specific activity of 225 units/mg of protein and a yield of 45% is obtained. The catalase content of the preparation is 0·08 units/mg of protein. If necessary GOD can be crystallized after purification.

2.6 Immobilization

Like glucose isomerase, GOD is a very convenient enzyme to use in enzyme immobilization studies and consequently there is an abundance of literature on immobilization of whole cells, pure enzyme or coimmobilization of two or more enzymes.

2.6.1 *Immobilized Cells*
Cells of *A. niger* NRRL 3, immobilized in calcium alginate gel, have yielded a 44% conversion of a 15% glucose solution within 24 h, using pure oxygen 93% conversion has been achieved.

GOD has been immobilized with glutaraldehyde and ovalbumin on the mycelium pellets of *Aspergillus* sp. Complete oxidation of glucose was attained in a batch system only under sufficient aeration (Fig. 3; Karube *et al.*, 1977).

Mycelia of *A. niger* which were permeabilized by treatment with iso-propanol were mixed with albumin and a catalase solution together with glutaraldehyde in iso-propanol. The coimmobilized GOD–catalase system was used in reactors for the preparation of gluconic acid from glucose or for deoxygenation of beer (Hartmeier & Doeppner, 1984).

2.6.2 *Immobilized Enzyme*
Immobilized GOD is commercially available and can be purchased, e.g. from Boehringer or Aldrich. A review of the first decade of immobilization was covered by Bucke (1983) in the first edition of this book. In the meantime GOD has been immobilized on numerous inorganic and organic support

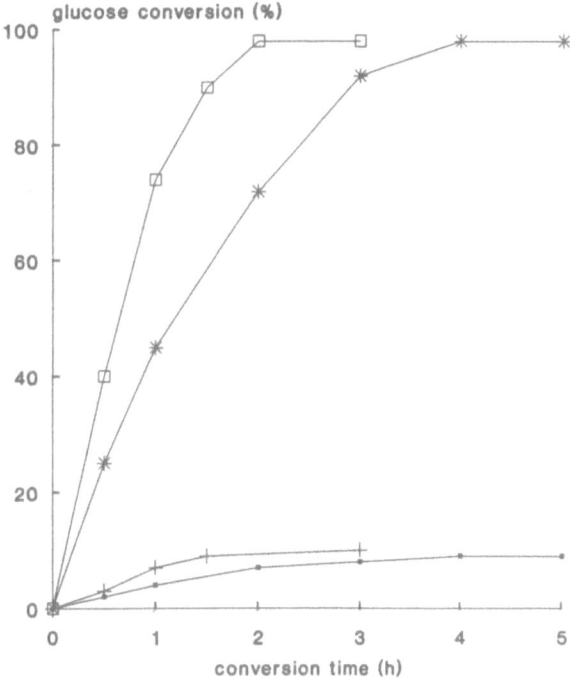

Fig. 3. Oxidation of glucose by immobilized glucose oxidase pellets (*Aspergillus* sp. No. 319) in a batch system (Karube *et al.*, 1977). For glucose conversion 1 g of wet glucose oxidase pellets was added to 50 ml of glucose solution at pH 5·0. Aeration was achieved by bubbling air through the system. With aeration: □—□, 0·5% glucose; *—*, 1·0% glucose. Without aeration: +—+, 0·5% glucose; ■—■, 1·0% glucose.

materials and on molecular sieves employing various methods of immobilization, e.g. immobilization by polyacrylamide coating obtained by electrochemically-initiated polymerization or by glutaraldehyde, by aminated polyamino acid, e.g. poly(Me glutamate) with glutaraldehyde, or by photopolymerization of polyvinylalcohol. Immobilization may be achieved by entrapment of GOD in dissolved nylon preparations, by embedding in silica gel, in alginate beads, or by dropwise addition of GOD solution in gelatine to xylene followed by crosslinking with glutaraldehyde. Supports for immobilization included nylon fibres or open nylon tubes treated with triethyloxonium-tetrafluoroborate, diaminohexane, and glutaraldehyde. Besides silica-based supports in the presence of crosslinking agents other carriers were also used, e.g. clay, ceramic, crab chitin, *p*-benzoquinone-impregnated graphite fixed with a nitrocellulose film, or graphite coated with TiO_2 or ZrO_2. Furthermore bentonites and zeolites, aluminium oxide, microfibrillar cellulose, low-porosity glassy carbon, porous (200–1000 Å pore size) metals, pore glass of different pore size, glass tubes coated by gel consisting of $Si(OH)_2$ in the presence of coupling agents (e.g. γ-aminopropyltriethoxysilane, γ-chloropropyltrimethoxysilane, etc.), or kieselguhr coupled via γ-aminopropyltri-ethoxysilanization and glutaraldehyde were also employed.

For analytical purposes immobilized enzyme membranes for tailor-made biosensors have been developed by entrapment or immobilization of GOD, e.g. in membranes of polyvinyl chloride, poly(2-hydroxyethyl methacrylate), polyacrylamide, collagen, polyvinylalcohol–collagen or polyvinylalcohol on cotton base, silk fibroin, nylon, etc.

In many cases the kinetic properties of the immobilized enzyme were close to those of the free enzyme, e.g. $K_m = 2\cdot0 \times 10^{-2}$ M for the immobilized enzyme and $K_m = 1\cdot93 \times 10^{-2}$ M for the free enzyme. In this case GOD was immobilized on Sepharose by photochemically (UV) initiated graft copolymerization with $FeCl_3$ as photocatalyst (D'Angiuro & Cremonesi, 1982). In most cases the reusability was fairly good. Usually the pH and temperature stability of the immobilized preparations were increased (Kozulic *et al.*, 1987; Kozhukharova *et al.*, 1988). GOD from *P. vitale* was immobilized alone and together with catalase in a gel of polyvinylalcohol. Compared to free enzyme the immobilized form showed increased thermal and pH stabilities. In the bioconversion of D-glucose to D-gluconic acid, the preparations were used repeatedly for a period of four months without any loss of the initial activity. Since the carbohydrate chains of GOD are not directly involved in catalysis, the enzyme was specifically crosslinked through these (Kozulic *et al.*, 1987). The method involves periodate oxidation of the carbohydrate aldehyde groups with adipic acid dihydrazide for crosslinking. The immobilized preparations retained a high level of the original activity (88–100%) depending on molar ratio of periodate to mannose. The minor decrease in specific activity was due to the periodate oxidation step. The preparations had high stability at increased temperatures (see Fig. 4) and up to a pH of 9·5–11.

Immobilized GOD was used in different reactor types mainly as a model system for studying reactor performances, or kinetic properties and stability of the enzyme. Chang *et al.* (1984) investigated a rotating packed disk reactor where the disks were packed with calcium alginate beads with immobilized GOD and catalase. The production rate of gluconic acid increased with the speed of rotation and the bulk flow rate. To study diffusional limitations of substrates and products through the membrane the same workers (Chang *et al.*, 1987) studied gluconic acid production in a dual hollow fibre reactor consisting of an outer silicone membrane (inner and outer diameter of 1·47 and 1·96 mm respectively) for O_2 supply and an inner polyamide membrane (porosity of 80–90%, i.d. 0·6 mm, o.d. 1·1 mm; nominal molecular weight cut off of 5×10^4) for substrate permeation. The reactor was operated in both diffusion and ultrafiltration modes. In the latter case, the conversion was much higher but the stability of the immobilized enzyme decreased. As the inlet glucose concentrations increased from 10 mmol to 500 mmol, the conversion decreased from 70% to 20%. An additional O_2 supply by the silicone tube oxygenator improved the glucose conversion when O_2 supply was rate-limiting, gluconic acid formed wtih oxygenator was 1·44 mmol/h and without 0·40 mmol/h.

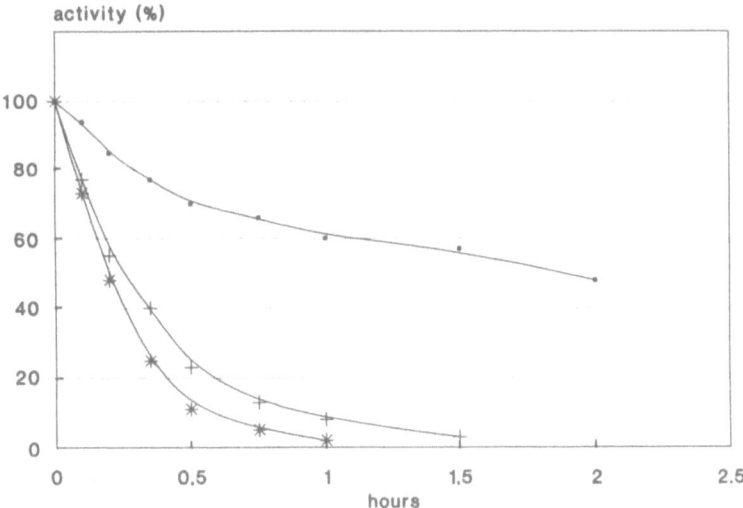

Fig. 4. Thermal stability of crosslinked glucose oxidase (Kozulic *et al.*, 1987). GOD was oxidized by periodate and crosslinked in 0·1 M sodium acetate buffer pH 4·6. The stability was measured as remaining activity at 65°C (native enzyme, *—*; oxidized enzyme, +—+; crosslinked enzyme; ■—■).

2.6.3 Multi-enzyme Systems

Besides the many examples of coimmobilization of GOD and catalase with the purpose of increasing GOD stability by decomposing H_2O_2, these systems are being developed mainly in order to use more complex systems than glucose, e.g. hydrolysed starch. As a model, maltose was converted to gluconic acid using a glucoamylase of *Rhizopus delemar* in combination with GOD and gluconolactonase of *A. niger* (Cho & Bailey, 1977). A second approach is the conversion of sucrose to fructose and gluconate, both of which can easily be separated from the reaction mixture. D'Souza & Nadkarni (1980) described a continuous conversion process carried out in a continuous-flow stirred tank reactor system for over 20 days. They used a multi-enzyme complex consisting of an invertase- and catalase-producing *Saccharomyces cerevisiae*, where GOD was fixed to the cell walls by concanavalin A. The cells were entrapped in polyacrylamide gel.

2.7 Uses of Glucose Oxidase

The most important application of GOD is the industrial-scale fermentation of gluconic acid. The market for gluconic acid and its salts, which are used for the detergent, textile, leather, photographic, pharmaceutical, food, feed, and concrete industries, is approximately 45 000 t annually worldwide (Bigelis, 1985). Gluconic acid, which usually is marketed as a 50% technical grade aqueous solution, finds application as a mild acidulant in leather tanning, metal processing, and foods. Sodium gluconate accounts for 85% of the market because of its metal sequestering qualities (mainly for the alkaline earth metals), which prevent the precipitation of lime soap scums in bottle and dish washing when added to the alkali at a rate of about one part gluconate to ten parts NaOH. When added to cement (0·02–0·2% relative to cement), sodium gluconate controls the setting time and increases the homogeneity of concrete and its resistance against water and frost. In human and veterinary medicine, calcium gluconate serves as a readily available form of calcium in cases of calcium deficiency and furthermore it is also used in cases of allergy. Further applications are in the tannery, metal and dye industries. Iron(II)gluconate finds application in cases of anaemia. Glucono-δ-lactone is mainly used in the food industry since it is non-acidic until dissolved in water, e.g. in baking powders, in bread mixes, sausage manufacturing, or in Japan as a coagulant of soybean protein.

GOD as a free enzyme has found several practical applications, usually in combination with catalase, examples include the stabilization of colour and flavour, in beer, fish meat during refrigeration, tinned foods, and soft drinks, by removal of oxygen, or the removal of glucose in the manufacture of egg powder, which prevents browning during dehydration due to Maillard reaction in the presence of glucose. The commercial use of GOD has often been hampered due to its high costs. As a result a few processes with immobilized enzyme have been developed up to the technical-scale (Hartmeier, 1979) for example, oxygen removal from beer with coimmobilized GOD and catalase. From the pilot plant trials it was found that the half-life of the enzyme system in continuous use was about three months. The initial oxygen content of beer (1·58–1·76 mg/litre) was reduced to 0·06–0·07 mg/litre. Another application is the production of fructose from sucrose hydrolysates, where glucose is oxidized and the resulting gluconic acid can be easily separated from the remaining fructose by ion-exchange or precipitation. In pilot-scale experiments with immobilized GOD and catalase it was estimated that 1 kg immobilized enzyme leads to 1250 kg pure fructose. But this method can hardly compete with other processes for the manufacture of fructose. Sankaran *et al.* (1989) developed a system for the repeated desugaring of the viscous egg melange by GOD and catalase immobilized to cotton cloth. This is an easily retrievable system for the production of egg powder.

High amounts of GOD have found application in medicine for the quantitative determination of glucose in body fluids, and recently in biotechnology, for analysing syrups from starch and cellulose hydrolysis. Immobilized GOD is used in devices for glucose measurement with the enzyme thermistor method (Owusu & Finch, 1986), e.g. the determination of lactose in concentrations of 0·05–10 mm/litre by a combination of lactase, GOD and catalase. The use of enzymes in biosensors was introduced by Clark & Lyons (1962) with a device in which GOD was retained between two Cuproplan membranes over an amperometric oxygen electrode, where oxygen consumption was measured. Another principle used GOD immobilized on a filter trap and the hydrogen peroxide produced is measured with a platinum electrode. The stability of this electrode exceeded 14 months, the range of determined glucose was 2×10^{-2} to 10^{-4} mol. Glucose electrodes based on both reaction types are on the market. In the meantime numerous types of enzyme electrodes have been developed differing physically as well as chemically. GOD is usually applied closely to the sensing part of the electrode by physical or chemical entrapment, or by immobilization. Automatic glucose analysers using glucose sensors were developed e.g. for fermentation process monitoring (Mizutani *et al.*, 1987). Immobilized GOD packed in small columns may be used in flow injection analysis of glucose. A further development is the use of multi-enzyme electrodes, e.g. with invertase, mutarotase and GOD coimmobilized in a collagen membrane, where saccharose concentration is measured as O_2 consumption according to the following reactions (Satoh *et al.*, 1976):

$$\text{saccharose} + H_2O \xrightarrow{\text{invertase}} \alpha\text{-D-glucose} + \text{D-fructose}$$

$$\alpha\text{-D-glucose} \xrightarrow{\text{mutarotase}} \beta\text{-D-glucose}$$

$$\beta\text{-D-glucose} + O_2 + H_2O \xrightarrow{\text{GOD}} \text{D-glucono-}\delta\text{-lactone} + H_2O_2$$

Other systems worth mentioning are a maltose sensor with coimmobilized α-glucosidase and GOD, or a bienzyme sensor consisting of glucoamylase and GOD for the determination of pullulan (Renneberg *et al.*, 1985).

3 GLUCOSE DEHYDROGENASES

3.1 Introduction

Gluconic acid is also the reaction product of a second group of enzymes, the glucose dehydrogenases (GDH). They all produce D-glucono-δ-lactone and a reduced acceptor. Beside the fungal process with glucose oxidase, the GDH-producing bacterium *Gluconobacter oxydans* has been applied for the technical-scale production of gluconic acid due to an increase in reaction rate and yield. Most of the other enzymes are till now without any commercial significance, so they will only be summarized in this context.

3.2 D-Glucose:(Acceptor) 1-Oxidoreductase (EC 1.1.99.10)

This enzyme occurs as a soluble GDH in *Aspergillus oryzae*. Being incapable of reacting with molecular O_2, it catalyses the oxidation of glucose by certain redox dyes and quinones (e.g. 2,6-dichlorophenolindophenol) according to the equation

$$\text{D-glucose} + \text{acceptor} \rightarrow \text{D-glucono-}\delta\text{-lactone} + \text{reduced acceptor}$$

This GDH containing one mole of FAD per mole of enzyme is a glycoprotein

(24% carbohydrate consisting of glucose, mannose and hexosamines). The relative molecular mass was estimated to be approx. 118 000 (Bak, 1967).

Another enzyme of this type is located in the cytoplasmic membrane of a *Pseudomonas* sp. as part of an electron transfer chain (Matsushita *et al.*, 1980). Besides redox dyes and quinones, coenzyme Q_1 was used as electron acceptor, thus leading to the conclusion that the coenzymes Q seem to be the natural acceptors.

3.3 D-Glucose:NAD$^+$ 1-Oxidoreductase (EC 1.1.1.118) and D-Glucose:NADP$^+$ 1-Oxidoreductase (EC 1.1.1.119)

The two NAD(P) dependant GDHs catalyse the reaction

$$\text{D-glucose} + \text{NAD(P)}^+ \rightleftharpoons \text{D-glucono-}\delta\text{-lactone} + \text{NAD(P)H} + \text{H}^+$$

The GDH (EC 1.1.1.119) also oxidizes D-mannose, 2-deoxy-D-glucose and 2-amino-2-deoxy-D-mannose. These glucose dehydrogenases are described as occurring in cyanobacteria, a group of photoautotrophic prokaryotes (Pritchard *et al.*, 1975) and mainly in those strains, which are capable of growing on glucose in the dark. Little attention has been paid to these enzymes due to the similarity of the general enzymologic properties (e.g. pH optimum, K_m values, substrate and coenzyme specificities) to those of the NAD(P) dependant GDHs of heterotrophic bacteria. Juhasz *et al.* (1987) succeeded in purifying the GDH of *Nostoc* sp. strain Mac, a filamentous cyanobacterium, more than 1100-fold to near homogeneity. It was shown that the enzyme can be modulated by various redox systems, e.g. thioredoxin. No thioredoxin sensitivity of any GDH from any source has so far been reported. The K_m value of the enzyme for glucose and NADP was determined both in the absence and presence of thioredoxin. No significant change was observed in the K_m value for glucose (14 mM, oxidized; 15 mM, reduced). The reduction of the enzyme with thioredoxin resulted in a 10-fold increase in the K_m value for NADP (0·024 mM, oxidized; 0·285 mM, reduced). It appears that thioredoxin may regulate the GDH activity, at least in part, by a thioredoxin induced shift of the K_m value for NADP. There is preliminary evidence that the apparent molecular mass of the enzyme (approx. 230 000) decreases upon reduction, this might indicate subunit interaction.

3.4 β-D-Glucose:NAD(P)$^+$ 1-Oxidoreductase (EC 1.1.1.47)

This type of enzyme has been isolated from a variety of sources, including mammalian liver, a blue-green alga, a number of bacteria, such as *Pseudomonas*, *Klebsiella*, *Vibrio*, *Serratia* and other oxidative species, an alkalophilic *Corynebacterium* (Kobayashi & Horikoshi, 1980), several archaebacteria such as the extreme halophile *Halobacterium saccharovorum*, the thermoacidophilic *Sulfolobus solfataricus* (Giardina *et al.*, 1986) and *Thermoplasma acidophilum* (Smith *et al.*, 1988). In various Bacillaceae GDH

(EC 1.1.1.47) is not found in vegetative cells but is synthesized in the forespore between 2–3 h after the onset of sporulation. Tochikubo & Yasuda (1985) proposed that GDH exists in the resting spores of *B. subtilis* in an inactive form (M_r about 55 000) and is restored *in vivo* by association to the active form (M_r about 120 000). Dipicolinic acid, which is accumulated in spores, strongly inhibits activation of GDH in resting spores (Tochikubo *et al.*, 1987). In germinating spores of *B. megaterium* only the glucose metabolism via the direct oxidative route of gluconate is operating so that NADH formed by the GDH reaction is solely responsible for aerobic supply of ATP at this growth stage (Sano *et al.*, 1988).

The GDH from *B. megaterium* and *B. subtilis* show high activities toward β-D-glucose and 2-deoxy-D-glucose. GDH from *B. megaterium* exists at pH 6·5 as a stable, active, tetrameric protein (M_r 118 000). By shifting the pH to 9, the enzyme is completely and reversibly dissociated into four inactive protomers, each consisting of 262 amino acids (Maurer & Pfleiderer, 1987). It has a relative molecular mass of 28 196 as deduced from the nucleotide sequence or 31 500 as determined from SDS-polyacrylamide gel electrophoresis. The K_m values of GDH from sporulating *B. subtilis* were 12·5 mM for glucose, 10·2 mM for 2-deoxyglucose and 85·0 mM for D-glucosamine. There was no activity on other carbohydrates tested. Crystallization and X-ray investigation on GDH from *B. megaterium* have been achieved by Pal *et al.* (1987). The complete amino acid sequence of the *B. megaterium* and *B. subtilis* enzyme have been determined (Fig. 5, Jany *et al.*, 1984) and a prediction of secondary, tertiary and quaternary structure has been proposed (Hönes *et al.*, 1987). The active tetrameric GDH from *B. megaterium* was rapidly inactivated upon reaction with tetranitromethane by nitration of tyrosine 254, which was essential for the catalytic activity.

The archaebacterial dehydrogenases from *Thermoplasma acidophilum* (Smith *et al.*, 1988) and from *Sulfolobus solfataricus* (Giardina *et al.*, 1986) can accept either NAD or NADP as cofactor, both native enzymes being tetramers, M_r 155 000 and 124 000, respectively and the subunits have a relative molecular mass of 38 000 and approximately 30 000 respectively. The *Sulfolobus* enzyme shows maximal activity at pH 9 and 77°C, with the K_m values were 8·0 mM with glucose as substrate and NAD$^+$ as cofactor, or 0·44 mM with NADP$^+$ as cofactor. The *Thermoplasma* GDH is extremely stable to high temperatures and organic solvents, K_m values of 10, 0·113 and about 30 mM were determined for glucose, NADP, and NAD, respectively. As a prerequisite for investigations on ligand binding and on the role of specific amino acid residues in catalysis, a gene encoding GDH of *B. megaterium* M1286 was isolated from a λ-EMBL3 phage library and cloned in *E. coli* (Heilmann *et al.*, 1988). The amino acid sequence as deduced from the coding region, consists of 261 amino acids and is different from the sequence described by Jany *et al.* (1984) (Fig. 5). By using this gene (*gdhA*) as a hybridization probe a second GDH gene (*gdhB*) was isolated and a further DNA region with extended sequence homology to the gene probe indicating a presumptive third gene was identified.

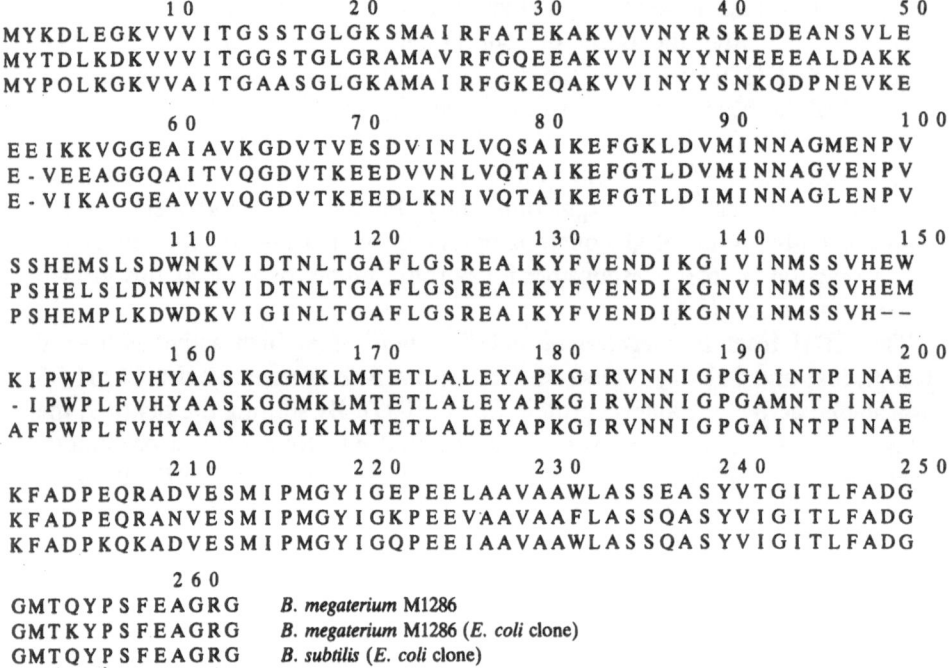

```
        1 0              2 0              3 0              4 0              5 0
MYKDLEGKVVVITGSSTGLGKSMAIRFATEKAKVVVNYRSKEDEANSVLE
MYTDLKDKVVVITGGSTGLGRAMAVRFGQEEAKVVINYYNNEEEALDAKK
MYPOLKGKVVAITGAASGLGKAMAIRFGKEQAKVVINYYSNKQDPNEVKE

        6 0              7 0              8 0              9 0              1 0 0
EEIKKVGGEAIAVKGDVTVESDVINLVQSAIKEFGKLDVMINNAGMENPV
E-VEEAGGQAITVQGDVTKEEDVVNLVQTAIKEFGTLDVMINNAGVENPV
E-VIKAGGEAVVVQGDVTKEEDLKNIVQTAIKEFGTLDIMINNAGLENPV

        1 1 0            1 2 0            1 3 0            1 4 0            1 5 0
SSHEMSLSDWNKVIDTNLTGAFLGSREAIKYFVENDIKGIVINMSSVHEW
PSHELSLDNWNKVIDTNLTGAFLGSREAIKYFVENDIKGNVINMSSVHEM
PSHEMPLKDWDKVIGINLTGAFLGSREAIKYFVENDIKGNVINMSSVH--

        1 6 0            1 7 0            1 8 0            1 9 0            2 0 0
KIPWPLFVHYAASKGGMKLMTETLALEYAPKGIRVNNIGPGAINTPINAE
-IPWPLFVHYAASKGGMKLMTETLALEYAPKGIRVNNIGPGAMNTPINAE
AFPWPLFVHYAASKGGIKLMTETLALEYAPKGIRVNNIGPGAINTPINAE

        2 1 0            2 2 0            2 3 0            2 4 0            2 5 0
KFADPEQRADVESMIPMGYIGEPEELAAVAAWLASSEASYVTGITLFADG
KFADPEQRANVESMIPMGYIGKPEEVAAVAAFLASSQASYVIGITLFADG
KFADPKQKADVESMIPMGYIGQPEEIAAVAAWLASSQASYVIGITLFADG

        2 6 0
GMTQYPSFEAGRG    B. megaterium M1286
GMTKYPSFEAGRG    B. megaterium M1286 (E. coli clone)
GMTQYPSFEAGRG    B. subtilis (E. coli clone)
```

Fig. 5. Homology alignment of glucose dehydrogenase of *Bacillus megaterium* and *B. subtilis* (Jany *et al.*, 1984; Hönes *et al.*, 1987; Heilmann *et al.*, 1988). The protein sequence of GDH from *B. megaterium* M11286 (top line) is compared to the DNA-derived sequence of *B. megaterium* M1286 (*gdhA*) cloned in *E. coli* (second line) and the corresponding sequence of GDH from *B. subtilis* (cloned in *E. coli*; third line). Gaps have been introduced to optimize homology.

In *Gluconobacter oxydans*, the commercial producer of gluconic acid, a GDH (EC 1.1.1.47) has proved to exist, but according to recently obtained results the main gluconic acid formation is catalysed by another type of GDH (EC 1.1.99.17, see Section 3.5).

GDH (EC 1.1.1.47) is on the market for the analytical determination of glucose. Production is achieved by using sporulating cells of *Bacillus cereus*, *B. subtilis* or *B. megaterium*. Fermentation and purification procedure including several steps of column chromatography are described in detail by Barker & Shirley (1980) and Ramaley & Vasantha (1983). Cloning of the *B. subtilis* gene in *E. coli* has resulted in an increased level of the enzyme (up to 2·5-fold) thus simplifying the purification procedure. The recovery rate of this process is given in Table 2 (Smith & Ramaley, 1988). Higher recovery rates of up to 83% could be achieved in the pilot-scale by application of phase partitioning in aqueous two phase systems (Schütte *et al.*, 1983).

GDH has been immobilized by common procedures for use in continuous flow analytical devices for glucose concentration measurement (e.g. Bissé *et al.*, 1981), or for the reduced cofactor production (NADH/NADPH) as a

Table 2
Purification of *B. subtilis* Glucose Dehydrogenase (Smith & Ramaley, 1988)

Procedure	Total[a] units	Specific[b] activity	Yield (%)
Cell extract	37 700	0·82	100
DEAE-Sephacel (batch adsorption and column chromatography)	31 500	1·93	84
Ammonium sulphate removal of contaminating proteins	28 400	20·35	75
Hydrophobic column chromatography (Phenyl-Sepharose)	23 600	150·00	63
Hydroxyapatite column	21 700	399·00	58
Chromatofocusing column chromatography	19 500	565·00	52
Hydrophobic column chromatography	17 200	543·00	46
Concentrated enzyme	16 500	540·00	44

[a] μmoles/min—a unit of activity is defined as the amount of enzyme needed to produce 1 μmole of NADH at 340 nm (22–25°C).
[b] Specific activity is defined as μmoles of NADH/min per mg protein.

coenzyme regeneration system for bioconversion steps (e.g. Wong & Drueckhammer, 1985).

3.5 D-Glucose:(Pyrroloquinoline-Quinone) 1-Oxidoreductase (EC 1.1.99.17)

This quinoprotein-containing GDH possesses a pyrroloquinoline-quinone (PQQ, methoxatin) as the prosthetic group. The enzyme has been studied extensively in *Acinetobacter calcoaceticus* (Geiger & Görisch, 1986; Dokter *et al.*, 1987), where the PQQ-dependent GDH is located at the outer side of the cytoplasmic membrane facing the periplasma. The same enzyme was shown to be responsible for gluconic acid production in *Pseudomonas aeruginosa* (van Schie *et al.*, 1985), *P. fluorescens* (Matsushita & Ameyama, 1982), *Klebsiella aerogenes* (Neijssel *et al.*, 1983), *Rhodobacter sphaeroides* (Niederpruem & Doudoroff, 1965), *Escherichia coli* (Ameyama *et al.*, 1986), *Agrobacterium radiobacter* (Linton *et al.*, 1987), and *Acinetobacter lwoffi, Azotobacter vinelandii, Agrobacterium* spp., *Rhizobium* spp. (van Schie *et al.*, 1987). In addition to membrane-bound PQQ-dependent GDH, *Acinetobacter calcoaceticus* contains a soluble enzyme. Recently Matsushita *et al.* (1989*a*) succeeded in purifying both enzyme types having M_r values of 83 000 and 55 000 (soluble form) and exhibiting totally different enzymatic properties. Only the membrane bound GDH is coupled to the respiratory chain via ubiquinone. By cloning the quinoprotein GDH it was found that the gene codes for a protein with an M_r of 83 000. Evidence exists for a second GDH, a dimer of two identical subunits with M_r 50 000 (Cleton-Jansen *et al.*, 1988). Besides the fungal process (see glucose oxidase, Section 2.2) gluconic acid is also commercially produced by a *Gluconobacter* sp. fermentation; most of this production is marketed as D-glucono-δ-lactone. Glucose metabolism in *G. oxydans* subsp. *suboxydans* proceeds extensively via the pentose phosphate pathway. Two enzymes exist for the direct oxidation of glucose to gluconic acid. The membrane-bound quinoprotein GDH (Ameyama *et al.*, 1981) and a soluble NADP-dependent GDH (EC 1.1.1.47; Adachi & Ameyama, 1982). In contrast to earlier results (Olijve & Kok, 1979*a*), Pronk *et al.* (1989) gave evidence that the quinoprotein GDH is the enzyme responsible for gluconic acid production. Activities measured as μmol/min per mg protein in cell-free extracts of *G. oxydans* were 4·0 for the PQQ-dependent GDH in relation to 0·15 for the NADP-dependent GDH. Thus gluconic acid production in this organism is an extracytoplasmic process. The PQQ-dependent GDH usually exists as the holo-form; in other bacteria such as *E. coli*, the enzyme is present in the apo-form and can be activated by PQQ (Ameyama *et al.*, 1985).

The purified enzyme has M_r 87 000 and it is specific for D-glucose and at a lower rate for maltose (5% of the glucose value). Other carbohydrates (D-mannose, D-galactose, L-rhamnose, D-mannitol, D-sorbitol, L-sorbose, D-ribose, D-xylose, lactose, sucrose) were not converted. NAD, NADP or oxygen were inactive as electron acceptors. The apparent K_m value for D-glucose was 2·8 mM for the crude membrane fraction and 7·7 mM for the purified enzyme. The GDH is active over a wide range from pH 2·0 through 10·0. The enzyme is contained within or associated tightly with the membrane and Matsushita *et al.* (1989*b*) recently demonstrated that the PQQ-dependent GDH of *G. suboxydans* donates electrons directly to ubiquinone in the respiratory chain.

The existing fermentation process is fairly efficient and medium and production conditions are given by Olijve & Kok (1979*b*). Efforts have been made to develop continuous production procedures of gluconic acid by immobilization. One example is the adsorption of *G. suboxydans* IFO 3290 cells on ceramic honeycomb monolith (Shiraishi *et al.*, 1989). Yields are not yet satisfactory because of the large amount (20–30%) of keto-gluconic acid formed as a by-product as a result of further oxidation of gluconic acid.

3.6 Other Glucose Dehydrogenases

Glucose–fructose oxidoreductase, a new NADP$^+$-dependent membrane-bound enzyme, in which the oxidation of glucose is coupled to the reduction of fructose to sorbitol, was isolated from an ethanol-producing *Zymomonas mobilis* (ATCC 29191) (Zachariou & Scopes, 1986; Chun & Rogers, 1988).

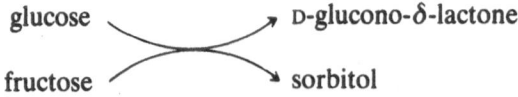

The K_m value for glucose is 2·8 mM, the pH optimum 6·5. Takahashi *et al.* (1987) described a novel NADP$^+$-dependent GDH from *Cryptococcus*

uniguttulatus, which is highly reactive and substrate specific toward glucose, but has very little activity toward 2-deoxy-glucose. This enzyme may be used for the quantitative determination of glucose.

4 PYRANOSE:OXYGEN 2-OXIDOREDUCTASE (EC 1.1.3.10)

While GOD and GDH are known to catalyse the oxidation of D-glucose at C-1, pyranose oxidase (glucose 2-oxidase, D-carbohydrate oxidase) has been shown to catalyse the oxidation of several carbohydrates (e.g. D-glucose, D-xylose, L-sorbose, and 1,5-D-gluconolactone) at the second carbon atom to yield 2-keto products and hydrogen peroxide. The reaction in the case of D-glucose is as follows:

$$\text{D-glucose} + O_2 \rightarrow \text{D-glucosone (D-arabino-L-hexosulose)} + H_2O_2$$

Glucose 2-oxidase (G2OD) was first mentioned by Bond *et al.* (1937) using plasmolysed mycelia of the ascomycetes *Aspergillus flavus-oryzae* and *A. parasiticus*. The capacity to produce G2OD is widespread among basidiomycetes. By screening 40 species of basidiomycetes Volc *et al.* (1985) found 15 producers, the best of them were *Oudemansiella mucida*, *Coriolopsis occidentalis*, *Fomes fomentarius*, and *Trametes* spp. Further well investigated glucosone producers are *Polyporus obtusus* (Ruelius *et al.*, 1968), *Coriolus versicolor* (Machida & Nakanishi, 1984), *Phanerochaete chrysosporium* (Eriksson *et al.*, 1986). The lignin-degrading basidiomycete *Oudemansiella mucida* produced in a laboratory fermenter 15 g D-glucosone per litre with D-glucose as substrate (50% conversion). The glucosone was metabolized after glucose exhaustion (Volc *et al.*, 1978). Other wood-rotting fungi with G2OD activity described by the authors are *Pleurotus ostreatus*, *Laetiporus sulphureus*, and *Phellinus abietis*. The metabolic pathway in which G2OD is involved has not yet been fully clarified.

The G2OD from *Coriolus versicolor* was purified by HPA-75 chromatography, Sepharose 4B and Sephadex G-100 gel filtration and hydroxyapatite chromatography (Machida & Nakanishi, 1984). The enzyme had M_r value of 220 000, consisting of identical subunits (M_r, 68 000), with FAD as cofactor. The highest activity was obtained with D-glucose as substrate and molecular oxygen as electron acceptor. The pH optimum was 6·2 at 50°C and the K_m value for glucose was 0·83 mM.

The G2OD from *Phanerochaete chrysosporium* was partially characterized by Eriksson *et al.* (1986). The intracellular enzyme appears in shake cultures on day 9 of incubation and peaks on day 11 (medium: D-glucose 2%, corn steep 0·7%, $MgSO_4 \cdot 7H_2O$, 0·15%, tap water, pH 5·5 with 6 M NaOH. Conidiospores ($2·6 \times 10^6$/ml) were used as inoculum. Glucose was the preferred substrate although D-gluconolactone, L-sorbose, and D-xylose were also oxidized (at 60%, 52%, and 37%, respectively). K_m values for glucose and xylose were 1·03 and 20 mM, respectively, the pH optimum was pH 7·5 but with a broad range of activity. Thus, the *Phanerochaete* G2OD showed the

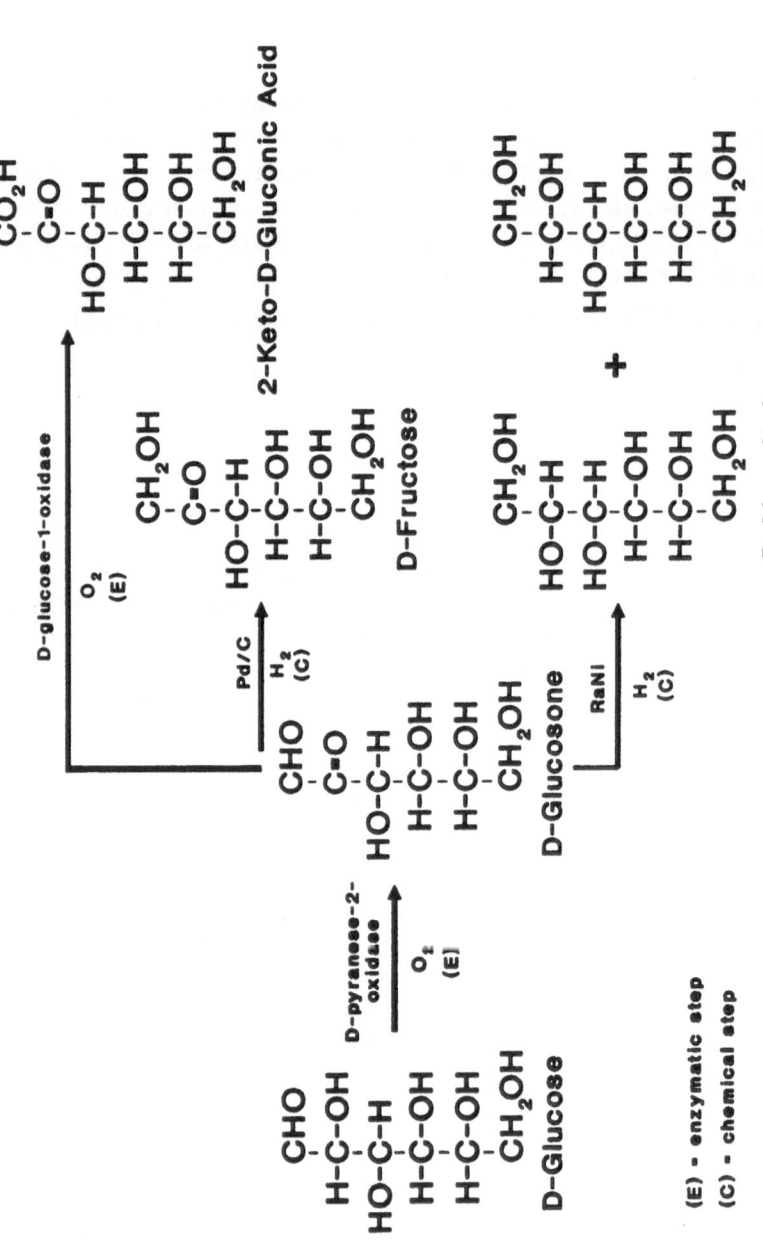

Fig. 6. Formation of glucosone by bioconversion of glucose with glucose-2-oxidase and further conversion products (Neidleman & Geigert, 1986).

same substrate specificity as the enzyme reported for *Polyporus obtusus* (Janssen & Ruelius, 1968).

G2OD has attracted considerable attention owing to its practical application, as D-glucosone is a key intermediate in the production of useful derivatives such as D-fructose, D-mannitol, D-sorbitol, and 2-keto-D-gluconic acid (Fig. 6). The commercial production of fructose involves first the hydrolysis of a polysaccharide such as starch and as a second step the isomerization of the so-produced glucose to form fructose. An alternative to this process may be the conversion of glucose to glucosone, which then is hydrogenated over a supported palladium catalyst to yield pure fructose. The most advanced process is claimed by the Cetus Corp. with immobilized pyranose 2-oxidase from a *Polyporus obtusus* mutant strain: The native enzyme is a tetramer (M_r 290 000; subunits M_r 69 000) with covalently bound flavin. The relative activities toward glucose, sorbose, xylose and gluconolactone substrates are 100:80:48:12 respectively, at 1% (w/v) substrate concentration. The pH optimum is 5·0 (Koths & Halenbeck, 1986). The yield of product is between 95–99% of the theoretical. The process may be carried out in a column reactor. The most effective way to get rid of produced H_2O_2 is its decomposition by catalase. Coupling to a second H_2O_2-consuming reaction, as the enzymatic halogenation to produce an intermediate halohydrin in the preparation of epoxides and glycols from alkenes, is not favourable due to the highly toxic by-products of this latter reaction.

A coupled enzymatic reaction of glucose 1-oxidase and glucose 2-oxidase yields 2-keto-D-gluconic acid, which may be converted to furfural, D-isoascorbic acid (food antioxidant), D-arabinose or D-ribulose (Neidleman *et al.*, 1981). A further interest exists in using G2OD for the determination of blood glucose levels (Taguchi *et al.*, 1985). D-glucose in blood exists as a mixture of α-D-glucose (37·3%) and β-D-glucose (62·7%). For determination, β-D-glucose oxidase (EC 1.1.3.4) is used, where mutarotase must be added to catalyse the interconversion of the anomeric forms of D-glucose. G2OD of *Coriolus versicolor* almost equally oxidizes both α- and β-D-glucose to the corresponding anomers of α,β-D-glucosone. The K_m values for these anomers (1·00 mM, α-D-glucose; 0·57 mM, β-D-glucose) were smaller than those for GOD (11·5 mM, β-D-glucose) from *Penicillium amagasakiense*.

5 GLUCOSE ISOMERASES

5.1 Introduction

In 1988 the main sweetener used in the world was sucrose which came from sugar beets (38.958.00 t) and sugar cane (77.786.00 t). In the US $3·6 \times 10^6$ tonnes of sucrose are used by consumers per year, the soft drink and food industry utilized $6·3 \times 10^6$ tonnes.

D-Glucose (dextrose) has 70–75% of the sweetening ability of sucrose.

However, D-fructose (levulose), the other monosaccharide moiety of sucrose, has twice the sweetening power of sucrose (Barker 1976). In addition, fructose plays an important role in the diet of diabetics as it is only slowly reabsorbed by the stomach and intestinal tract and does not influence the blood glucose level. Fructose was originally produced from invert sugar solutions using the calcium fructonate method. This involves the mixing of calcium hydroxide with an invert sugar solution, treatment with CO_2, separation of the $CaCO_3$, followed by vacuum evaporation and crystallization. Since 1964 fructose has been produced on an industrial scale using cationic resins (Lauer, 1980).

Glucose is produced from starch by means of acid hydrolysis or enzymatic hydrolysis. As sugar beets are available only 100 days per year, wheat, corn or mannitol are used as starting materials. Studies have been carried out on the chemical conversion of glucose to fructose using alkaline conditions and high temperatures. However, these methods turned out to be non-selective, and non-metabolizable sugars such as psicose (Fig. 7) are formed in addition to formate and various coloured substances. Therefore, chemically-obtained fructose syrup has not been employed commercially.

Fructose is produced biochemically from inulin, sucrose, or glucose which is obtained from starch (Fig. 8). Today, however, due to inulin shortages and high prices, the inulin technique is no longer used (Kierstan, 1980). Although the scientific discussion to convert glucose into fructose is much older, the first patent 'On production of fructose from glucose through the action of xylose isomerase' was published by Marshall (1960). Marshall & Kooi (1957) discovered that in *Pseudomonas hydrophila* isomerization takes place without phosphorylation. In a cell with 'normal' metabolism carbohydrate isomerization occurs following a phosphorylation step. A glucose isomerase technique

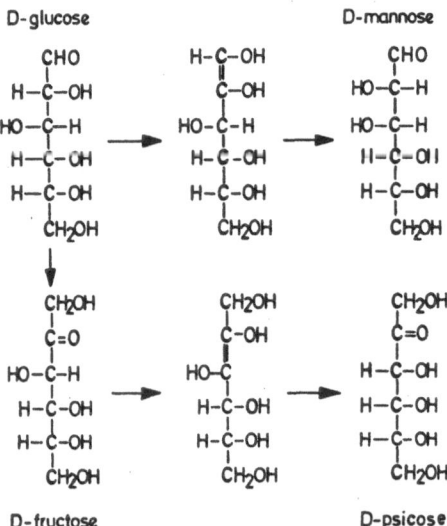

Fig. 7. Chemical isomerization of glucose.

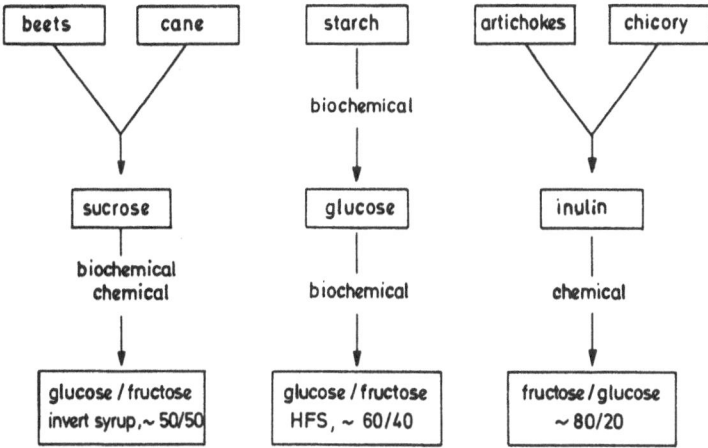

Fig. 8. Production of fructose syrup.

for the production of glucose–fructose-syrup (high fructose syrup, HFS, containing about 42% fructose) was first developed in Japan and later in the US. The enzymes were expensive, because of the necessity to induce the isomerase production by feeding of xylose. The enzymes needed cobalt or arsenate as cofactors, had poor thermostability and the immobilization methods were uneffective. Some of the microorganisms for the production of glucose isomerase were non-GRAS. But microorganisms and process additives have to be considered as GRAS (Generally Recognized As Safe) under the Federal Food, Drug and Cosmetic Act, if they are used for food production (FDA, 1983). The preparations are not allowed to be mutagenic, nor to provoke chromosomal changes. They should not contain antimicrobial and teratogenic activity (Ashby *et al.*, 1987). In feeding and injection tests no signs of infections or toxicosis are allowed (Porter *et al.*, 1984).

In the field of industrial enzymes the glucose isomerase process shows the largest expansion in the market today. Initially soluble enzymes were used and later immobilized enzymes were developed. This optimized technique is the largest commercial application of an immobilized enzyme. Latest developments include increasing the yield of fructose and also separation of fructose from fructose-bearing solutions. In 1978 the US produced 22% of its sweeteners from corn. By 1983 the percentage increased to 33%. Reviews on this topic were written by Barker & Shirley (1980), Chen (1980*a,b*), Bucke (1983), Crueger & Crueger (1984), Verhoff *et al.* (1985) and Crueger & Crueger (1990).

5.2 Historic Perspective

Various types of enzymes are able to convert glucose to fructose (Fig. 9). The first type is a glucose-phosphate isomerase (D-glucose-6-phosphate-ketol-isomerase, EC 5.3.1.9). Producers of this enzyme are *Esherichia intermedia*,

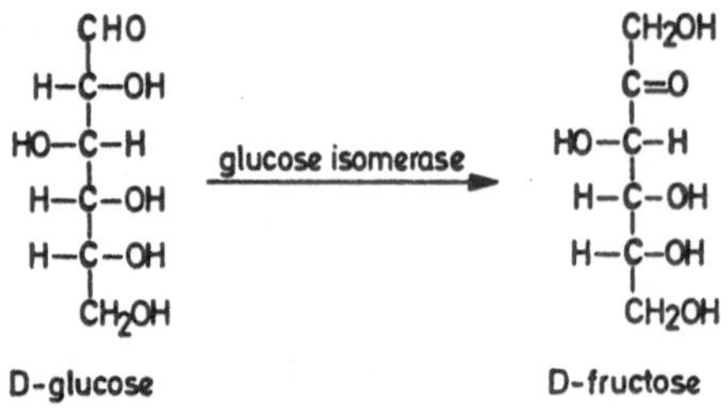

Fig. 9. Reaction of glucose isomerase.

E. freundii, Aerobacter aerogenes, and *A. cloacae*. The enzymes need arsenate to form a glucose–arsenate complex which is isomerized as follows:

$$\text{Glucose} + \text{arsenate} \leftrightarrow \text{glucose–arsenate}$$

$$\text{Glucose–arsenate} + \text{enzyme} \leftrightarrow \text{glucose–arsenate–enzyme}$$

$$\text{Glucose–arsenate–enzyme} \leftrightarrow \text{fructose} + \text{arsenate} + \text{enzyme}$$

Only for some of the enzymes is xylose needed as an inducer (Natake & Yoshimura, 1963, 1964; Natake, 1966, 1968). The enzymes have pH optima at 7·0 and temperature optima at 50°C. However, due to their requirement of arsenate these enzymes are not used in production processes.

A glucose isomerase (D-glucose ketol-isomerase, EC 5.3.1.18) has been characterized which is linked to NAD^+ and produced by *Bacillus megaterium* (Takasaki & Tanabe, 1962, 1963). Its pH optimum is 7·8 and the temperature optimum at 35°C.

Various heterolactic bacteria (*Lactobacillus brevis, L. fermenti, L. pentoaceticus, L. buchneri, L. mannitopolus, L. gayonii*) produce glucose isomerases (Yamanaka, 1962, 1963a, b, 1965, 1968; Shukla & Prabhu, 1985a, b). These enzymes require D-xylose as an inducer as well as manganese ions. In Japan these enzymes have been thoroughly studied for their utility in production of fructose. A disadvantage of these enzymes is their relative instability at higher temperatures.

5.3 Enzyme Sources

Reviews of the different technical glucose isomerase processes are written by Verhoff *et al.* (1985) and Jensen & Rugh (1987). The only commercially employed enzymes are D-xylose isomerases (D-xylose ketol-isomerase, EC 5.3.1.5). Their advantages are:

—a low pH optimum (less secondary reactions);
—high specific activity;
—a high temperature optimum (less contaminations);
—cofactors (ATP, NAD^+) are not needed.

The enzymes can be concentrated by ultrafiltration with a 100 000 M_r cut off membrane (Johnson *et al.*, 1986). They have been purified to homogeneity. Verhoff *et al.* (1983) described the purification and characterization of glucose isomerase from *Flavobacterium arborescens*. The molecular mass of native enzyme is 180 000, with four ionically associated subunits of 44 700 daltons each. These data are generally in accordance with the enzyme of other strains, which have molar masses in the range of 120 000–175 000 daltons (Drocourt *et al.*, 1988). No disulphide bonds have been detected. Detailed information about the purification of genetically overproduced *E. coli* isomerase is given by Tucker *et al.* (1988). The enzyme of *Flavobacterium arborescens* has a Michaelis–Menton constant of 0·51 mmol for xylose and 0·11 mol for glucose, so that xylose as substrate is strongly preferred.

Determination of the crystal structure of glucose isomerase also serves as a guide to the engineering of a microbial strain carrying a gene which has been adapted to give more convenient properties to the enzyme. The C-alpha backbones of the glucose isomerase of *Streptomyces rubiginosus* and *Arthrobacter* sp. were determined by X-ray crystallography. Each molecule is a tetramer of 8-stranded alpha/beta barrels, and the mode of association of the tetramers is identical in each case. The *Arthrobacter* electron density shows four additional amino acids at the carboxyl terminus. There is also an insertion of eight amino acids within the molecule (Henrick *et al.*, 1987; Shaw *et al.*, 1987). The assay methods for glucose isomerases are summarized by Montenecourt *et al.* (1985).

Many microorganisms have been screened for production strains. A typical screening is described by Suekane & Iizuka (1982). The screening was performed using 46 species and 63 strains of *Streptomyces*. The microorganisms were inoculated into a medium containing starch, peptone, meat extract, yeast extract and salts, and cultured at 28°C with shaking for 24 h. After harvesting by centrifugation, the cells were washed and treated in suspension by sonic

Table 3

Composition (w/v%) of Media used for Screening for Glucose Isomerase (Suekane & Iizuka, 1981)

Bacteria, yeasts, actinomycetes		Acetic acid bacteria		Lactic acid bacteria	
D-Xylose	2	D-Xylose	1	D-Xylose	1
$MgSO_4 \cdot 7H_2O$	0·0025	D-Glucose	0·3	D-Glucose	0·1
$(NH_4)_2HPO_4$	0·6	Autolyzed yeast solution	20	Yeast extract	0·4
KH_2PO_4	0·2	$MgSO_4 \cdot 7H_2O$		Nutrient broth	1
pH 6·8		$(NH_4)_2HPO_4$		CH_3COONa	1
		KH_2PO_4		$MgSO_4 \cdot 7H_2O$	0·02
		pH 6·8		NaCl	0·001
				$FeSO_4 \cdot 7H_2O$	0·001
				$MnSO_4 \cdot 4H_2O$	0·001
				pH not adjusted	

Table 4

Distribution of Glucose Isomerase Activity Within Bacteria (Suekane & Iizuka, 1981)

Microorganism (genera)	Number of tested strains	Number of strains to grow on xylose	Number of strains with glucose isomerase activity	
			Medium	Minimal
Pseudomonas	210	80	—	80
Xanthomonas	7	2	1	1
Acetobacter	30	10	7	3
Gluconobacter	10	4	3	1
Aeromonas	3	—		
Protaminobacter	1	—		
Bacerium	5	4	1	3
Kluyvera	5	2	2	
Vibrio	2	—		
Agrobacterium	3	2	2	
Alcaligenes	4	—		
Achromobacter	14	4	2	2
Flavobacterium	14	6	4	2
Escherichia	16	13	13	
Aerobacter	14	14	13	1
Erwinia	7	6	6	
Serratia	23	3	3	
Proteus	1	—		
Micrococcus	31	10	8	2
Staphylococcus	2	—		
Sarcina	17	9	4	5
Brevibacterium	29	7	3	4
Streptococcus	10	6	5	1
Leuconostoc	7	5	4	1
Lactobacillus	21	14	12	2
Propionibacterium	2	2	2	
Corynebacterium	16	—		
Microbacterium	2	—		
Arthrobacter	2	—		
Bacillus	95	13	13	
Streptomyces	124	88	63	25

oscillation. Centrifugation gave a supernatant which was used as a crude intracellular enzyme solution. Enzyme activities were studied by conversion of glucose to fructose in a reaction mixture. Suekane & Iizuka (1981) tested 603 strains of bacteria on xylose media (Table 3). The results show that isomerase activity may be found widely among the various microorganisms (Table 4). Thirteen of 16 strains of *Escherichia,* and all 14 *Aerobacter* strains showed isomerase activity. On the other hand, only 80 of 210 tested *Pseudomonas* strains were able to grow on xylose and showed only minor activity. Seventy per cent of the tested actinomycetes, and forty yeast strains out of 56 gave a positive result: 24 strains showed medium and 16 strains fairly weak activities.

Bok *et al.* (1984) found in their screening for acidophilic *Streptomyces* producers that nearly half of the strains have high activity within 24 h.

Yield improvement and cost reduction are the main purposes in industrial development. The screening for new efficient strains continues. Industry directed its research to the following topics: the enzymes should be constitutive (no xylose needed), non-toxic cofactors should only be required, the microorganisms should be GRAS, the fermentation method should be safe and not expensive.

Among the strains with xylose isomerase activity the following are of commercial interest, today:

Bacillus coagulans (Novo Industries A/S)
Streptomyces murinus (Novo Industries A/S)
Arthrobacter sp. (Reynolds Tobacco Comp., ICI Americas Inc.)
Actinoplanes missouriensis (Gist-Brocades NV, Anheuser-Busch Inc.)
Streptomyces olivaceus (Miles Laboratories Inc.)
Streptomyces olivochromogenes (CPC Intern. Inc.)
Streptomyces phaeochromogenes (Nagase/Denki Kagaku Kogyo)
Streptomyces rubiginosus (Miles-Kali Chemie, Finnsugar/Fermco)

The characteristics of different glucose isomerases are reviewed by Roels & van Tilburg (1979) Chen (1980*a*, *b*), and Makkee *et al.*, (1985*a*). Some trade names of glucose isomerase are:

Maxazyme® (Gist Brocades NV)
Sweetzyme® (Novo Industries)
Optisweet 22® (Miles-Kali Chemie)
Sweetase® (Nagase/Denki Kagaku Kogyo)
Takasweet® (Miles Laboratories Inc.)

5.4 Genetics and Regulation

Very few results have been published about genetic strain improvement (Table 5). As for other industrial and scientific research, nitrous acid, ethylene imine,

Table 5

Increase of Isomerase Activity by Genetic Manipulation

Strain	Mutagenic agent	Increase (%)	Reference
Streptomyces ATCC 21175	Ethylene imine	62	Crueger & Crueger, 1984
S. olivaceus	UV	16	Crueger & Crueger, 1984
S. olivochromogenes	UV	50	Crueger & Crueger, 1984
S. nigrificans	UV	198	Crueger & Crueger, 1984
S. rubiginosus	Mutagens + UV, gamma-rays	17-fold	Antrim & Auterinen, 1986
E. coli	Cloning	12-fold	Lastick *et al.*, 1986*a*
E. coli	Cloning	20-fold	Mullings, 1985

UV irradiation and X-ray treatment are used for mutagenic treatment (Djejeva *et al.*, 1986*a*; Ellaiah *et al.*, 1987). Bacteria used in the production for glucose isomerase have been mutated to obtain mutants which are constitutive in respect of induction by xylose (Roth *et al.*, 1987). Mutants of *Streptomyces chrysomallus* were obtained by a chemostat selection procedure that produced an isomerase that was inducible by glucose. The chemostat was inoculated with a preculture in a glucose-containing medium. After the population had reached steady state, mutagenesis was performed using nitrosomethylurea. At the same time a shift to xylose medium was carried out. During the transient phase after medium shift, mutants with constitutive glucose isomerase gained a growth advantage. A further shift to xylose-limited medium intensified the selection pressure. Clones showing altered colony morphology produced glucose isomerase in the presence of xylose or glucose. However the glucose isomerase of the mutants is not completely constitutive. Hafner (1985) claimed a thermostable glucose isomerase producer *Streptomyces thermoviolaceous* fermented at 42°C which was stable at 80–90°C. Horwath (1986) described a thermostable (>90°C) enzyme from *Arthrobacter* sp.

Glucose analogues like 2-deoxyglucose or β-D-glucose oxime are used to select constitutive mutants (Miles Laboratories, 1983; Verhoff *et al.*, 1985). β-D-glucose oxime is a good competitive inhibitor of glucose isomerase from *S. albus* and *F. arborescens*. Its binding to glucose isomerase was demonstrated. Glucose oxime effect *in vivo* is very complex. Both the initial rate and the extent of cell growth are affected. The cells cultivated in the presence of the glucose oxime exhibit much higher glucose isomerase activity than do cells grown in the absence of the inhibitor. The increase in activity is due to the elevated enzyme synthesis rather than to enzyme stimulation, though glucose oxime itself is not an inducer. Glucose oxime also increased β-galactosidase activity in cells (Table 6). It is suggested that glucose oxime interferes with the carbohydrate catabolite repression in the organisms and thereby induces the optimal enzyme biosyntheses involved in catabolic metabolism (Boguslawski & Whitted, 1984).

There are some data about the molecular basis of xylose metabolism in bacteria. For the non-industrial strain *Salmonella typhimurium* it was shown

Table 6

Effect of Glucose Oxime on Glucose Isomerase
and β-Galactosidase

Enzyme	Glucose oxime (%)	Activity
Glucose isomerase	—	15·3
	0·1	32·2
β-Galactosidase	—	0·47
	0·1	0·88

that xylose metabolism is controlled by an operon-like cluster of loci consisting of the structural gene xy1T (xylose transport), xy1B (xylosekinase) and xy1A (xylose isomerase). The genes are under the control of a regulatory gene xy1R. In xylose free media the xy1R gene product represses the operon. Feeding of xylose results in a complex-binding of xylose and the xy1R gene product, which activates the transcription of the operon. In *Bacillus subtilis* expression of the xyl operon is repressed at the level of transcription and is again induced by xylose (Gärtner *et al.*, 1988).

The nucleoside sequences of the glucose isomerase genes have been determined. An open reading frame of 1167 nucleosides encoding a protein sequence of 388 amino acids was identified for *S. violaceoniger* (Drocourt *et al.*, 1988). *Ampullariella* strains have a 1182 nucleotide open reading frame as coding sequence for the enzyme, but has limited amino acid sequence homology to other available sequences from *E. coli*, *B. subtilis* and *S. violaceoniger* (Saari *et al.*, 1987). Wilhelm & Hollenberg (1985) demonstrated an extensive homology of nucleoside sequences between *Bacillus* spp. and *E. coli*.

Although much work has been published about cloning the xylose isomerase gene and demonstrating overproduction, no production strains are in use which has been optimized by genetic engineering. For cloning, DNA from *S. violaceoniger* was extracted and ligated with plasmid pUT 206. Enzyme deficient strains were transformed with the plasmid and positive glucose isomerase transformants selected (Viot *et al.*, 1987). Using *Saccharomyces cerevisiae* expression vectors, the *E. coli* isomerase gene has been introduced into *S. cerevisiae*. The transformants produced an amount of enzyme parallel to the copy number of the plasmid, but comparable to *E. coli* the yeast protein was 1000 fold less active (Sarthy *et al.*, 1987). Studies to get expression of *B. subtilis* and *A. missouriensis* xylose isomerase genes in yeast were carried out by Amore *et al.* (1989). About 5% of cellular protein of transformants consisted of xylose isomerase, but the enzyme was inactive and partly insoluble. By cloning with a regulatory and promoter sequence deficient fragment ligated into an overexpression plasmid and transformed to *E. coli*. Lastick *et al.* (1986*a*, *b*) succeeded in producing an isomerase up to 38% of total protein in the cell lysate. In a review Workman *et al.* (1986) summarized the cloning strategy for glucose isomerase in *Bacillus* spp. comprising competent cell transformation, protoplast transformation, plasmid deletion and rearrangement. In an early paper Luenser (1983) discussed the application of genetic engineering to yield improvements. Cetus (1988) claimed to increase the glucose isomerase stability by substituting an amino acid at a preselected substitution side in the protein.

5.5 Production of Glucose Isomerases

In order to reach optimal yields, the inducer, xylose, must be added to production media (Fig. 10). Xylose is very expensive and so it has been

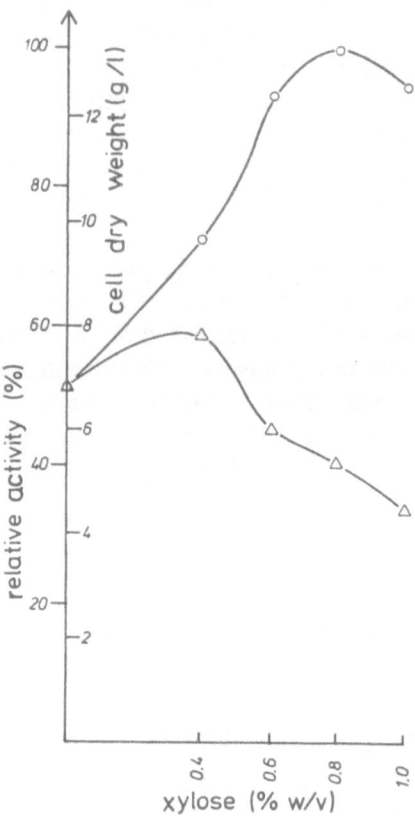

Fig. 10. Induction of glucose isomerase formation by xylose (Vaheri & Kauppinen, 1977). O—O cell dry weight, △—△ glucose isomerase activity.

replaced by xylan hydrolysate and wheat bran. Drazic *et al.* (1980) showed that during *S. bambergiensis* fermentation, xylose, to some extent, may be substituted by starch, glucose, sorbitol or glycerol, but at least 25% of the carbohydrate concentration should be xylose. A typical medium consists of (Park *et al.*, 1980):

—3·0% wheat bran acid hydrolysate,
—2·0% corn steep liquor,
—0·1% $MgSO_4 \cdot 7H_2O$,
—0·01% $CoCl_2 \cdot 6H_2O$.

On the other hand, constitutive mutants are also being used for production purposes. Shieh (1977) described a process with *A. missouriensis*; similarly Long (1978) has worked with an *Arthrobacter* sp. Table 7 presents data from a fermentation with *Arthrobacter* nov. sp. NRRL B-3728. *F. arborescens* ATCC 4358 mutants are able to produce more glucose isomerase than the parent strain growing in lactose medium instead of xylose (Lee, 1981). The regulation of glucose isomerase production is influenced by catabolic repression. In batch

Table 7
Arthrobacter-Production Medium (Long, 1978)

Medium	Seed culture (%)	Production culture (%)
Dextrose	2·0	2·0
Meat protein	0·3	0·6
Yeast extract	0·15	0·15
$(NH_4)_2HPO_4$	0·6	0·6
KH_2PO_4	0·2	0·2
$MgSO_4{\cdot}7H_2O$	0·01	0·01

and continuous fermentation glucose acts as a repressor. *B. coagulans* does not produce glucose isomerase during the log phase. When glucose is used up in the medium, growth comes to a halt. Then, following a typical diauxic pattern, catabolism of other carbohydrates starts and enzyme production increases. In batch fermentation with *Bacillus* spp. strains, maximum enzyme titres are obtained within 24 h.

Not only yeast extract and meat protein but other cheaper protein sources may replace those which were originally used in the production medium. The nitrogen source is a critical point which has to be optimized for the different strains.

Table 8 shows a comparison of the different nitrogen sources. For many strains corn steep liquor (CSL) seems to be most efficient and therefore is most commonly used. Due to the corn quality (germination potency, age, origin) and the manufacturing process (9–20% lactic acid), corn steep liquor is very variable, therefore enzyme yields with CSL are hardly reproducible. Pretreatment by heating (50°C), pH adjustment (pH 7) and centrifugation of the sludge by a decanter may be used, but less yield often results due the pretreatment conditions. Some bacteria are able to utilize inorganic nitrogen [NH_4Cl, $(NH_4)_2HPO_4$, or $(NH_4)_2SO_4$]; however, these are not optimally suitable on the production scale.

In most cases mineral salts such as $CoCl_2{\cdot}6H_2O$, $MgSO_4{\cdot}7H_2O$, and $MnSO_4{\cdot}4H_2O$ need to be added to the production culture medium. *F.*

Table 8
Influence of the Nitrogen Source on Glucose Isomerase Production

Nitrogen source	Activity (%)
Corn steep liquor	100
Casein hydrolysate	73
Soya pepton	66
Yeast extract	58
Malt extract	24

arborescens needs 0·1 mmol cobalt and 0·1 mmol magnesium for optimal activity; these metals have a synergistic effect. But due to cobalt content in complex media there is no need for feeding cobalt during the fermentation of the strain. For *S. kanamyceticus* Debnath & Majumdar (1987) estimated the optimal mineral concentration of 0·07% $MgSO_4$, 0·05% K_2HOP_4, Fe^{2+} 10 mg/litre, Mn^{2+} 3 mg/litre, Zn^{2+} 3 mg/litre, but Cu^{2+}, Co^{2+} and Ca^{2+} have inhibitory effects on enzyme production. It is important to reduce the cobalt content because of its potential to cause environmental hazards from the fermentation waste of the spent media and because of health problems related to the cobalt content of the high fructose syrup. *S. fradiae* SCFS needs cobalt chloride for optimal production (Kowser & Joseph, 1982). High peptone feeding causes low enzyme yield, due to chelating of cobalt by the peptone, so that the ions are not available for the strain. *Arthrobacter* sp. and *S. olivaceus* (Reynolds, 1973) as well as some mutants of *S. olivochromogenes* (CPC International Inc., 1975) do not require cobalt for optimal production.

B. coagulans enzyme may be produced in a continuous fermentation. In this case a specific growth rate of $\mu = 0·1–0·4/h$ is achieved. With strains of *Mycobacterium smegmatis* and *Arthrobacter* nov.sp. Meers (1981) showed, that in continuous culture carbon-limitation causes enhanced specific enzyme yield and carbon conversion efficiency for glucose isomerase production. Not only glucose limitation but also oxygen limitation is optimal at the production scale (Diers, 1976). Microanaerobic conditions within the cells stabilize the system. *Streptomyces* spp. strains only produce isomerase under batch conditions.

Continuous fermentation processes have never reached production scale for glucose isomerase but scale up experiments for feeding and repeated fill and draw procedures are on the way.

Table 9 shows the fermentation conditions relating to a 120 m³ scale. Only few data are available. The Miles process utilizing *S. olivaceus* has been described by Verhoff *et al.* (1985). Master slants are prepared by incubation of the lyophilized strains on sporulation agar slants with the addition of 1% 2-deoxyglucose to check for mutants. The check for stability is very important, because the high-producing strains tend to deteriorate. Significant variations

Table 9

Fermentation Conditions for Glucose Isomerase Production (Batch Fermentation)

Strain	pH	Time (h)	Temp. (°C)	Activity[a] (U/ml)
Bacillus coagulans	6·8	24	40	
Streptomyces olivaceus	7·0	24	30	
Streptomyces albus	7·5	40	30	33·3
Arthrobacter sp.	6·9	55	30	
Actinoplanes missouriensis	7·0	72	30	67·5

[a] U/ml amount of enzyme which produces 1 µmol/min fructose.

can be observed (Djejeva *et al.*, 1986*b*; Prapulla *et al.*, 1987). The incubation time for the Miles process is 7–9 days at 30°C. The master slants, stored at 4°C, are good for preparing the working slants on the same agar. Shake flasks, 24–48 h incubation at 30°C with spores from the working slants make up the inoculum (5%) for the preseed tanks (first fermentation stage). The medium of the preseed tank is as follows: glucose, soy protein, yeast extract, phosphate, soft water. The seed tank (corn starch, corn steep liquor, yeast extract, silicon antifoam agent) is inoculated with 5% of the preseed tank. The production bioreactor is presterilized and filled with continuous sterilized medium (sterilized 1–2 min at 150°C). The medium is the same as that of the seed culture, but with addition of dextrose. The fermentation is controlled online for pH, pO_2, CO_2 and O_2 in the off-gas, and offline for sterility, biomass concentration and enzyme activity. The fermentation time is 1–2 days, followed by cooling and pumping to the downstream process unit.

5.6 Immobilization of Glucose Isomerase

The first step in the downstream processing is the concentration of the biomass (von Hemfort & Kohlstette, 1984).

Enzyme immobilization can either be accomplished with the enzyme or with the glucose isomerase-bearing cells by the following procedures:

—covalent binding to an insoluble carrier,
—adsorption to an insoluble carrier,
—entrapping in a matrix,
—immobilization within the cells.

There is still a considerable amount of experimental work being undertaken on the optimization of the downstream process and the immobilization procedure. For example, Chung *et al.* (1987) reported a dual hollow fibre bioreactor for whole cell immobilization of *S. griseus*. The strain proliferates in the bioreactor and in a second step glucose is converted to fructose. But most of the new systems are only of interest for research studies, one example is the paper of Katwa & Rao (1983) discussing the immobilization on cyanogen bromide-activated Sepharose. Due to the hazards involved this process will never be used in production scale.

The first small industrial-scale production plant (1967–1970) in the US used a Japanese process with *S. wedmorensis* ATCC 21230 (AIST, 1965; Takasaki *et al.*, 1969). The enzyme activity has been immobilized within the cells by heat treatment (15 min, 65°C). In the repeated batch process for isomerization (20 h per batch) the enzymatic activity had a half-life time of approximately 170 h.

Since 1974 crosslinked glucose isomerase preparations for batch isomerization processes are available. This process is still in use for HFS production. One example of this procedure is the old Novo process with *B. coagulans* for production of Sweetzyme S (Fig. 11, Jensen & Rugh, 1987). The enzymatic activity recovery is 50–60%. When isomerization has been finished the

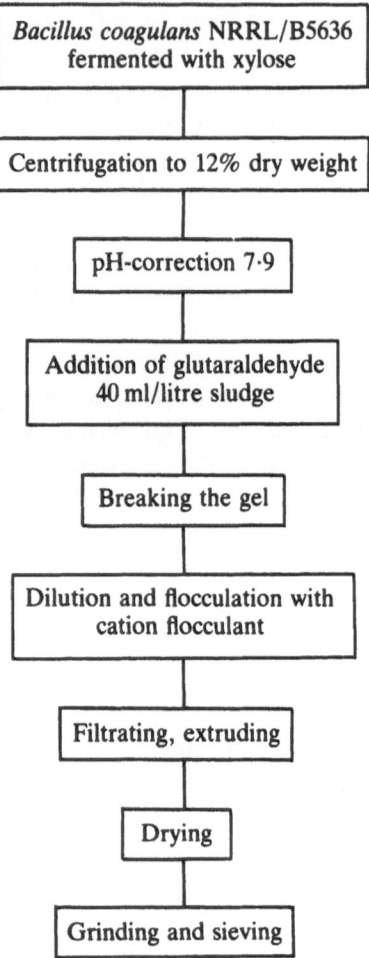

Fig. 11. Preparation of glutaraldehyde crosslinked glucose isomerase (Sweetzyme S) for batch isomerization (Jensen & Rugh, 1987).

particles are allowed to settle at the bottom of the bioreactor, the HFS is drawn off. When Co^{2+} and Mg^{2+} are added, 5% of enzymatic activity is lost per batch during an isomerization time of 20 h at pH 6·5–7·0 and a temperature of 60–65°C.

The next step in the development (1976) was the production of immobilized glucose isomerase for fixed bed operation. Novo changed the process from the immobilization of the *B. coagulans* cells to producing the immobilized enzyme. After fermentation and concentration the cells were disrupted by pressing the sludge with 300–350 kg/cm² through a Manton-Gaulin homogenizer valve. By cavitation the enzyme is released and subsequently immobilized by crosslinking with glutaraldehyde. The particle sizes are in the range 0·3–1·0 mm. With this material Co^{2+} is no longer needed for the isomerization. The optimization of the process over the years within the Novo company is shown in Table 10.

Table 10
Historical Development of the Glucose Isomerase Processes at Novo

Isomerization parameter	Sweetzyme S 1976	Sweetzyme Q 1984	Sweetzyme T 1987
Organism/enzyme	B. coagulans	B. coagulans— enzyme	S. murinus— enzyme
Temperature (°C)	65	60	60
Activity (IGIU/g)	175–225	225–275	275–325
Activity half-life (days)	20	50–65	90–115
Lifetime to 10% residual activity (days)	65	165–215	160–210
Productivity (kg/kg enzyme)	1 750	3 000–4 000	8 000–9 000

To get better productivity, higher activity and better handling characteristics with these enzymes Sweetzyme Q was recently improved to Sweetzyme T by changing the microbial production strain from *B. coagulans* to *S. murinus* (Jörgensen *et al.*, 1988).

The Gist Brocades process to produce Maxazyme has been described by Hupkes & van Tilburg (1976) and van Tilburg (1983). After the fermentation of *A. missouriensis* the pH is raised to 8·6 and the temperature to 72°C. The killed cells are concentrated by centrifugation. At 40°C gelatin is added to a final concentration of 8%. The process is summarized in Fig. 12.

The Sweetase production developed by Nagase starts with the fermentation of *S. phaechromogenes*. The fermentation broth is heat-treated and concentrated by filtration. The sludge is immobilized on a water-insoluble anion-exchange resin, containing quaternized nitrogen in pyridine rings. The immobilized enzyme resin is granulated and dried (Denki Kagaku, 1974).

Miles Kali produces a glucose isomerase from *S. rubiginosus*. The fermentation broth is concentrated, the mycelium disrupted and the enzyme partially purified. Optisweet 22 is produced by adsorbing the enzyme to the support (10–70 ppm SiO_2) and then immobilizing by crosslinking with glutaraldehyde (Weidenbach *et al.*, 1984). A review of the enzyme immobilization on inorganic supports is given by Weetall (1985).

Finnsugar/Fermco offers a regenerable immobilized glucose isomerase from *S. rubiginosus* (Antrim & Auterinen, 1986). The enzyme is extracted from the biomass, the debris are filtered off and the enzyme is further purified by ultrafiltration and crystallization to 80% purity. The enzyme is immobilized by electrostatically binding to a carrier (extruded mixture of 30% DEAE-cellulose, 20% titanium dioxide and 50% food grade polystyrene, particle size 0·4–0·8 mm). The immobilization takes place over 3–5 h in a column or in a batch tank by adsorbing on DEAE-cellulose. Due to the adsorption the half-life of the enzyme is lower, but the carrier may be reused by removing the old inactive enzyme and regenerating the DEAE with sodium hydroxide.

Table 11 shows a comparison of commercial glucose isomerase preparations. Amoco (Chao *et al.*, 1986) presented data of a procedure where spherical

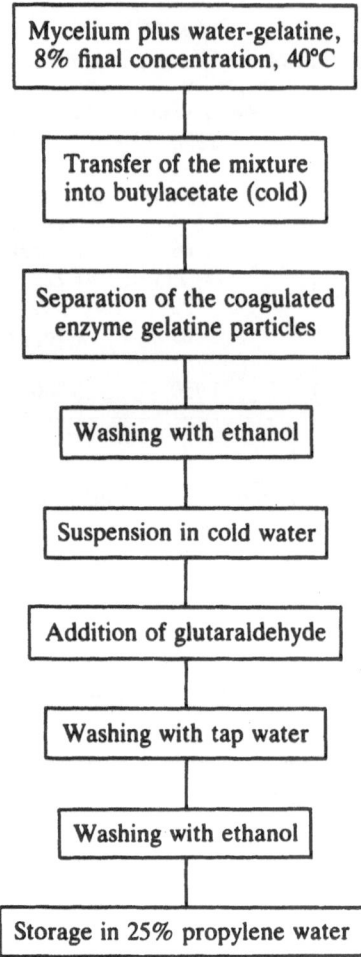

Fig. 12. Immobilization steps for *Actinoplanes missouriensis* (Hupkes & van Tilburg, 1976).

beads of κ-carrageenan containing entrapped cells were prepared using a two step process. A lot of development has been done to increase the knowledge about fluid dynamics within the columns (Verhoff & Furjanic, 1983).

5.7 Fructose Production Processes

Today the use of glucose to produce high fructose syrup is by far the largest industrial business with immobilized enzymes. Reviews of the industrial process are from Fullbrook (1984), Barker & Petch (1985), Harder & Wilms (1985) and Jörgensen *et al.* (1988).

5.7.1 *Degradation of Starch to Glucose Syrup*
To produce a hydrolysate, starch has to be converted first to a paste at 65–90°C. During this process the viscosity increases significantly. α-Amylases

Table 11

Comparison of Some Commercial Isomerase Preparations

Microorganism	pH optimum	Temperature optimum (°C)	Metal requirement	
			Cobalt M	Magnesium M
Bacillus coagulans (Jensen & Rugh, 1987)	7·5	60	–	1×10^{-1}
Actinoplanes missouriensis (Gist Brocades, 1979)	7·5	58–60	3×10^{-4}	3×10^{-3}
Streptomyces murinus (Jörgensen *et al.*, 1988)	7·5	60	–	+
Streptomyces rubiginosus (Antrim & Auterinen, 1986)	7·8	55–57	–	$1·5 \times 10^{-3}$
Streptomyces rubiginosus (Weidenbach *et al.*, 1984)	7·5	58–60	–	120 ppm

(hydrolysing α-1,4 glycosidic bonds, mainly from *Bacillus amyloliquefaciens*, *B. licheniformis*) and isoamylases (α-1,6-splitting enzymes, acting on amylopectin) break down the starch to glucose oligomers (7–10 glucose units). The use of isoamylases increases the yield of glucose by 1–1·5%. The viscosity is reduced by the amylase treatment.

In the next step, the starch solution, which has been liquefied is saccharified with glucoamylase at pH 4·0–5·0 and a temperature of 55–60°C without an intermediate purification. But the final glucose syrup has to be purified by filtration, carbon treatment and a strong cation and a weak anion exchanger, because the solution contains impurities like amino acids, lipids and ions, which may reduce the glucose isomerase activity or may plug the columns. Calcium as an enzyme inhibitor which would bond strongly to the enzyme, has to be purified out of the solution below 1 ppm. A glucose syrup up to 98 DE (dextrose equivalent) is thus achieved.

5.7.2 Isomerization of Glucose

Enzyme stability is one of the factors affecting productivity. The influence of temperature, ionic strength, pH and the presence of poisonous impurities has been investigated. The stability of the immobilized enzyme may be improved by strong diffusion resistance. The substrate stabilizes the glucose isomerase (Chen & Wu, 1987). Some characteristics of immobilized enzymes on the market for glucose treatment in a fixed bed reactor are summarized in Table 12.

The substrate conditions for the isomerization in a fixed bed reactor are an economic compromise between enzyme usage and the desired fructose level. The conversion rate from starch to glucose with 93–96% glucose on a dry solids basis is normal. Substrate solids for industrial isomerization are in the

Table 12
Immobilized Glucose Isomerase Preparations for Industrial Use (Jensen & Rugh, 1987)

Manufacturer	Immobilization method	Initial flow rate (g syrup Ds per litre reactor per hour at 60°C)	Activity half-life at optimal conditions (h)	Recommended operating pH range
Gist Brocades (Maxazyme)	Spherical particles of gelatin-entrapped glutaraldehyde crosslinked cells of *Actinoplanes missouriensis*	700–900	1 500	6·8–7·5
Godo Shusei (Godo-AGI)	Granulate of chitosan-treated glutaraldehyde crosslinked *Streptomyces griseofuscus* cells	750–1 000	850–1 200	8·0–8·3
Miles (Takasweet)	Spheronized extrudate of polyamine flocculated, glutaraldehyde crosslinked *Flavobacterium arborescens* cells	1 000–1 200	1 450–1 800	7·2–7·6
Nagase (Sweetase)	Granulate of heat-treated *S. phaechromogenes* cells adsorbed on insoluble anion-exchange resin containing a pyridine ring	800–1 000	1 300	7·3–8·0
Novo (Sweetzyme)	Cylindrical, extruded particles of glutaraldehyde cross-linked cell homogenate of: *Bacillus coagulans*	700–900	1 500	7·0–8·0
	Streptomyces murinus	800–1 100	2 500	7·0–8·0
Finnsugar (Spezyme IGI)	Pure GI from *S. rubiginosus* adsorbed on DEAE-cellulose embedded in polystyrene with TiO_2	1 000–2 400	1 200	7·7
Miles Kali-Chemie (Optisweet 22)	Spherical SiO_2 particles with adsorbed, glutaraldehyde crosslinked purified GI from *S. rubiginosus*	5 500	1 200	7·5

scale 40–48%DS (dry substance). Above 40% the activity decreases 1% for each percent increase of glucose.

Magnesium concentration of 1·5 mmol/litre stabilizes and activates the enzyme 15–20-fold greater than the enzyme inhibitor calcium. Oxygen in the syrup inactivates the enzyme and is responsible for increased formation of secondary products during isomerization. Therefore a low oxygen tension has to be achieved by feeding of sodium metabisulphite ($Na_2S_2O_5$, 2 mmol/litre). Temperature has an influence on conversion rate, activity, and total productivity (operating time to 10% final activity). At 55°C the fructose content at equilibrium is 50%, at 60°C, 50·7% fructose, at 70°C, 52·4% fructose, at 80°C, 53·9% and at 90°C, 55·6% fructose. But increasing temperature decreases stability and the enzyme half-life and therefore productivity. Most industrial plants run at 58–60°C, a temperature with low risk for microbial contamination. The pH values for the optimal activity and the optimal productivity are different (Jörgensen *et al.*, 1988; Fig. 13), so that the decision for optimal production pH depends on the expected operation time (Fig. 14).

Production plants are constructed as in the arrangement published by Oestergaard & Knudsen (1976; Fig. 15). This plant has a capacity of 100 t/day using two parallel lines with three reactors in series. Nowaday up to eight reactors run parallel. To keep a constant throughput, the reactors have to start in time intervals. The isomerization time is 0·8–4 h.

After isomerization the high fructose syrup is treated with carbon, strong

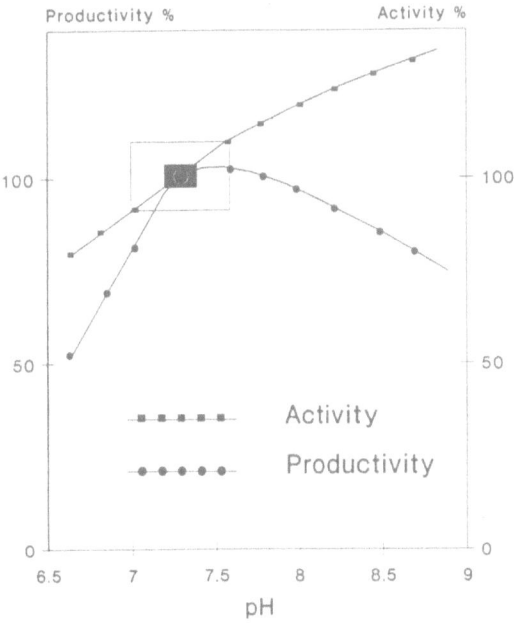

Fig. 13. pH values for optimal productivity and optimal activity (Jörgensen *et al.*, 1988).

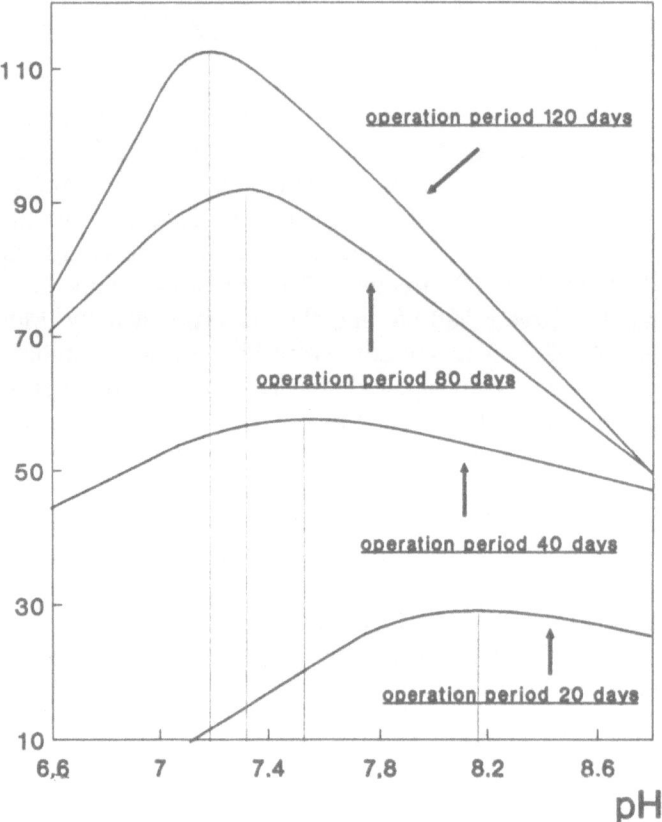

Fig. 14. Effect of pH on productivity in relation to operation time (Harder & Wilms, 1985).

acid cation-exchange resins and slightly alkaline anion-exchange resins, and then concentrated in order to reach a stable state (71% dry weight). The following characteristics are typical of a high fructose syrup (HFS):

Dry substance	71%
pH	4–5
Ash	0·05–0·1%
Fructose	42% (dry substances)
Glucose	53%
Oligosaccharide	5%
Psicose	<0·1%

One gram of HFS contains 4·1 kcal = 17·2 kJ.

The high fructose syrup is directly sold to the food industry or concentrated to reach a stable state to ~70% dry weight.

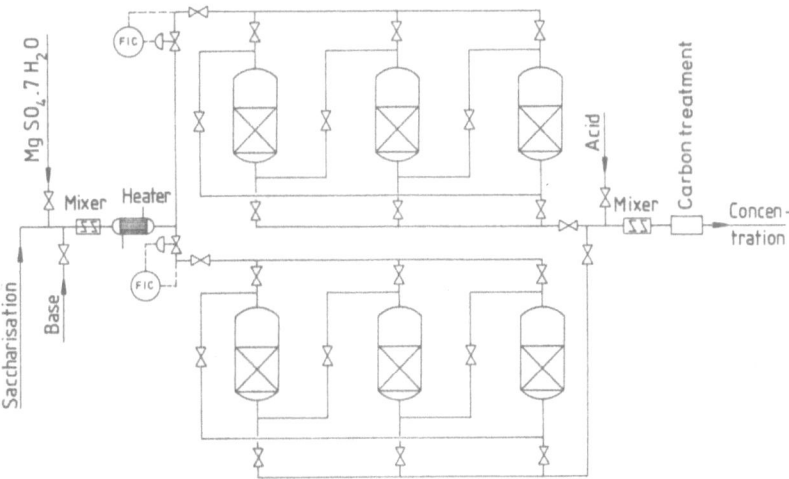

Fig. 15. Reactor arrangement for a 100 tonne/day production plant (Oestergaard & Knudsen, 1976).

Today 60% of the HFS production is a 55% fructose syrup, which is produced starting from the 42% solution by adsorption of the fructose on either zeolites or calcium salts of strong acid cation-exchange resins and elution by water (Keller *et al.*, 1981). Another interesting method is to dilute the 42% fructose solution with ethanol (Visuri & Klibanov, 1987). That ethanol/water mixture is passed through a glucose isomerase column to enhance fructose content to the desired level. Ethanol can be recovered by distillation. This approach can be used for the production of syrup containing 55% fructose.

5.8 Economics

In 1985 the US production of HFS was about 6.5×10^6 t. Due to free competitions HFS is sold 15% cheaper than sugar. The consumption per person in 1985 was (kg/year):

	Sucrose	Dextrose	HFS
US	31	2	18
Canada	34	1	12
Japan	21	1	7

The US reduced importation of sucrose by two million tonnes per annum as a result of HFS production. In the US HFS is produced by the following companies: American Maize Products Comp., Amstar Corp., Archer Daniels Midland Corn Sweeteners, Cargill Inc., Clinton Corn Processing Comp., CPC International Inc., The Hubinger Comp., and A. E. Staley Manufacturing Co.

In Europe, especially in the EC, the sugar industry has a strong political influence. The result is a production rate of HFS of no more than

300 000 tonnes per annum. There are plants in Germany, Spain, UK, The Netherlands, Belgium, France, Italy, Ireland, Yugoslavia, Austria and Hungary.

The syrup is a direct substitute for sucrose in carbonated and still soft drinks, canned fruits, lactic acid beverages, juice, bread, icecream, frozen candies, and processed milk products.

In addition to a small increase in HFS production over the next years developments in this field will most likely proceed along the following lines:

1. Reduction of the isomerization costs. Amplification of the glucose isomerase genes in producer organisms will lead to an increase of fermentation productivity.
2. Higher fructose levels. Higher isomerization yields may be achieved with temperature-stabilized glucose isomerase by raising the working temperature. Temperatures of about 90°C lead to an increase in fructose concentration. Barker *et al.* (1983) produced a fructose syrup containing more than 90% fructose by forming fructose–oxyanion complexes with germanate.
3. Combination of saccharification and isomerization. Several production processes exist which use starch as the initial substrate (Hebeda & Leach, 1975; Naarden International, 1976; Walon, 1977; Katwa & Rao, 1983). A combination of α-amylase, glucoamylase, and glucose isomerase may yield a 39–45% fructose syrup in a single step. Gunei Chemical Co. has installed a 70,000 t/per annum production plant.
4. An interesting development is the production of mannitol. A process involving both a bio- and chemo-catalysts has been applied to the conversion of HFS into D-mannitol for use as sugar-free sweetener and in pharmacutical preparations. To investigate this, the use of immobilized glucose isomerase and different hydrogenation catalysts for example, 20% Cu/silica was evaluated. Good mannitol yields (62–66%) were obtained (Makkee *et al.*, 1985*b*).
5. Large-scale production of fructose. As discussed, modern chromatography techniques using ion-exchange resins to separate fructose and glucose have proved the best methods on a technical scale (Lauer, 1980). In France a product with the following specification is on the market:

Fructose syrup (levulys 70/95)

Dry weight (%)	70
Fructose (%)	95
Glucose (%)	4·5–5·0
$(\alpha)_D^{20}$	−85°
Calcium (mg/kg)	100
pH	4·0

This material is sold in a crystalline form.

REFERENCES

Adachi, O. & Ameyama, M. (1982). *Methods in Enzymology*, **89**, 159.

AIST (1965). Japanese Patent 27525.

Ameyama, M., Shinagawa, E., Matsushita, K. & Adachi, O. (1981). *Agricultural and Biological Chemistry* **45**, 851.

Ameyama, M., Nonobe, M., Hayashi, M., Shinagawa, E., Matsushita, K. & Adachi, O. (1985). *Agricultural and Biological Chemistry* **49**, 1227.

Ameyama, M., Nonobe, M., Shinagawa, E., Matsushita, K., Takimoto, K. & Adachi, O. (1986). *Agricultural and Biological Chemistry*, **50**, 49.

Amore, R., Wilhelm, M. & Hollenberg, C. P. (1989). *Applied Microbiology and Biotechnology*, **30**, 351.

Antrim, R. L. & Auterinen, A. L. (1986). *Starch/Stärke*, **38**, 132.

Ashby, R., Hjortkjaer, R. K., Stavnsbjerg, M., Gurtler, H., Pedersen, P. B., Bootmann, J., Hodson-Walker, G., Tesh, J. M. & Willoughby, C. R. (1987). *Toxicology Letters*, **36**, 23.

Bak, T.-G. (1967). *Biochimica et Biophysica Acta*, **139**, 277.

Barker, S. A. (1976). *Process Biochemistry*, **11**, Dec., 25.

Barker, S. A. & Petch, G. S. (1985). *Biotechnology Series*, **5**, 93.

Barker, S. A. & Shirley, J. A. (1980). In *Economic Microbiology*, ed. A. H. Rose. Academic Press, London, p. 171.

Barker, S. A., Pelmore, H. & Somers, P. (1983). *Enzyme and Microbial Technology*, **5**, 121.

Bentley, R. & Bhate, D. S. (1960a). *Journal of Biological Chemistry*, **235**, 1219.

Bentley, R. & Bhate, D. S. (1960b). *Journal of Biological Chemistry*, **235**, 1225.

Bigelis, R. (1985). In *Gene Manipulation in Fungi*, eds J. W. Bennett & L. L. Lasure. Academic Press, New York, p. 357.

Bissé, E., Scholer, A. & Vonderschmitt, D. J. (1981). *Journal of Applied Biochemistry*, **3**, 176.

Boguslawski, G. & Whitted, B. E. (1984). *Journal of Applied Biochemistry*, **6**, 289.

Bok, S. H., Seidman, M. & Wopat, P. W. (1984). *Applied and Environmental Microbiology*, **47**, 1213.

Bond, R. C., Knight, E. C. & Walker, T. K. (1937). *Biochemical Journal*, **31**, 1033.

Bucke, C. (1983). In *Microbial Enzymes and Biotechnology*, ed. W. M. Fogarty. Applied Science Publ., London, p. 93.

Cetus (1988). US Patent 171693.

Chang, H. N., Joo, I. S. & Ghim, Y. S. (1984). *Biotechnology Letters*, **6**, 487.

Chang, H. N., Kyung, Y. S. & Chung, B. H. (1987). *Biotechnology and Bioengineering*, **29**, 552.

Chao, K. C., Haugen, M. M. & Royer, G. P. (1986). *Biotechnology and Bioengineering*, **28**, 1289.

Chen, W. (1980a). *Process Biochemistry*, **15**, June–July, 30.

Chen, W. (1980b). *Process Biochemistry*, **15**, Aug.–Sept., 36.

Chen, K. C. & Wu, J. Y. (1987). *Biotechnology and Bioengineering*, **30**, 817.

Cho, Y. K. & Bailey, J. E. (1977). *Biotechnology and Bioengineering* **19**, 185.

Chun, U. H. & Rogers, P. L. (1988). *Applied Microbiology and Biotechnology* **29**, 19.

Chung, B. H., Chang, H. N. & Kho, Y. H. (1987). *Journal of Fermentation Technology*, **65**, 575.

Clark, L. C. Jr. & Lyons, C. (1962). *Annals of the New York Academy of Sciences*, **102**, 29.

Cleton-Jansen, A.-M., Goosen, N., Wenzel, T. J. & van de Putte, P. (1988). *Journal of Bacteriology*, **170**, 2121.

CPC International Inc. (1975). British Patent 1411764.

Crueger, A. & Crueger, W. (1984). In *Biotechnology 6a*, eds H.-J. Rehm & G. Reed.

Verlag Chemie, Weinheim, p. 441.

Crueger, W. & Crueger, A. (1990). *Biotechnology, a Textbook of Industrial Microbiology*, Sinauer Associated Inc., Sunderland, Ma.

D'Angiuro, L. & Cremonesi, P. (1982). *Biotechnology and Bioengineering*, **24**, 207.

Debnath, M. & Majumdar, S. K. (1987). *Applied Microbiology and Biotechnology*, **26**, 189.

Denki Kagaku, K. K. (1974). Japanese Patent 2414694.

Diers, I. (1976). In *Continuous Culture 6: Applications in New Fields*, ed. A. C. R. Dean. Ellis Horwood Ltd, Chichester, p. 208.

Van Dijken, J. P. & Veenhuis, M. (1980). *European Journal of Applied Microbiology and Biotechnology*, **2**, 275.

Djejeva, G., Stoeva, N. & Stoychev, M. (1986a). *Acta Microbiologica Bulgarica*, **19**, 32.

Djejeva, G., Grigorova, J., Raykovska, V. & Stoychev, M. (1986b). *Acta Microbiologica Bulgarica*, **19**, 37.

Doeppner, T. & Hartmeier, W. (1984). *Starch/Stärke*, **36**, 283.

Dokter, P., Pronk, J. T., van Schie, B. J., van Dijken, J. P. & Duine, J. A. (1987). *FEMS Microbiology Letters*, **43**, 195.

Drazic, M., Golubic, Z. & Cizmek, S. (1980). *Periodicum Biologorum*, **82**, 481.

Drocourt, D., Bejar, S., Calmels, T., Reynes, J. P. & Tiraby, G. (1988). *Nucleic Acids Research*, **16**, 9337.

D'Souza, S. F. & Nadkarni, G. B. (1980). *Biotechnology and Bioengineering*, **22**, 2179.

Ellaiah, P., Sambamurthy, K. & Ramesh, K. V. R. N. S. (1987). *Indian Journal of Pharmaceutical Sciences*, **49**, 127.

Eriksson, K.-E., Pettersson, B., Volc, J. & Musilek, V. (1986). *Applied Microbiology and Biotechnology*, **23**, 257.

FDA (1983). United States Food and Drug Administration. *Federal Register, 8. Fe.*, **48**, 5715.

Fiedurek, J., Rogalski, J., Ilczuk, Z. & Leonwicz, A. (1986). *Enzyme and Microbial Technology*, **8**, 734.

Fullbrook, P. D. (1984). The enzymic production of glucose syrups. In *Glucose Syrups*, eds S. Z. Dziedzic & H. W. Kearsley. Elsevier Applied Science, London, pp. 65–115.

Fujimoto, A., Ingram, P. & Smith, R. A. (1965). *Biochimica et Biophysica Acta*, **96**, 91.

Gärtner, D., Geissendörfer, M. & Hillen, W. (1988). *Journal of Bacteriology*, **170**, 3102.

Geiger, O. & Görisch, H. (1986), *Biochemistry*, **25**, 6043.

Giardina, P., De Biasi, M.-G., De Rosa, M., Gambacorta, A. & Buonocore, V. (1986). *Biochemical Journal* **239**, 517.

Gibson, Q. H., Swoboda, B. E. P. & Massey, V. (1974). *Journal of Biological Chemistry* **239**, 3927.

Gist Brocades (1979). Technical data sheet Mgi-01-01/79.02.En05.

Hafner, E. W. (1985). US Patent 4551430.

Harder, A. & Wilms, J. (1985). *European Congress of Biotechnology*, S **IV**, 345.

Hartmeier, W. (1979). *Biotechnology Letters*, **1**, 21.

Hartmeier, W. & Doeppner, T. (1984). Ger. Offen. DE 3301992.

Hayashi, S. & Nakamura, S. (1981). *Biochimica et Biophysica Acta*, **657**, 40.

Hebeda, R. E. & Leach, H. W. (1975). US Patent 3922201.

Heilmann, H. J., Mägert, H. J. & Gassen, H. G. (1988). *European Journal of Biochemistry*, **174**, 485.

Von Hemfort, H. & Kohlstette, W. (1984). *Starch/Stärke*, **36**, 109.

Henrick, K., Blow, D. M., Carrel, H. L. & Glusker, J. P. (1987). *Protein Engineering*, **1**, 467.

Hönes, J., Jany, K.-D., Pfleiderer, G. & Wagner, A. F. V. (1987). *FEBS Letters*, **212**, 193.

Horwath, R. O. (1986). US Patent 4593001.

Hupkes, J. V. & van Tilburg, R. (1976). *Starch-Stärke*, **28**, 356.

Janssen, F. W. & Ruelius, H. W. (1968). *Biochimica et Biophysica Acta*, **167**, 501.

Jany, K.-D., Ulmer, W., Fröschle, M. & Pfleiderer, G. (1984). *FEBS Letters*, **165**, 6.

Jeanes, A. R. (1978). *Dextran Bibliography*. United States Department of Agriculture, Research Science, Miscellaneous Publication 1355.

Jensen, V. J. & Rugh, S. (1987). *Methods in Enzymology*, **136**(C), 356.

Johnson, R. A., Antrim, R. L. & Lloyd, N. E. (1986). European Patent 186871.

Jörgensen, O. B., Karlson, L. G., Nielsen, N. B., Pedersen, S. & Rugh, S. (1988). *Starch/Stärke*, **40**, 307.

Juhasz, A., Gsizmadia, V., Borbely, G. & Udvardy, J. (1987). *Biochimica et Biophysica Acta*, **916**, 119.

Karube, I., Hirano, K.-J. & Suzuki, S. (1977). *Biotechnology and Bioengineering*, **19**, 1233.

Katwa, L. C. & Rao, M. R. R. (1983). *Biotechnology Letters*, **5**, 191.

Keller, H. W., Reents, A. C. & Larawey, J. W. (1981). *Starch/Stärke*, **33**, 55.

Kelley, R. L., Ramasamy, K. & Reddy, C. A. (1986). *Archiv für Mikrobiologie*, **144**, 254.

Keyes, M. H. (1980). US Patent 4194067.

Kierstan, M. (1980). *Process Biochemistry*, **15**, May, 2.

Kobayashi, Y. & Horikoshi, K. (1980). *Agricultural and Biological Chemistry*, **44**, 2261.

Koths, K. E. & Halenbeck, R. F. (1986). US Patent 4569913.

Kowser, N. & Joseph, R. (1982). *Starch/Stärke*, **34**, 171.

Kozhukharova, A., Kirova, N., Popova, Y., Batsalova, K. & Kunchev, K. (1988). *Biotechnology and Bioengineering*, **32**, 245.

Kozulic, B., Leustek, I., Pavlovic, B., Mildner, P. & Barbaric, S. (1987). *Biochemistry and Biotechnology*, **15**, 265.

Kucera, J., Hradil, J., Stamberg, J., Rada, V. & Tomisek, J. (1983). CS Patent 206922.

Kusai, K., Sekuzu, K., Okunuki, K., Hagihara, B., Nakai, M. & Yamauchi, S. (1960). *Biochemica et Biophysica Acta*, **40**, 555.

Lastick, S. M., Tucker, M. Y., Mackedonski, V. & Grohmann, K. (1986*a*). *Biotechnology Letters*, **8**, 1.

Lastick, S. M., Tucker, M. Y. & Grohmann, K. (1986*b*). Abstract Papers American Chemical Society 192. Meeting, BTEC 52.

Lauer, K. (1980). *Starch/Stärke*, **32**, 11.

Lee, C. K. (1981). US Patent 4283496.

Linton, J. D., Woodard, S. & Gouldney, D. G. (1987). *Applied Microbiology and Biotechnology*, **25**, 357.

Lockwood, L. B. (1975). In *The Filamentous Fungi*, Vol. I, eds J. E. Smith & D. R. Berry. Edward Arnold, London, p. 140.

Lockwood, L. B. (1979). In *Microbial Technology*, ed. H. J. Peppler, Academic Press, New York, p. 355.

Long, M. E. (1978). US Patent 29692.

Luenser, S. J. (1983). *Developments in Industrial Microbiology*, **24**, 79.

Machida, Y. & Nakanishi, T. (1984). *Agricultural and Biological Chemistry*, **48**, 2463.

Makkee, M., Kieboom, A. P. G. & van Bekkum, H. (1985*a*). *Starch/Stärke*, **37**, 231.

Makkee, M., Kieboom, A. P. G. & van Bekkum, H. (1985*b*). *Carbohydrate Research*, **138**, 237.

Markwell, J., Frakes, L. G., Brott, E. C., Osterman, J. & Wagner, F. W. (1989). *Applied Microbiology and Biotechnology*, **30**, 166.

Marshall, R. O. (1960). US Patent 2950288.

Marshall, R. O. & Kooi, E. R. (1957). *Science*, **125**, 648.

Matsushita, K. & Ameyama, M. (1982). *Methods in Enzymology,* **89,** 149.

Matsushita, K., Ohno, Y., Shinagawa, E., Adachi, O. & Ameyama, M. (1980). *Agricultural and Biological Chemistry,* **44,** 1505.

Matsushita, K., Shinagawa, E., Adachi, O. & Ameyama, M. (1989a). *Biochemistry,* **28,** 6276.

Matsushita, K., Shinagawa, E., Adachi, O. & Ameyama, M. (1989b). *Journal of Biochemistry,* **105,** 633.

Maurer, E. & Pfleiderer, G. (1987). *Zeitschrift für Naturforschung, Section C: Biosciences,* **42,** 907.

Meers, J. L. (1981). *Biotechnology Letters,* **3,** 136.

Miall, L. M. (1978). In *Primary Products of Metabolism,* ed. A. H. Rose. Academic Press, London, p. 99.

Miles Laboratories (1983). US Patent 465243.

Milsom, P. E. & Meers, J. L. (1985). In *Comprehensive Biotechnology,* Vol. 3, ed. M. Moo-Young. Pergamon Press, Oxford, p. 681.

Mischak, H., Kubicek, C. P. & Röhr, M. (1985). *Applied Microbiology and Biotechnology,* **21,** 27.

Mizutani, S., Iijima, S., Morikawa, M., Shimizu, K., Matsubara, M., Ogawa, Y., Izumi, R., Matsumoto, K. & Kobayashi, T. (1987). *Journal of Fermentation Technology,* **65,** 325.

Molliard, M. (1922). *Compte Rendue Academic Science,* **174,** 881.

Montenecourt, B. S., Carroll, J. O. & Lanzilotta, R. P. (1985). In *Comprehensive Biotechnology,* ed. M. Moo-Young. Pergamon Press, Oxford, p. 329.

Müller, D. (1928). *Biochemische Zeitschrift,* **199,** 136.

Mullings, R. (1985). *Enzyme and Microbial Technology,* **7,** 592.

Naarden International N. V. (1976). British Patent 1456262.

Nakajima, Y. (1988). *Journal of the Japanese Society of Starch Science,* **35,** 131.

Nakamatsu, T., Akamatsu, T., Miyajima, R. & Shiio, I. (1975). *Agricultural and Biological Chemistry,* **39,** 1803.

Natake, M. (1966). *Agricultural and Biological Chemistry,* **30,** 887.

Natake, M. (1968). *Agricultural and Biological Chemistry,* **32,** 303.

Natake, M. & Yoshimura, S. (1963). *Agricultural and Biological Chemistry,* **27,** 342.

Natake, M. & Yoshimura, S. (1964). *Agricultural and Biological Chemistry,* **28,** 510.

Neidleman, S. L. & Geigert, J. (1986). *The preparation of D-glucosone: a case history favoring enzymatic over chemical synthesis.* Online International Inc., New York, p. 49.

Neidleman, S. L., Amon, W. F. & Geigert, J. (1981). European Patent 42221.

Neijssel, O. M., Tempest, D. W., Postma, P. W. & Duine, J. A. (1983). *FEMS Microbiology Letters,* **20,** 35.

Niederpruem, D. J. & Doudoroff, M. (1965). *Journal of Bacteriology,* **89,** 697.

Oestergaard, F. & Knudsen, S. L. (1976). *Starch/Stärke,* **28,** 350.

Olijve, W. & Kok, J. J. (1979a). *Archiv für Mikrobiologie,* **121,** 291.

Olijve, W. & Kok, J. J. (1979b). *Archiv für Mikrobiologie,* **121,** 283.

O'Malley, J. J. & Weaver, J. L. (1972). *Biochemistry,* **11,** 3527.

Owusu, R. K. & Finch, A. (1986). *Biochimica et Biophysica Acta,* **872,** 83.

Pal, G. P., Jany, K. D. & Saenger, W. (1987). *European Journal of Biochemistry,* **167,** 123.

Park, Y. H., Chung, T. W. & Han, M. H. (1980). *Enzyme and Microbial Technology,* **2,** 227.

Porter, E. V. & Chassy, B. M. (1982). *Methods in Enzymology,* **90,** 25.

Porter, M. C., Hartnagel, R. E., Kowalski, R. L., Clemens, G. R., Jasty, V., Bare, J. J. & Boguslawski, G. (1984). *Journal of Food Protection,* **47,** 359.

Prapulla, S. G., Thakur, M. S., Jaleel, S. A., Srikanta, S., Prasad, M. S., Devi, P. N., Ghildyal, N. P. & Lonsane, B. K. (1987). *Chemie, Mikrobiologie, Technologie der Lebensmittel,* **10,** 168.

Pritchard, R., Beauclerk, A. & Smith, A. J. (1975). *Biochemical Society Transactions*, **3**, 384.

Pronk, J. T., Levering, P. R., Olijve, W. & van Dijken, J. P. (1989).˙ *Enzyme and Microbial Technology*, **11**, 160.

Ramaley, R. F. & Vasantha, N. (1983). *Journal of Biological Chemistry*, **258**, 12558.

Renneberg, R., Kaiser, G., Scheller, F. & Tsujisaka, Y. (1985). *Biotechnology Letters*, **7**, 809.

Reynolds, R. J. (1973). British Patent 1328980.

Roels, J. A. & van Tilburg, R. (1979). In *Immobilized Microbial Cells*, ed. K. Venkatsubramanian. American Chemical Society, Washington, p. 147.

Rogalski, J., Fiedurek, J., Szczordrak, J., Kapusta, K. & Leonowicz, A. (1988). *Enzyme and Microbial Technology*, **10**, 508.

Rosenberg, S., Frederick, K., Bauer, M., Tung, W., Chamberlain, S. & Masiarz, F. (1989). *Journal of Cellular Biochemistry*, **13A**, 77.

Roth, M., Neigenfind, M., Haenel, F. & Bormann, E. J. (1987). *Biotechnology Letters*, **9**, 855.

Ruelius, H. W., Kervin, R. M. & Janssen, F. W. (1968). *Biochimica et Biophysica Acta*, **167**, 493.

Saari, G. C., Kumar, A. A., Kawasaki, G. H., Insley, M. Y. & O'Hara, P. J. (1987). *Journal of Bacteriology*, **169**, 612.

Sankaran, K., Godbole, S. S. & D'Souza, S. F. (1989). *Enzyme and Microbial Technology*, **11**, 617.

Sano, K., Otani, M. & Umezawa, C. (1988). *Biochemical and Biophysical Research Communications*, **151**, 48.

Sarthy, A. V., McConaughy, B. L., Lobo, Z., Sundstorm, J. A., Furlong, C. E. & Hall, D. B. (1987). *Applied and Environmental Microbiology*, **53**, 1996.

Satoh, I., Karube, I. & Suzuki, S. (1976). *Biotechnology and Bioengineering*, **18**, 269.

Van Schie, B. J., Hellingwerf, K. J. & van Dijken, J. P. (1985). *Journal of Bacteriology*, **163**, 493.

Van Schie, B. J., De Mooy, O. H., Linton, J. D., van Dijken, J. P. & Kuenen, J. G. (1987). *Journal of General Microbiology*, **133**, 867.

Schütte, H., Kroner, K. H., Hummel, W. & Kula, M.-R. (1983). In *Biochemical Engineering III. Annals of the New York Academy of Sciences*, **413**, 270.

Shaw, P. C., Anderton, T., Rangarajan, M. & Hartley, B. S. (1987). *Protein Engineering*, **1**, 264.

Shieh, K. K. (1977). US Patent 40003793.

Shiraishi, F., Kawakami, K., Tamura, A., Tsuruda, S. & Kusunoki, K. (1989). *Applied Microbiology and Biotechnology*, **31**, 445.

Shukla, G. & Prabhu, K. (1985a). *Enzyme and Microbial Technology*, **7**, 499.

Shukla, G. & Prabhu, K. (1985b). *Journal of Basic Microbiology*, **25**, 273.

Smith, E. P. & Ramaley, R. F. (1988). *Preparative Biochemistry*, **18**, 165.

Smith, L. D., Bungard, S. J., Danson, M. J. & Hough, D. W. (1988). *Biochemical Society Transactions*, **16**, 864.

Sojka, W., Dauth, C., Tschiersch, B. & Schwabe, K. (1984). Ger. (East) DD 207930.

Spencer, J. F. T. & Gorin, P. A. J. (1965). *Progress in Industrial Microbiology*, **7**, 177.

Stärk, J. (1976). US Patent 3930953.

Suekane, M. & Iizuka, H. (1981). *Zeitschrift für Allgemeine Mikrobiologie*, **21**, 457.

Suekane, M. & Iizuka, H. (1982). *Zeitschrift für Allgemeine Mikrobiologie*, **22**, 577.

Szymona, M. & Ostrowski, W. (1964). *Biochimica et Biophysica Acta*, **85**, 283.

Taguchi, T., Ohwaki, K. & Okuda, J. (1985). *Journal of Applied Biochemistry*, **7**, 289.

Takahashi, M., Sunto, S., Imamura, S., Takada, M. & Tagata, S. (1987). DE Patent 3714544.

Takasaki, Y. & Tanabe, O. (1962). *Hakko Kyokaishi*, **20**, 449.

Takasaki, Y. & Tanabe, O. (1963). *Kogyo Bijutsuin, Hakko Kenkyusho Kenkyo Hokuko*, **23**, 41.

Takasaki, Y., Kosugi, Y. & Kanbayashi, A. (1969). In *Fermentation Advances,* ed. D. Perlman. Academic Press, New York, p. 561.

Van Tilburg, R. (1983). Engineering aspects of biocatalysis in industrial starch conversion technology. PhD Thesis, Delft University of Technology, The Netherlands.

Tochikubo, K. & Yasuda, Y. (1985). *Microbiology and Immunology,* **29,** 213.

Tochikubo, K., Yasuda, Y. & Kozuka, S. (1987). *Microbiology and Immunology,* **31,** 95.

Tominaga, Y. & Kelly, S. (1984). *Journal of Protein Chemistry,* **3,** 49.

Tsuga, H. & Mitsuda, H. (1973). *Journal of Biochemistry,* **73,** 199.

Tucker, M. Y., Tucker, M. P., Himmel, M. E., Grohmann, K. & Lastick, S. M. (1988). *Biotechnology Letters,* **10,** 79.

Vaheri, M. & Kauppinen, V. (1977). *Process Biochemistry,* **12,** July–Aug., 5.

Verhoff, F. H. & Furjanic, J. J. (1983). *Industrial and Engineering Chemistry, Process Design and Development* **22,** 192.

Verhoff, F. H., Boguslawski, G., Lantero, O. J., Schlager, S. T. & Jao, Y. C. (1983). *Journal of Applied Biochemistry,* **5,** 186.

Verhoff, F. H., Boguslawski, G., Lantero, O. J., Schlager, S. T. & Jao, Y. C. (1985). In *Comprehensive Biotechnology III,* ed. M. Moo-Young. Pergamon Press, Oxford, p. 837.

Viot, D., Bouquelet, S., Tiraby, G., Marcel, T., Sicard, G. & Verwaerde, F. (1987). *Annals of the New York Academy of Sciences,* **501,** 92.

Visuri, K. & Klibanov, A. M. (1987). *Biotechnology and Bioengineering,* **30,** 917.

Volc, J., Sedmera, P. & Musilek, V. (1978). *Folia Microbiologica,* **23,** 292.

Volc, J., Denisova, N. P., Nerud, F. & Musilek, V. (1985). *Folia Microbiologica,* **30,** 141.

Walon, R. G. P. (1977). US Patent 4009074.

Ward, G. E. (1967). In *Microbial Technology,* ed. H. J. Peppler. Reinhold Publ. Comp., London, p. 200.

Weetall, H. H. (1985). *Trends in Biotechnology,* **3,** 276.

Weidenbach, G., Bonse, D. & Richter, G. (1984). *Starch/Stärke,* **36,** 412.

Wilhelm, M. & Hollenberg, C. P. (1985). *Nucleic Acids Research,* **13,** 5717.

Wong, C. H. & Drueckhammer, D. G. (1985). *Bio/Technology,* **3,** 649.

Workman, W. E., McLinden, J. H. & Dean, D. H. (1986). *Critical Reviews in Biotechnology,* **3,** 199.

Yamanaka, K. (1962). *Agricultural and Biological Chemistry,* **26,** 167.

Yamanaka, K. (1963a). *Agricultural and Biological Chemistry,* **27,** 265.

Yamanaka, K. (1963b). *Agricultural and Biological Chemistry,* **27,** 271.

Yamanaka, K. (1965). Japanese Patent 20230/65.

Yamanaka, K. (1968). *Biochimica et Biophysica Acta,* **151,** 670.

Ye, W. N., Combes, D. & Monsan, P. (1988). *Enzyme and Microbial Technology,* **10,** 498.

Zachariou, M. & Scopes, R. K. (1986). *Journal of Bacteriology,* **167,** 863.

Zamfirescu, I., Gomoiu, I., Gozia, O., Ciopraga, J. & Schell, H. D. (1983). *Studii si Cercetari de Biochimie,* **26,** 129.

Ziffer, J., Gaffney, A. S., Rothenberg, S. & Cairney, T. J. (1971). British Patent 1249347.

Chapter 6

MICROBIAL PROTEINASES AND BIOTECHNOLOGY

HELLE OUTTRUP & C. O. L. BOYCE

Novo Nordisk A/S, Novo Allé, DK-2880, Bagsvaerd, Denmark

CONTENTS

1 INTRODUCTION

It should be stressed at the outset that the term, microbial proteinase, refers only to proteinases with microbial origins. Hence, in this chapter the term will try to stay within this definition even though, today, proteinases from other organisms, e.g. human, have been expressed by microorganisms via genetic engineering.

The vastness of the subject, microbial proteinases, makes it nearly impossible to thoroughly review all aspects in one article. Three excellent reviews (Aunstrup, 1980; Ward, 1983; Kalisz, 1988) precede this chapter; each developing different aspects of the story about microbial proteinases along with some common elements. Aunstrup's article focused on microbial selection and fermentation technology of microbial proteinases. Ward's chapter in the previous edition of this book and Kalisz' article described the many sources of microbial proteinases and discussed the possible functional role of these in nature. All give detailed accounts of the classification of proteinases; e.g. serine, metallo- and sulfhydryl proteinases; so we will not repeat these definitions and explanations here.

Therefore, this chapter will serve as an update on the literature since about 1982 and a focus on the industrially important proteinases. It will review the impact that genetic and 'protein' engineering has had, as well as, its potential and limitation for commercial successes. It will also update the information about significant industrial applications of microbial proteinases and put into perspective the individual contributions to the industrial enzyme industry. In this connection, it will describe the commercial importance, recent technical developments and the future of the major microbial proteinase products.

2 PROTEIN ENGINEERING—PRESENT STATUS

2.1 Introduction

During the last decade, advances in molecular biology, genetic engineering and computer hardware and software have enhanced the scientist's ability to determine amino acid sequence and 3-dimensional structure of proteins. Confirmation of the amino acid sequence and 3-dimensional structure by high-resolution, X-ray crystallography remains the rate limiting step.

These new technological tools for the protein scientist have made possible the new discipline called, protein engineering. This field might be considered the prize for all the structure–function research on proteins over the decades. Today, scientists are able to predict 3-D structure and function based on a given amino acid sequence. Hence protein scientists are now able to look at altering the amino acid sequence in order to improve the function of a specific protein, such as a proteolytic enzyme.

The four requirements for a protein engineer to carry out this work are:

—A cloned, sequenced and expressed gene coding for the specific protein.
—A system for site-specific mutagenesis.
—A 3-D structure of the protein based on X-ray crystallography.
—A computer modeling program to predict 3-D structures from the amino acid sequences.

In deciding which amino acid to change scientists make great use of the available data from sequence analysis of homologous proteins. For example, when scientists compare this information about structure to the differences in function/performance of homologous proteins, they gain insight as to which substitutions in amino acids are likely to bring about the desired properties. Still, each guess must be tried and confirmed through experimentation. While one has better tools today to formulate a narrower range of possibilities, one still is faced with the arduous task of cultivating many mutants and evaluating the result of each.

2.2 Serine Proteinases

2.2.1 Introduction
Alkaline serine proteinases are used as detergent additives and, thus, represent the largest volume of microbial enzymes in the industrial sector. The two main enzymes, subtilisin Novo (BPN') and subtilisin Carlsberg, are produced by *Bacillus amyloliquefaciens* and *Bacillus licheniformis*, respectively. Just as these were the first enzymes to be produced on an industrial scale, so also they are among the first targets for protein engineering.

Because of their industrial importance, a large body of published and unpublished information exists about these two enzymes. The amino acid sequences were reported more than 20 years ago (Markland & Smith, 1967; Smith *et al.*, 1968) a three-dimensional structure of subtilisin BPN' from X-ray crystallography was published shortly after (Wright *et al.*, 1969).

A refined three-dimensional structure for subtilisin Carlsberg has only recently been described (Bode *et al.*, 1986, 1987; McPhalen *et al.*, 1986; Neidhart & Petsko, 1988). This lag between the earlier work and now was because the derivatized crystals of the enzyme were extremely sensitive to X-ray radiation (Neidhart & Petsko, 1988). If it were not for the commercial importance of this enzyme, researchers might have abandoned it for an easier model.

The final requirements for protein engineers were supplied when the gene for subtilisin BPN' was identified, cloned and expressed for sequencing (Wells *et al.*, 1983). Later a 'cassette system' for site specific mutagenesis for *Bacillus subtilis* was developed which made possible specific changes in the amino acid sequence of the enzyme and to produce enough material for functionality testing (Wells *et al.*, 1985).

2.2.2 Active Site Elements
The catalytic area in the serine proteinases consists of four main elements:

—the 'catalytic triad' (Ser 221, His 64 and Asp 32 for subtilisin BPN');
—an oxyanion binding site (Asn 155);
—the 'main chain' amide of Ser 221;
—other binding determinants for specific substrates.

The amino acid sequence of 8 subtilisin-type proteinases and 3 proteinases

```
BPN'                 1          10              20        30        37
BPN'        * * * * * * * A Q S * V P Y G V S Q I K * * * * A P A L H S Q G Y T G S N V K V A V I D S G I D S
amylosacch  * * * * * * * A Q S * V P Y G I S Q I K * * * * A P A L H S Q G Y T G S N V K V A V I D S G I D S
l 168       * * * * * * * A Q S * V P Y G I S Q I K * * * * A P A L H S Q G Y T G S N V K V A V I D S G I D S
B mesent    * * * * * * * A Q S * V P Y G I S Q I K * * * * A P A L H S Q G Y T G S N V K V A V I D S G I D S
DY          * * * * * * * A Q T * V P Y G I P L I K * * * * A D K V Q A Q G Y K G A N V K V G I I D T G I A A
Carlsberg   * * * * * * * A Q T * V P Y G I P L I K * * * * A D K V Q A Q G F K G A N V K V A V L D T G I Q A
Carlsberg   * * * * * * * A Q T * V P Y G I P L I K * * * * A D K V Q A Q G F K G A N V K V A V L D T G I Q A
Savinase    * * * * * * * A Q S * V P W G I S R V Q * * * * A P A A H N R G L T G S G V K V A V L D T G I * S
Esperase    * * * * * * * Q T * V P W G I S F I N * * * * T Q Q A H N R G I F G N G A R V A V L D T G I * A
Thermitase  Y T P N D P Y F S S * R Q Y G P Q K I Q * * * * A P Q A W * D I A E G S G A K I A I V D T G V Q S
ProtK       * * * * * * A A Q T N A P W G L A R I S S T S P G T S T Y Y Y D E S A G Q G S C V Y V I D T G I E A
Aqualysin   * * * * * * A T Q S P A P W G L D R I D Q R D L P L S N S Y T Y T A T G R G V N V Y V I D T G I R T
```

```
BPN'        38 40              50              60        70        80    83
BPN'        S H P D L * * K V A G G A S M V P S E T N P F * Q D N N S H G T H V A G T V A A L * N N S I G V L G
amylosacch  S H P D L * * N V R G G A S F V P S E T N P Y * Q D G S S H G T H V A G T I A A L * N N S I G V L G
l 168       S H P D L * * N V R G G A S F V P S E T N P Y * Q D G S S H G T H V A G T I A A L * N N S I G V L G
B mesent    S H P D L * * N V R G G A S F V P S E T N P Y * Q D G S S H G T H V A G T I A A L * N N S I G V L G
DY          S H T D L * * K V V G G A S F V S G E S * Y N * T D G N G H G T H V A G T V A A L * D N T T G V L G
Carlsberg   S H P D L * * N V V G G A S F V A G E A * Y N * T D G N G H G T H V A G T V A A L * D N T T G V L G
Carlsberg   S H P D L * * N V V G G A D F V A G E A * Y N * T D G N G H G T H V A G T V A A L * D N T T G V L G
Savinase    T H P D L * * N I R G G A S F V P G E P * S T * Q D G N G H G T H V A G T I A A L * N N S I G V L G
Esperase    S H P D L * * R I A G G A S F I S S E P * S Y * H D N N G H G T H V A G T I A A L * N N S I G V L G
Thermitase  N H P D L A G K V V G G W D F V D N D S T P * * Q N G N G H G T H C A G I A A A V T N N S T G I A G
ProtK       S H P E F * * * * E G R A Q M V K T Y Y Y S S * R D G N G H G T H C A G T V G S * R * * * * * T Y G
Aqualysin   T H R E F * * * * G G R A R V G Y D A L G G N G Q D C N G H G T H V A G T I G G V * * * * * * T Y G
```

```
BPN'        84      90        100         110             120         126
BPN'        V A P S A S L Y A V K V L G A D G S G Q Y S W I I N G I E W * A I A * N N M D * * * * V I N M S L
amylosacch  V A P S A S L Y A V K V L D S T G S G Q Y S W I I N G I E W * A I S * N N M D * * * * V I N M S L
l 168       V S P S A S L Y A V K V L D S T G S G Q Y S W I I N G I E W * A I S * N N M D * * * * V I N M S L
B mesent    V A P S A S L Y A V K V L D S T G S G Q Y S W I I N G I E W * A I S * N N M D * * * * V I N M S L
DY          V A P N V S L Y A I K V L N S S G S G T Y S A I V S G I E W * A T Q * N G L D * * * * V I N M S L
Carlsberg   V A P S V S L Y A V K V L N S S G S G S Y S G I V S G I E W * A T T * N G M D * * * * V I N M S L
Carlsberg   V A P S V S L Y A V K V L N S S G S G T Y S G I V S G I E W * A T T * N G M D * * * * V I N M S L
Savinase    V A P S A E L Y A V K V L G A S G S G S V S S I A Q G L E W * A G N * N G M H * * * * V A N L S L
Esperase    V A P S A D L Y A V K V L D R N G S G S L A S V A Q G I E W * A I N * N N M H * * * * I I N M S L
Thermitase  T A P K A S I L A V R V L D N S G S G T W T A V A N G I T Y * A A D * Q G A K * * * * V I S L S L
ProtK       V A K K T Q L F G V K V L D D N G S G Q Y S T I I A G M D F V A S D K N N R N C P K G V V A S L S L
Aqualysin   V A K A V N L Y A V R V L D C N G S G S T S G V I A G V D W V * T * R N H R R P A * * * V A N M S L
```

Fig. 1. Amino acid sequences of different subtilisins and related proteinases aligned and numbered according to the sequence of subtilisin BPN' (Sven Branner, pers. comm.).

Footnote.

BPN' = subtilisin BPN' from *Bacillus amyloliquefaciens* (Wells *et al.*, 1983).

amylosacch = *Bacillus amylosacchariticus* subtilisin (Kurihara *et al.*, 1972).

l 168 = subtilisin E from *Bacillus subtilis* strain 168 (Stahl & Ferrari, 1984).

B mesent = *Bacillus mesentericus* subtilisin (Svendsen, 1986).

DY = subtilisin DY (Nedkov *et al.*, 1985).

Carlsberg = *Bacillus licheniformis* subtilisin (Smith *et al.*, 1968; Jacobs *et al.*, 1985).

Savinase = serine proteinase of an alkaline *Bacillus* species (Hastrup *et al.*, 1989).

Esperase = serine proteinase of an alkaline *Bacillus* species (Hastrup *et al.*, 1989).

Thermitase = serine proteinase from *Thermoactinomyces vulgaris* (Meloun *et al.*, 1985; Dauter *et al.*, 1988).

ProtK = proteinase K from *Tritirachium album* (Betzel *et al.*, 1988).

Aqualysin = *Thermus aquaticus* proteinase (Kwon *et al.*, 1988).

```
BPN'       127 130        140        150        160        170    176
BPN'       GGPSGSAALKAAVDKAVASGVVVVA AAGNEGTSGSSSTVGYPGKYPSVIA
amylosacch GGPSGSTALKTVVDKAVSSGIVVAA AAGNEGSSGSSSTVGYPAKYPSTIA
1168       GGPTGSTALKTVVDKAVSSGIVVAA AAGNEGSSGSTSTVGYPAKYPSTIA
B mesent   GGPTGSTALKTVVDKAVSSGIVVAA AAGNEGSSGSTSTVGYPAKYPSTIA
DY         GGPSGSTALKQAVDKAYASGIVVVA AAGNSGSSGSQNTIGYPAKYDSVIA
Carlsberg  GGASGSTAMKQAVDNAYARGVVVVA AAGNSGNSGSTNTIGYPAKYDSVIA
Carlsberg  GGPSGSTAMKQAVDNAYARGVVVVA AAGNSGSSGNTNTIGYPAKYDSVIA
Savinase   GSPSPSATLEQAVNSATSRGVLVVA ASGNSGA*GSIS***YPARYANAMA
Esperase   GSTSGSSTLELAVNRANNAGILLVG AAGNTGR*QGVN***YPARYSGVMA
Thermitase GGTVGNSGLQQAVNYAWNKGSVVVA AAGNAGNTAPN****YPAYYSNAIA
ProtK      GGGYSSSVNSAAA*RLQSSGVMVAV AAGNNNADARNYS***PASEPSVCT
Aqualysin  GGGV*STALDNAVKNSIAAGVVYAV AAGNDNANACNYS***PARVAEALT


BPN'       177 180        190        200        210        220    224
BPN'       VGAVDSSNQRASFSSVGPELDVMAPGVSIQSTLPGN*K*YGAYNGTSMAS
amylosacch VGAVNSSNQRASFSSAGSELDVMAPGVSIQSTLPGG*T*YGAYNGTSMAT
1168       VGAVNSSNQRASFSSAGSELDVMAPGVSIQSTLPGG*T*YGAYNGTSMAT
B mesent   VGAVNSANQRASFSSAGSELDVMAPGVSIQSTLPGG*T*YGAYNGTSMAT
DY         VGAVDSNKNRASFSSVGAELEVMAPGVSVYSTYPSN*T*YTSLNGTSMAS
Carlsberg  VGAVDSNSNRASFSSVGAELEVMAPGAGVYSTYPTN*T*YATLNGTSMAS
Carlsberg  VGAVDSNSNRASFSSVGAELEVMAPGAGVYSTYPTS*T*YATLNGTSMAS
Savinase   VGATDQNNNRASFSQYGAGLDIVAPGVNVQSTYPGS*T*YASLNGTSMAT
Esperase   VAAVDQNGQRASFSTYGPEIEISAPGVNVNSTYTGN*R*YVSLSGTSMAT
Thermitase VASTDQNDNKSSFSTYGSVVDVAAPGSWIYSTYPTS*T*YASLSGTSMAT
ProtK      VGASDRYDRRSSFSNYGSVLDIFGPGTSILSTWIGG*S*TRSISGTSMAT
Aqualysin  VGATTSSDARASFSNYGSCVDLFAPGASIPSAWYTSDTATQTLNGTSMAT


BPN'       225 230        240        250        260        270    275
BPN'       PHVAGAAALILSKHPNWTNTQVRSSLENTTTKLGDSFYY*GKGLINVQAAAQ
amylosacch PHVAGAAALILSKHPTWTNAQVRDRLESTATYLGDSFYY*GKGLINVQAAAQ
1168       PHVAGAAALILSKHPTWTNAQVRDRLESTATYLGNSFYY*GKGLINVQAAAQ
B mesent   PHVAGAAALILSKHPTWTNAQVRDRLESTATYLGSSFYY*GKGLINVQAAAQ
DY         PHVAGAAALILSKYPTLSASQVRNRLSSTATNLGDSFYY*GKGLINVEAAAQ
Carlsberg  PHVAGAAALILSKHPNLSASQVRNRLSSTATYLGSSFYY*GKGLINVEAAAQ
Carlsberg  PHVAGAAALILSKHPNLSASQVRNRLSSTATYLGSSFYY*GKGLINVEAAAQ
Savinase   PHVAGAAALVKQKNPSWSNVQIRNHLKNTATSLGSTNLY*GSGLVNAEAATR
Esperase   PHVAGVAALVKSRYPSYTNNQIRQRINOTATYLGSPSLY*GNGLVHAGPATQ
Thermitase PHVAGVAGLLASQGRS**ASNIRAAIENTADKISGTGTYWAKGRVNAYKAVQY
ProtK      PHVAGLAAYLMTLGKTTAASACR*YIADTANKGDLSNIPFGTVNLLAYNNVGA
Aqualysin  PHVAGVAALYLEQNPSATPASVASAILNGATTGRLSGIGSGSPNRLLYSLLSSGSG
```

Fig. 1.—(*Continued*)

from other microbial sources which are completely unrelated appear in Fig. 1. These sequences have been aligned and the amino acids numbered according to the system for subtilisin BPN' (S. Branner, 1988, personal communication).

From this alignment it is easy to see that a great deal of homology exists among these serine proteinases. That is, the amino acids of the 'catalytic triad', the oxyanion binding site, and surrounding sequences have been conserved. Therefore the differences in the substrate specificity and catalytic activity among these enzymes must reside outside the homologous sequences. The same is true for the differences in the enzymes' stability toward the destabilizing action of hot washing temperatures, high pH and oxidizing agents. Therefore, finding the sites in the non-homologous sequences responsible for the functional differences of these enzymes is the new challenge which has been set.

2.2.3 Engineering the Subtilisins

2.2.3.1 SUBSTRATE SPECIFICITY

All of the subtilisins exhibit endoproteolytic attack on proteins and peptides; some show esterase activity toward specific esters. All are inactivated by serine reagents, such as, di-isopropylfluorophosphate (DFP) and phenylmethanesulfonyl fluoride (PMSF). Likewise, all remain stable over a broad pH range (5–6 at the low end; 11–12 at the top). While all are stable at temperatures below 50°C, many show adequate activity between 60°–70°C.

Studies with synthetic tetrapeptides have demonstrated the subtle differences in specificity between the subtilisins (Wells *et al.*, 1987*a*). The tetrapeptides used for these studies possessed the general formula, *N*-succinyl–Ala–Ala–Pro–P_1-*p*- nitroanilide. Different amino acids can be substituted at the P_1-residue so as to evaluate the cleavage of the peptide bond at the carboxy-side. Phe at this site is the 'preferred' substrate for subtilisins. The differences can also be dramatic; e.g. subtilisin Carlsberg has up to 50 times greater activity than subtilisin BPN' for certain amino acids in the P_1 position.

An example of the difference in specificity can be seen by analyzing the difference in reactivity of subtilisin BPN' toward tetrapeptides with Glu, Gln, Met, or Lys substituted at the P_1-residue. The Phe as the P_1-residue is thought to interact with the amino acids numbered 156 and 166 in the subtilisin BPN' protein. If the P_1 residue were oppositely charged, molecular modeling predicted that salt bridges probably would form with these sites (Wells *et al.*, 1987*a*).

In one study mutants of subtilisin BPN' were generated in which the amino acid at sites 156 and/or 166 were substituted with others bearing a net charge from −2 to neutral to +1. As the net charge became more positive, the reactivity toward the tetrapeptides series changed. That is, with a net charge of +1, the reactivity of the tetrapeptide with Glu as the P_1-residue was 10-fold greater than its neutral homolog, Gln. Likewise, the one with Lys as the P_1-residue was 10 times less than its neutral homolog, Met (Wells *et al.*, 1987*a*).

Further, the study showed nearly additive effects for combined mutations at sites 156 and 166. For example, when using the tetrapeptide with Glu as the P_1-residue, the Gly 166→ Lys subtilisin mutant possessed a 500-fold increase in activity versus the wild-type enzyme. In addition, the double mutant, Glu 156→ Gln plus Gly 166→ Lys, exhibited a 1900-fold jump in activity (Wells *et al.*, 1987*a*).

Subtilisin Carlsberg and BPN' differ in 85 amino acids (plus a deletion) out of 275; and, as mentioned, these have a 50-fold difference in their activity toward certain of the synthetic tetrapeptides. It is remarkable that substitution of only 3 out of 85 non-homologous amino acids can be so significant. That is, these changes confer on subtilisin BPN' nearly the same substrate specificity as subtilisin Carlsberg.

Molecular modeling suggested that the amino acids at sites 156 or 217 could make direct contact with either the P_1-residue or the *p*-nitroanilide moiety, respectively in the synthetic tetrapeptide. The studies with mutants showed that the simultaneous substitutions at Glu 156→ Ser, Tyr 217→ Leu, and Gly 169→ Ala were sufficient to favorably alter the substrate specificity of subtilisin BPN' (Wells *et al.*, 1987*b*).

In other work with subtilisin E, which is produced by *Bacillus subtilis* 168 similar results were observed. The substitution, Ile 31→ Leu, significantly increased the specific activity toward casein, as well as small peptides (Takagi *et al.*, 1988). Interestingly, in wild-type subtilisin Carlsberg, leucine already is present at site 31 (Fig. 1).

2.2.3.2 OXIDATIVE STABILITY

Detergents often contain oxidizing chemicals as bleaching agents which can inactivate serine proteases, such as subtilisin. Hence, resistance to oxidation is of special interest to companies making detergents.

Met 222, which neighbors the active site Ser 221, is known to be susceptible to oxidation. When it is oxidized, subtilisins lose some of their activity (Stauffer & Etson, 1969). Thus, a protein engineering strategy was undertaken to evaluate if substitution of Met 222 with another amino acid could instill resistance to oxidizing agents, like hydrogen peroxide. Nineteen mutants of subtilisin BPN' were made where Met 222 was substituted with another amino acid. These were evaluated for their catalytic activity and resistance to hydrogen peroxide. In these experiments a synthetic tetrapeptide with Phe as the P_1 residue was used as substrate. When Ser, Ala or Leu were substituted for Met 222, the mutant enzymes were resistant to peroxide inactivation for over one hour (Estell *et al.*, 1985). Interestingly, only the Met 222→ Cys mutation showed an increase in catalytic activity; all the rest produced a decrease in activity. Of course, this Cys-mutant was rapidly inactivated in the presence of 1 M hydrogen peroxide.

2.2.3.3 HEAT STABILITY

Wash temperatures above 60°C are not recommended for detergents containing enzymes. Even at this temperature some inactivation occurs (Aunstrup, 1980); so an alkaline proteinase with greater temperature stability is an economic and technical asset to detergent manufacturers.

In pursuing a protein engineering route to this goal mixed results were obtained. The strategy was to introduce two cysteine residues at sites 22 and 87 of the subtilisin BPN' to create a disulfide bond. It was considered that the bond would prevent the peptide chain from denaturing until higher temperatures were reached. One group reported an increase of 3·1°C for the temperature for irreversible inactivation (Pantoliano *et al.*, 1987) while another group reported a decrease in thermostability compared to wild-type subtilisin BPN' (Wells & Powers, 1986).

A more conventional mutagenesis approach has definitely produced a subtilisin with improved thermostability. The mutant enzyme possessed an

Asn 218→Ser substitution; one that would hardly have been predicted by computer modeling (Bryan *et al.*, 1986). This 'direct' approach involved random mutagenesis coupled to a phenotypic screen. Mutated *Bacillus* colonies were grown on nitrocellulose filters to which the excreted proteinase would bind. The filters with bound subtilisin mutants were transferred to assay plates under extreme heat conditions in order to identify thermostable mutants (Bryan *et al.*, 1986).

2.2.3.4 ALKALINE pH STABILITY
The alkalophilic *Bacilli* are known to produce proteinases similar to subtilisin Carlsberg and BPN', but which exhibit maximum enzymatic activity at pH values up to 12 instead of 11 for the former (Aunstrup *et al.*, 1972; Aunstrup, 1980). Hence, there was good reason to believe that a more alkaline stable mutant of the subtilisins could be produced. Through a course of conventional mutagenesis and prototype screening, a mutant enzyme with improved alkaline stability was identified (Cunningham & Wells, 1987). The double mutant, Ile 107→ Val plus Lys 213→ Arg, exhibited the desired improvement. However, a triple mutant (double mutant plus Met 50→ Phe) was twice as stable as wild-type subtilisin at pH 12.

2.2.4 Conclusions
The replacement of amino acids at specific sites within proteins and the subsequent measurement of functional differences has increased our understanding of the relationship of enzyme structure and function. While computer modeling and refined 3-D X-ray crystallography have been valuable tools, they do not make the task of generating and evaluating 19 different permutations of a single amino acid substitution any easier.

Protein engineering will not make traditional mutagenesis and prototype screening obsolete. The value of these conventional microbiological techniques is clearly demonstrated in the work to find more thermo- and alkaline stable subtilisins. Also, the refined 3-dimensional X-ray crystallographic data have not explained the difference in specificity toward the synthetic tetrapeptides that the subtilisins Carlsberg and BPN' exhibit (Neidhart & Petsko, 1988). Thus, while protein engineering allows scientists to make predictable changes, some of the results still remain unpredictable.

2.3 Aspartic Proteinases

2.3.1 Introduction
The second major industrial proteinase group are the renneting enzymes for cheesemaking. These enzymes are basically acid proteinases which have an aspartic acid residue as a key catalytic moiety in the active site (Foltmann & Pedersen, 1977). In contrast to the serine proteinases, aspartic proteinases show a high degree of substrate specificity which means that they exhibit a low degree of general proteolytic activity on most proteins. For cheesemaking a

renneting enzyme should only cleave the Phe 105–Met 106 peptide bond in the κ-casein protein chain (Delfour *et al.*, 1965).

In cheesemaking calf-chymosin is the preferred enzyme; however, it commands a premium price because of its source: the fourth stomach of the suckling animal. While porcine pepsin has been used as an economical adjunct, its higher degree of proteolytic activity against casein is undesirable. Thus, several microbial sources of milk-clotting enzyme were sought. Microbial sources of rennets with nearly the same activity as calf-chymosin include: *Rhizomucor miehei*, *Mucor pusillus* Lindt and *Endothia parasitica*. The first two organisms are described as thermophilic fungi while the third is a yeast (Morihara, 1974). All of these are produced on an industrial scale as alternatives for calf-chymosin but mainly out of tradition these have not penetrated the renneting enzyme market to the degree anticipated.

The known architecture of the active site of aspartic proteinases comes mostly from work with pepsin A. It is the only one which has been characterized via a 3-dimensional X-ray study (James & Sielecki, 1986). The amino acid sequences and the coding genes have been identified and published for the microbial aspartic proteinases as well as for calf-chymosin and porcine pepsin. Figure 2 presents the amino acid sequences for these enzymes which have been aligned to demonstrate their homology (Foltmann, 1988).

The functioning and structure of the active site of aspartic proteinases has been described as a 'cleft' and 'flap'. One hypothesis for the catalytic mechanism suggests that within the flap portion the residue, Tyr 75, appears important for binding the substrate. Thus, this residue is partly responsible for ensuring that a substrate peptide bond comes in close proximity to the aspartic acid complex in the cleft portion.

The two aspartic acid residues (Nos 32 and 215, according to the numbering for pepsin A) are hydrogen bonded through their carboxylic acid side-chains. This complex, in turn, polarizes a molecule of water which inserts via nucleophilic attack, into a nearby peptide bond of the substrate (Pearl, 1985).

Researchers have found that the genes for the fungal and mammalian renneting enzymes actually code for a zymogen, not the active enzyme. That is, these enzymes are produced as an inactive precursor, which must be modified by proteinases before gaining activity. The prochymosin and pre-prochymosin moieties have been cloned by several groups (Harris *et al.*, 1982; Moir *et al.*, 1982; Hidaka *et al.*, 1986). However, expression and post-translational maturation of these moieties in genetically engineered microbes have met with problems.

In Escherichia coli the cloned prochymosin is packaged into inclusion bodies. These have to be solubilized; the prochymosin isolated, purified and treated with acid in order to have active genetic-engineered chymosin. Solubilization of the inclusion bodies is accomplished with a urea treatment. Dialysis under alkaline conditions is necessary to ensure proper folding of the prochymosin (Beppu, 1988). Thus, while it may be feasible to get microorganisms to produce proteins from higher life forms, the importance of post-

```
        1         10          20          30          40
PA     IGDEPLENYLD—TEYFGTIGIGTPAQDFTVIFDTGSSNLWVPSVYC
BC     GEVASVPLTNYLD—SQYFGKIYLGTPPQEFTVLFDTGSSDFWVPSIYC
MM  AAADGSVDTPGYYDFDLEEYAIPVSIGTPGQDFLLLFDTGSSDTWVPHKGC
MP   AEGDGSVDTPGLYDFDLEEYAIPVSIGTPGQDFYLLFDTGSSDTWVPHKGC
EP    STGSATTTPIDSLD—DAYITPVQIGTPAQTLNLDFDTGSSDLWVFSSET
```

```
     46    50          60          70          80          90
PA   —SSLACSDHNQFNPDDSSTFEA—TSQELSITYGTGS—MTGILGYDTVQV————
BC   —KSNACKNHQRFDPRKSSTFQN—LGKPLSIHYGTGS—MQGILGYDTVTV————
MM   TKSEGCVGSRFFDPSASSAFKA—TNYNLNITYGTGGANGLYFEDSIAIGDITV
MP   DNSEGCVGKRFFDPSSSSTFKE—TDYNLNITYGTGGANGIYFRTSITVGGATV
EP   TASE—VDGQTIYTPSKSTTAKLLSGATWSISYGDGSSSSGDVYTDTVSV————
```

```
           100         110         120         130         140
PA   GGISDTNQIFGLSETEPGSFLYYAPFDGILGLAYPSISASGATPV———FDNLWD
BC   SNIVDIQQTVGLSTQEPGDVFTYAEFDGILGMAYPSLASEYSIPV———FDNMMN
MM   TKQILAYVDNVRGPTAEQSPNADIFLDGLFGAAYPDNTAMEAEYGSTYNTVHVN
MP   KQQTLAYVDNVSGPTAEQSPDSELFLDGIFGAAYPDNTAMEAEYGDTYNTVHVN
EP   GGLTVTGQAVESAKKVSSSFTEDSTIDGLLGLAFSTLNTVSPTQQKTFFDNAKA
```

```
           150         160         170         180
PA   ———QGLVSQDLFSVYLSSNDDSGSVVLLGGIDSSYYTGSLNWVPV————SVEGYW
BC   ———RHLVAQDLFSVYMDR—DGQESMLTLGAIDPSYYTGSLHWVPV————TVQQYW
MM   LYKQGLISSPLFSVYMNT—NSGTGEVVFGGVNNTLLSGDIAYTDVMSRYGGYYFW
MP   LYKQGLISSPVFSVYMNT—NDGGGQVVFGGVNNTLLGGDIQYTDVLKSRGGYFFW
EP   S————LDSPVFTADLGY——HAPGTYNFGFIDTTAYTGSITYTAVS———TKQGFW
```

```
     191       200         210         220         230         240
PA   QITLDSITMDGETI—ACSGGCQAIVDTGTSLLTGPTSAIAINIQSDIGASENS—
BC   QFTVDSVTISGVVV—ACEGGCQAILDTGTSKLVGPSSDIL—NIQQAIGATQNQ—
MM   DAPVTGITVDGSAAVRFSRPQAFTIDTGTNFFIMPSSAASKIVKAALPDATETQ
MP   DAPVTGVKIDGSDAVSFDGAQAFTIDTGTNFFIAPSSFAEKVVKAALPDATESQ
EP   EWTSTGYAV—GSGTFK—STSIDGIADTGTTLLYLPATVVSAYWAQVSGAKSSSS
```

```
         250         260         270                    280
PA   DGEMVISCSSIDSLPDIVFTIDGVQYPLSPSAYILQ———————————DDDSCTS
BC   YGEFDIDCDNLSYMPTVVFEINGKMYPLTPSAYTSQ——————————DQGFCTS
MM   QG-WVVPCASYQNSKSTISIVMQKSGSSSDTIEISVPVSKM-LPVDQSNETCMF
MP   QG-YTVPCSKYQDSKTTFSLDLQKSGSSSDTIDVSVPISKMLLPVDKSGETCMF
EP   VGGYVFPCSA——TLPSFTFGVGSARIVIPGDYIDFG———————PISTGSSSCFG
```

```
       290         300         310         320       327
PA   GFEGMDVPTSSGELWILGDVFIRQYYTVFDRAN-NKVGLAPVA
BC   GFQS————ENHSQKWILGDVFIREYYSVFDRAN-NLVGLAKAI
MM   IIILP—————NGGNQYIVGNLFLRFFVNVYDFGN—NRIGFAPLASAYENE
MP   IVLP—————DGGNQFIVGNLFLRFFVNVYDFGK—NRIGFAPLASGYENN
EP   GIQSS————AGIGINIFGDVALKAAFVVFNGATTPTLGFASK
```

Fig. 2. Amino acid sequences of aspartic proteinases with milk-clotting activity aligned and numbered according to the sequence of porcine pepsin A (Foltmann, 1988).

Footnote.
PA = porcine pepsin A (Tang *et al.*, 1973).
BC = bovine chymosin (Hidaka *et al.*, 1986).
MM = aspartic proteinase from *Rhizomucor miehei* (Boel *et al.*, 1986).
MP = aspartic proteinase from *Mucor pusillus* (Tonouchi *et al.*, 1986).
EP = *Endothia parasitica* proteinase (Barkholt, 1987).

translational modifications should not be overlooked. Microbial hosts for genetically engineered proteins may lack the cellular equipment to duplicate the 'maturation' which routinely occurs in the original organism. In this case, these problems could mean the difference between commercial success and laboratory curiosity.

Production of cloned-chymosin has been successful with the production of fully active enzyme from *Aspergillus nidulans* (Cullen *et al.*, 1987) and *A. awamori* (Hayenga *et al.*, 1988). Also, cloned-chymosin from *Saccharomyces cerevisiae* has been commercialized by Gist-Brocades.

2.3.2 Engineering of Calf-Chymosin

Site-directed mutagenesis has helped to elucidate the structure–function relationship of certain residues in calf-chymosin (Beppu, 1988). Modifications at three sites in the chymosin chain have produced interesting effects.

When Tyr 77 was replaced with Phe, a significant increase in clotting activity relative to general proteolytic activity occurred. The Tyr 77 residue of chymosin corresponds to the Tyr 75 substrate binding site in the 'flap' region of the active site of pepsin A. Similarly, when the dipeptide, Ser–Arg, is inserted after Phe 110 or when Phe is substituted at Val 113, significant increases of milk-clotting activity occur. Finally, when Glu or Leu are substituted for Lys 221, the pH optimum for the proteolytic activity against denatured hemoglobin is lowered (Beppu, 1988).

2.3.3 Engineering of Microbial Rennets

While microbial rennets are a lot less proteolytic than pepsin, they are a little more active than the calf-chymosin. This deficiency plus the conservative nature of the dairy tradition explain the preference for calf-chymosin by the cheesemaker. Hence, microbial rennets are an attractive target for the protein engineer. The goal of course, is to erase this minor flaw in order to create a microbial rennet with exactly the same action profile as calf-chymosin.

The amount of work in this area lags behind that for the serine proteinases. The gene coding for the proteinase from *Rhizomucor miehei* has been cloned (Boel *et al.*, 1986) and successfully expressed in *Aspergillus oryzae* (Christensen *et al.*, 1988). Although the enzyme was slightly over-glycosylated compared to the wild-type, the specific activity of the proteinase was unaltered. No reports of site-directed mutagenesis on this enzyme have appeared to date. On the other hand, the closely related proteinase from *Mucor pusillus* has been cloned for the purpose of protein engineering. The researchers are interested in finding out if His 48 and His 155 are involved in the catalysis (Beppu, 1988).

Although protein engineering of the aspartic proteinases is still in its infancy, progress is being made. More information is needed about the structure–function relationship of more sites in the enzyme protein. Besides the interest in determining the exact relationships within the 'cleft' and 'flap' regions, the economic reward of a precise calf-chymosin substitute will fuel the fire of various research teams in this endeavor.

2.4 Metallo-proteinases

The third ranked group of industrially important proteinases are the sulfhydryl proteinases from plant sources, such as papain and bromelain. Unlike the aspartate proteinases, there are no commercially viable microbial alternatives. While at least one microbial sulfhydryl proteinase, clostripain, has been reported, it is from a non-food-grade source, *Clostridium histolyticum* (Gros & Labanesse, 1960).

Thus, the metallo-proteinases are the next important class of microbial proteinases in the industrial sector. As implied in their group name, the metallo-proteinases require a metal cation, usually Zn^{2+}, residing in the active site. In the enzyme from *Bacillus thermoproteolyticus*, thermolysin, the Zn-cation is bound by three ligands to His 142, His 146 and Glu 166. Also characteristic of metallo-proteinases, Glu 143 participates in catalysis by promoting the attack of a water molecule on the carbonyl carbon of the peptide bond in question (Kester & Matthews, 1977).

Figure 3 presents the amino acid sequences of four metallo-proteinases. The key active site residues described for thermolysin are conserved throughout. Also noteworthy is the high degree of homology (85%) between the sequence for thermolysin and the less thermostable proteinase from *Bacillus stearothermophilus*.

Because of this homology and the fact that a tertiary structure for thermolysin has been proposed (Matthews *et al.*, 1972) researchers have assumed a similar structure for the proteinase from *B. stearothermophilus*. Also because of the difference in these enzymes' thermostability, this pair of metallo-proteinases made an attractive subject for further study (Imanaka *et al.*, 1986). Molecular modeling suggested that the Gly 141 → Ala substitution might increase the thermostability of the *B. stearothermophilus* proteinase. The insertion of an Ala for the Gly was thought to increase the internal hydrophobicity and thus, stabilize the α-helix to a greater degree. Experiments confirmed the molecular modeling hypothesis.

Further, molecular modeling predicted that the single substitution, Thr 63 → Ser would decrease thermostability. This mutant confirmed researchers' predictions. However, the double mutant, Thr 63 → Ser and Gly 141 → Ala, did not reverse the effect. That is, the double mutant had the same degree of thermolability as the destabilized single mutant at site 63. Only the triple mutant (with the addition of a Gly 58 → Ala substitution) showed an increase in thermostability (Imanaka *et al.*, 1986).

The neutral proteinases from *Bacillus subtilis* and *B. amyloliquefaciens* require Ca^{2+}-cations, which are bound by four residues: Asp 138, Asp 185, Glu 190 and Asp 191. These sites are conserved in the thermolabile, as well as the thermostable proteinases, as depicted in Fig. 3. Interestingly, another residue, Glu 177, appears only in the thermostable proteinases. Its appearance gives rise to the unconfirmed hypothesis that deletion of this residue accounts for the poor thermostability of the metallo-proteinases from *B. subtilis* and *B. amyloliquefaciens* (Imanaka *et al.*, 1986).

```
       1            10            20            30            40
TH  I TGTS TVGVGRGVLDGQKN I NTTYSTYY - - - YLQDNTR - -GDG I FTYDAKYRT-TL
BS  VAGAS TVGVGRGVLGDQKY I NTTYSS YYGYY YLQDNTR - -GSG I FTYDGRNRT-VL
SU  AAA-T-GSGTTLKGATVPLN - - I SYEGGKYVLRDLSKPTGTQ I I TYDLQNRQSRL
AM  AAT-T-GTGTTLKGKTVSLN - - I SSESGKYVLRDLSKPTGTQ I I TYDLQNREYNL

       51           60            70            80            90           100
TH  PGSLWADADNQFFAS YDAP AVDAHYYAGVTYDYYKNVHNRL S YDGNNAA I RS SVHY
BS  PGSLWTDGDNQFTAS YDAAAVDAHYYAFVVYDYYKNVHGRL S YDGS NAA I RSTVHY
SU  PGTLVS STTKTFTSS SQRAAVDAHYNLGKVYDYFYSNKFRNS YDNKGSK I VSSVHY
AM  PGTLVS STTNQFTTS SQRAAVDAHYNLGKVYDYFYQKFNRNS YDNKGGK I VSSVHY

       110          120           130           140           150          160
TH  SQGYNNAFWNGS EMVYGDGDGQTF I PLSGG I DVVAHELTHAVTDYT AGL I YQNES G
BS  GRGYNNAFWNGSQMVYGDGDGQTFLPFSGG I DVVGHELTHAVTDYT AGLVYQNES G
SU  GTQYNNAAWTGDQM I YGDGDGS FFS PLSGSLDVTAHEMTHGVTQET ANL I YENQPG
AM  GSRANNAAW I GDQM I YGDGDGS FFS PLSGSMDVTAHEMTHGVTQET ANLNYENQPG

       170          180           190           200           210
TH  A I NEA I SD I FGTLVEFTYAKNPDWE I GEDVYT PG I SGDSLRSMSD PAKYGDPDHY S
BS  A I NEAMSD I FGTLVEFYANRNPDWE I GED I YT PGVAGDALRSMSD PAKYGDPDHY S
SU  ALNES FSDVFG - - - - - YFNDTEDWD I GED I -T - -VSQPALRSLSN PTKYNQPDNY A
AM  ALNES FSDVFG - - - - - YFNDTEDWD I GED I -T - -VSQPALRSLSN PTKYGQPDNF K

       220          230           240           250           260
TH  K - RYT - - -GTQDNGGVH I NS G I I NKAAYL I SQGGTHYGVSVVG I GRDKLGK I FYRA
BS  K - RYT - - -GTQDNGGVHTNS G I I NKAAYLLSQGGVHYGVSVVG I GRDKMGK I FYRA
SU  NYRNL PNTDEGDYGGVHTNS G I PNKAAYNT I TK - - - - - - - - -LGVSKSQQ I YYRA
AM  NYKNL PNTDAGDYGGVHTNS G I PNKAAYNT I TK - - - - - - - - - I GVNKAEQ I YYRA

       271          280           290           300           310          316
TH  LTQYLT PTSNFSQLRAAAVQSAYDLYGST SQEVAS VKQAFDAVGVK
BS  LVYYLT PTSNFSQLRAAC VQAAADLYGST SQEVNS VKQAFNAVGVY
SU  LTTYLT PSS TFKDAKAAL I QSARDLYGSTD - -AAKVEAAWNAVGL
AM  LTVYLT PSS TFKDAKAAL I QSARDLYGSQD - -AASVEAAWNAVGL
```

Fig. 3. Amino acid sequences of metallo-proteinases from four different *Bacillus* species (Imanaka *et al.*, 1986) aligned and numbered according to the sequence of thermolysin.

Footnote.

TH = thermolysin (Titani *et al.*, 1972).

BS = neutral metallo-proteinase from *Bacillus stearothermophilus* (Takagi *et al.*, 1985).

SU = neutral proteinase from *Bacillus subtilis* (Yang *et al.*, 1984).

AM = *Bacillus amyloliquefaciens* neutral proteinase (Vasantha *et al.*, 1984).

3 MICROBIAL PROTEINASE MARKETS

In discussing the industrial sector, reference is made to enzymes which are sold in bulk (kg) quantities and which are used as additives or processing aids for consumer products or ingredients thereof. Enzymes which are parts of clinical diagnostic systems or which are sold in small quantities (usually at premium prices) and which are intended mainly for medical or research purposes are excluded.

It is of interest to consider the few proteinases which are major items of commerce in the industrial enzyme sector versus the many which have been reported in the scientific literature. Previous reviews on proteinases (Aunstrup, 1980; Ward, 1983; Kalisz, 1988) describe many interesting proteinases from many different sources. A simple library database search will turn up

Table 1

Industrial Proteinase Markets. Estimated Annual Sales Turnover
and Market Share for the Major Proteinases in the Industrial
Sector. Data used by permission (Hepner & Male, 1986)

	Sales (million US$)	*Share of industrial proteinase market(%)*
Bacterial proteinase®	145	60
Animal rennet®	50	21
Microbial rennet®	12	5
Papain®	15	6
Pancreatin®	12	5
Bromelain®	5	2
Fungal proteinase®	3	1
Total:	242	100

additional citations about new proteases from new or mutant sources. Some of
these references will describe various novel applications of proteinases to
modify proteins, many hinting or predicting commercial success. Sadly, the
'transfer coefficient' from research-paper discussion to the industrial sector for
enzymes is quite small.

Industrial proteinases as a class represent the largest category of industrial
enzymes. While industrial enzymologists consider the 'starch-modifying' enzymes as the second largest class, it should be pointed out that this is a more
heterogeneous grouping of amylases with glucose (xylose) isomerase. Table 1
and Fig. 4 present a breakdown of the market shares for industrial proteinases
as of 1986 (Hepner & Male, 1986). At that time this sector was valued at

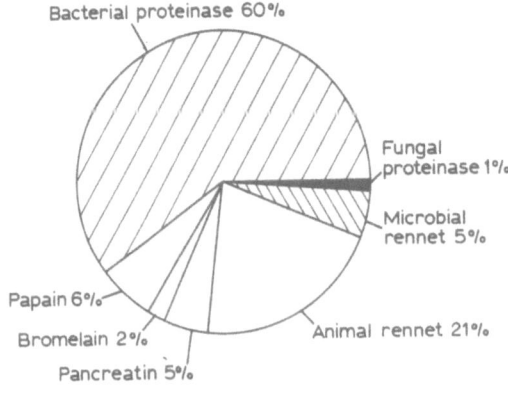

Fig. 4. Piechart of the industrial proteinase market. Market shares for each proteinase
are given as percent of a total estimated sales turnover of US$242 m. Microbial
proteinases are depicted in pie slices with shading. Data used by permission (Hepner &
Male, 1986).

Table 2

Markets for Microbial Proteinases. Estimates of Annual Sales Turnover and Market Share for the Major Applications of Microbial Proteinases. Data used with permission (Hepner & Male, 1986)

	Sales (million US$)	*Share of microbial proteinase market (%)*
Detergent proteinases	140	89·2
Microbial rennets	12	7·6
Baking proteinases	3	1·9
Leather	1	0·6
Miscellaneous	1	0·7
Totals	157	100·0

US$242 m compared to a US$420 m market for all industrial enzymes. These data clearly show that bacterial proteinases are the most significant segment, representing 60% of the total industrial proteinase turnover. Microbial rennets and fungal proteinases play a minor role in comparison: 5% and 1%, respectively (Hepner & Male, 1986).

It is surprising to see the small market share for papain and pancreatin (trypsin being the main proteinase component). Research and patent literature about these two enzymes is disproportionately large compared to their market values.

So how are proteinases from microbial sources employed in the industrial sector? The applications appear lopsided. That is, the data in Table 2 and Fig. 5 show that proteinases for detergents are the dominant use with sales turnover of US$140 m/year or over 89% of the total microbial proteinase use. In a very distant second place are the microbial rennets which account for a little under 8% of the microbial proteinase market. Microbial rennets' failure to penetrate the dairy rennet market possibly illustrates the importance of customer

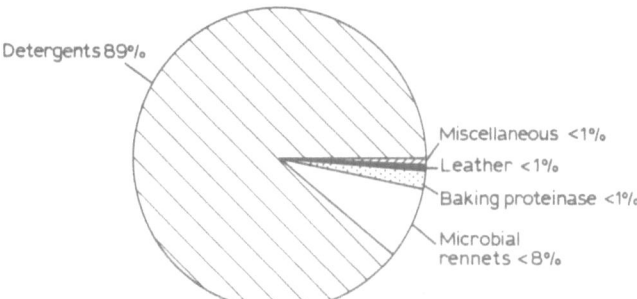

Fig. 5. Piechart for microbial proteinase markets. Market shares for each major application given in percent of an estimated annual sales turnover of US$157 m. Data used by permission (Hepner & Male, 1986).

preference and perceptions over lower pricing and technical specification for a biochemical product.

The markets for microbial proteinases for the baking and leather industries have a different character. That is, bakeries and tanneries usually purchase a processing aid, not just an enzyme. In most cases the enzyme is offered in a proprietary blend with other ingredients which impart additional functions. Thus, it is difficult to accurately identify the actual quantities of enzymes used and/or allocate a portion of the sales value of these blends.

In the following sections on the major applications of microbial proteinases, an attempt will be made to comment on what enzymological and marketing developments have occurred since the previous edition of this book. Consequently, repetition of the detailed descriptions of these applications, which have appeared elsewhere (Aunstrup, 1980; Godfrey & Reichelt, 1983; Ward, 1983; Boyce, 1984; Kalisz, 1988) will be avoided.

4 INDUSTRIAL APPLICATIONS

4.1 Detergents

4.1.1 Marketing Aspects

The use of microbial proteinases as active ingredients for laundry detergents is the largest single application of enzymes. Microbial proteinases can be found in many types and many brands of detergents for use in the home and commercial establishments. Despite earlier reports of enzymes in detergents, significant commercialization took hold in the late 1960s, suffered a decline in the early 1970s but has gained impressive ground in the 1980s.

Penetration of the home laundry detergent market has been fueled in recent years by the consumer's desire to wash at lower temperatures and to avoid environment-polluting products containing phosphates. For example, in 1988 the market share for phosphate-free detergents was about 70% in West Germany and 60% in the US (Rohe, 1988). This increase in proteinase usage demands a lot from enzyme suppliers. That is, manufacturers require microbial proteinases which are compatible with the new surfactants and detergent adjuncts and which remain active longer on the supermarket shelf.

Enzymes have been a valuable component of detergents in the low phosphate detergents and in those formulated for warm and cold water washing temperatures. That is, earlier formulations designed for hot washing temperatures contained sodium phosphate, tribasic (a very effective soil loosening agent) and sodium perborate a bleaching agent that effectively becomes activated at high temperatures (greater than 60°C). Because of environmental pressure to reduce phosphate pollution and the increased use of polyester fabrics, these compounds were reduced or taken out of detergents. The resulting products were inferior to those 'from the good old days'. Microbial proteinases via their environmentally-safe action, increase performance for the newer formulations.

Another reason for the increased usage of enzymes in detergents has been the development of better enzymes. Because efficacy testing can be run at laboratory-scale and in the same manner as the detergent manufacturers, enzyme suppliers have themselves been able to devise effective screening programs for new product prototypes. This situation is in contrast to that of other applications, for example, corn (maize) wet milling where the difference in scale between research experiment and commercial use is very much larger and which can reverse the final outcome of a new enzyme's performance.

Whatever edge the detergent enzyme supplier might have, the demands on R & D groups for new and better microbial proteinases has taxed this advantage to the limit. The phenomenal growth in heavy-duty liquid detergents in recent years switched pH optimum specifications and the liquid environment demanded more stable proteinases in order to meet shelf-life requirements. Also as new surfactants, builders and extenders are introduced or as changes in the petrochemicals industry rearranges ingredient costs, more compatible enzymes have been sought.

Another hidden demand on the R & D groups, of both enzyme supplier and detergent manufacturer is the different way people do their laundry around the world. In the US preheated water is delivered to the machine for a 10–20 min wash and rinse. On the other hand, in many places in Europe only cold water is delivered to the machine; it is heated with the clothes to a set temperature and then undergoes a wash cycle up to 1·5 h. It seems clear then that a single type of microbial proteinase could not satisfy all the needs for the plethora of detergent products and the conditions under which the consumer uses them.

In Table 3 it can be seen that two microbial proteinases dominate detergent products today. While these appear as seven products from four suppliers, the microbial sources tend to be either *Bacillus licheniformis* or the alkalophilic *Bacilli*. In the next section some of the new sources of proteinases being offered to detergent manufacturers will be discussed.

Despite optimistic market reports and patent applications for new detergent proteinases, why have so few products appeared in the ingredient marketplace? One reason may be that detergent enzymes are low-priced, sold in ton quantities and evaluated on a cost–performance basis. It will be interesting to see if the new enzymes can 'pass muster' as these two 'old workhorses' have;

Table 3
Major Detergent Proteinases

Trade name	Producer	Source
Alcalase	Novo-Nordisk	
Maxatase	Ibis	*B. licheniformis*
Optimase	Miles-Kali Chemie	
Esperase	Novo-Nordisk	
Savinase	Novo-Nordisk	
Maxacal	Ibis	Alkalophilic *Bacilli*
Kazusase	Showa Denko	

and what impact these will have on future relationships in this highly competitive market.

4.1.2 Technical Aspects

The use of microbial enzymes in home laundry detergent is a fairly unique application because it is one of the few industrial uses of enzymes where the public consumer receives an active enzyme. (Digestive-aids enzymes being the other.) With all other major applications the enzyme functions as a processing aid. Therefore, the consumer receives a product, such as bread or corn syrup, where the enzyme is inactive or has been removed. In our discussion of technical aspects we will focus on the factors that help or hinder the detergent manufacturer delivering a product containing enzyme active enough to 'deliver the punch' in the washing machine.

4.1.2.1 PROTEINASE EFFICACY

First, one might wonder how proteinases can be so effective in removing so many different types of laundry dirts or soilings. Aunstrup's review (1980) of the mechanism of laundry soils and proteinase efficacy invokes two modes of protein involvement in laundry soilings. One is the obvious protein-based stains, such as blood, where the protein is the stain (Smulders, 1984). Therefore, if hemoglobin is hydrolyzed, it becomes solubilized and washes away.

The other is less obvious; where protein functions as a binder to which other compounds adhere. For example, the notorious 'ring around the collar' in US television commercials might technically be described as soot particles and sebum adhering to a base of proteinaceous skin-cell debris deposited on a shirt collar. Therefore, hydrolyze the proteinaceous skin-cell debris and the 'dirt' can more easily be washed away by the detergent surfactants.

Cutler (1972) points out that people lose and regenerate 6–14 grams of skin cells per day; most of this being deposited on our clothing as dirt. Thus, it is easy to understand how proteinases have so much substrate available, even in a load of white undershirts which have been protected from the dirt that comes in contact with our top-clothing.

With consumer products the deciding test of an 'improved' product are home-use test where consumers use the new and a reference product. Unless consumers can perceive a difference, smart detergent manufacturers will not add an extra ingredient. However, home-use tests are expensive and take a lot of time. They are certainly not the most efficient way to screen new ingredients or the numerous reformulated prototypes that are evaluated by product development personnel. Likewise, microbiologists and biochemists screening for new and improved enzymes for detergent cannot rely on this method for their work.

Happily, although these may not be conclusive, researchers in both arenas have alternatives known as model wash tests and use the Terg-O-Meters or Launder-O-Meter machines. Both are brand names of machines which can simulate the washing process under controlled conditions for several samples at one time. Therefore, developmental scientists have a basis for comparing the performance of several detergent formulations under consideration.

Performance with these laboratory-scale wash tests is measured on the basis of how much of a uniform stain has been removed from the cloth patches by each detergent prototype. Stain removal is correlated by measuring how much light is reflected by the washed patches. The greater the amount of light reflected from the patches, the greater the degree of stain removal. The nice thing about this methodology is that scientists outside the detergent industry can operate them and have a basis for discussion with detergent manufacturers using methods recognized in the industry. There remains only one caveat: a good Terg-O-Meter result does not guarantee financial success of a new proteinase. This relies on a successful home-use test, an acceptable price and the absence of negative consumer marketing reactions.

4.1.2.2 PROTEINASE STABILITY

The environment for a microbial proteinase in a detergent is considerably harsh and destabilizing. As observed previously, alkaline serine proteases from *Bacillus* species are and have been the enzyme (group) of choice as detergent additives. Subtilisin Carlsberg produced by *B. licheniformis* was the first microbial proteinase used in detergents because it is relatively stable at high pH and in the presence of highly concentrated phosphates.

The second generation of detergent proteinases was selected to withstand even higher pH and temperatures in the presence of not only sodium tripolyphosphate, but perborates (oxidizing bleach compounds) and sequestering agents (called 'builders'). Interestingly, these highly alkalophilic *Bacilli* were found growing at pH values of 9·7 in media buffered with sodium sesquicarbonate (Aunstrup *et al.*, 1972).

The removal of phosphates from detergents for environmental purposes, has caused great changes in the overall composition of detergents; and hence, the conditions in which microbial proteinases must maintain activity for an acceptable shelf-life. The new surfactants and bleaching agents that are employed in the better phosphate-free products are well-known serine proteinase inhibitors. The anionic surfactants, especially alklyl benzene sulfonates, cause unfolding and self-digestion of proteinases. Luckily, the new 'builders', zeolite, silicates, nitrilotriacetic acid (NTA) and ethylenediamine tetraacetic acid (EDTA) are not known to be seriously destabilizing (Smith *et al.*, 1987).

With the fantastic upsurge in consumer use of liquid detergents, enzyme suppliers were faced with a new dilemma: a liquid, aqueous environment. The shelf-life of a proteinase as a dry powder or as a granulate is much longer than one in a liquid. Further, with a broad range of water as a diluent in the diverse liquid detergent products in the market today, improved shelf-life stability continues to be a key goal for enzyme suppliers and detergent manufacturers.

Another hurdle facing enzyme suppliers is to find proteinases which exhibit

Table 4
New Detergent Proteinases (from alkalophilic *Bacilli*)

Code name	Molecular weight (Da)	Isoelectric point	Reference
Ya	21 000	10·1	Takeuchi *et al.*, (1986)
Yb	40 000	5·1	Takeuchi *et al.*, (1986)
GX 6644	26 500	9·5	Durham (1987)
GX 6638-I	27 500	5·2	Durham *et al.*, (1987)
GX 6638-II	36 000	4·2	Durham *et al.*, (1987)
GX 6638-III	22 000	9·2	Stellwag *et al.*, (1987)
LA 1-1	Not reported	Not reported	Daido Nippon (1985*a*)
ES 2-5	Not reported	Not reported	Daido Nippon (1985*b*)
NF-1	30 000	Not reported	Osaka Prefect (1986)

significant activity at cold water wash temperatures, say between 15–25°C. The alkaline serine proteinases are not known to be inhibited by cold temperatures. However, like most catalytic processes, the rate of hydrolysis is much slower as temperature drops; colloquially, 'they just work slower when conditions are cooler'.

Tables 4 and 5 list new microbial proteinases isolated from alkalophilic *Bacilli* and other microbial sources, respectively. Some of these have been patented for use in detergents. Presuming that some of these possess demonstrative cleaning advantages over the subtilisins from *B. licheniformis*, *B. subtilis*, and the alkalophilic *Bacilli*, the keys to any future success will lie in a successful consumer-use test, economic yield in production (acceptable price to the detergent manufacturers) and the absence of any other negative consumer marketing reactions.

Those who have claimed that the industrial enzyme market for detergent proteinases lacks excitement and is becoming a boring, 'commodity' sales market, must have been on the 'outside' these past 10–20 years. It is hoped that the impression provided here is that detergents and, thus, microbial

Table 5
New Detergent Proteinases (from new microbial sources)

Microbe	Molecular weight	Isoelectric point	Reference
Bacillus Thermoruber	39 000	5·3	Manachini *et al.* (1988)
Bacillus APG 501	19 000	6·96	Gunze (1987)
Flavobacterium arborescens	19 400	Not reported	Boguslawski & Boyer (1984)
Fusarium sp.	26 500	11	Isono *et al.* (1972)
Conidiolobus sp.	22 000	8·4	Srinavasan *et al.* (1983)
Thermomonospora fusca	Not reported	Not reported	Gusek & Kinsella (1987)
Streptomyces sp.	Not reported	Not reported	KAO Corp. (1986)

proteinases is an ever changing and evolving technical effort. While the concept of an enzyme in a detergent is a simple and long known concept, the challenge to advance the technology still places great demands on both basic and applied science.

4.2 Cheesemaking

Calf-chymosin (rennin) has been used for ages to clot or curdle milk during cheesemaking. Because of its source, calf-chymosin commands a premium price. One answer has been to use bovine or porcine pepsin as an adjunct. Because of pepsin's higher proteolytic action and because of bans on the use of rennets from animal sources by some religious groups, microbial rennets emerged as an acceptable substitute for many during the 1970s.

Microbial rennets have increased their market penetration from about 10% in 1978 (Aunstrup, 1980) to about 19% in 1986 (Hepner & Male, 1986). While some of this increase may be due to an increased familiarity with microbial rennets, undoubtedly the chemical modification of microbial rennets to increase their thermolability had a positive effect. Despite the efforts to minimize the differences between microbial and calf-chymosin, differences still exist, making the latter the preferred enzyme for cheesemaking. Some of these differences include:

—The ratio of general proteolytic to milk-clotting activity is still a little higher for microbial rennets. This means there is more hydrolysis of cheese protein during ripening (the tertiary phase). The result is a faster ripening cheese and/or a softer curd (Aunstrup, 1980).
—Microbials have higher heat stability, even after chemical modification, which means increased proteolysis in processes involving scalded milk or pasteurization.
—Differences in the change of milk-clotting activity versus varied processing conditions, such as temperature, pH and calcium concentration. This results in the inconvenience of adjusting the processing parameters if one switches from calf-chymosin to microbial rennet.

Reduction in the performance gap between microbial rennets and calf-chymosin will be the next technological advance. It is because of these differences that calf-chymosin was cloned in microbes. As seen above, this has only been partially successful. Thus, while microbes can be made to produce the basic protein, uneconomic yields and/or an inactive enzyme may be the real shortcomings. Despite the promising reports in the trade and scientific press during the past several years, cloned-chymosin has yet to make a significant appearance in the market-place. Part of this delay is due to the above problems; another factor is regulatory pressure.

The high concern over experimentation and commercialization of gene-spliced organisms and their fermentation products is one example. Also, in one case, the production microbe is not a food-approved organism. While the delay has not rested easily with those that have invested in this research, it may have produced an unspoken benefit: the delay has bought time for these competitors to work out their own particular problems connected with yield. The net benefit may be that it will result in better product launches.

When cloned-chymosin does appear, it will be an exciting market to observe. For example, will the producers try to compete directly with calf-chymosin as was the case with microbial rennet. If so, one can expect that it will carry a premium price but at a discount compared to calf-chymosin. The other alternative is to offer it at the considerably lower prices of microbial rennets and pitch it as a 'same-as-the-real-thing' product. The deciding factor may be: will the R & D and post-fermentation processing costs permit such a strategy?

4.3 Leather Processing

Two operations in converting animal skins to leather use proteases, at least, to some extent: unhairing of hides and bating. Use of microbial enzymes in these steps can be characterized as, 'limited', for both, but for different reasons.

4.3.1 Unhairing

For unhairing of hides microbial serine proteases are competing against inexpensive chemicals: lime and sodium sulfide. The alkaline swelling and dehairing is very effective and reasonably fast (Aunstrup, 1980). Microbial proteases from alkalophilic *Bacilli* technically can replace the chemical treatment, but at an increased cost. The emphasis on controlling pollution and worker safety (hydrogen sulfide fumes) has pressured the leather industry in some countries to find replacement processes or to relocate to countries with more relaxed environmental rules/enforcement. The first option has led to an opening for microbial proteases; however the second option has been the more popular elective.

4.3.2 Bating

Leather bating is the process which gives leather a degree of flexibility and suppleness. Erroneously, consumers often refer to this quality as leather 'softness'. For example, leather for gloves has undergone extensive bating while leather for shoe-soles has had little or no treatment.

Bating has traditionally been an enzymatic process, but the main source of protease has been pancreatic glands. Hence, trypsin or the cruder preparation, pancreatin, has been the main agent for bating leather.

Bacterial serine proteinases have attempted to displace animal trypsin as an economical substitute, but with little success. In the 1970s trials with these alkaline proteinases produced poor results due to the greater proteolytic activity of the serine proteinases. Only more recently have improved microbial proteinase products met with success. Thus today, proprietary mixtures of chemicals and proteinases from *Aspergillus oryzae*, *B. amyloliquefaciens* or *B. licheniformis* can be found in the bating house supply room.

4.3.3 Future Developments

Besides being a source of trypsin/pancreatin, pancreas glands are also a source of insulin for diabetes therapy. Some insulin extraction processes inactivate trypsin and the other enzymes found in the glands; in other processes, trypsin and insulin are co-products.

The introduction of cloned-human insulin could have an interesting secondary effect on the availability and use of trypsin and/or pancreatin. One scenario predicts that because pancreas glands no longer are needed to produce extracted insulin, more are available for trypsin production. Also since trypsin would be the primary end-product, the extraction process can be streamlined for it. Thus, with a more plentiful supply of trypsin and/or pancreatin, lower pricing may diminish the advantage of using a microbial proteinase as an economical substitute.

Another scenario argues the opposite: that prices for trypsin and/or pancreatin are likely to rise; increasing the pressure to choose a microbial proteinase agent. This view sees gland extraction as an expensive process in which the sale of both insulin and trypsin/pancreatin is required to offset costs. Thus, with the loss of animal-insulin revenues, the price of trypsin will necessarily rise.

So is this then an invitation to clone trypsin into a fermentable microbe? Probably not: profits from sale of industrial enzymes are much lower than for pharmaceuticals, such as insulin.

Therefore, while this application of microbial proteinases pales in comparison with those of detergents and cheesemaking, it deserves scrutiny for its spectator value. That is, use of proteinases for leather processing may rest on the border between industrial breakthrough via new technologies and pure hype which envelop the introduction of new technologies, such as biotechnology.

4.4 Brewing

Brewers have two applications involving proteinases: chill proofing and low/no malt brewing. Chill proofing, the more significant application involves papain, a plant proteinase; not a microbial one. From time to time proteinases already in the industrial market-place have tried to compete for this application, but without success.

Chill proofing is a treatment which prevents the development of a haze in some finished beers after extended storage at cold temperatures. The hazes are thought to be a complex of polyphenols, carbohydrates and proteins. The mechanism of action of papain in preventing haze development is unclear, but some have proposed a hypothesis involving a coprecipitation reaction similar to the formation of curds in cheesemaking (Godfrey & Reichelt, 1983).

Papain's superiority to current microbial proteinases may lie in its high degree of proteolytic action in the pH 5-7 range. This pH is too low for the alkaline serine proteinases and the activity required is greater than that of the industrial neutral metallo-proteinases. While one could theoretically isolate, gene-splice, or protein-engineer a microbial proteinase to be like papain, the payback in terms of sales to chill proofing probably cannot justify the effort.

The second application of proteinases in brewing is the addition of proteinases to the mash tun in order to increase the amount of free amino nitrogen (FAN) for the fermentation. Several situations could require this: a malt batch with low proteinase activity, a low malt grist (high use of adjuncts) or formulation, and replacement of malt with barley which lacks proteolytic activity. Supplementation with neutral metallo-proteinase from *B. amyloliquefaciens* has been the preferred remedy, at least for improving low-proteinase malts. The main advantages of this enzyme is that it is not affected by the serine proteinase inhibitors in barley extracts and its pH optimum is closer to that of the wort.

Low and, even, no malt brewing processes have been reported at the European Brewing Convention by several researchers (Proceedings, 1973). For these processes, supplementation with exogeneous enzymes replaces those normally provided by the malt in the formula. In these newer (and to conservative brewmasters, heretical!) formulations, exogenous proteinases are required because endogenous proteinases in the mashed grains is too low or absent.

Proteinases hydrolyze the grain proteins in the mash tun so that the peptides and amino acids can be assimilated by the yeast during fermentation. Worts (fermentable grain extract) with insufficient FAN will not reach the required alcohol content early enough in fermentation and thus incur a greater risk of spoiling (Beckerich & Denault, 1987).

An additional benefit of adding exogenous proteinase to high barley brews is that the proteolytic action activates a portion of the latent β-amylase present in barley extracts. The elevated β-amylase activity increases the yield of fermentable sugars in the wort which can undergo conversion to alcohol. While there is an economic incentive to low/no malt brewing with exogenous enzymes, conservatism in the industry and differing regulatory requirements from state to state will impede the growth of this use of microbial proteinases.

4.5 Baking

Microbial proteinases are used for two applications in the bakery: to modify high gluten doughs and for cracker and biscuit production. For both, fungal acid proteinases have been the dominant product in this sector. Even so, the sales turnover of enzyme alone is fairly small. Like the leather industry, enzymes usually are sold in proprietary mixtures which impart a product attribute or aid in processing; so the level of fungal acid proteinase is probably very much smaller than actual tonnage of material sold.

In the first application, fungal proteinase is added to doughs made with flour having an abnormally high gluten content, resulting in an unusually stiff dough. A stiff dough will not assume the desired shape after the molding machines and this subjects the mixing equipment to excessive wear. This situation arises only when the wheat crop has been affected by bad weather. Therefore, some characterize this application as a 'band-aid': only used when it is needed. This also means that enzyme volumes used will be very low. In this application the proteinase breaks down some of the gluten which forms an elastic web within the dough. A little hydrolysis converts a stiff, rubbery dough into one which is more pliable, easier to stretch and mold.

The function of fungal acid proteinase in the manufacture of crackers and biscuits is similar to the first. In this situation the gluten is heavily hydrolyzed so that the dough does not hold the gas bubbles produced during yeast fermentation. The result is an unleavened dough which, after baking, develops a crunchy, crispy texture.

4.6 Miscellaneous Applications

Previous reviews have made reference to additional industrial uses of microbial and other proteinases. While these may be technically interesting, these are not commercial successes in terms of microbial enzyme sales. For example, in the case of meat tenderization, the principal enzymes are papain and bromelain, two plant sulfhydryl proteinases. Interestingly, these enzymes are not the ideal ones for the application, leaving an opening for a superior microbial proteinase.

The shortcoming with papain is that it is more active on the muscle fiber tissue than on the connective tissue in tough meat. The connective tissue is the more appropriate target for hydrolysis if tender meat is the goal. Over-hydrolysis of a tough piece of meat with papain results in a texture similar to mixing ground meat in a nylon-mesh produce bag—a mouthful of mushy meat and string!

There may, therefore, be a future opportunity for a collagenase to selectively attack the connective tissue. Unfortunately, most of the collagenases reported to date come from non-food-approved organisms. Possibly genetic engineering of an effective collagenase into an appropriate microorganism might work. However, because meat tenderization can be considered a 'niche' application, the final payback may not justify the development expenses.

Another area touched on in previous papers is protein hydrolysis. Outside of hydrolyzed milk, soy and corn proteins for infant formulas, this area has never expanded as predicted. In any case the real technical interest rests with the properties of the modified proteins, not the enzymes used to do it.

5 CONCLUDING REMARKS

Although no new significant industrial applications have appeared during the last decade the microbial proteinases have been subject of intensive studies especially on the molecular level. The economical impact of the detergent proteinases has provided a data base of well characterized serine proteinases from different microbial sources, with different degrees of molecular homology and different catalytic properties. This spectrum of natural variants combined with the increasing knowledge of structure–function relationships of the proteolytic enzymes has formed a good basis for the protein engineering of 'tailor'-made proteinases for future industrial applications.

ACKNOWLEDGEMENTS

The authors want to thank Sven Branner for the alignment of the amino acid sequences of the serine proteinases, Sven Hastrup, Tove Christensen and Birthe Stentebjerg-Olesen for valuable critical assessment of the manuscript and Kim Lan Chong and Hanne Hart Hansen for typing part of the manuscript.

REFERENCES

Aunstrup, K. (1980). In *Economic Microbiology 5*, ed. A. H. Rose. Academic Press, New York, p. 49.

Aunstrup, K., Outtrup, H., Andresen, O. & Dambmann, C. (1972). In *Fermentation Technology Today*, ed. G. Terui. Kyoto, Japan, p. 299.

Barkholt, V. (1987). *European Journal of Biochemistry*, **167**, 327.

Beckerich, R. P. & Denault, L. J. (1987). In *Enzymes and Their Role in Cereal Technology*, eds J. Kruger, D. Lineback & C. E. Stauffer. American Association of Cereal Chemists, St. Paul, MN, p. 335.

Beppu, T. (1988). *The 18th Linderstroem-Lang Conference*, Elsinore, Denmark, 4–8 July.

Betzel, C., Bellemann, M., Pal, G. P., Bajorath, J., Saenger, W. & Wilson, K. S. (1988). *Proteins: Structure, Function and Genetics*, **4**, 157.

Bode, W., Papamokos, E., Musil, D., Seemueller, U. & Fritz, H. (1986). *The EMBO Journal*, **5**(4), 813.

Bode, W., Papamokos, E. & Musil, D. (1987). *European Journal of Biochemistry*, **166**, 673.

Boel, E., Beck, A.-M., Randrup, K., Draeger, B., Fiil, N. P. & Foltman, B. (1986). *Proteins: Structure, Function and Genetics*, **1**, 363.

Boguslawski, G. & Boyer, E. W. (1984). European Patent Application EP 0104554.

Boyce, C. O. L. (1984). *Novo's Handbook of Practical Biotechnology*. Novo Industri A/S, Denmark.

Bryan, P. N., Rollence, M. L., Pantoliano, M. W., Wood, J. F., Finzel, B. C., Gilliland, G. L., Howard, A. J. & Poulos, T. L. (1986). *Proteins: Structure, Function and Genetics*, **1**, 326.

Christensen, T., Woeldike, H., Boel, E., Mortensen, S. B., Hjortshoej, K., Thim, L. & Hansen, M. T. (1988). *Biotechnology*, **6**, 1419.

Cullen, D., Gray, G. L., Wilson, L. J., Hayenga, K. J., Lamsa, M. H., Rey, M. W., Norton, S. & Berka, R. M. (1987). *Biotechnology*, **5**, 369.

Cunningham, B. C. & Wells, J. A. (1987). *Protein Engineering*, **1**(4), 319.

Cutler, W. G. (1972). *Detergency: Theory and Test Methods, Vol 5*, Part I. Marcel Dekker, New York.

Daido Nippon (1985*a*). Japanese Patent Publication No. 60203184.
Daido Nippon (1985*b*). Japanese Patent Publication No. 60203185.
Dauter, Z., Betzel, C., Ingelmann, M., Höhne, W. E. & Wilson, K. S. (1988). *FEBS Letters*, **236**, 171.
Delfour, A., Jolles, J., Alais, C. & Jolles, P. (1965). *Biochemical and Biophysical Research Communications*, **19**, 452.
Durham, D. R. (1987). *Journal of Applied Bacteriology*, **63**, 381.
Durham, D. R., Stewart, D. B. & Stellwag, E. J. (1987). *Journal of Bacteriology*, **169**, 2762.
Estell, D. A., Graycar, T. P. & Wells, J. A. (1985). *Journal of Biological Chemistry*, **260**, 6518.
Foltmann, B. (1988). *The 18th Linderstroem-Lang Conference*, Elsinore, Denmark, 4–8 July.
Foltmann, B. & Pedersen, V. B. (1977). In *Acid Proteases: Structure, Function and Biology*, ed. J. Tang, Plenum Press, New York.
Godfrey, T. & Reichelt, J. (1983). *Industrial Enzymology*. Macmillan Publishers Ltd, Surrey, UK.
Gros, P. & Labanesse, B. (1960). *Bulletin of the Society of Chemistry and Biology*, **42**, 559.
Gunze (1987). Japanese Patent Publication No. 62158483.
Gusek, T. W. & Kinsella, J. E. (1987). *Biochemical Journal*, **246**, 511.
Harris, T. J. R., Lowe, P. A., Lyons, A., Thomas, P. G., Eaton, M. A. W., Millican, T. A., Patel, T. P., Bose, C. C., Carey, N. H. & Doel, M. T. (1982). *Nucleic Acids Research*, **10**, 2177.
Hastrup, S., Branner, S., Norris, F., Petersen, S. B., Nørskov-Lauritsen, L., Jensen, V. & Aaslyng, D. (1989). International Patent Publication No. WO89/06279.
Hayenga, K. J., Crabb, D., Carlomagno, L., Arnold, R., Heinsohn, H. & Lawlis, B. (1988). *The 18th Linderstroem-Lang Conference*, Elsinore, Denmark, 4–8 July.
Hepner, L. & Male, C. (1986). *Report: Industrial Enzymes by 1990*. L. Hepner & Assoc., London.
Hidaka, M., Sasaki, K., Vozumi, T. & Beppu, T. (1986). *Gene*, **43**, 197.
Imanaka, T., Shibazaki, M. & Takagi, M. (1986). *Nature*, **324**, 695.
Isono, M., Tomoda, K., Miyata, K., Maejima, K. & Tsubaki, K. (1972). United States Patent No. 3,652,399.
Jacobs, M., Eliasson, M., Uhlen, M. & Flock, J. (1985). *Nucleic Acids Research*, **13**, 8913.
James, M. N. G. & Sielecki, A. R. (1986). *Nature*, **319**, 33.
Kalisz, H. M. (1988). *Advances in Biochemical Engineering Biotechnology*, **36**, 1.
KAO Corp. (1986). Japanese Patent Publication No. 61012282.
Kester, W. R. & Matthews, B. W. (1977). *Journal of Biological Chemistry*, **252**, 7704.
Kurihara, M., Markland, F. S. & Smith, E. L. (1972). *Journal of Biological Chemistry*, **247**, 5619.
Kwon, S.-T., Terada, I., Matsuzawa, H. & Ohta, T. (1988). *European Journal of Biochemistry*, **173**, 491.
Manachini, P. L., Fortina, M. G. & Parini, C. (1988). *Applied Microbiological Biotechnology*, **28**, 409.
Markland, F. S. & Smith, E. L. (1967). *Journal of Biological Chemistry*, **242**, 5198.
Matthews, B. W. Jansonius, J. N., Colman, P. M., Schoenborn, B. P. & Dupourque, D. (1972). *Nature and New Biology*, **238**, 37.
McPhalen, C. A., Schnebli, H. P. & James, M. N. G. (1986). *FEBS Letters*, **188**(1), 55.
Meloun, B., Baudys, M., Kostka, V., Hausdorf, G., Frömmel, C. & Höhne, W. E., (1985). *FEBS Letters*, **183**, 195.
Moir, D., Mao, J.-I., Schumm, J. W., Vovis, G. F., Alford, B. L. & Taunton-Rigby, A. T. (1982). *Gene*, **19**, 127.
Morihara, K. (1974). *Advances in Enzymology*, **41**, 179.

Nedkov, P., Oberthur, W. & Braunitzer, G. (1985). *Biologische Chemie Hoppe-Seyler,* **366,** 421.

Neidhart, D. J. & Petsko, G. A. (1988). *Protein Engineering,* **2**(4), 271.

Osaka Prefect (1986). Japanese Patent Publication No. 61124381.

Pantoliano, M. W., Ladner, R. C., Bryan, P. N., Rollence, M. L., Wood, J. F. & Poulos, T. L. (1987). *Biochemistry,* **26,** 2077.

Pearl, L. (1985). In *Aspartic Proteinases and Their Inhibitors,* ed. V. Kostka. Walter de Gruyter, Berlin, New York.

Proceedings of the 14th Congress of the European Brewing Convention (1973), Elsevier, Amsterdam.

Rohe, D. (1988). *Chemische Industrie,* **4,** 28.

Smith, A., Smith, E. & Mackirdy, I. (1987). *Proceedings of the 2nd World Conference on Detergents.* American Oil Chemist's Society, Montreux, 1986.

Smith, E. L., DeLange, R. J., Evans, W. H., Landon, M. & Markland, F. S. (1968). *Journal of Biological Chemistry,* **243,** 2184.

Smulders, E. J. (1984). *Trends in Laundry Detergents.* Henkel, KGaA, Düsseldorf, FRG.

Srinavasan, M. C., Vartak, H. G., Powar, V. K. & Sutar, I. I. (1983). *Biotechnology Letters,* **5,** 285.

Stahl, M. L. & Ferrari, E. (1984). *Journal of Bacteriology,* **158,** 411.

Stauffer, C. E. & Etson, D. (1969). *Journal of Biological Chemistry,* **244,** 5333.

Stellwag, E. J., Swann, W. E., Stewart, D. B., Durham, D. R. & Nolf, C. A. (1987). European Patent Application EP 220921A.

Svendsen, I. (1986). *FEBS Letters,* **196,** 229.

Takagi, H., Morinaga, Y., Ikemura, H. & Inouye, M. (1988). *Journal of Biological Chemistry,* **263,** 19592.

Takagi, M., Imanaka, T. & Aiba, S. (1985). *Journal of Bacteriology,* **163,** 824.

Takeuchi, K., Nishino, T. Odera, M., Shimogaki, H. & Negi, T. (1986). European Patent Application EP 0204342.

Tang, J., Sepulveda, P., Marciniszyn, J., Chen, K. C. S., Huang, W.-Y., Tao, N., Liu, D. & Lanier, J. P. (1973). *Proceedings of the National Academy of Science, USA,* **70,** 3437.

Titani, K., Hermodson, M. A., Ericsson, L. H., Walsh, K. A. & Neurath, H., (1972). *Nature New Biology,* **238,** 35.

Tonouchi, N., Shoun, H., Uozumi, T. & Beppu, T. (1986). *Nucleic Acids Research,* **14,** 7557.

Vasantha, N., Thompson, L. D., Rhodes, C., Banner, C., Nagle, J. & Filpula, D. (1984). *Journal of Bacteriology,* **159,** 811.

Ward, O. P. (1983). *Microbial Enzymes & Biotechnology,* ed. W. M. Fogarty. Elsevier, London, p. 251.

Wells, J. A. & Powers, D. B. (1986). *Journal of Biological Chemistry,* **261,** 6564.

Wells, J. A., Ferrari, E., Henner, D. J., Estell, D. A. & Chen, E. Y. (1983). *Nucleic Acids Research* **11,** 7911.

Wells, J. A., Vasser, M. & Powers, D. B. (1985). *Gene,* **34,** 315.

Wells, J. A., Powers, D. B., Bott, R. R., Graycar, T. P. & Estell, D. A. (1987*a*). *Proceedings of the National Academy of Science, USA,* **84,** 1219.

Wells, J. A., Cunningham, B. C., Graycar, T. P. & Estell, D. A. (1987*b*). *Proceedings of the National Academy of Science, USA,* **84,** 5167.

Wright, C. S., Alden, R. A. & Kraut, J. (1969). *Nature* (London), **221,** 235.

Yang, M. Y., Ferrari, E. & Henner, D. J. (1984). *Journal of Bacteriology,* **160,** 15.

Chapter 7

MICROBIAL LIPASES

SVEN ERIK GODTFREDSEN

Novo Nordisk A/S, Novo Allé, DK-2880, Bagsvaerd, Denmark

CONTENTS

1 INTRODUCTION

In comparison to the proteases and carbohydrases the lipases are today of little importance in an industrial context. This is due, in part, to the fact that few important industrial application areas for lipases have been identified until the last ten years, and to the fact that the production of lipases in industrial quantities and yields has proven technologically difficult.

Recently this situation has begun to change, and some important industrial applications for lipases have now been identified e.g. within the areas of oleochemistry, organic synthesis, formulation of detergent compositions, and nutrition. The expectation is that lipases will be as important industrially in the future as the proteases and carbohydrates, not only as industrial enzymes as such but also as catalysts for preparation of industrially important materials.

In the past, most studies on lipases and their mode of action have been directed towards mammalian enzymes such as pancreatic lipase, lipoprotein lipase, and phospholipase A_2. These works are well described and reviewed in several classical papers such as those of Desnuelle (1961), Brockerhoff & Jensen (1974), and Borgström & Brockman (1984). The intriguing nature of lipases as expressed in their dependence on an interface for their action and the likely occurrence on their surface of structures, different from the active site, for their binding to the interface is excellently accounted for in these early reviews. Also well outlined is the early work on plant lipases, fungal lipases, the cutinases, and bacterial lipases (Borgström & Brockman, 1984).

The present chapter provides a quick reference to recent papers on microbial lipases regarding their occurrence, production and uses. It will provide a quick overview of microbial lipases and their industrial applications in addition to a general overview of the rapidly developing field of microbial lipases.

2 SCREENING, CULTIVATION AND OCCURRENCE OF LIPASE-PRODUCING MICROORGANISMS

Screening and isolation of microorganisms for lipase activity from soil samples or other biological materials is relatively easy and most frequently carried out employing agar plates containing triglycerides. Lipase-catalysed hydrolysis gives rise either to clearing zones, or precipitates of e.g. salts of fatty acids (Sztajer & Zboinska, 1982; Gowland et al. 1987; Whitesides & Ladner, 1988). Screening systems making use of chromogenic substrates have also been described and found useful for isolation of lipase-producing fungi (Yeoh et al., 1986), e.g. plate assays based on an agar medium containing trioleylglycerol and the fluorescent dye rhodamine B (Kouker & Jaeger, 1987). The lipase activity of isolated microorganisms, cell-free extracts, and purified enzymes are most commonly assayed using an emulsion of tributyrin as substrate in a pH-stat equipment.

The commonly used methods may be modulated to direct screening towards microorganisms generating lipases of certain desired characteristics as, for example, a high activity towards long chain alcohols or a high stability and activity at elevated temperatures (El-Hoseiny, 1986; Gowland et al., 1987). These proporties may be of importance particularly in regard to the use of lipases in organic synthesis and for modification or hydrolysis of high-melting fats and oils. Also, it is usually possible to apply the commonly used screening systems in a semi-quantitative fashion and thus make use of these in the course of mutation programmes being carried out for the purpose of furnishing high yielding mutants of lipase-producing microorganisms. Information in the literature on mutation of lipase producers is, however, quite limited.

Like screening, cultivation of lipase-producing microorganisms does not usually present great difficulty. Yields of lipases reported in the literature are, however, quite low i.e. usually in the mg/litre range. Improvements in lipase fermentation yields by medium and cultivation conditions are described in several reports (Lawrence et al., 1967; Eitenmiller et al. 1970; Chander & Klostermeyer, 1983; Bloquel & Veillet-Poncet, 1984). It appears difficult however to find microorganisms and fermentation conditions that furnish lipases in yields sufficiently high to make the industrial exploitation of the enzymes feasible (even though it is easy to find lipase-producing microorganisms). Yield improvements by classical mutation and fermentation development have, however, been carried out to make some industrial uses of lipases economically feasible but the real breakthrough in regard to yield improve-

Table 1
Sources of Microbial Lipases

Microorganism	Reference
Absidia corymbifera	98
Absidia hyalospora	(36)
Acinetobacter calcoaceticus	20, 106
Acinetobacter pseudoalcaligenes	19
Achromobacter sp.	(77)
Achromobacter lipolyticum	107, 106
Alcaligenes sp.	(77)
Alcaligenes denitrificans	[86]
Amylomyces rouxii	56
Arthrobacter sp.	(77)
Aspergillus awamori	119
Aspergillus flavus	56, 118
Aspergillus fumigatus	98
Aspergillus japonicus	[116]
Aspergillus niger	119, (114), (36), (52), 22, [90], (47), (71), (74), 112, (102), (66), (78), (91), (92)
Aspergillus oryzae	119, 56
Bacillus cereus	107
Bacillus megaterium	(36)
Bacillus stearothermophilus	18, 31
Candida antarctica	(53), 77
Candida auricularia	77
Candida curvata	107
Candida cylindraceae	(57), (100), (64), (99), 21, (6), (72), (55), (102), (38), 110, (66), (84), (78), (12)
Candida deformans	107
Candida foliorum	77
Candida humicula	77
Candida lipolytica	107
Candida rugosa	107
Candida tsukubaensis	77
Chaetomium thermofile	98
Chromobacterium chocolatum	(36)
Chromobacterium viscosum	(100), (77), (38)
Corynebacterium acnes	107, 106
Flavobacterium arborescens	(36)
Fusarium oxysporum	[51], (50)
Fusarium solari	107
Geotricum candidum	(100), (114), [90], [113], 112, 117, 14, (105), (91)
Humicula grisea	1
Humicula insulens	1
Humicula lanuginosa	4, [67], 68, 40, 79
Lactobacillus sp.	107
Leishmania donovani	15
Malbrancheae pulcella	87
Micrococcus freudenreichii	63

Table 1—*contd.*

Microorganism	Reference
Mucor sp.	80, 82
Mucor javanicus	[42]
Mucor lipolyticus	[81]
Mucor miehei	(101), 39, (52), (72), (38), (102), <u>110</u>
Mucor pusillus	87, 106
Myxococcus xantus	107
Penicillium crustosum	89
Penicillium cyclopium	(114), [90], [43], (112), (91), 42
Penicillium roquefortii	17
Phycomyces nitens	(38)
Pichia miso	(88)
Propionibacterium acnes	107, 106
Propionibacterium granulosum	107, 106
Proteus sp.	107
Pseudomonas sp.	<u>110</u>
Pseudomonas aeruginosa	(35), (34), [104], 106, [86]
Pseudomonas fluorescens	(77), [96], (38), 106, (78)
Pseudomonas fragi	[59]
Pseudomonas pseudoalcaligenes	<u>19</u>
Pseudomonas stutzeri	<u>19</u>
Rhizopus sp.	(99)
Rhizopus arrhizus	(26), (100), (54), <u>110</u>, (66)
Rhizopus chinensis	(61), (60)
Rhizopus delemar	(26), (35), (77), (109), (114), 23, 45, [90], 112, (102), (91), (38), [46]
Rhizopus japonicus	(35), (74), (77), (38)
Rhizopus microsporus	107
Rhizopus niveus	(100), (71)
Rhizopus nodosus	107
Rhizopus oligosporus	56
Rhizopus oryzae	(100)
Rhodotorula minuta	(24)
Rhodotorula pilimonae	107
Saccharomyces fragilis	(36), 63
Saccharomyces lipolytica	107
Schizosaccharomyces pombe	(36)
Sporotrichum thermophile	1
Staphylococcus aureus	107, 106
Staphylococcus carnosus	32
Streptococcus lactis	107, 106
Streptomyces coelicolor	106
Streptomyces fradiae	106
Streptomyces panayensis	(36)
Talaromyces thermophilus	87
Thielavia minor	98
Torula thermophila	1

[ ]; ( ); —; deal with characteristics of enzymes, use of enzymes in synthesis and use of enzymes in detergents, respectively.

ments has come from cloning of the lipase genes, rather than from traditional screening, mutation, medium and fermentation developments.

Microorganisms producing lipases are very widespread and have been the subject of several reviews (Johri *et al.*, 1985; Stuer *et al.*, 1986). Table 1 summarizes the microbial sources of lipases and provides reference to papers dealing with the source of the enzymes, their use in synthesis and fat and oil processing, their purification and characterization, and their possible use as detergent additives.

Those lipases currently available industrially and their manufacturers are listed in Table 2. It also provides reference to papers dealing with applications

Table 2
Industrial Sources of Lipases

Lipase	Source	Reference
Achromobacter sp.	Meito Sangyo	77
Alcaligenes sp.	Meito	36
Arthrobacter sp.	Sumitomi	77
Aspergillus niger	Amano	2, 3, 52, 74, 102,
	Novo	66
	Röhm	28, 37
Candida cylindraceae	Amano	100
	Enzyme Dev. Co.	66, 102
	Meito	28, 64, 21, 72,
		38, 110
Chromobacterium viscosum	US Biochemicals	100
	Toyo Jozo	36, 38, 77
Humicula lanuginosa	Amano	100, 36
	Novo	40
Mucor miehei	Amano	2, 101, 102, 110
	Gist	102
	Röhm	28
	Novo	39, 52, 94, 102,
		38, 110
Phycomyces nitens	Takeda	38
Pseudomonas sp.	Amano	100, 36, 110
Pseudomonas aeruginosa	Amano	34
Pseudomonas fluorescens	Amano	38, 41
Rhizopus sp.	Serva	100
	Nagase	3
Rhizopus arrhizus	Precibio (France)	26
	Boehringer-Mannheim	100
	Rapidase.	28
	Gist-Brocades	28, 110
Rhizopus delemar	Chemical Dynamics	100
	Tanabe	38, 77, 102
Rhizopus japonicus	Amano	34
	Nagase	34
	Osaka Saiken Lab.	38
Rhizopus niveus	Amano	100
Rhizopus oryzae	Amano	100

Table 3
Industrial Application Areas for Microbial Lipases

Industry	Effect	Product
Dairy food	Hydrolysis of milk fat[5]	Flavour agents
	Cheese ripening[83,93]	Cheese
	Modification of butter fat[52]	Butter
Bakery food	Flavour improvement and shelf life pro-longation	Bakery products
Beverages	Improved aroma	Beverages
Food dressing	Quality improvement	Mayonnaise, dressings and whippings
Health food	Transesterification	Health food
Meat and fish	Flavour development and fat removal	Meat and fish products
Fat and oils	Transesterification	Cocoa butter, margarine
	Hydrolysis	Fatty acids, glycerol, mono- and diglycerides
Chemical	Enantioselectivity	Chirale building blocks and chemicals
	Synthesis	Chemicals
Pharmaceutical	Transesterification	Specialty lipids
	Hydrolysis	Digestive aids
Cosmetics	Synthesis	Emulsifiers, moisturing agents
Leather	Hydrolysis	Leather products
Paper	Hydrolysis	Paper products
Cleaning	Hydrolysis	Removal of cleaning agents e.g. surfactants

of the commercially available lipases. Besides the enzymes listed, the detergent lipase 'Lipolase' as well as the *Mucor miehei* (Huge-Jensen *et al.*, 1989) lipase (both being produced using *Aspergillus oryzae* as a host) are available commercially. In addition a list of the important application areas of these lipases and future lipases is given in Table 3 (Borgström & Brockman, 1984).

3 CHARACTERIZATION OF MICROBIAL LIPASES

As indicated in Table 1 many of the known lipases have been subjected to purification and characterization, usually in regard to their pH-activity profiles, stability and activity at elevated temperatures, their positional specificity in triglyceride hydrolysis, and their fatty acid specificity. In some studies the lipases have been purified to homogeneity and crystallized. Some of the most detailed studies reported are those dealing with the lipases produced by *Penicillium* (Tsujisaka *et al.*, 1973; Iwai *et al.*, 1975; Bloquel & Veillet-Poncet, 1984; Borgström & Brockman, 1984), *Rhizopus delemar* (Fukumoto *et al.*, 1964; Tsujisaka *et al.*, 1973; Iwai & Tsujisaka, 1974*a*, *b*) *Aspergillus niger* (Fukumoto *et al.*, 1963; Tsujisaka *et al.*, 1973; Höfelmann *et al.*, 1985), *Aspergillus japonicus* (Vora *et al.*, 1988), *Humicula lanuginosa* (Liu *et al.*, 1973*a*), *Geotricum candidum* (Tsujisaka *et al.*, 1973, 1977), *Mucor lipolyticus* (Nagaoka & Yamada, 1969, 1973), *Mucor japonicus* (Ishihara *et al.*, 1975), *Acinetobacter calcoaceticus* (Fischer & Kleber, 1987), *Pseudomonas fluorescens* (Roussis *et al.*, 1988), and *Pseudomonas aeruginosa* (Sugihara *et al.*, 1988).

The positional specificity which is an important characteristic in regard to some industrial and analytical uses of lipases has been determined in the course of the characterization of most of the lipases listed as has the substrate specificity of many of the enzymes (Okumura *et al.*, 1976; Joshi & Mathur, 1985; Joshi & Dhar, 1987). *A. niger*, *R. delemar*, *M. Miehei* and *H. lanuginosa* appear to be attractive sources of 1,3-specific lipases while the lipases produced by *G. candidum* and *P. cyclopium* have been shown to be positionally unspecific. The lipase of *G. candidum* has been found to preferentially cleave certain unsaturated fatty acids in triglyceride hydrolysis while, in contrast, the lipase of *Fusarium oxysporum* appears to preferentially cleave saturated fatty acids. The selectivity of the lipases is most commonly studied using natural substrates for the enzymes but may also be tested conveniently by using pseudolipids as substrates (Sonnet & Antonia, 1988). Some lipases have recently been further characterized in regard to their inhibition by other proteins which, interestingly, appear to exert their action on lipase activity by modifying the quality of the substrate–water interface which is important in the catalytic process (Gargouri *et al.*, 1984, 1986).

So far, few lipases only have been cloned and sequenced. The *Alcaligenes denitrificans* gene has been cloned into *E. coli* (Odera *et al.*, 1986) like the *Pseudomonas fragii* lipase gene which has been sequenced (Kugimiya *et al.*, 1986). Also cloned and sequenced are the lipase genes of *M. miehei* (Huge-Jensen *et al.*, 1989), *Staphylococcus hyicus* (Götz *et al.*, 1985), *Staphylococcus aureus*, *H. lanuginosa* and from non-microbial sources the lipase gene of rat lingual lipase, rat hepatic lipase, porcine pancreatic lipase, human lipoprotein lipase, and, the related enzyme, human lecithin–cholesterol acyltransferase (Tsujisaka & Iwai, 1984). A comparison of these known lipase structures is summarized in Fig. 1. It appears that a serine residue located in a stretch of hydrophobic amino acid residues is totally conserved among the lipases characterized.

No studies on high resolution three dimensional structures of microbial triglyceride lipases have been published. Two such structures are, however, being refined and finalized at the 2 Å level at present (Huge-Jensen *et al.*, 1989). These latter works on cloning and structure elucidation of microbial lipases probably represent the most important, recent developments in the

PFL	R	V	N	L	I	G	H	S	Q	G	A	L	T	A
SAL	K	V	H	L	V	G	H	S	M	G	G	Q	T	I
SHL	P	V	H	F	I	G	H	S	M	G	G	Q	T	I
RML	K	V	A	V	T	G	H	S	L	G	G	A	T	A
RLL	K	I	H	Y	V	G	H	S	Q	G	T	T	I	G
RHL	K	V	H	L	I	G	Y	S	L	G	A	H	V	S
PPL	N	V	H	V	I	G	H	S	L	G	S	H	A	A
HLPL	N	V	H	L	L	G	Y	S	L	G	A	H	A	A
HLCAT	P	V	F	L	I	G	H	S	L	G	C	L	H	L

PFL   = *Pseudomonas fragi* lipase
SAL   = *Staphylococcus aureus* lipase
SHL   = *Staphylococcus hyicus* lipase
RML   = *Rhizomucor miehei* lipase
RLL   = Rat lingual lipase
RHL   = Rat hepatic lipase
PPL   = Porcine pancreatic lipase
HLPL   = Human lipoprotein lipase
HLCAT Human lecithin-cholesterol acyltransferase

Fig. 1. Homology between substrate binding regions in lipases.

field. By cloning the lipase genes into hosts such as *Aspergillus oryzae* which is well suited to industrial fermentation of enzymes a quantum jump has been achieved in regard to the yields obtained in lipase fermentations. Moreover, detailed knowledge of lipase structures will make possible design of enzymes well suited for future industrial applications. Protease-resistant lipases for the detergent industry and lipases of high selectivity towards substrates of interest are some obvious examples of such possible developments. There can be little doubt that the breakthrough in cloning and structure determination of microbial lipases will have profound consequences in the future.

4 APPLICATION OF MICROBIAL LIPASES

The potential applications of microbial lipases have been the subject of several recent reviews (Werdelmann & Schmid, 1982; Poserske, 1984; Ratledge, 1984; Macrae & Hammond, 1985; Nielsen, 1985). A broad account on the biotechnology of fats has been provided by Werdelmann & Schmid (1982), while an excellent review on the present and future applications of lipases has also been published (Macrae & Hammond, 1985).

The major target in the fat and oil industry is the production of new triglyceride types with desirable melting properties using lipase-catalysed interesterification of readily available triglycerides (Tanaka *et al.*, 1981; Chakrabarty, 1985; Macrae, 1983, 1984, 1985; Macrae & How, 1986) (Fig. 2). This technological potential of lipases was first envisioned by Macrae and the Unilever research team as was the possibility of applying lipases in aqueous free environments for industrial purposes.

One of the crucial issues for the actual realization of the Unilever process

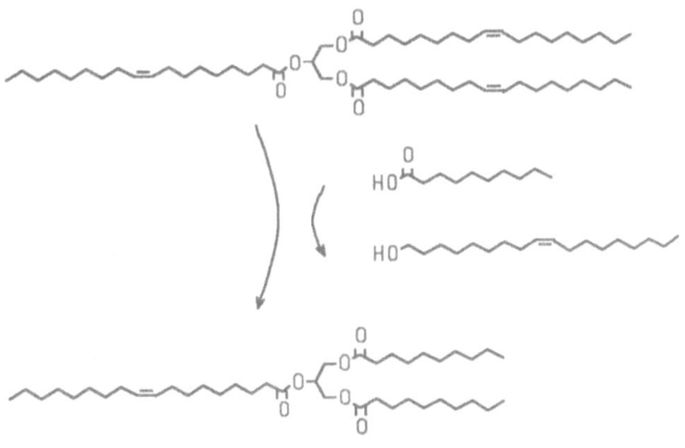

Fig. 2. Lipase catalyzed acidolysis for 1,3-specific acyl exchange of triglycerides.

has been development of immobilized preparations of lipases exhibiting the desired specificities. One such product, Lipozyme, is available today in industrial quantities and new improved immobilized lipase preparations are being developed at present. Many recent papers deal with various aspects of improved biocatalysts for the Unilever process including new carrier types (Brady *et al.*, 1986; Sztajer *et al.*, 1988), chemically modified lipases soluble in organic solvents (Inoda *et al.*, 1984), and lipases entrapped in hydrophobic polymers (Fukui & Tanaka, 1982; Kyotani *et al.*, 1988; Yeoh *et al.*, 1986). Also, several recent papers deal with improvements of the actual process design by employing loop reactor systems (Knox & Cliffe, 1984), membrane reactor types (Hoq *et al.*, 1985), batch reactor systems (Kyotani *et al.*, 1988a), micro-emulsion techniques (Bello *et al.*, 1986), and packed bed reactor systems (Macrae, PCT Appl. 83/03844). The latter is now recognized as being the preferred method for interesterification of triglycerides. All in all, the technological problems related to use of microbial lipases for triglyceride modification have been solved to the point that interesterification of triglycerides must be considered as an industrially feasible process. The actual application of the process for triglyceride modification is, today, a strategic issue and will depend on the developments of competing technologies such as plant genetics and the availability of natural materials capable of matching the performance of biosynthesized materials. In all probability the interesterification process will first be utilized for preparation of very valuable materials for use in the pharmaceutical area.

The potential application of lipases for fat hydrolysis (Linfield *et al.*, 1984; Kohr *et al.*, 1986) has likewise been the subject of several studies and is now considered technologically feasible. This technology is, however, also subject to competition from other technologies such as steam hydrolysis. Application of microbial lipases for fat hydrolysis is, therefore, likely to be put into practice mainly for treatment of very valuable materials such as hydrolysis of triglycerides containing valuable and sensitive polyunsaturated fatty acids which are gaining increasing importance in the area of nutrition and as pharmaceutical products. Likewise, the use of lipases for glyceride synthesis (Jensen *et al.*, 1978; Tanaka *et al.*, 1981; Vermiere *et al.*, 1987) and oil processing (Fullbrook, 1983) is likely to be realized only when application for high value materials emerge.

Within the field of organic synthesis an increasing interest in application of lipases is apparent from the literature. In this field the lipases are now utilized experimentally for synthesis of simple achirale esters, for synthesis and production (Vora *et al.*, 1988) of chirale building blocks, for synthesis of chirale esters and for regioselective esterification of polyfunctional alcohols.

The simple esterification reaction illustrated in Fig. 3 (Lazar, 1985) exemplifies how lipases may advantageously be used in the synthesis of achirale esters carrying sensitive groupings such as (in this instance), an allylic alcohol function. The *Candida cylindraceae* lipase is used in the example quoted, but probably the best lipase presently available for ester synthesis is that produced by *Candida antarctica* (Michiyo, 1986). Using this lipase immobilized on an inert support esterification can be carried out simply by mixing the reactants in the presence of the enzyme and removing the water generated in the course of the esterification process. The quality of esters synthesized in this fashion is often superior to that of esters prepared using conventional methods. Since the output of reactors operated using enzymes as catalysts can be greater than their chemical counterparts there is every reason to apply enzymes for ester synthesis. Probably this is the chemical area where lipases will first gain industrial importance.

The example illustrated in Fig. 4 (Kirk *et al.*, 1989) indicates how lipases can be applied for regioselective esterification of polyfunctional alcohols. The process shown is carried out simply by mixing the reactants in the presence of the enzyme and removing water from the reaction mixture in the course of the reaction (Kirk *et al.*, 1989). The process shown is also an example of how new, potentially useful chemicals may become abundantly available because of the developments of lipase technologies. The glycoside esters shown are excellent surfactants and can be prepared at a low cost using the lipase-based process.

The processes illustrated in Figs 5 (Sih, 1986) and 6 (Hirohara *et al.*, 1985) make use of the enantioselectivity of lipases in hydrolytic reactions. The process in Fig. 5 leads to an optically active building block for captopril while

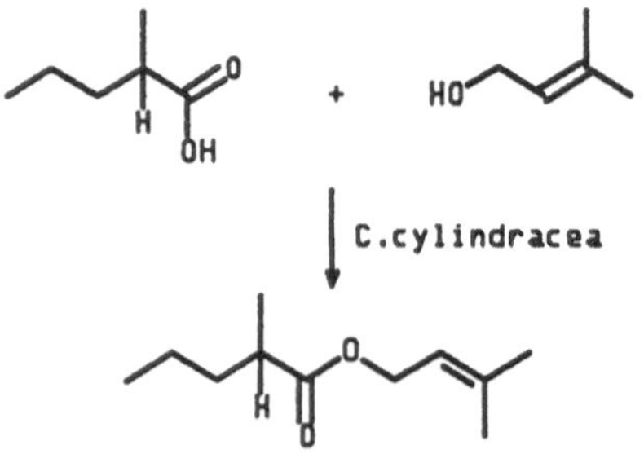

Fig. 3. Lipase catalyzed synthesis of
a simple, sensitive ester.

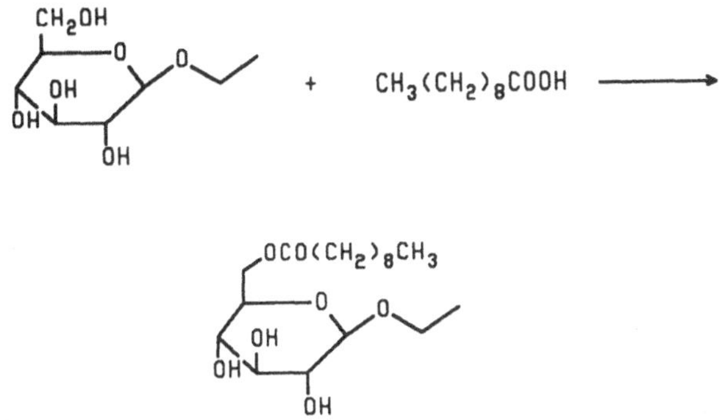

Fig. 4. The *Candida antarctica* lipase catalyzed synthesis of glycoside esters. The glyco-lipids are formed in nearly quantitative yields in the solvent free process (Kirk *et al.*).

the product of the process shown in Fig. 6 is applied as a building block for the synthesis of a pyrethroid insecticide (Satoshi *et al.*, 1988). This latter process is particularly elegant since the recycling problem inherent in synthesis of chirale materials by enantioselective hydrolysis of racemates finds an easy solution. This problem may in some cases be solved even more elegantly by using prochirale starting materials for the lipase-catalysed hydrolysis. An example is shown in Fig. 7 (Naumura *et al.*, 1988) where one of the chirale compounds could be prepared as the predominant product from the prochirale starting material. In some applications it has been possible to make use of lipases for synthesis of chirale esters from racemic starting materials. An example of this mode of lipase application in organic synthesis is shown in Fig. 8 (Koshiro *et al.*, 1985).

The substrate types amenable for lipase-catalysed synthesis and hydrolysis are very varied indeed, cf. some recent examples summarized in Figs 9 and 10. Reviews on the applications of lipases are currently provided by the Warwick Biotransformation Club of The Laboratory of the Government Chemist. Unfortunately, little knowledge is available for a rational approach to the use of lipases for synthesis. Most workers simply test lipases available until a

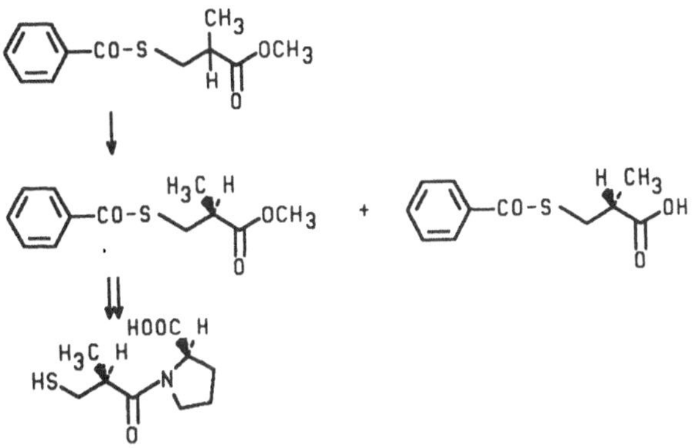

Fig. 5. The use of lipases for synthesis of an optically active building block for captopril production.

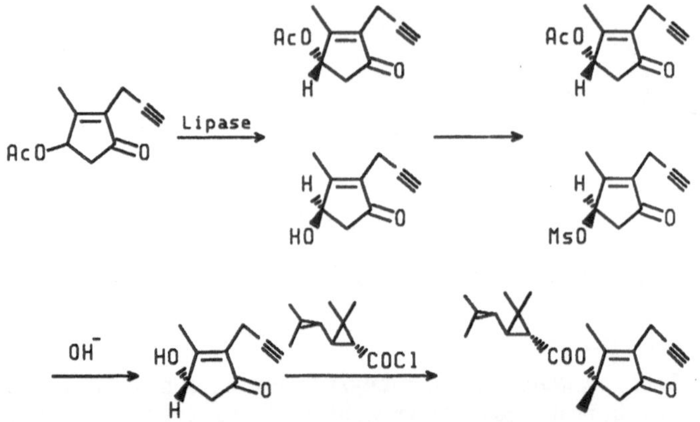

Fig. 6. Application of lipases for synthesis of optically pure building blocks for insecticide production.

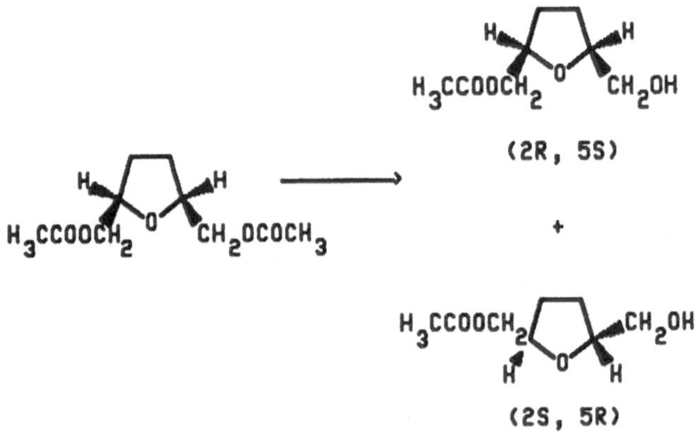

Fig. 7. Conversion of a pro-chirale substrate into optically active products by lipase catalyzed hydrolysis. One of the products indicated was generated predominantly.

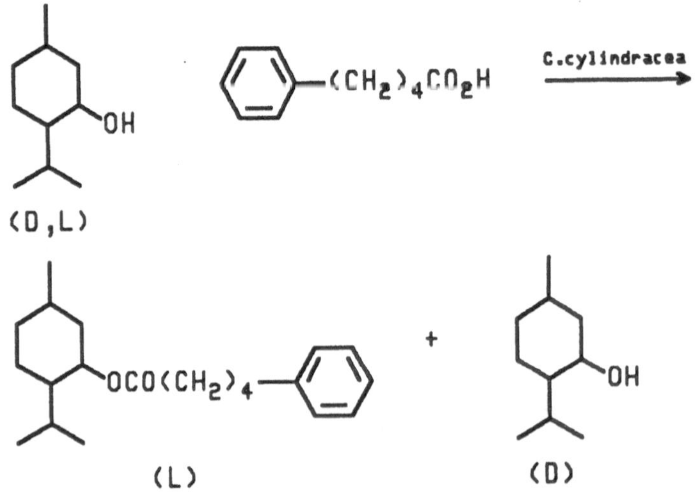

Fig. 8. One of the classical examples of application of lipases for synthesis of optically pure esters by enantioselective esterification of a racemic alcohol.

suitable enzyme is identified. Presumably this is going to change as three dimensional structures of the lipases become publicly available and as computer graphic equipment becomes more widespread. This development is probably going to take place quite fast. One could well imagine that the producers of industrial lipases in the future will make available together with their enzyme products the corresponding three dimensional structures in a form amenable for easy analysis by the potential user.

Substrate   Reference

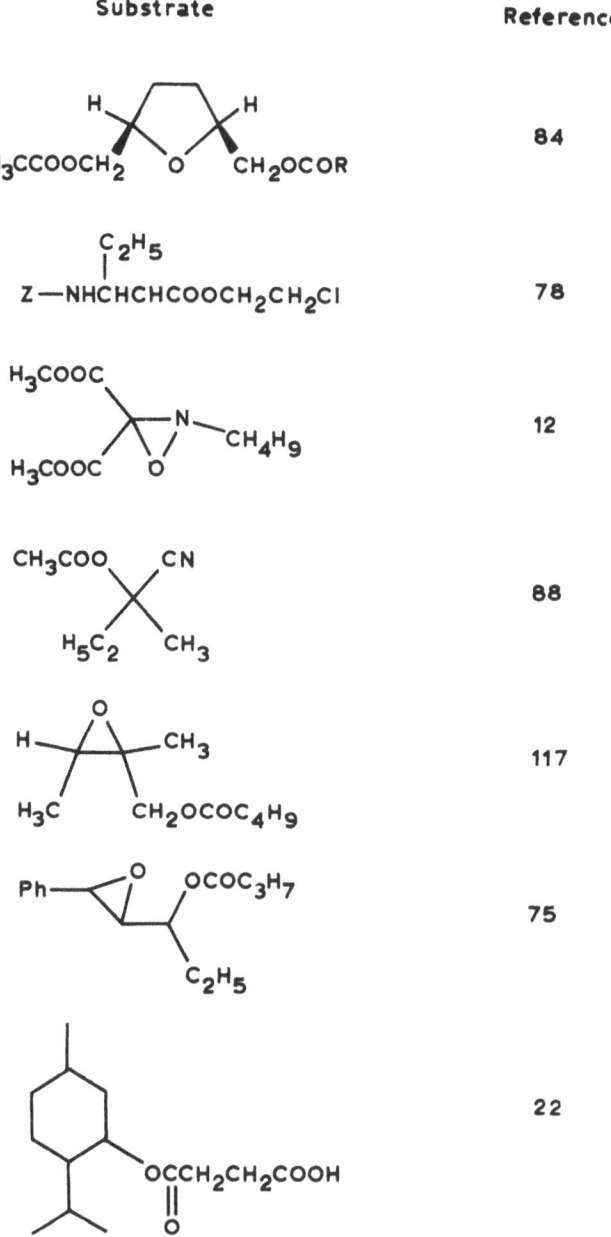

84

78

12

88

117

75

22

Fig. 9. Examples of substrates applied in lipase catalyzed organic synthesis.

Substrate   Reference

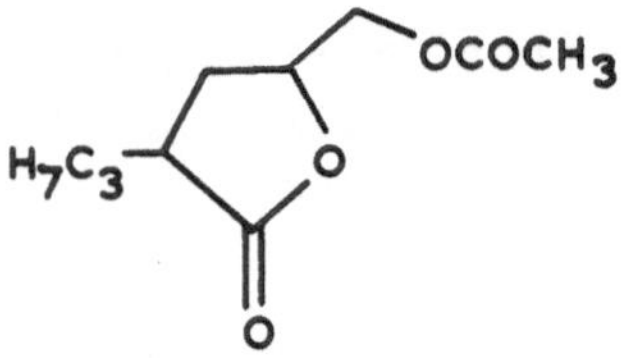

34

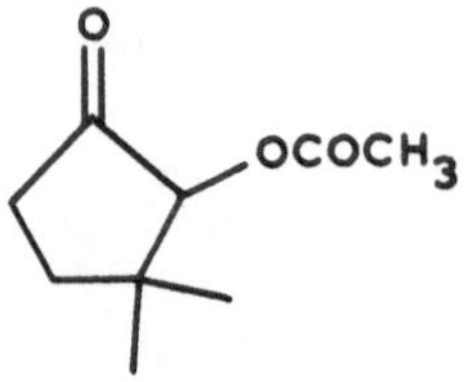

29

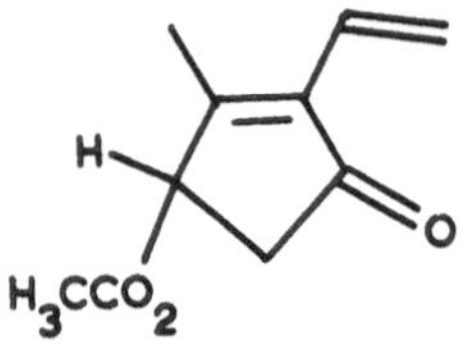

77

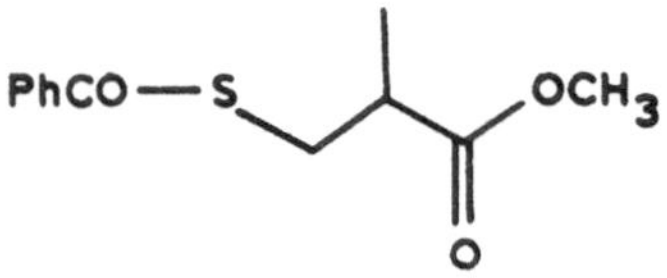

100

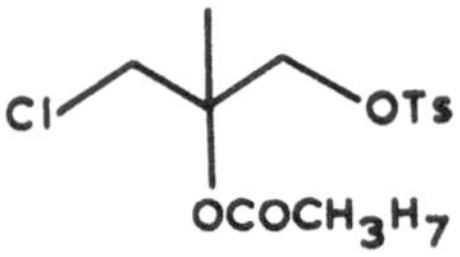

34,35

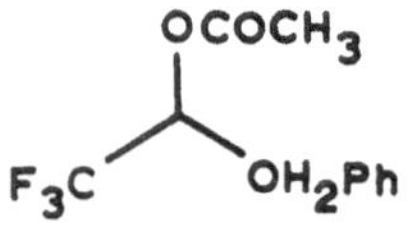

65

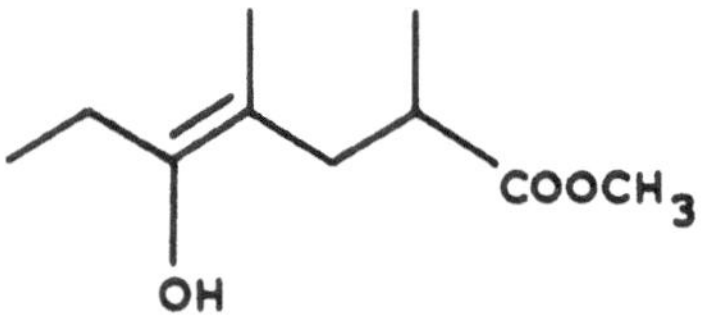

2

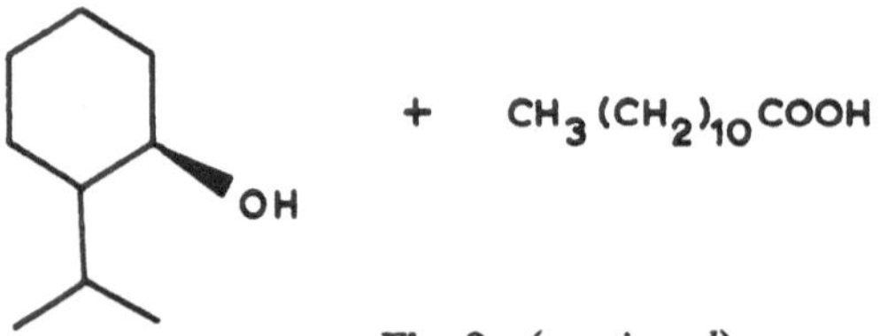

62

Fig. 9—*(continued)*

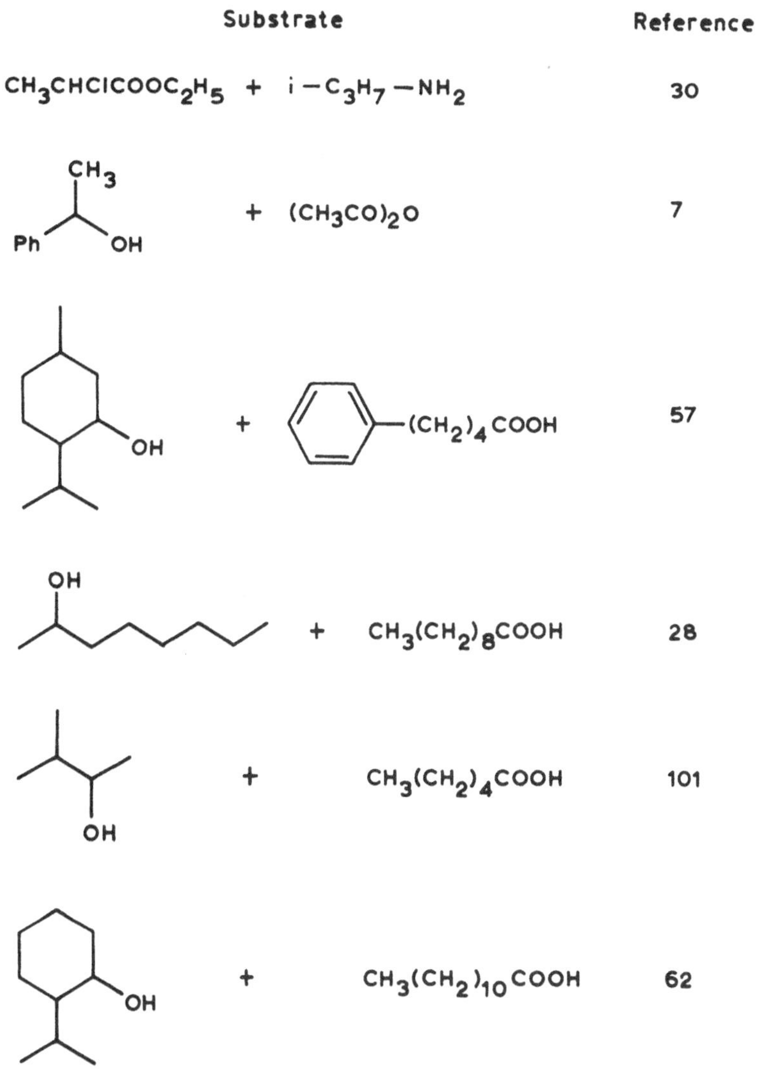

Substrate   Reference

$CH_3CHClCOOC_2H_5$ + $i-C_3H_7-NH_2$ 30

(structure: Ph–C(CH$_3$)(OH)–) + $(CH_3CO)_2O$ 7

(menthol-type structure with OH) + (phenyl)$-(CH_2)_4COOH$ 57

(branched alcohol with OH) + $CH_3(CH_2)_8COOH$ 28

(branched alcohol with OH) + $CH_3(CH_2)_4COOH$ 101

(cyclohexyl isopropyl structure with OH) + $CH_3(CH_2)_{10}COOH$ 62

Fig. 10. Examples of substrate types applied in lipase catalyzed syntheses.

Application of lipase as detergent additives to facilitate removal of fats from fabrics in the course of washing procedures is today the most important application of lipases. Early studies in this area indicated that lipases would under proper conditions enhance the fat-removing potency of detergent compositions (Farin *et al.*, 1986; Fujii *et al.*, 1986; Thornton *et al.*, 1988) although, in some studies, the effects achieved by addition of lipases could be matched by increasing the content of non-ionic surfactants in the detergent composition (Andree *et al.*, 1980). For this and other reasons the price for a detergent lipase had to be at the level of other detergent enzymes. Moreover, a useful detergent lipase had to be stable in the presence of proteases, at least during a washing period, and the lipase had to be active under basic conditions and in the presence of surfactants. These latter compounds are often referred to as lipase inhibitors but do in fact influence lipase activity by decreasing surface tension thereby influencing the quality of the oil–water interface required for lipase action. Even so, lipases can be made to function under washing conditions (Huge-Jensen & Gormsen, 1989) and, in fact, a detergent lipase, Lipolase, was marketed in 1988. This product is currently produced using an *Aspergillus oryzae* strain transformed with the detergent lipase gene. The development of Lipolase is, in many respects, the most important recent development in the field of microbial lipases. Due to this product microbial lipases will now gain industrial importance and, due to the technologies for manufacturing the product, other industrial applications of microbial lipases are now economically feasible.

REFERENCES

1. Adams, P. R. & Deploey, J. J. (1978). *Mycologia,* **70,** 906.
2. Akita, H., Matsukura, H. & Oishi, T. (1986). *Tetrahedron Letters,* **27,** 5241.
3. Andree, H., Müller, W.-R. & Schmid, R. D. (1980). *Journal of Applied Biochemistry* **2,** 218.
4. Arima, K., Liu, W.-H. & Beppu, T. (1972). *Agricultural and Biological Chemistry,* **36,** 1913.
5. Arnold, R. G., Shahani, K. M. & Dwivedi, B. K. (1975). *Journal of Food Science,* **50,** 1127.
6. Bello, M., Pievic, M., Adenier, H., Thomas, D. & Legoy, M.-D. (1986). *Comptes Rendues de l'Academie des Sciences de Paris,* **303,** 187.
7. Bianchi, D., Cesti, P. & Battistel, E. (1988). *Journal of Organic Chemistry* **53,** 5531.
8. Bloquel, R. & Veillet-Poncet, L. (1984). *Microbiologie-Aliments-Nutrition,* **2,** 179.
9. Borgström, B. & Brockman, H. L. (eds) (1984). *Lipases.* Elsevier, Amsterdam.
10. Brady, C. D., Metcalfe, L. D. & Slaboszewar, Frank, D. (1986). United States Patent No. 4,629,742.
11. Brockerhoff, H. & Jensen, R. G. (1974). *Lipolytic Enzymes,* Academic Press, New York.
12. Bucciarelli, M., Forni, A., Moretti, I. & Prati, F. (1988). *Journal of the Chemical Society, Chemical Communications,* 1614.
13. Chakrabarty, M. M. (1985). *Journal of the Indian Chemical Society,* **62,** 1.
14. Chander, H. & Klostermeyer, H. (1983). *Milchwissenschaft,* **38,** 410.
15. Chandhuri, G., Pal, S. & Banerjee, A. B. (1986). *IRCS Medical Science,* **14,** 1091.
16. Desnuelle, P. (1961). *Advances in Enzymology,* **23,** 129.
17. Eitenmiller, R. R., Vakil, J. R. & Shahani, K. M. (1970). *Journal of Food Science,* **35,** 130.
18. El-Hoseiny, M. M. (1986). *Journal of Microbiology,* **21,** 81.
19. Farin, F., Labout, J. J. M. & Verschoor, G. J. (1986). European Patent Application No. 0 218.272.
20. Fischer, B. E. & Kleber, H.-P. (1987). *Journal of Basic Microbiology,* **27,** 427.
21. Fujii, T., Tatara, T. & Minagawa, M. (1986). *Journal of the American Oil Chemist's Society,* **63,** 796.

22. Fukui, S. & Tanaka, A. (1982). In *Enzyme Engineering,* Vol. 6, eds I. Chibata, S. Fukui & L. B. Wingard, Jr. Plenum Press, New York, p. 191.
23. Fukumoto, J., Iwai, M. & Tsujisaka, Y. (1964). *Journal of General and Applied Microbiology,* **10,** 257.
24. Fukumoto, J., Iwai, M. & Tsujisaka, Y. (1963). *Journal of General and Applied Microbiology,* **9,** 353.
25. Fullbrook, P. D. (1983). *Journal of the American Oil Chemist's Society,* **60,** 476.
26. Gargouri, Y., Piéroni, G., Rivière, C., Sarda, L. & Verger, R. (1986). In *Enzymes of Lipid Metabolism II,* NATO ASI Series, A **116,** 23.
27. Gargouri, Y., Julien, R., Pieroni, G., Verger, R. & Sarda, L. (1984). *Journal of Lipid Research,* **25,** 1214.
28. Gerlach, D., Missel, C. & Schreier, P. (1988). *Zeitschrift für Lebensmittel Untersuchung und Forschung,* **186,** 315.
29. Glänzer, B. I., Faber, K. & Griengl, H. (1988). *Enzyme and Microbial Technology,* **10,** 689.
30. Gotor, V., Brieva, R. & Rebolledo, F. (1988). *Tetrahedron Letters,* **29,** 6973.
31. Gowland, P., Kernick, M. & Sundaram, T. K. (1987). *FEMS Microbiology Letters,* **48,** 339.
32. Gustafson, C., Franzén, L. & Tagesson, C. (1988). *Scandinavian Journal of Gastroenterology,* **23,** 413.
33. Götz, F., Popp, F., Korn, E. & Schleifer, K. H. (1985). *Nucleic Acids Research,* **13,** 5895.
34. Hamaguchi, S., Ohashi, T. & Watanabe, K. (1986). *Agricultural and Biological Chemistry,* **50,** 375.
35. Hamaguchi, S., Ohashi, T. & Watanabe, K. (1986). *Agricultural and Biological Chemistry,* **50,** 1629.
36. Hirohara, H., Mitsuda, S., Ando, E. & Komaki, R. (1985). *Studies in Organic Chemistry,* **22,** 119.
37. Hoq, M. M., Tagami, H., Yamane, T. & Shimizu, S. (1985). *Agricultural and Biological Chemistry,* **49,** 335.
38. Höfelmann, M., Hartmann, J., Zink, A. & Schreier, P. (1985). *Journal of Food Science,* **50,** 1721.
39. Huge-Jensen, B., Boel, E., Thim, L., Christensen, M., Andreasen, F. & Christensen, T. (1989). Paper presented at '15. Nordisk Lipidsymposium' in Rebild, Denmark, 1989.
40. Huge-Jensen, I. B. & Gormsen, E. (1989). United States Patent No. 4,810,414.
41. Inada, Y., Nishimura, H., Takahashi, K., Yoshimoto, T., Saha, A. R. & Saito, Y. (1984). *Biochemical and Biophysical Research Communications,* **122,** 845.
42. Ishihara, H., Okuyama, H., Ikezawa, H. & Tejima, S. (1975). *Biochimical et Biophysica Acta,* **388,** 413.
43. Iwai, M., Okumura, S. & Tsujisaka, Y. (1975). *Agricultural and Biological Chemistry,* **39,** 1063.
44. Iwai, M., Okumura, S. & Tsujisaka, Y. (1980). *Agricultural and Biological Chemistry,* **44,** 2731.
45. Iwai, M. & Tsujisaka, Y. (1974a). *Agricultural and Biological Chemistry,* **38,** 1241.
46. Iwai, M. & Tsujisaka, Y. (1974b). *Agricultural and Biological Chemistry,* **38,** 1249.
47. Iwai, M., Tsujisaka, Y. & Fukumoto, J. (1964). *Journal of General and Applied Microbiology,* **10,** 13.
48. Jensen, R. G., Gerrior, S. A., Hagerty, M. M. & McMahon, K. E. (1978). *Journal of the American Oil Chemist's Society,* **55,** 422.
49. Johri, B. N., Jain, S. & Chouhan, S. (1985). *Proceedings of the Indian Academy of Sciences (Plant Sci.),* **94,** 175.

50. Joshi, S. & Dhar, D. N. (1987). *Acta Microbiologica Hungarica*, **34**, 111.
51. Joshi, S. & Mathur, J. M. S. (1985). *Indian Journal of Microbiology*, **25**, 76.
52. Kalo, P. (1988). *Meijeritieteellinen Aikakauskirja*, **46**, 36.
53. Kirk, O., Björkling, F. & Godtfredsen, S. E. (1989). *Journal of the Chemical Society, Chemical Communications*, 939.
54. Knox, T. & Cliffe, K. R. (1984). *Process Biochemistry*, **19**, 188.
55. Kohr, H. T., Tan, N. H. & Chua, C. L. (1986). *Journal of the American Oil Chemist's Society*, **63**, 538.
56. Koritala, S., Hesseltine, C. W., Pryde, E. H. & Mounts, T. L. (1987). *Journal of the American Oil Chemist's Society*, **64**, 509.
57. Koshiro, S., Sonomoto, K., Tanaka, A. & Fukui, S. (1985). *Journal of Biotechnology*, **2**, 47.
58. Kouker, G. & Jaeger, K.-E. (1987). *Applied Environmental Microbiology*, **53**, 211.
59. Kugimiya, W., Otani, Y., Hashimoto, Y. & Takagi, Y. (1986). *Biochemical and Biophysical Research Communications*, **141**, 185.
60. Kyotani, S., Fukuda, H., Morikawa, H. & Yamane, T. (1988*a*). *Journal of Fermentation Technology* **66**, 71.
61. Kyotani, S., Fukuda, H., Nojima, Y. & Yamane, T. (1988*b*). *Journal of Fermentation Technology*, **66**, 567.
62. Langrand, G., Secchi, M., Buono, G., Baratti, J. & Triantaphylides, C. (1985). *Tetrahedron Letters*, **26**, 1857.
63. Lawrence, R. C., Fryer, T. F. & Reiter, B. (1967). *Journal of General Microbiology*, **48**, 401.
64. Lazar, G. (1985). *Fette Seifen und Anstrichmittel*, **87**, 394.
65. Lin, J. T., Yamazaki, T. & Kitazume, T. (1987). *Journal of Organic Chemistry*, **52**, 3211.
66. Linfield, W. M., Barauskas, R. A., Sivieri, L., Serota, S. & Stevenson, SR, R. S. (1984). *Journal of the American Oil Chemist's Society*, **61**, 191.
67. Liu, W.-H., Beppu, T. & Arima, K. (1973*a*). *Agricultural and Biological Chemistry*, **37**, 157.
68. Liu, W.-H., Beppu, T. & Arima, K. (1973*b*). *Agricultural and Biological Chemistry*, **37**, 2487.
69. Macrae, A. R. (1984). *American Oil Chemical Society Monograph*, **11**, 189.
70. Macrae, A. R. (1983). *Journal of the American Oil Chemist's Society*, **60**, 291.
71. Macrae, A. R. (1985). *Philosophical Transactions of the Royal Society of London B*, **310**, 227.
72. Macrae, A. R. (1986). European Patent Application No. 0 274 798.
73. Macrae, A. R. & How, P. (1986). United States Patent No. 4,719,178.
74. Macrae, A. R. & Hammond, R. C. (1985). *Biotechnology and Genetic Engineering Reviews*, **3**, 193.
75. Marples, B. A. & Roger-Evans, M. (1989). *Tetrahedron Letters*, **30**, 261.
76. Michiyo, I. (1986). PCT Patent Application No. WO 88/02775.
77. Mitsuda, S., Umemura, T. & Hirohara, H. (1988). *Applied Microbiology and Biotechnology*, **29**, 310.
78. Miyazawa, T., Takitani, T., Ueji, S., Yamada, T. & Kuwata, T. (1988). *Journal of the Chemical Society, Chemical Communications*, 1214.
79. Morinaga, T., Kanda, S. & Nomi, R. (1986). *Journal of Fermentation Technology*, **64**, 451.
80. Nagaoka, K. & Yamada, Y. (1969). *Agricultural and Biological Chemistry*, **33**, 986.
81. Nagaoka, K. & Yamada, Y. (1973). *Agricultural and Biological Chemistry*, **37**, 2791.

82. Nagaoka, K., Yamada, Y. & Koaze, Y. (1969). *Agricultural and Biological Chemistry*, **33**, 299.
83. Nasr, M. (1983). *Egyptian Journal of Dairy Science*, **11**, 309.
84. Naumura, K., Takahashi, N. & Chikamatsu, H. (1988). *Chemical Letters*, 1717.
85. Nielsen, T. (1985). *Fette, Seifen und Anstrichmittel*, **87**, 15.
86. Odera, M., Takeuchi, K. & Tho-E, A. (1986). *Journal of Fermentation Technology*, **64**, 363.
87. Ogundero, V. W. (1980). *Mycologia*, **72**, 118.
88. Ohta, H., Kimura, Y. & Sugano, Y. (1988). *Tetrahedron Letters*, **29**, 6957.
89. Oi, S., Sawada, A. & Satomura, Y. (1967). *Agricultural and Biological Chemistry*, **31**, 1357.
90. Okumura, S., Iwai, M. & Tsujisaka, Y. (1976). *Agricultural and Biological Chemistry*, **40**, 655.
91. Okumura, S., Iwai, M. & Tsujisaka, Y. (1979). *Biochimica et Biophysica Acta*, **575**, 156.
92. Okumura, S., Iwai, M. & Tominaga, Y. (1984). *Agricultural and Biological Chemistry*, **48**, 2805.
93. Peppler, H., Dooley, J. G. & Huang, H. T. (1975). *Journal of Dairy Science*, **59**, 859.
94. Posorske, L. H. (1984). *Journal of the American Oil Chemist's Society*, **61**, 1758.
95. Ratledge, C. (1984). *Fette Seifen und Anstrichmittel*, **86**, 379.
96. Roussis, I. G., Karabalis, I., Papadopoulou, C. & Drainas, C. (1988). *Lebensmittelwissenschaft und Technologie*, **21**, 188.
97. Satoshi, M., Umemura, T. & Hirohara, H. (1988). *Applied Microbiology and Biotechnology*, **29**, 310.
98. Satyanarayana, T. & Jori, B. N. (1981). *Current Science*, **50**, 680.
99. Seino, H., Uchibori, T., Nishitani, T. & Inamasu, S. (1984). *Journal of the American Oil Chemist's Society*, **61**, 1761.
100. Sih, C. J. (1986). PCT Patent Application No. WO 87/05328.
101. Sonnet, P. E. (1987). *Journal of Organic Chemistry*, **52**, 3477.
102. Sonnet, P. & Antonia, E. (1988). *Journal of Agricultural and Food Chemistry*, **36**, 856.
103. Stuer, W., Jaeger, K. E. & Winkler, U. K. (1986). *Journal of Bacteriology*, **168**, 1070.
104. Sugihara, A., Shimada, Y. & Tominaga, Y. (1988). *Agricultural and Biological Chemistry*, **52**, 1589.
105. Sztajer, H., Maliszewska, I. & Wieczorek, J. (1988). *Enzyme and Microbial Technology*, **10**, 492.
106. Sztajer, H. & Zboinska, E. (1982). *Acta Biotechnology*, **8**, 169.
107. Tahoun, M. K., El-Kady, M. & Wahba, A. (1985). *Microbios Letters*, **28**, 133.
108. Tanaka, T., Ono, E., Ishihara, M., Yamanaka, S. & Takinami, K. (1981). *Agricultural and Biological Chemistry*, **45**, 2387.
109. Tatara, T., Fujii, T., Kawase, T. & Minagawa, M. (1985). *Journal of the American Oil Chemist's Society*, **62**, 1053.
110. Thornton, J., Howard, S. P. & Buckley, J. T. (1988). *Biochimica et Biophysica Acta*, **959**, 153.
111. Tsujisaka, Y. & Iwai, M. (1984). *Kagaku to Kogyo. Osaka*, **58**, 60.
112. Tsujisaka, Y., Iwai, M. & Tominaga, Y. (1973). *Agricultural and Biological Chemistry*, **37**, 1457.
113. Tsujisaka, Y., Okumura, S. & Iwai, M. (1977). *Biochimica et Biophysica Acta*, **489**, 415.
114. Vermiere A., Pille, S., Himpe, J. & Vandamme, E. (1987). *Rijksuniversitet Gent*, **52**, 1853.

115. Vora, K. A., Bhandara, S. S., Pradhan, R. S., Amin, A. R. & Modi, V. V. (1988). *Biotechnology and Applied Biochemistry,* **10,** 465.
116. Werdelmann, B. W. & Schmid, R. D. (1982). *Fette Seifen und Anstrichmittel,* **84,** 436.
117. Whitesides, G. H. & Ladner, W. (1988). United States Patent No. 4,732,853.
118. Yeoh, H. H., Wong, F. M. & Lim, G. (1986). *Mycologia,* **78,** 298.
119. Yokozeki, K., Tanaka, T., Yamanaka, S., Takinami, T., Hirose, Y., Sonomoto, K., Tanaka, A. & Fukui, S. (1982). *Enzyme Engineering,* **6,** 151.

Chapter 8

ENZYMES OF ALKALOPHILES

KOKI HORIKOSHI*

*Department of Bio-engineering, Tokyo Institute of Technology,
O-okayama, Meguro-ku, Tokyo 152, Japan*

CONTENTS

* Present address: Department of Applied Microbiology, The Institute of Physical and
Chemical Research, Wako-shi, Saitama 351-01, Japan.

1 INTRODUCTION

Studies of alkalophiles have led to the discovery of many new types of enzymes which exhibit unique properties in many respects. Work with these microorganisms, has led to the establishment of an 'Alkaline Fermentation Process'. By this process, about 35 new kinds of enzymes were isolated and purified. Some of them were genetically investigated, and their genes were cloned and sequenced. One of the most important factors in the production of these alkaline enzymes is pH control of the culture media as well as other factors, such as carbon source or nitrogen source. For laboratory scale, the culture media commonly contain 1% sodium carbonate or sodium bicarbonate, which is usually sufficient to maintain the pH above 8 without readjustment during cultivation (Table 1).

2 ALKALINE PROTEASES

The first production of alkaline protease from alkalophilic bacteria was reported by Horikoshi (1971a), and next by Aunstrup *et al.* (1972). An alkaline protease from an alkalophilic *Bacillus* sp. No. 221 was isolated from soil, which is different from the subtilisin group. The optimum pH of the enzyme was 11·5, with 75% of the activity remaining at pH 13·0 (Fig. 1). The enzyme was completely inhibited by diisopropylphosphofluoridate (DFP) or 6 M urea, but not by (ethylenediaminetetraacetic acid) EDTA or *p*-chloromercuribenzoate. The relative molecular mass of the enzyme was 30 000 which is slightly higher than those of other reported alkaline proteases. Calcium ions

Table 1
Basal Media for Alkalophilic Microorganisms

Ingredient	Horikoshi-I (g/litre)	Horikoshi-II (g/litre)
Glucose	10	—
Soluble starch	—	10
Polypeptone	5	5
Yeast extract	5	5
KH_2PO_4	1	1
$MgSO_4 7H_2O$	0·2	0·2
Sodium carbonate	10	10
Agar	20	20

Sodium carbonate must be sterilized separately.

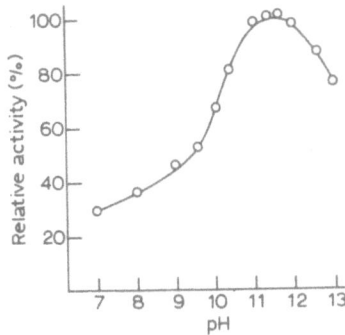

Fig. 1. pH-activity curve of protease from alkalophilic *Bacillus* sp. No. 221.

affected both activity and stability of the enzyme. The addition of 5 mM calcium ions reflected a 70% increase in activity at the optimum temperature (60°C).

Subsequently two further *Bacillus* species, AB42 and PB12, were reported which also produced alkaline protease (Aunstrup *et al.*, 1972). They had relative molecular masses of about 20 000 and 26 000 and both had an isoelectric point of 11·0. They exhibited activity in the pH range 9·0–12·0 and the temperature optimum of the strains was 60°C for AB42 and 50°C for PB12. Both enzymes were inhibited by phenylmethan sulfonyl fluoride indicating that they are serine enzymes.

Recently, *Bacillus* sp. No. ES 2-2-5, No. LA 1-1-1 (Nomoto *et al.*, 1984), *Bacillus* sp. NKS-21 (FERM BP-93) (Tsuchida *et al.*, 1986) and *Bacillus* sp. B21-2 were isolated from soil (Fujiwara & Yamamoto, 1987). The four proteases produced by these organisms had pH optima, like most other bacterial alkaline proteases, at 10–12.

Takami *et al.* (1989) isolated a new alkaline protease, which was extremely thermostable, from alkalophilic *Bacillus* sp. No. AH-101. A protease from alkalophilic *Bacillus* sp. No. AH-101 was purified from the culture fluid by only two steps, passing through a DEAE-Toyopearl 650M column followed by CM-Toyopearl 650M column chromatography. The relative molecular mass was about 29 000–30 000 and the sedimentation constant was 3·0S. The enzyme was most active toward casein at pH 12–13 and stable to 10 min incubation at 60°C in the range pH 5–13. Calcium ions were effective in stabilizing the enzyme molecule especially at higher temperatures. The optimum temperature was about 80°C in the presence of 5 mM calcium ions. The enzyme was stable at 30–70°C in the presence of 5 mM calcium ions. The enzyme was completely inactivated by phenylmethan sulfonyl fluoride, but little affected by EDTA, urea, sodium dodecyl sulfate (SDS) and sodium dodecylbenzene sulfonate (DBS). These results indicate that the enzyme is more stable against both temperature and high alkaline conditions in the presence of detergents than other proteases so far reported. This enzyme is therefore a good candidate for use as a detergent additive, although its yield has not been sufficient enough for an industrial scale plant.

Tsai *et al.* (1983) isolated an alkaline elastase from an alkalophilic *Bacillus* sp. Ya–B, which showed a marked preference for elastin over other proteins. The relative molecular mass was estimated to be 25 000 and the specific activity of the purified enzyme was 12 400 for casein and 2440 for elastin at pH 11·75, which is the optimum pH for enzyme activity. The isoelectric point was 10·6 and the enzyme was stable in the pH range 5·0–10·0. This elastase, which is a serine enzyme, was strongly inhibited by the alkaline proteinase inhibitor and *Streptomyces* subtilisin inhibitor, but was not inhibited by metalloproteinase inhibitor nor elastatinal. The N-terminal sequence of the enzyme was determined and compared with those of subtilisin BPN', subtilisin Carlsberg and *Bacillus* No. 221 protease. Extensive sequence homology was exhibited among these three enzymes.

2.1 Industrial Applications

2.1.1 Detergent Additives
With one notable exception, the main industrial applications for alkalophilic enzymes are in the detergent industry. Detergent enzymes account for approximately 25% of total world wide enzyme production and the alkalophilic or alkalotolerant proteins represent a good example of a successful commercial bio-product. Detergents usually have a pH in solution between 8 and 10·5. For an enzyme to be useful as a detergent additive it must be active in solution at alkaline pH values and stable in the presence of detergent additives such as bleaching agents, bleach activators, surfactants, perfumes and so on. Furthermore an enzyme must exhibit long term stability in the detergent product.

Proteolytic enzymes, classified as serine proteases, are widely used in detergent compositions. Not all of these are produced by alkalophilic bacteria. The most widely studied serine proteases are the so-called subtilisins produced by neutrophilic *B. subtilis* strains. Recently, several alkaline proteases have been produced by alkalophilic *Bacillus* strains and these are commercially available from companies such as Novo Industries, Showa-Denko Co., etc.

2.1.2 Other Applications
Use of alkaline enzymes has been made in the hide-dehairing process where dehairing is carried out at pH values between 8 and 10. In the old dehairing process, the hides were placed in a bath containing calcium hydroxide and sodium sulfide at pH around 12. This process had the disadvantages of being detrimental to the hairs and causing swelling of the skins, resulting in difficulties in further processing. Suitable enzymes are now commercially available from several companies.

An interesting application of alkaline protease was reported by Fujiwara & Yamamoto (1987). B21-2 protease could be used to decompose the gelatinous coating of X-ray films, from which silver was recovered.

3 STARCH-DEGRADING ENZYMES

There are many types of starch-degrading enzymes, such as amylase, cyclomaltodextrin glucanotransferase, pullulanase, α-glucosidase, etc. (Fogarty, 1983). The enzymes most extensively studied generally have pH optima for activity in the acid or neutral range. Horikoshi (1971*b*) was the first to detect alkaline amylase. The enzyme was produced in the Horikoshi-II medium by the cultivation of alkalophilic *Bacillus* sp. No. A-40-2. Then several types of alkaline starch-degrading enzymes were discovered by cultivating alkalophilic microorganisms. No alkaline amylases produced by neutrophilic microorganisms have been so far reported.

3.1 Amylases

Various types of alkaline amylases have been discovered (Yamamoto *et al.*, 1972). They have been classified into four types according to their pH-activity curves (Fig. 2). The type-I curve has only one peak at pH 10·5; the type-II curve has two peaks at pH 4·0–4·5 and 9·0–10·0; the type-III curve has three peaks at pH 4·5, 7 and 9·5–10·0; the type-IV curve has one peak at pH 4·0 with a shoulder at pH 10·0. Two or three peaks in these curves suggested the existence of multiple components in the enzymes. However, experiments on cloning the amylase gene revealed that one enzyme (not a mixture) which was coded in one gene still exhibited three peaks for activity.

The characteristics of the four types of amylases are summarized in Table 2. Type-I are completely adsorbed by DEAE-cellulose at pH 9·0, but the other three types are not adsorbed. Calcium ions have a protective action against heat inactivation of these amylases, although the extent of the protective action is different among the four amylases. Of the four amylases, type-I amylases are the most thermolabile. They were completely inactivated by heating for 15 min at 55°C in the presence of 10 mM Ca^{2+}. Type-II and type-IV amylases are relatively more stable than type-I amylases. They retained about 70–95% of their original activity after the same heat treatment. Type-III

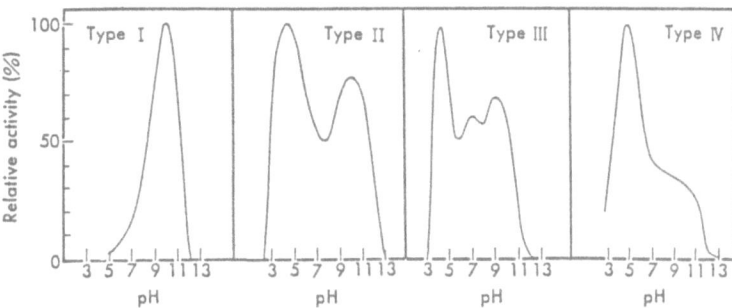

Fig. 2. Four types of pH-activity curves of alkaline amylases of alkalophilic *Bacillus* spp.

Table 2

Properties of Alkaline Amylases of Alkalophilic *Bacillus* spp.

Type	Strain No.	Optimum pH	Stable pH^a	Protection by calcium	Product
I	A-40-2	10·5	7·0–9·5	+	Glucose and malto-oligosaccharides
	A-59	10·5	7·0–9·5	+	
	27-1	10·5	7·0–9·0	+	
	124-1	10·5	7·0–9·0	+	
II	135	4·0; 10·0	7·0–9·0	+	Cyclodextrins $\beta > \alpha$
	169	4·0; 10·0	7·0–9·0	+	
III	38-2	4·5; 9·0	5·0–10·5	+	Cyclodextrins $\alpha:\beta:\gamma = 1:15:4$
IV	17-1	4·5	6·5–10·0	+	Cyclodextrins $\alpha:\beta:\gamma = 1:45:15$
	13	4·5	6·5–10·0	+	

^a pH range where 50% of activity remains after incubation at 50°C for 15 min.
α: α-cyclodextrin; β: β-cyclodextrin; γ: γ-cyclodextrin; + : positive.

amylases are the most stable. They retained about 50% of their activity at pH 4·5 or pH 10·0 even when heated for 40 min at 65°C. Type-III amylase (No. 38-2 enzyme) and type-IV amylase (No. 17 and No. 13 enzymes) have high cyclomaltodextrin glucanotransferase activities which convert starch to cyclodextrins. Cyclomaltodextrin glucanotransferase (CGTase) catalyzes the degradation of starch to form cyclodextrins which are composed of six to eight glucose units linked by an α-1,4-bond. The corresponding cyclodextrins formed are named as α-, β-, and γ-CD, respectively.

Strain No. A-40-2 (ATCC 21592) was selected from about 300 colonies of bacteria grown in Horikoshi-II medium (Horikoshi, 1971*b*). The enzyme is most active at pH 10·0–10·5 and retains 50% of its activity between pH 9·0 and 11·5. The enzyme is not inhibited by 10 mM EDTA at 30°C and is completely inactivated by 8 M urea. But about 95% of the activity is recovered by removing the urea by dialysis. The enzyme can hydrolyze 70% of starch to yield glucose, maltose and maltotriose. Therefore, the enzyme is a type of saccharifying α-amylase. Alanine, DL-norvaline and D-methionine were effective additives to media in the production of the alkaline amylase. The addition of 0·5% DL-norvaline and 0·5% D-methionine to the culture medium increased amylase production 1·7-fold, while they repressed microbial growth (Ikura & Horikoshi, 1987).

Alkalophilic *Bacillus* sp. No. A-59 (ATCC 21591) isolated from soil also produces a type-I alkaline amylase in the Horikoshi-II medium. The enzyme preparation was separated into two fractions, major alkaline amylase having a relative molecular mass of 50 000 and minor alkaline amylase (M_r 70 000). The ratio between them was about 10 to 1 and other properties were not significantly different from the alkaline amylase of alkalophilic *Bacillus* sp. No. A-40-2.

With *Bacillus* sp. No. A-59 (ATCC 21591) Kelly *et al.* (1983) showed it produced three alkaline enzymes associated with degradation of starch: α-amylase, pullulanase and α-glucosidase.

Recently, another alkalophilic *Bacillus* sp. NCIB 11203 which was isolated from soil produced several enzymes in the culture broth such as alkaline amylase, alkaline protease and alkaline phosphatase (Kelly *et al.*, 1987). In the culture broth two extracellular α-glucosidases were detected. The enzyme was purified by fractionation with ammonium sulfate and chromatography on DEAE-Biogel A. Both enzymes had pH optima at 7·0, which is distinct from enzymes of most other alkalophilic bacilli but similar to that of *Bacillus* sp. No. A-59.

3.2 CGTases

3.2.1 CGTase of Bacillus *sp. No. 38-2*

Bacillus sp. No. 38-2 was selected as the best enzyme producer of CGTase from about 1000 strains examined. The enzymes of *Bacillus* sp. No. 38-2 and *Bacillus* sp. No. 17-1 were purified in almost the same manner—starch adsorption and DEAE-cellulose chromatography followed by gel filtration on a Sephadex G-100 column. Then the enzyme fraction was applied to a preparative gel electrophoresis apparatus. The crude enzyme of *Bacillus* No. 38-2 was a mixture of three enzymes: acid CGTase having optimum pH for enzyme activity at 4·6; neutral CGTase, pH 7 and alkaline CGTase, pH 8·5.

Recently Mäkelä *et al.* (1988) purified the CGTase of alkalophilic *Bacillus* sp. No. 38-2 by a two-step procedure involving affinity chromatography followed by HPLC anion exchange. The HPLC anion exchange chromatogram exhibited at least six active fractions from the sample purified by affinity chromatography. But their enzymatic properties were identical except for isoelectric point (pI = 4·9–4·6). The main fraction, comprising 70% of the total activity, had a relative molecular mass of 70 500 (pI = 4·9).

In order to investigate the genetic information of the enzyme/s, a gene of CGTase of alkalophilic *Bacillus* No. 38-2 was cloned by a shot-gun cloning method using pBR322. A transformant carrying a plasmid pSC8, contained a 5·3 kbp DNA fragment which had the CGTase activity. There was a single open reading frame of 2136 bp which encoded a polypeptide of 712 amino acids. The nucleotide sequence and the amino acid sequence of this CGTase has strong homology with those of the CGTase of *B. macerans*. At the amino acid level, 448 (63%) of the aligned amino acids are identical and the overall homology of the aligned nucleotide sequences was about 64% (Hamamoto *et al.*, 1987; Kaneko *et al.*, 1988).

The plasmid-borne CGTase hydrolyzed potato starch and the major product was β-CD ($10:70:20 = \alpha\text{-}:\beta\text{-}:\gamma\text{-CDs}$). Why was it possible to isolate three CGTases from the culture fluid? There are three possibilities: (1) artifact formation during purification process, (2) three genes in the chromosomal DNA, but not found or (3) one gene makes several gene products in *Bacillus* sp. No. 38-2, although in *E. coli* HB101 system it exhibited one protein band. Attempts to clone acid-, neutral-, or alkaline-CGTase gene/s from *Bacillus* sp. No. 38-2, have not been successful.

3.2.2 CGTase of Bacillus *sp. No. 1011*

Independently, Kimura *et al.* (1987*a*, *b*) cloned the gene for β-CGTase from an alkalophilic bacterium, *Bacillus* sp. No. 1011 in an *Escherichia coli* phage lambda D69. The alkalophilic bacterium, *Bacillus* sp. No. 1011, was isolated from soil by the method described by Horikoshi (1971*b*). Amino acid sequence analysis suggests that β-CGTase may have two protein domains; the one in the N-terminal side cleaves the α-1,4-glycosidic bond in starch, and the other in the COOH-terminal side catalyzes other activities, such as cyclization, etc., although this hypothesis is tentative.

3.3 Maltohexaose-Forming Enzymes

3.3.1 The Enzyme of Bacillus *sp. No. 707*

From the beginning of 1970s, many bacterial strains, which produce amylases catalyzing the degradation of starch to malto-oligosaccharides (G*n*-amylase), have been isolated and their enzymes characterized. In order to obtain hyperproducers of the enzymes for industrial applications, an alkalophilic bacterium, *Bacillus* sp. No. 707 was isolated and the gene for maltohexaose-producing amylase (G6-amylase) was cloned (Kimura *et al.*, 1988). The gene for G6-amylase in *Bacillus* No. 707 was cloned in an *E. coli* phage lambda D69. The major hydrolysis product from soluble starch by the enzyme from *Bacillus* sp. No. 707 was G4. By contrast, the product of the enzyme from *B. subtilis* (pTUB812) and *E. coli* (pTU306) was G6. The content of G6 in the hydrolysate was approximately 50–60%.

3.3.2 The Enzyme of Bacillus *sp-H-167*

Independently, (Hayashi *et al.*, 1988*a*, *b*) have isolated alkalophilic *Bacillus* spp. producing maltohexaose-forming enzymes. One of them, alkalophilic *Bacillus* sp. H-167 produced three α-amylases that yielded maltohexaose as the main product from starch. All the enzymes were most active at pH 10·5 and were stable in the range pH 7·5–12·0 at 50°C. And all the enzymes produced maltohexaose in the early stage of hydrolysis. The maximum yield was about 25–30%.

In order to obtain genetic information on the multi-enzyme systems and to elucidate the regulation of their formation, we have cloned the gene of G6-amylase of *Bacillus* sp. H-167 and expressed it in *E. coli* (Shirokizawa *et al.*, 1989). The pH-activity profile of the plasmid-borne amylase was essentially the same as that of the amylases from *Bacillus* sp. H-167.

3.4 Pullulanase

During the course of screening of alkaline amylases from alkalophilic *Bacillus* spp, Nakamura *et al.* (1975) discovered, *Bacillus* sp. No. 202-1, which produced an extracellular pullulanase in an alkaline medium of pH 10.

The relative molecular mass of the alkaline pullulanase is 92 000. The isoelectric point of the enzyme is below 2·5. The enzyme has an optimum pH for activity at 8·5–9·0, and is stable for 24 h at pH 6·5–11·0 at 4°C. The enzyme is most active at 55°C, and is stable up to 50°C for 15 min in the absence of substrate. The enzyme is inhibited by Hg^{2+} and Zn^{2+}, but not inhibited by sulfhydryl reagents such as monoiodoacetate and *p*-chloromercuribenzoate or by chelating agents such as EDTA. These results would indicate that no sulfhydryl or serine residues are involved in the catalytic site of the enzyme.

Kelly *et al.* (1983) found alkalophilic *Bacillus* sp. No. A-59 (ATCC 21591) produced three enzymes, α-amylase, pullulanase and α-glucosidase in the culture broth. These three enzymes were separately produced and the levels of α-glucosidase and pullulanase reach maxima after 24 h cultivation at an initial pH of 9·7. Although this pullulanase was not purified, it had a pH optimum at 7·0. Therefore, this enzyme differs from other enzymes of alkalophilic *Bacillus* strains which have optima between pH 9·5 and 11·5.

4 CELLULASES OF ALKALOPHILIC *BACILLUS* SPECIES

Cellulases which are commercially available display optimum activity over a pH range from 4–6. Prior to the mid-1980s enzymes with an alkaline optimum pH for activity (pH 10 or higher) had not been reported. Horikoshi *et al.* (1984) and Fukumori *et al.* (1985) reported that newly isolated bacteria (*Bacillus* sp. No. N-4 and No. 1139) produced extracellular car-boxymethylcellulases (CMCases) in alkaline growth media. One of these bacteria, alkalophilic *Bacillus* sp. N-4, produced multi-CMCases which were active over a broad pH range (pH 5–10). These CMCases have been partially purified and characterized. Another bacterium, *Bacillus* sp. No. 1139 produced one CMCase which was purified and had an optimum pH for activity at pH 9·0.

4.1 Cellulases of *Bacillus* sp. No. N-4

An alkalophile, *Bacillus* sp. No. N-4 (ATCC 21833) isolated from soil was shown to be a facultative, aerobic, spore-forming (spherical, terminal, 1·4 μm), Gram-positive, motile, and rod-shaped bacterium (0·3–0·4 μm × 2·0–3·0 μm). The isolate, therefore, is quite similar to *Bacillus pasteurii*, although its optimum pH for growth is higher than that of *B. pasteurii*.

The crude enzyme showed very weak activity toward Avicel and very strong activity on CMC at pH 6·7 and 10·0. The enzyme preparation had a broad pH activity curve in the range pH 5–10 as shown in Fig. 3. The enzyme was very stable at pH values ranging from 6–10. The CMCases were separated by passing through a Sephadex G-150 column followed by hydroxy apatite column chromatography. Two alkaline CMCases (enzymes E1 and E2) having an optimal pH for enzyme activity at pH 10·0 were partially purified from the crude enzyme preparation. In comparison with other enzymes, cellulases E1 and E2 have high pH optima and are the first reports of alkaline CMCases. Such multiplicity of CMCase components has also been reported by many investigators. However, no crucial experiment on the multiplicity of cellulase molecules have been studied. To remove the effect of proteolytic degradation of the cellulase molecule, we have cloned three cellulase genes {pNK1(celB), pNK2(celA) and pNK3(celC)} of *Bacillus* sp. No. N-4 in *Escherichia coli* HB101 with pBR322 (Fukumori *et al.*, 1986*a, b*, 1987*a, b*, 1989). These three genes were found in different positions on the chromosomal DNA, although very high homology was revealed among these genes. The plasmid-borne CMCases had broad pH activity curves (pH of 5–10·5) except celC (pH 9·0) and were stable up to 75°C. It is of interest that the plasmid-encoded cellulases have very broad pH activity curves, as was observed for the enzymes of the parental strain.

The *celC* protein had a relative molecular mass of about 100 000. The deduced amino acid sequence exhibited strong homology with cellulases from *Bacillus* species, especially the *celF* from *Bacillus* sp. No. 1139. The *celC* had an optimum pH for enzyme activity in the range of pH 9–10 which is similar to that of *celF*.

4.2   A Cellulase from Alkalophilic *Bacillus* sp. no. 1139

Another alkalophilic *Bacillus* sp. No. 1139 was selected from about 1200 colonies (Fukumori *et al.*, 1985, 1986*a*). The purified CMCase was electrophoretically homogeneous, with an estimated relative molecular mass (SDS-PAGE) of 92 000. The isoelectric point was pH 3·1. The CMCase was most active at pH 9·0, and still retained some activity at pH 10·5. The enzyme was stable over the range of pH 6–11 (24 h at 4°C) and up to 40°C for 10 min. The enzyme hydrolyzed cellotriose or cellotetraose but not cellobiose. Cellotriose was converted to cellobiose which was the main product. To obtain further information, the alkalophilic *Bacillus* sp. No. 1139 CMCase gene (pFK1) was

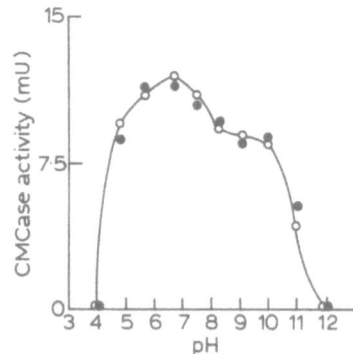

Fig.   3. Effect   of   pH   on   cellulase activity of N-4 cellulase preparation produced from alkalophilic *Bacillus* sp. N-4.

cloned in *E. coli* HB101 using the plasmid pBR322. The optimum pH for the cellulase activity (celF) of *E. coli* (pFKl) was 9·0, the same as that observed for *Bacillus* sp. 1139.

4.3 Cellulases as Laundry Detergent Additives

Protease in laundry detergents can hydrolyze proteinaceous soils stacked on textiles and improve washing efficiency. However, sebum soil other than protein cannot be hydrolyzed at all.

Cotton absorbs sweat and sebum very well, but it is very difficult for conventional laundry detergents to remove sebum soil on cotton fabrics at relatively lower temperatures.

Soil is thought to be adsorbed or stacked on cotton fibers. Kao Corp. Japan (H. Hurato, personal communication) reported that soils on cotton fibers were trapped in amorphous hydrated cellulose (interlamellar space) and in the presence of cellulase, a part of this hydrated cellulose is modified and subsequently soils are easily removed by a detergent. However, several conditions must be met: (1) cellulases must have wide pH activity range and preferably be alkaline cellulases, (2) they must have high activity and stability under high alkaline conditions in the presence of detergent and (3) they must have high stability over a wide temperature range. ˙

Saito & Ito (personal communication) had mixed our alkaline cellulases with their laundry detergents and studied the change of washing effect on washing cotton underwear. One of these enzymes produced by an alkalophilic *Bacillus* sp. N-1 showed the best result. However, production of the enzyme by the organism was not sufficient from the industrial point of view. Ito *et al.* (1989) in collaborative study isolated alkalophilic *Bacillus* sp. KSM-635 from soil, which closely resembled *Bacillus* sp. No. 1139 and succeeded in producing a suitable alkaline cellulase in an industrial scale plant.

5 β-1,3-GLUCANASES

A polysaccharide β-1,3-glucan occurs in some microorganisms (fungi and yeast) and higher plants as a component of cell walls, as cytoplasmic reserve materials, and as extracellular products. A typical β-1,3-glucan is laminaran found in *Laminaria* seaweed species.

In 1958, Horikoshi & Iida reported that *Bacillus circulans* isolated from a contaminated culture vessel, strongly lyzed the cell walls and spore coats of fungi during the cultivation of *Aspergillus oryzae* (Horikoshi & Iida, 1958, 1959, 1964, Horikoshi *et al.*, 1963). A β-1,3-glucanase was isolated and purified from the culture fluid. This was the first paper showing that β-1,3-glucanase is a key enzyme in lyzing fungal cell walls.

Most β-1,3-glucanase so far reported have pH optima on the acid side, but have no activities at pH 8. Alkaline β-1,3-glucanases, which are active at pH values above 8, were recovered from the culture broths of an alkalophilic soil isolate *Bacillus* sp. No. K-12-5 and from *Bacillus* sp. No. 221 which was also an alkaline protease-producer (Horikoshi & Atsukawa, 1973*a*, 1975). The β-1,3-glucanase from both strains were purified by the conventional procedures including column chromatography on DEAE-cellulose at pH 8·0, precipitation with 70% saturated ammonium sulfate, and gel filtration on Sephadex G-100 or G-75. No. 221 enzyme was crystallized from the eluate of a Sephadex G-75 column with 30% saturated ammonium sulfate. The enzymatic properties are summarized in Table 3, together with those from *B. circulans* IAM 1165 and *Basidiomycete* sp. QM 806 for comparison.

No enzymes having optima for activity at pH 10 or capable of withstanding 10 min at 80°C had been reported until the discovery of alkaline *Bacillus* sp. No. AG-430 (Y. Nogi & K. Horikoshi, unpublished data). The enzyme was purified by DEAE-Sepharose CL-6B column chromatography followed by passing through a Sephadex G-75 column. Its relative molecular mass, estimated by SDS-PAGE method, was approximately 35 000 and its isoelectric point was pH 3·8. The enzyme was most active at pH 9–10 (Fig. 4) and most stable in the range of pH 4–12 at 40°C for 1 h incubation. The enzyme was most active at 65°C in the presence of laminaran as a substrate. It is, however, very interesting that in the absence of the substrate the enzyme was very stable in 0·01 M phosphate buffer (pH 7·0). About 50% of the original activity still remained after 50 min incubation in boiling water bath. As its thermal stability was decreased in the presence of some ions such as Ca^{2+}, Mg^{2+}, etc., it is

Table 3

Physicochemical Properties of β-1,3-Glucanases

Property	β-1,3-glucanase				
	IAM 1165	*No. 221*	*No. K-12-5*	*AG-430*	*QM 806*
Relative molecular mass	28 000	36 000	40 000	35 000	57 000
Sedimentation constant	3·0	3·2	3·6	—	3·7
Isoelectric point	5·4	4·1	3·5	3·8	—
Optimum pH for enzyme action	5·8	8·0	6·0–8·0	9·0–10·0	4·6–6·0
Stable temperature (°C)	up to 65	75	65	100[a]	—

[a] In the presence of laminarin as the substrate, the enzyme was active up to 65°C.
IAM 1165: *Bacillus circulans* IAM1165 (Horikoshi & Iida, 1958).
No. 221: Alkalophilic *Bacillus* sp. No. 221.
No. K-12-5: Alkalophilic *Bacillus* sp. No. K-12-5.
AG-430: Alkalophilic *Bacillus* sp. AG-430.
QM 806: *Basidiomycetes* sp. QM 806.

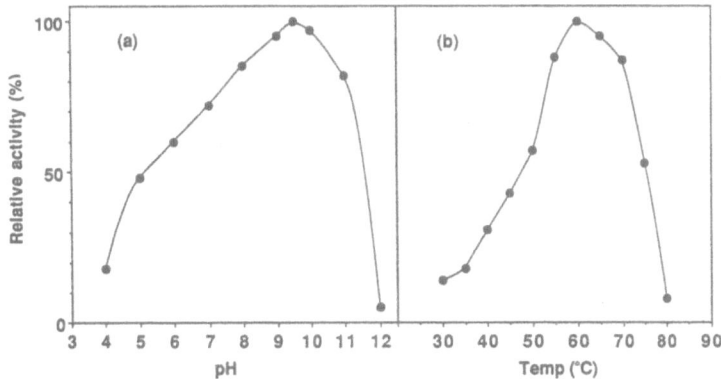

Fig. 4. pH-activity profile (a) and temperature-activity curve (b) of β-1,3-glucanases from *Bacillus* sp. AG-430.

suspected that the enzyme exhibits reversible thermal denaturation. The enzyme was catalytically inactive at temperatures above 80°C, but returned to an active structure after the sample was cooled in an ice bath. These results indicate that the β-1,3-glucanase of alkalophilic *Bacillus* sp. AG-430 is the most thermostable β-1,3-glucanase so far reported. The end-products in the enzymatic digest were glucose, laminaribiose and laminaritriose, and this enzyme is a type of endo-β-1,3-glucanase.

6 MANNAN-DEGRADING ENZYMES

β-Mannosidase and β-mannanase cleave the β-D-mannans to yield D-mannose and manno-oligosaccharides, respectively. Several bacterial β-mannanases have been reported, but there is no information so far available about β-mannosidase and β-mannanase formation by alkalophilic microbes. Recently, Akino *et al.* (1987, 1988*a, b*) isolated an alkalophilic *Bacillus* sp. AM-001 which produced high amounts of the cell-associated β-mannosidase or β-mannanase.

Three extracellular β-mannanases (M-I, M-II and M-III) were purified from the crude enzyme preparation by ammonium sulfate precipitation (80% saturation) followed by a DEAE-Toyopearl 650M and hydroxyapatite column chromatography. The relative molecular masses estimated by SDS-PAGE were 58 500 for M-I, 59 500 for M-II and 42 000 for M-III. β-Mannanase M-I and M-II were most active at pH 9·0 and M-III had an optimal enzyme activity at pH 8·5. Although M-III enzyme was relatively more stable than the others, there are no significant differences among these three mannanases except in relative molecular masses. These enzymes hydrolyzed β-1,4-manno-oligosaccharides and the major components in the digests were di-, tri- and tetra-saccharides. The mannanase gene was cloned into *E. coli* using pUC19 as a vector and sequenced.

The β-mannosidase was also purified from the cell extract treated with 0·1% (w/v) Triton X-100. Its relative molecular mass was estimated to be 94 000 and it had an isoelectric point of pH 5·5. The purified enzyme was most active at pH 6·0 and stable at 40°C for 30 min in the pH range of 6·5–8·0.

7 LIPASES

Lipases occur in microorganisms, plants and animals. Much attention has been devoted in recent years to lipase and certain lipases have been utilized as detergent additives. Furthermore, it has been found that some lipases exhibit fatty acid transfer activity. The pH optima of microbial lipases are generally in the range of 5–8; a few exceptions are lipases from *Penicillium crustosum* and *Mucor lipolyticus* which have pH optima around 9·0.

7.1 *Achromobacter* sp. Lipase

Attempts to isolate alkalophilic bacteria from soil samples which are capable of producing alkaline lipases have been made. Iodate No. 865, (Meito Sangyo Co., Japan, personal communication) which was identified as an *Achromobacter* species, produced a large amount of extracellular lipase in an alkaline medium (pH 10). The optimum pH of the purified lipase was 10·0; the enzyme retains 50% of the maximum activity between pH 6·0 and 11·8. However, the inhibition of the lipase by detergents is markedly reduced by the addition of 0·025% surfactants such as Tween 80 and Span 80.

7.2 *Pseudomonas* spp. Lipase

Watanabe *et al.* (1977) conducted an extensive screening for alkaline lipase-producing microorganisms from soil and water samples. Of 1606 strains isolated, two bacterial strains, 26·1B and 22·39B, were selected as potent producers of alkaline lipase, and were identified as *Pseudomonas nitroreducens* nov. var. *thermotolerans* and *P. fragi*, respectively. The optimum pH of the two lipases was 9·5. Both enzymes were inhibited by bile salts such as sodium cholate, sodium deoxycholate, and sodium taurocholate at a concentration of 0·25%.

Recently a Japanase company has produced a laundry detergent containing a fungal lipase produced by recombinant DNA technology. The fungus used is not alkalophilic, but an alkalophile must be a good candidate for this field.

8 β-LACTAMASE

In 1976, an alkalophilic bacterium, *Bacillus cereus* No. 170, was found to produce extracellular β-lactamase (Sunaga *et al.*, 1976, 1979). Three fractions showing penicillinase activity were observed in the culture broth of *Bacillus cereus* No. 170 induced by the addition of benzyl-penicillin. Relative molecular masses of penicillinases found in fractions I and II were approximately 26 000 and the fraction III penicillinase was approximately 24 000. Their pH optima for activity were 6·0–6·5 and the stable pH ranges were broad, ranging from about 7·0 to 10·0. Fraction III enzyme was the most thermostable and was strongly activated by the addition of Zn^{2+} (5 mM) and should be a class B β-lactamase according to the classification system proposed by Ambler (1980). However, its yield was too low to study further. The β-lactamase gene of alkalophilic *Bacillus* sp. No. 170 was cloned, strongly expressed in *E. coli*, and purified (Kato *et al.*, 1983, 1985).

9 XYLANASES

Xylan biomass is generally abundant in annual crops, particularly in agricultural residues such as rice straw, barley straw, and corn cobs. From the industrial point of view, alkaline xylanases would be interesting to study, because the enzyme can readily hydrolyze xylan which is soluble in alkaline solution.

9.1 Mesophilic Xylanases of Alkalophilic *Bacillus* species

Two alkalophilic bacteria, *Bacillus* sp. No. C-59-2 and No. C-11, which produce xylanase have been isolated (Horikoshi & Atsukawa, 1973*b*; Ikura & Horikoshi, 1977). Both organisms could grow well and produce xylanase in alkaline media at pH 10. The purified enzyme exhibited a broad optimum pH range for activity from 6·0–8·0. Maximum hydrolysis of xylan was about 40% at either pH 6·0 or pH 9·0. The optimum pH for activity of xylanase from *Bacillus* sp. No. C-11 was 7·0, and approximately 37% of the activity remains even at pH 10·0. Xylanases from alkalophiles retain considerably higher activity in the alkaline pH range than do other bacterial xylanases. Recently, it was found that several alkalophiles produced multi-component xylanases, such as *Bacillus* sp. No. C-125, etc. In the culture fluid, a major component, xylanase N and a minor component, xylanase A (approximately 1/10 of the major component) were detected. Xylanase N was most active at pH 6·0–7·0 but xylanase A had a broad pH activity curve (pH 6–10) and was still active at pH 12 (Honda *et al.*, 1985*a*, *b*).

9.2 Thermophilic Xylanases from Alkalophilic *Bacillus* species

Four isolates (W1, W2, W3 and W4) of alkalophilic thermophilic bacteria which produced xylanase were obtained from soils using a xylan medium (Okazaki *et al.*, 1984, 1985). The pH optimum for enzyme activity of strains W1 and W3 was 6·0 and for those strains W2 and W4 it was between 6 and 7. The enzymes were stable between pH 4·5 and 10·5 at 45°C for 1 h. The optimum temperatures of xylanases of W1 and W3 were 65°C and those of W2 and W4 were 70°C. The enzymes were stable up to 60°C and the addition of 5 mM $CaCl_2$ had no effect on their thermal stabilities. All xylanases hydrolyzed xylan to yield xylose and xylobiose. The degree of hydrolysis of xylan was about 70% after 24 h incubation.

10   PECTINASES

The pectinases are classified into three groups according to their mode of action towards the substrate: (1) esterases; (2) hydrolases; and (3) lyases. Among the pectinases, the most widely distributed enzymes in microorganisms, are endo-polygalacturonase and endo-polygalacturonate lyases. The pH optima of endo-polygalacturonases generally range from 4·0–5·0 and those of endo-polygalacturonate lyases range from 7·0–10·0.

10.1   *Bacillus* sp. No. P-4-N Polygalacturonase

The first paper on alkaline endo-polygalacturonase produced by alkalophilic *Bacillus* sp. No. P-4-N was that of Horikoshi (1972). The enzyme was purified about 300-fold from the culture fluid by column chromatography on DEAE-cellulose and Sephadex G-100 and G-200. The pH optimum of the enzyme was 10·0 on pectic acid. This is one of the unique properties of this enzyme in contrast with other polygalacturonases. In general, hydrolytic pectinases do not require calcium. The activation of the enzyme by calcium is also a unique feature. The enzyme was most stable at pH 6·5 and up to 70°C in the presence of Ca^{2+} but was inactivated completely at 80–90°C.

10.2   *Bacillus* sp. RK9 Polygalacturonate Lyase

Kelly & Fogarty (1978) reported that *Bacillus* sp. RK9 isolated from garden soil could grow well in alkaline media at pH 9·7 and produced endo-polygalacturonate lyase. The endo-polygalacturonate lyase of RK9 was purified 163-fold from culture fluid by precipitation with 50–90% saturated $(NH_4)_2SO_4$ and DEAE-cellulose column chromatography. The pH optimum of the enzyme was 10·0 towards acid-soluble pectic acid. The enzyme activity was significantly affected by the buffer system used for the reaction. The optimum temperature of the enzyme was 60°C. The enzyme retained 100% of its activity after incubation for 1 h at 37°C and pH 11·0.

10.3   Industrial Applications

10.3.1   *Production of Japanese Paper*
The first application of alkaline pectinase-producing bacteria in retting of Mitsumata bast (*Edgeworthia papyrifera*) was reported by Yoshihara &

Kobayashi (1982). They isolated about 800 alkalophilic bacteria from soil, sewage, and decomposed manure in Japan and Thailand. *Bacillus* sp. GIR-277 had strong macerating activity toward Mitsumata bast. The retted basts were harvested and Japanese paper was prepared by the method described in Japanese Industrial Standard P8209. The overall yield of pulp was about 70%. The strength of the unbeaten pulp resulting from bacterial retting was as high as that by the conventional soda ash-cooking method. The paper sheets were very uniform and soft to touch.

10.3.2 *Treatment of Pectic Waste Water with an Alkalophilic* Bacillus *sp.*

The waste water from citrus processing industry contains pectinaceous materials that are hardly decomposed by microbes during the activated-sludge treatment. Tanabe *et al.* (1987) had tried to develop a new waste treatment by using alkalophilic microorganisms. They isolated an alkalophilic *Bacillus* sp. GIR 621 from soil in Thailand which produced an extracellular endo-pectate lyase in alkaline media of pH 10·0 (Tanabe *et al.*, 1987). The treatment with the strain GIR 621-7 proved to be useful as a pretreatment of the waste water to remove pectic substances. The endo-pectate lyase from GIR 621-7 was purified and characterized as follows: relative molecular mass, 33 000; isoelectric point, pH 8·8; optimal pH, 9·0; optimal temperature, 55–60°C; preferred concentration of Ca^{2+}; 0·4 mM.

11 OTHER ENZYMES

11.1 Catalase

Microbial catalases are intracellular enzymes. Kurono & Horikoshi (1973) reported that an alkalophile *Bacillus* sp. No. Ku-1 isolated from soil, produced alkaline catalase in the Horikoshi-1 medium. The unique feature of this organism was the extracellular production of catalase. The pH optimum of the enzyme is 10·0; the activity decreases sharply at pH values above 10, and practically no activity is detected at pH 11·0. The enzyme is stable for 10 min between pH 7·0 and 8·5 at 60°C, but completely inactivated at 70°C.

11.2 Alkaline Phosphatase

In *E. coli* alkaline phosphatase is located in the periplasmic space as a soluble enzyme. In Gram-positive bacteria such as *B. subtilis* and *B. licheniformis*, the alkaline phosphatase appears to be membrane-bound.

Bacillus sp. RK11 (IMD No. 278), a soil isolate, produced an extracellular alkaline phosphatase. In the absence of Mn^{2+} in a complex medium, neither alkaline phosphatase production nor sporulation were detected. No other divalent metal could be substituted for Mn^{2+} (Kelly *et al.*, 1984).

Recently Nomoto *et al.* (1988) isolated alkalophilic *Bacillus* sp. OK-1, which excreted alkaline phosphatase into the Horikoshi-II medium. The enzyme was purified by DEAE-cellulose column chromatography followed by gel filtration. The relative molecular mass was estimated as 108 000 and was composed of two subunits having the same value of 54 000. The purified enzyme had a pH optimum of 11 and was fairly stable between pH 5 and 12. This enzyme was inactivated by EDTA and recovered by the addition of Ca^{2+} ions.

Although Kelly *et al.* (1984) did not report that the presence of any metal ion was responsible or associated with alkaline phosphatase activity, it is of interest that alkaline phosphatase which was thought to be a typical intracellular enzyme, was secreted by an alkalophilic *Bacillus* sp. into the culture liquor.

REFERENCES

Akino, T., Nakamura, N. & Horikoshi, K. (1987). *Applied Microbiology and Biotechnology*, **26**, 323.

Akino, T., Nakamura, N. & Horikoshi, K. (1988a). *Agricultural and Biological Chemistry*, **52**, 773.

Akino, T., Nakamura, N. & Horikoshi, K. (1988b). *Agricultural and Biological Chemistry*, **52**, 1459.

Ambler, R. P. (1980). *Philosophical Transactions of the Royal Society, London*, **B189**, 321.

Aunstrup, K., Outtrup, H., Andresen, O. & Dambmann, C. (1972). *Proteases from alkalophilic* Bacillus *sp. Fermentation Technology Today*, Proceedings of the 4th International Fermentation Symposium. Society Ferment. Technol., Osaka, Japan, p. 299.

Fogarty, W. M. (1983). In *Microbial Enzymes—Biotechnology*, ed. W. M. Fogarty. Applied Science Publishers, London, p. 1.

Fujiwara, N. & Yamamoto, K. (1987). *Journal of Fermentation Technology*, **65**, 53.

Fukumori, F., Kudo, T. & Horikoshi, K. (1985). *Journal of General Microbiology*, **131**, 3339.

Fukumori, F., Kudo, T., Narahashi, Y. & Horikoshi, K. (1986a). *Journal of General Microbiology*, **132**, 2329.

Fukumori, F., Sashihara, N., Kudo, T. & Horikoshi, K. (1986b). *Journal of Bacteriology*, **168**, 479.

Fukumori, F., Kudo, T. & Horikoshi, K. (1987a). *FEMS Microbiology Letters*, **40**, 311.

Fukumori, F., Ohnishi, F., Kudo, T. & Horikoshi, K. (1987b). *FEMS Microbiology Letters*, **48**, 65.

Fukumori, F., Kudo, T., Sashihara, N., Nagata, Y., Ito, K. & Horikoshi, K. (1989). *Gene*, **76**, 289.

Hamamoto, T., Kaneko, T. & Horikoshi, K. (1987). *Agricultural and Biological Chemistry*, **51**, 2019.

Hayashi, T., Akiba, T. & Horikoshi, K. (1988a). *Agricultural and Biological Chemistry*, **52**, 443.

Hayashi, T., Akiba, T. & Horikoshi, K. (1988b). *Applied Microbiology* and *Biotechnology*, **28**, 281.

Honda, H., Kudo, T., Ikura, Y. & Horikoshi, K. (1985a). *Canadian Journal of Microbiology*, **31**, 538.

Honda, H., Kudo, T. & Horikoshi, K. (1985b). *Journal of Bacteriology*, **161**, 784.

Horikoshi, K. (1971a). *Agricultural and Biological Chemistry*, **35**, 1407.

Horikoshi, K. (1971*b*). *Agricultural and Biological Chemistry*, **35**, 1783.

Horikoshi, K. (1972). *Agricultural and Biological Chemistry*, **36**, 285.

Horikoshi, K. & Atsukawa, Y. (1973*a*). *Agricultural and Biological Chemistry*, **37**, 1449.

Horikoshi, K. & Atsukawa, Y. (1973*b*). *Agricultural and Biological Chemistry*, **37**, 2097.

Horikoshi, K. & Atsukawa, Y. (1975). *Biochimica et Biophysica Acta*, **384**, 477.

Horikoshi, K. & Iida, S. (1958). *Nature*, **181**, 917.

Horikoshi, K. & Iida, S. (1959). *Nature*, **183**, 186.

Horikoshi, K. & Iida, S. (1964). *Biochimica et Biophysica Acta*, **83**, 197.

Horikoshi, K., Koffler, H. & Arima, K. (1963). *Biochimica et Biophysica Acta*, **73**, 267.

Horikoshi, K., Nakao, M., Kurono, Y. & Sashihara, N. (1984). *Canadian Journal of Microbiology*, **39**, 774.

Ikura, Y. & Horikoshi, K. (1977). *Agricultural and Biological Chemistry*, **41**, 1373.

Ikura, Y. & Horikoshi, K. (1987). *Journal of Fermentation Technology*, **65**, 707.

Ito, S., Shikata, S., Ozaki, K., Kawai, S., Ohta, Y. & Satoh, T. (1989). *Agricultural and Biological Chemistry*, **53**, 1275.

Kaneko, T., Hamamoto, T. & Horikoshi, K. (1988). *Journal of General Microbiology*, **134**, 97.

Kato, C., Kudo, T., Watanabe, K. & Horikoshi, K. (1983). *European Journal of Applied Microbiology and Biotechnology*, **18**, 339.

Kato, C., Kudo, T., Watanabe, K. & Horikoshi, K. (1985). *Journal of General Microbiology*, **131**, 3317.

Kelly, C. T. & Fogarty, W. M. (1978). *Canadian Journal of Microbiology*, **24**, 1164.

Kelly, C. T., O'Reilly, F. & Fogarty, W. M. (1983). *FEMS Microbiology Letters*, **20**, 55.

Kelly, C. T., Nash, A. M. & Fogarty, W. M. (1984). *Applied Microbiology and Biotechnology*, **19**, 61.

Kelly, C. T., Brennon, P. A. & Fogarty, W. M. (1987). *Biotechnology Letters*, **9**, 125.

Kimura, K., Takano, T. & Yamane, K. (1987*a*). *Applied Microbiology and Biotechnology*, **26**, 147.

Kimura, K., Kataoka, S., Ishii, Y., Takano, T. & Yamane, K. (1987*b*). *Journal of Bacteriology*, **169**, 4399.

Kimura, K., Tsukamoto, A., Ishii, Y., Takano, T. & Yamane, K. (1988). *Applied Microbiology and Biotechnology*, **27**, 372.

Kurono, Y. & Horikoshi, K. (1973). *Agricultural and Biological Chemistry*, **37**, 2565.

Mäkelä, M., Mattsson, P., Schinina, M. E. & Koppela, T. (1988). *Biotechnology and Applied Biochemistry*, **10**, 414.

Nakamura, N., Watanabe, K. & Horikoshi, K. (1975). *Biochimica et Biophysica Acta*, **397**, 188.

Nomoto, M., Lee, T.-C., Su, C.-S., Liao, C.-W., Yen, T.-M. & Yang, C.-P. (1984). *Agricultural and Biological Chemistry*, **48**, 1627.

Nomoto, M., Ohsawa, M., Wang, H.-L., Chen, C.-C. & Yeh, K.-W. (1988). *Agricultural and Biological Chemistry*, **52**, 1643.

Okazaki, W., Akiba, T., Horikoshi, K. & Akahoshi, R. (1984). *Applied Microbiology and Biotechnology*, **19**, 335.

Okazaki, W., Akiba, T., Horikoshi, K. & Akahoshi, R. (1985). *Agricultural and Biological Chemistry*, **49**, 2033.

Shirokizawa, O., Akiba, T. & Horikoshi, K. (1989). *Agricultural and Biological Chemistry*, **53**, 491.

Sunaga, T., Akiba, T. & Horikoshi, K. (1976). *Agricultural and Biological Chemistry*, **40**, 1363.

Sunaga, T., Akiba, T. & Horikoshi, K. (1979). *Agricultural and Biological Chemistry*, **43**, 477.

Takami, H., Akiba, T. & Horikoshi, K. (1989). *Applied Microbiology and Biotechnology*, **30**, 120.

Tanabe, H., Yoshihara, K., Tamura, K., Kobayashi, Y., Akamatsu, I., Niyomwan, N. & Footrakul, P. (1987). *Journal of Fermentation Technology*, **64**, 243.

Tsai, Y., Yamasaki, M., Yamamoto-Suzuki, Y. & Tamura, G. (1983). *Biochemistry International*, **7**, 577.

Tsuchida, O., Yamagata, Y., Ishizuka, T., Arai, Y., Yamada, J., Takeuchi, M. & Ichishima, E. (1986). *Current Microbiology*, **14**, 7.

Watanabe, N., Ota, Y., Minoda, Y. & Yamada, K. (1977). *Agricultural and Biological Chemistry*, **41**, 1353.

Yamamoto, M., Tanaka, Y. & Horikoshi, K. (1972). *Agricultural and Biological Chemistry*, **36**, 1819.

Yoshihara, K. & Kobayashi, Y. (1982). *Agricultural and Biological Chemistry*, **46**, 109.

Chapter 9

ENZYMATIC AND MICROBIAL TRANSFORMATION OF ANTIBIOTICS AND STEROIDS

JOSEPH O'SULLIVAN

The Squibb Institute for Medical Research, Princeton, New Jersey 08543-4000, USA

CONTENTS

1 INTRODUCTION

The medical and economic importance of antibiotics and steroids can hardly be overstated. Since their introduction about fifty years ago, antibiotics have transformed the way medicine is practiced; they have affected life expectancy and life quality in a significant way and, in consequence, have had a profound effect on society. This is also true for steroids, their widespread use has been a major factor in the control of pain and their application in controlling the reproductive cycle has had far reaching effects on population growth in the economically developed countries.

The study of antibiotics has helped open up our understanding of fundamental cellular processes such as protein synthesis, nucleic acid and general cellular metabolism as well as in the elucidation of cell wall and membrane structures.

The importance of antibiotics, therefore, extends well beyond their clinical application. It is the clinic however, which offers the greater challenge to industrial scientists: the challenge to overcome the problems of drug resistance, to provide new drugs for new diseases and to improve old drugs. It is clear that these challenges can best be met by continually developing methods to find novel chemical structures, by studying the potential of enzymatic and microbial transformations and by applying the new technologies in recombinant DNA and immunology to open up new areas of cellular metabolism for exploration.

In this chapter the medically and economically most important antibiotics will be described. The estimated worldwide revenues for antibacterials in 1987 was 13 614 million dollars, of which cephalosporins contributed 45%, penicillins 19%, macrolides, aminocyclitols and tetracyclines, 5% each; the synthetic antibacterials, quinolones and sulphonamides, contributed another 10% and the remainder could be accounted for by antitubercular agents and several minor compounds. Because of the vast body of literature available on antibiotics and steroids, it has not been possible to deal with those agents having activity against fungi, viruses or mammalian cells nor with the many antibacterials of minor medical and industrial interest.

2 ANTIBIOTICS

2.1 β-Lactam Antibiotics

The β-lactam antibiotics comprise six classes of naturally occurring compounds; penicillins, cephalosporins, clavams, carbapenems, nocardicins and monobactams (Fig. 1). With the exception of nocardicins, representatives of all the other classes have found clinical utility and make β-lactams the single most important group of antibacterials commercially.

2.1.1 Penicillins
Penicillins are *N*-acyl derivatives of 6-aminopenicillanic acid (6-APA) and are produced in two types of fermentation. In the first type, fungi such as *Penicillium chrysogenum*—but also some ten other genera (Elander, 1983; Kitano, 1983)—produce penicillins with non-polar side chains. The nature of the non-polar side chain can be determined by the addition of a suitable precursor (Behrens, 1948). For example, if phenylacetic acid is added to fermentations of *P. chrysogenum*, benzylpenicillin (Penicillin G) is produced (Fig. 2). Other monosubstituted acetic acids can also be used, thus, phenoxyacetic acid results in phenoxymethyl penicillin or penicillin V production. A number of questions relating to the transacylation of isopenicillin N to give penicillins with non-polar side chains (Fig. 2) still exist but several reports suggest that a single enzyme is involved which removes the L-α-aminoadipyl side chain of isopenicillin N giving rise to 6-APA, probably in an enzyme

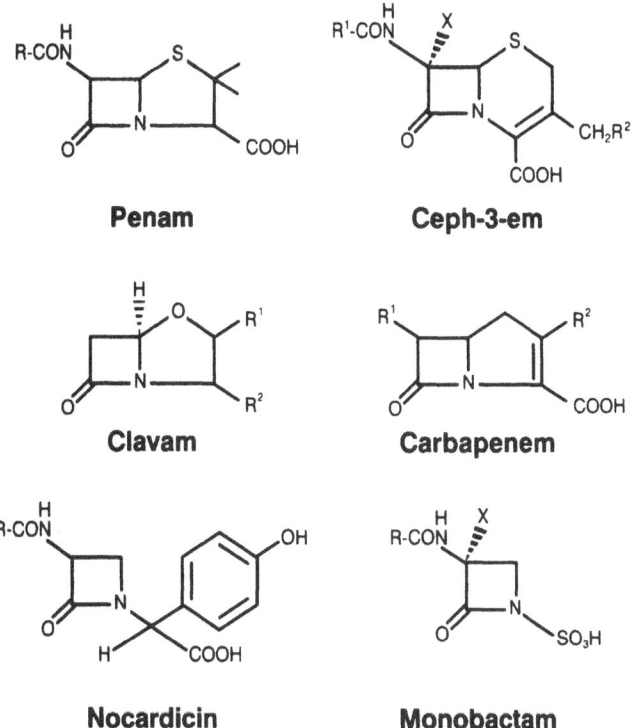

Fig. 1. The β-lactam group of antibiotics.

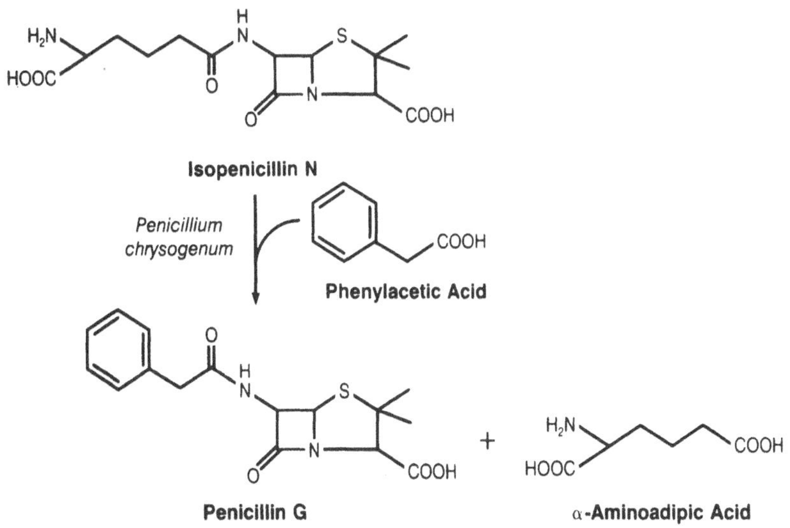

Fig. 2. The production of penicillin G by *Penicillium chrysogenum*.

bound state; the 6-APA is then acylated with the activated side chain to give the penicillin. (Spencer & Maung, 1970; Loder, 1972; Fawcett *et al.*, 1975).

The second type of penicillin fermentation leads to the production of penicillin N and then usually to cephalosporins. Eight fungal genera, of which the best known is *Cephalosporium acremonium* (*Acremonium chrysogenum*) produce penicillin N and cephalosporins (Kitano, 1983) as do numerous *Streptomyces* species and the bacterial species, *Flavobacterium* and *Xanthomanas*, (Singh *et al.*, 1982), although it has not been formally demonstrated that the bacteria produce penicillin N.

The penicillin fermentation of the first type, is of major industrial importance since all the clinically used penicillins start out as penicillins with non-polar side chains which are then cleaved by penicillin acylases to give 6-APA (Fig. 3). It is estimated that 15–20 000 tons of penicillin are produced each year and the bulk of this is converted to 6-APA (Lowe, 1988). The 6-APA is then modified to the penicillin of choice, examples of which are, methicillin, nafcillin, ampicillin and amoxicillin. The structure–activity relationships of penicillins have been reviewed by Hoover (1983).

There are basically three types of enzymes which engage penicillins: penicillin acylases which remove the side chain (Fig. 3) or in the reverse reaction, add the activated side chain to 6-APA to produce penicillin; secondly β-lactamases which hydrolyze the cyclic amide bond of β-lactam compounds and thirdly, the penicillin-binding proteins which possess carboxypeptidase or transpeptidase activity and through whose inhibition penicillins exert their action.

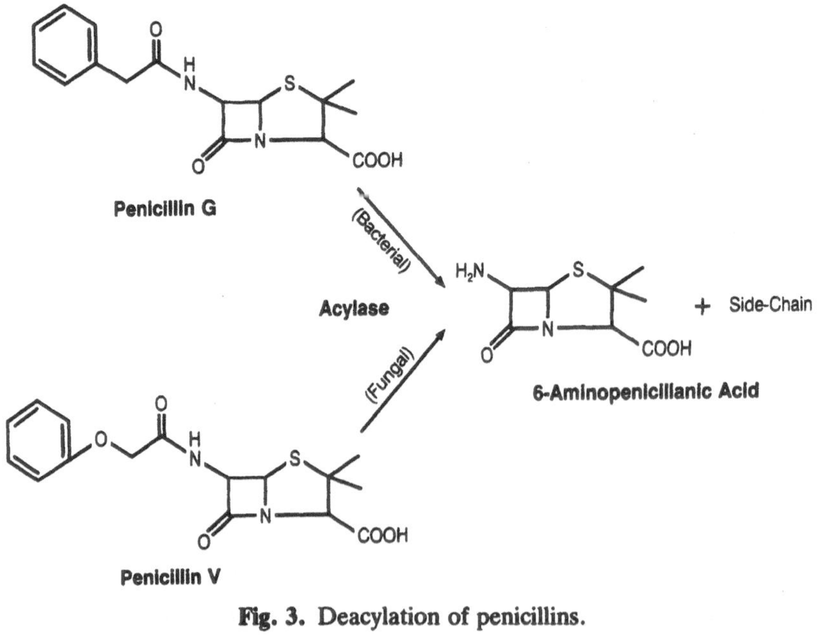

Fig. 3. Deacylation of penicillins.

2.1.1.1 ACYLASES

The enzymatic cleavage of penicillin G or V is of substantial interest. Soon after the discovery of 6-APA (Bachelor *et al.*, 1959) numerous investigators reported the existence of microbial enzymes which would deacylate penicillins (Kaufmann & Bauer, 1960; Rolinson *et al.*, 1960). Later it was found that the ability to deacylate penicillins was relatively widespread and that penicillin G was the preferred substrate for bacterial acylases while penicillin V was the better substrate for fungal acylases (Fig. 3). A third acylase, ampicillin acylase, has also been described. Acylase-producing organisms have been extensively studied and much strain development has been carried out; acylases are found both intra- and extracellularly and the choice of a particular enzyme or enzyme source will depend on several industry-specific conditions. Savidge (1984) lists ten major penicillin-producing companies each of which uses a different microbial source of acylase. The gene for penicillin G acylase from *Escherichia coli* was cloned (Mayer *et al.*, 1979) and expressed in high copy in *E. coli*. Uninduced levels over a thousand fold higher than the parent were achieved. Phenylacetic acid can be used effectively to induce acylase in several organisms (Kaufmann & Bauer, 1960; Klein & Wagner, 1980). Although penicillin acylase can be used as a pure enzyme in solution or impure and immobilized in its producing cell (Sato *et al.*, 1976), the most successful application of penicillin acylases is in a purified form immobilized by either adsorption, adsorption and cross-linking, covalent attachment to resins or lastly, by physical entrapment. Immobilization procedures have been reviewed by Abbot (1976) and by Vandamme (1988) and a discussion of the benefits of using immobilized β-lactam acylases has been presented by Lowe (1988). He draws special attention to the safety features of immobilized enzyme usage since they lead to the elimination of hazardous reagents such as solvents and other chemicals. They are also efficient because of their reduced energy use, increased activity and reduced side reactions over whole cells and finally, extensive reuse is possible. The reversibility of the acylase reaction in Fig. 3 has been referred to above. This was first demonstrated by Robinson *et al.* (1960) and Kaufmann & Bauer (1960) who showed that it proceeded best at pH 4–6 when the methyl ester of the D-side chain enantiomer was in excess. Immobilized *E. coli* was shown to produce ampicillin and amoxicillin in over 50% yield from 6-APA and the methyl esters of D-phenylglycine and *p*-hydroxy-phenylglycine, respectively (Marconi *et al.*, 1975). Other enzyme sources have been explored for this reverse reaction and conversion rates of 63% for ampicillin (Nara *et al.*, 1971) and 90% for amoxicillin (Kawamori *et al.*, 1983) have been achieved. However, these processes are not as efficient as chemical semisynthesis and are unlikely to find extensive industrial application.

2.1.1.2 β-LACTAMASES

The second major group of enzymes which interact with penicillins are the β-lactamases. These were discovered almost as soon as the penicillin structure became known (Abraham & Chain, 1940) and they have been an ominous presence throughout the history of the development of penicillins and other β-lactams. The first major problem encountered with penicillin usage was the occurrence of penicillin-resistant *Staphylococcus aureus* due to β-lactamase production. By the mid 1950s, the incidence of these organisms in some hospitals was as high as 80%. This led to the search for β-lactams which were resistant to attack by β-lactamases and with the discovery of cephalosporins (Newton & Abraham, 1956) this was achieved in part. However, the identification of 6-APA (Batchelor *et al.*, 1959) presented the possibility of generating semisynthetic penicillins and led to the development of β-lactamase stable penicillins such as methicillin, nafcillin, cloxacillin and others. With the development of broad-spectrum β-lactams another problem arose, the production of β-lactamase by Gram-negative bacteria. This problem, in turn, was overcome by advances in cephalosporin activity due to the discovery of cephalosporins with a methoxyl group in the C-7 position (X in Fig. 1) (Nagarajan *et al.*, 1971). These cephamycins had broad-spectrum Gram-negative activity and were stable to β-lactamases (Stapley *et al.*, 1972). In the mid-1970s clavulanic acid was found in broths of *Streptomyces clavuligerus* (Howarth *et al.*, 1976) and it could inhibit a broad range of β-lactamases. This discovery was soon followed by other novel β-lactam compounds such as the carbapenems, nocardicins and monobactams. Thus, each challenge offered by β-lactamases was met by either new natural product discovery or by semisynthetic manipulation of existing structures. β-Lactamases continue to proliferate however, and enzymes with novel properties continue to arise, thereby, extending the challenge to develop new antibiotics. β-Lactamases owe their success in part, to the transposable nature of β-lactamase genes (Hedges *et al.*, 1974). These genes may be located chromosomally or extrachromosomally and can be mobilized within an organism by transposition and between organisms by mechanisms such as transduction, conjugation or transformation. β-Lactamases can be classified on the basis of microbial origin, size, substrate and inhibition profiles, isoelectric point and, increasingly, by sequence. Numerous attempts to classify this group of enzymes have been made (Richmond & Sykes, 1973; Sykes & Matthew, 1976; Bush, 1988) and as new data is gathered, especially gene sequence data, additional schemes will be offered. β-Lactamases are produced by most bacteria, including blue-green algae (Kushner & Breuil, 1977) but have also been reported from yeast (Mehta & Nash, 1978) and in human kidney (Kropp *et al.*, 1982). β-Lactamases have been the subject of numerous reviews, for example, Medeiros & Jacoby (1986), Amyes (1987) and Sanders (1987).

2.1.1.3 PENICILLIN-BINDING PROTEINS

The third class of proteins which interact with penicillins comprise a group of membrane-bound proteins with molecular weights from about 40 000–91 000 daltons and are commonly referred to as penicillin-binding proteins (PBP) (Blumberg & Strominger, 1974; Spratt 1975; Georgopapadakou & Liu, 1980). PBPs are involved in the biosynthesis of the bacterial cell wall and it is by interaction with PBPs that penicillins and other β-lactam antibiotics exert their action. There are several PBPs in bacterial membranes and significant differences in the type of PBP between bacterial groups and in their response to penicillins. Specific PBPs have been assigned to specific cell functions. Thus, in *E. coli*, PBP1a and 1b are involved in cell elongation, PBP2 in cell shape and PBP3 in septation (Spratt, 1977). Binding of penicillin to PBP 4, 5 and 6 in *E. coli* led to no discernible morphological change. PBPs have basically two types of enzymatic activity, DD-carboxypeptidase activity (PBP 4, 5 and 6) and peptidoglycan transpeptidase activity (PBP1), although the high molecular weight PBPs (e.g. 1a) may also have transglycosylase activity (Nakagawa *et al.*, 1984). The PBPs all bind penicillin covalently and some of them exhibit β-lactamase-like activity. Recent analysis of the active site of PBPs would indicate considerable homology with β-lactamases suggesting that both groups of proteins may be derived from a common evolutionary origin (Ghuysen, 1988; Nicholas & Strominger, 1988; Spratt & Cromie, 1988).

2.1.2 *Cephalosporins and Cephamycins (Cephems)*
The cephem nucleus (Fig. 1) consists of a β-lactam ring fused to a dihydrothiazine ring. Most cephalosporins and cephamycins typically, have a δ-linked D-α-aminoadipoyl side chain and cephamycins have a methoxyl group at the 7α-position of the β-lactam ring. The R^2 group in Fig. 1 can be —H in deacetoxycephalosporin, —OH in deacetylcephalosporin, —OCOCH$_3$, in cephalosporin C, —OCONH$_2$ in cephamycin A and also cinnamoyl groups in certain cephamycins (Stapley *et al.*, 1972), heterocyclic thiols in organomycins (Osono *et al.*, 1980) and very complex structures in certain bacterially-produced cephalosporins (Singh *et al.*, 1984; O'Sullivan & Sykes, 1986).

Cephalosporins are produced by fungi, streptomycetes and bacteria, whereas cephamycins are produced only by streptomycetes. *Flavobacterium* produces a 7-α-formamidocephalosporin with the complex R^2 substituent referred to above.

Cephalosporin C was discovered at a time when penicillin-resistant *Staphylococcus aureus* was becoming a clinical problem and many of the developments which occurred with penicillins have been paralleled by similar developments in the cephems. Clinically, cephems have found very wide usage and may be thought of as progressing through three generations of semi-synthetic modification, examples of which are: cephalothin, cephalexin and cephaloridine, first generation, cefoxitin and cefuroxime, second and cefotaxime, third generation. Each generation extended the antibacterial spectrum of the cephems (Hoover, 1983).

2.1.2.1 ENZYMES REACTING WITH CEPHEMS
Figure 4 shows the sites of enzymic action on the cephem structure. The cephalosporins are biosynthetically derived from penicillin N by the action of a thiazolidine ring expanding enzyme, deacetoxycephalosporin C synthase found in crude extracts of *C. acremonium* (Kohsaka & Demain, 1976). It was shown that deacetoxycephalosporin C was converted to deacetylcephalosporin C (R^2

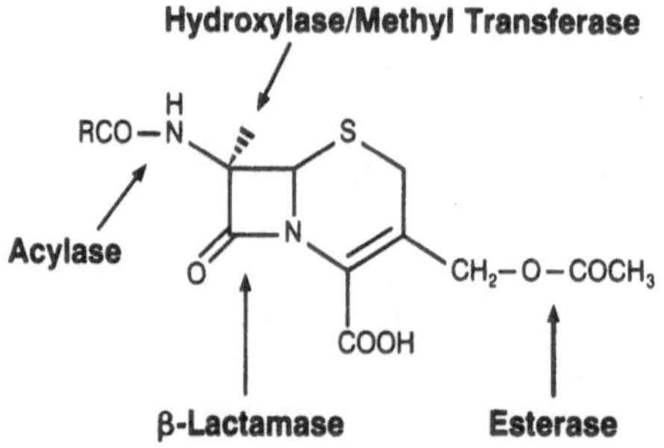

Fig. 4. Enzymatic modifications of cephems.

in Fig. 1 = OH) by a dioxygenase (Turner *et al.*, 1978). It was later found that the synthase and the dioxygenase had the same co-factor requirements and were inseparable by chromatography or DEAE-Sephacryl. Since they catalyzed sequential biosynthetic reactions, Scheidegger *et al.* (1984) proposed that both reactions were catalyzed by the same protein. This proved not to be the case in *Streptomyces* where the activities were clearly separable (Jensen *et al.*, 1985). A partial amino acid sequence of the synthase–dioxygenase was accomplished and was used to clone the gene from *C. acremonium* with expression of the cloned gene in *E. coli* from which active extracts were made (Samson *et al.*, 1985). The sequence of part of this enzyme was very homologous (50%) with a region of the enzyme isopenicillin N synthase which appeared to be involved in substrate binding (Samson *et al.*, 1987); isopenicillin N synthase is the earlier enzyme in the pathway that cyclizes the tripeptide α-aminoadipyl–cysteinyl–valine to isopenicillin N (Konomi *et al.*, 1979; O'Sullivan *et al.*, 1979).

Two other biosynthetic steps are of interest here, the acetylation of deacetylcephalosporin to give cephalosporin C (and the reverse reaction) and the introduction of the methoxyl group in cephamycins. The involvement of an acetylase in the terminal step of cephalosporin C biosynthesis was noted by Fujisawa *et al.* (1973); it had been shown earlier by Jeffery *et al.* (1961) that the acetyl group could be removed and this process was of some commercial importance because semisynthetic modifications could be more easily carried out with the free hydroxyl group of deacetylcephalosporin C. Since that report, numerous sources of the acetylesterase have been found, including bacteria, plants and mammals. Abbot & Fukuda (1975*a*, *b*) have described, in some detail, an enzyme from *Bacillus subtilis* with a molecular weight of 190 000 which has a temperature optimum of 40–50°C and a pH optimum of 7·0. The purified enzyme was used in a filtration apparatus with a 10 000 dalton cut off. When the substrate was hydrolyzed and then filtered off, the enzyme was reusable up to twenty times over an eleven day period.

The introduction of the methoxyl group on the β-lactam ring of cephalosporins at the C-7 position was first studied enzymatically by O'Sullivan & Abraham (1980). They showed that a cell-free preparation of *Streptomyces clavuligerus* converted cephalosporin C to 7α-methoxy cephalosporin C in the presence of O_2, Fe^{2+}, α-oxoglutarate, *S*-adenosylmethionine and a reducing agent. They proposed that the C-7 position was first hydroxylated and then methylated and later Hood *et al.* (1984) succeeded in isolating the 7α-hydroxyl intermediate. Hood and coworkers studied the substrate specificity of the oxygenation and methylation steps and concluded that the methylation was less specific. Enzymes from different *Streptomyces* species had different substrate specificities. The *S. lipmanii* and *S. clavuligerus* oxygenases were markedly affected by alteration of the substituent at C-3 in the cephalosporins whereas *S. wadayanensis* oxygenase was less affected. Other substrate changes were also studied.

2.1.2.2 SIDE-CHAIN MODIFICATION

Just as the production of 6-APA had opened up a host of semisynthetic possibilities for penicillin so too did the generation of 7-aminocephalosporanic acid (7-ACA) for cephalosporin production. The first demonstration of 7-ACA was carried out chemically by Loder *et al.* (1961) when they also showed that acylation of the 7-ACA led to big increases in bioactivity. Since that time, people have been looking for enzymes which would accomplish the deacylation of cephalosporins; most of these efforts have been unsuccessful.

What has been possible is a two step procedure where D-amino acid oxidase from *Gliocladium deliquescens* (Banyu, 1972) or from *Trigonopsis vulgaris* (Fildes *et al.*, 1974) converts the amino acid to a keto acid; the keto acid can be converted by decarboxylation in the presence of hydrogen peroxide to give glutaryl-7ACA. Ichikawa *et al.* (1981) described an acylase from *Pseudomonas* SY-77-1 which was active against the keto acid and the glutaryl derivative with the production of 7-ACA. Another two step method of producing 7-ACA is a procedure where the amino group of the D-α-aminodipyl side chain is acylated with isobutoxychloroformate then treated with *Paecilomyces* which hydrolyzed the side chain and the acyl group (Kawate *et al.*, 1978). Matsuda *et al.* (1987) have cloned and characterized the genes for two distinct cephalosporin acylases from *Pseudomonas*. Sequence analysis of these genes and comparing them with the sequence for the penicillin G acylase gene from *E. coli* (Schumacher *et al.*, 1986) revealed striking similarity between them; both acylases consist of two nonidentical subunits processed from a common precursor encoded by a single open reading frame. Analysis of the active site of these enzymes may well lead to procedures which will more easily result in efficient deacylation of cephalosporin C. Semisynthetic cephalosporins are readily deacylated by penicillin acylases and indeed the reverse, synthetic reaction has also been carried out. Kim *et al.* (1983) described the synthesis of cephalexin using immobilized *Xanthomonas citri* cells with an 83% yield; however, it is unlikely that enzymatic procedures will find extensive industrial usage in the present economic environment.

2.1.2.3 β-LACTAMASES AND PBPS

The last groups of enzymes to be considered in connection with cephalosporins are the β-lactamases (cephalosporinases) and the PBPs. Much of the discussion of these enzymes in connection with penicillins applies equally to cephalosporins. Depending on the classification criteria used, cephalosporinases may end up being grouped apart from penicillinases and there are certainly β-lactamases which show a substrate preference for cephalosporins. But when amino acid sequence around the active site, isoelectric point, substrate profiles and inhibition profiles, microbial origins, metal requirements, etc. are considered, they do not form a very homogeneous group. Likewise, the interaction of cephalosporins with PBPs is in many respects similar to that for penicillins and other β-lactams.

2.1.3 Clavulanic Acid

Some other β-lactam compounds in clinical use are clavulanic acid, thienamycin (a carbapenem) and aztreonam (a monobactam). Clavulanic acid is produced by several *Streptomyces* species and was originally isolated in a screen to detect inhibitors of β-lactamase (Brown *et al.*, 1976). Several other clavams have been isolated since then (see review by O'Sullivan & Sykes, 1986). Clavulanic acid is weakly antibacterial but is a potent inhibitor of β-lactamase produced by staphylococci and of the plasmid encoded β-lactamases of several Gram-negative bacteria. The interaction with the *E. coli* R-TEM enzyme has been studied in detail (Charnas *et al.*, 1978; Fisher *et al.*, 1978; Labia & Peduzzi, 1978) and consists of three stages, a normal hydrolysis of the β-lactam ring followed by a transient intermediate that slowly regenerates free enzyme and then an enzyme–clavulanate complex formation which results in irreversible inactivation of the β-lactamases (Fisher *et al.*, 1980). Clavulanic acid binds well to PBP2 of *E. coli,* moderately to PBPs 1, 4, 5 and 6 and has low affinity for PBP3 (Spratt *et al.*, 1977).

2.1.4 Carbapenems

Carbapenems consist of a β-lactam ring fused to an unsaturated five membered ring (Fig. 1). Unlike all other bicyclic β-lactams they do not contain sulphur or oxygen in the ring system but frequently the R^2 group contains sulfur, presumably derived from cysteamine. Often the R^1 group has a sulfonic acid component and contains one to three carbon atoms. The hydrogen atoms on the β-lactam ring may be *cis* or *trans* and this feature together with the nature and sterochemistry of the side chains has a profound effect on the type and extent of activity of the various carbapenems. There are in excess of thirty different carbapenems produced by a variety of *Streptomyces* species (Cassidy, 1981; O'Sullivan & Sykes, 1986) and a single example of a carbapenem produced by *Serratia* and *Erwinia* (Parker *et al.*, 1982) where R^1 and R^2 are hydrogen atoms. The first example of a carbapenem was thienamycin ($R^1 = $ —CH(CH$_3$)OH; $R^2 = $ —SCH$_2$CH$_2$NH$_2$) produced by *S. cattleya* (Albers-Schönberg *et al.*, 1978) which, in addition, produces four thienamycin derivatives. The differences between the numerous carbapenems described is often very slight and they are frequently produced in groups such as the thienamycins, the olivanic acids produced by *S. olivaceus* (Hood *et al.*, 1979), the PS series of compounds produced by *S. cremeus auratilis* (Okamura *et al.*, 1978) and several others (see reviews by Cassidy, 1981; O'Sullivan & Sykes, 1986). The carbapenems may be potent antibacterial agents or potent β-lactamase inhibitors or both. The *trans* isomers are more resistant to attack by β-lactamases than the *cis* isomers and sulfated forms are more resistant than unsulfated forms (Hoover, 1983; Basker *et al.*, 1980). Carbapenems are reversible inactivators of β-lactamases for the most part (Okamura *et al.*, 1980; Charnas & Knowles, 1981; Easton & Knowles, 1982), but irreversible inactivation of *Citrobacter freundii* and *Proteus vulgaris* β-lactamases has been demonstrated (Yamaguchi *et al.*, 1984). When *N*-formyl thienamycin was administered to human subjects it was found to be hydrolyzed in the kidney; the enzyme responsible was a dipeptidase and in order to overcome this problem, a peptidase inhibitor, cilastatin, has to be coadministered with the *N*-formylthienamycin (Norrby *et al.*, 1983).

The interaction of carbapenems with PBPs has not been extensively studied but the concentration of thienamycin which leads to 50% reduction of ^{14}C-benzylpenicillin binding in *E. coli* has been determined by Spratt *et al.* (1977). It had a high affinity for PBPs 1 (1·1 μg/ml), 2 (0·06 μg/ml), 4 (0·15 and 5·5 μg/ml), 5 (1 μg/ml) and 6 (3·8 μg/ml) and a lower affinity for PBP 3 (31 μg/ml). The minimum inhibitory concentration was 0·1 μg/ml, at which concentration *E. coli* KN126 converted to large osmotically stable round cells; at values above 0·6 μg/ml rapid lysis occurred. Details of other interactions of carbapenems with enzymes have come largely from biosynthetic studies. The coproduction of carbapenems with other β-lactams such as penams and cephems has raised the question of whether common mechanisms for β-lactam ring formation exist. Further, the structural relatedness of the carbapenems suggest a common pathway and with these considerations in mind Williamson *et al.* (1985) proposed a scheme for carbapenem biosynthesis. They postulate an initial condensation of actylcoenzyme A with γ-glutamylphosphate which could then be enzymatically cyclized to give the carbapen-2-em-3-carboxylic acid produced by *Serratia* and *Erwinia*. An intermediate in this process could also be methylated and successively modified to account for most of the chiral and side chain variations found in the carbapenems. The proposal of this biosynthetic scheme was greatly aided by studies with blocked mutants (Brown *et al.*, 1977; Rosi *et al.*, 1981). Nozaki *et al.* (1984) also studied blocked mutants of *S. griseus* subsp. *cryophilus* which produces eight 5,6 *cis*-carbanems and established a sequence for their biosynthesis. Kubo *et al.* (1984) isolated a specific acylase which removes the pathothenyl group of OA-6129 carbapenems and is believed to play a central role in the biosynthesis of carbapenems (Fig. 5). In addition to the depantothenylation of OA-6129, the enzyme catalyzed the acyl exchange of OA-6129 carbapenems with acyl coenzyme A. It can deacylate *N*-acetyl L-amino acids but not the carbapenem PS-5 ($R^2 = SCH_2CH_2NHCOCH_3$; $R^1 = CH_2CH_3$) and it can acylate 6-APA and certain deacylpenams; the enzyme can be found in *S. fuloviridis*, *S. catteya*, *S. cremeus* and *S. argenteolus*.

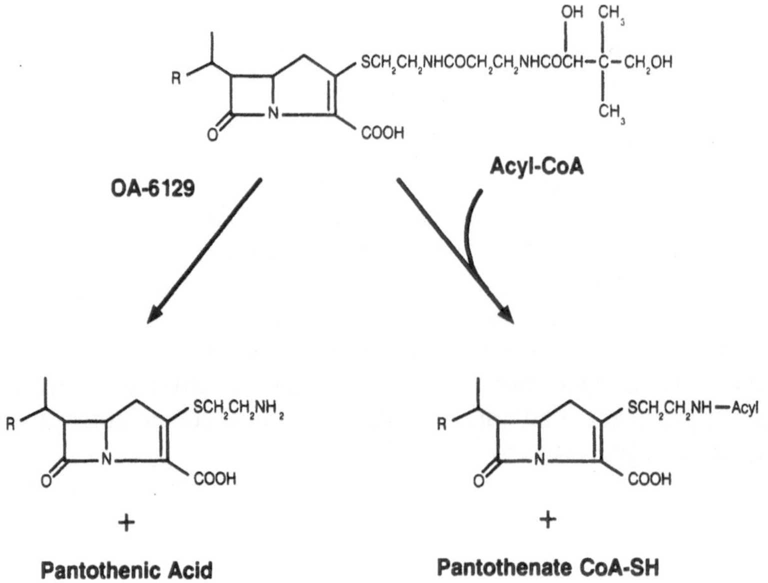

Fig. 5. The depantothenylation of OA-6129 and the acyl exchange of OA-6129 with acyl coenzyme A by an acylase (R = H, OH or OSO_3H).

2.1.5 Monobactams

Compounds having the monobactam structure (Fig. 1) were first reported by Japanese workers (Imada *et al.*, 1981) who identified sulfazecin (X = OCH_3, R = γ-D-glu-D-ala) and isosulfazecin (X = OCH_3, R = γ-D-glu L-ala) produced by *Pseudomonas acidophila* and *P. mesoacidophila*. Independently, and using a different detection procedure, monobactams were discovered from *Gluconobacter* sp., *Acetobacter* sp. and *Chomobacterium violaceum* (Sykes *et al.*, 1981). These monocyclic β-lactams are distinctive in having a sulfonic acid residue attached to the nitrogen of the ring. Since the early reports, several additional monobactams have been discovered (see reviews by O'Sullivan & Sykes, 1986; Parker *et al.*, 1986). Since the monobactams were produced in very low quantities in fermentations and since their synthesis was uncompli-cated, synthetic routes for the production of a wide variety of monobactam structures have been developed (Slusarchyk *et al.*, 1984). It was clear from biosynthetic studies that nutritional modification would not lead to various *N*-acyl derivatives (O'Sullivan *et al.*, 1982) and, indeed, the naturally produced monobactams were not hydrolyzed by penicillin acylases. However, penicillin G acylase from *E. coli* could acylate the 3-aminomonobactam nucleus to give active products (O'Sullivan & Aklonis, 1984). The acylation was optimal at pH 4·5 whereas deacylation of the product was optimal at pH 7·5. Monosubsti-tuted acetic acids were suitable as acyl groups but if the α-carbon was substituted, as for example, in phenylglycine, the acylation did not take place. Monobactams are like other β-lactams in that they interact with β-lactamases and PBPs. The natural products are not very good antibiotics, with minimal inhibitory concentrations against a panel of bacteria ranging from about 10 µg/ml to well over 100 µg/ml and this is consistent with the binding of monobactams to the essential PBPs of *E. coli* and *Staphylococcus aureus*, which is also poor (Wells *et al.*, 1982*a*, *b*). The simplest natural monobactam, SQ 26 180 (*N*-acetyl-derivative of (*R*)-3-amino-3-methoxymonobactamic acid) was relatively stable to β-lactamase action, with little affinity for the penicillinase-type β-lactamases, K_i, TEM-2 and that produced by *S. aureus* or to the β-lactamase found in the producing organism *C. violaceum*. SQ 26 180 is a reversible competitive inhibitor of the Class 1 P-99 cephalosporinase of *Enterobacter cloacae* with a K_i of 80 µM and it was also found to inhibit DD-carboxypeptides from *Streptomyces* R61 with an I_{50} of 3 µM (Wells *et al.*, 1982*a*). Sulfazecin had moderate activity against Gram-negative bacteria (Imada *et al.*, 1981; Sykes *et al.*, 1981) and while it was hydrolyzed by cephalosporinase, it was much less labile than benzylpenicillin, cephalosporin C or cephamycin C (Kintaka *et al.*, 1981) presumably due to the presence of a 3 α-methoxy group (Imada *et al.*, 1981). Since the monobactam nucleus has proven amenable to chemical synthesis and modification, several analogues have been made with excellent activity against Gram-negative bacteria and azthreonam is currently in clinical use.

2.2 Tetracyclines

Tetracyclines are broad-spectrum antibacterials produced chiefly by *Streptomyces* species; *S. aureofaciens* produces tetracycline and chlortetracycline (Minieri *et al.*, 1954) and *S. rimosus* produces oxytetracycline (Finlay *et al.*, 1950). Synthetic modifications of tetracyclines have been carried out and structures of two of these, minocycline and doxycycline, are given in Fig. 6. The biosynthesis of tetracyclines has been exhaustively studied: malonyl-CoA condenses with malonate semiamide-CoA to give a 19-carbon chain which is then aromatized to yield a hypothetical tricyclic nonaketide (McCormick *et al.*, 1965). This is then reduced at C-8 and methylated at C-6 followed by ring closure to give 6-methylpretetramide which is believed to be the starting intermediate in the biosynthesis of tetracyclines (McCormick *et al.*, 1965). The 6-methylpretetramide is converted *in vivo* to tetracycline and chlortetracycline. Use of mutants blocked in the biosynthetic pathway has been of great value in elucidating the pathway; a mutant of *S. aureofaciens* produced 6-demethyltetracycline (McCormick *et al.*, 1957). In chlortetracycline biosynthesis the 6-methylpretetramide is hydroxylated at C-4, then further oxidized to a ketone and later transformed to chlortetracyline (McCormick *et al.*, 1965). The amino group at C-4 is introduced by transamination and the reaction product can be recovered from cell-free extracts of *S. aureofaciens* (Miller *et al.*, 1964). Methylation of the C-4 amino group proceeds in two steps by transmethylation from methionine and these reactions can be measured in cell-free systems using radiolabelled *S*-adenosyl methionine (Miller & Nash, 1975*a*). Hydroxylation at C-6 has been measured in cell-free extracts of *S.*

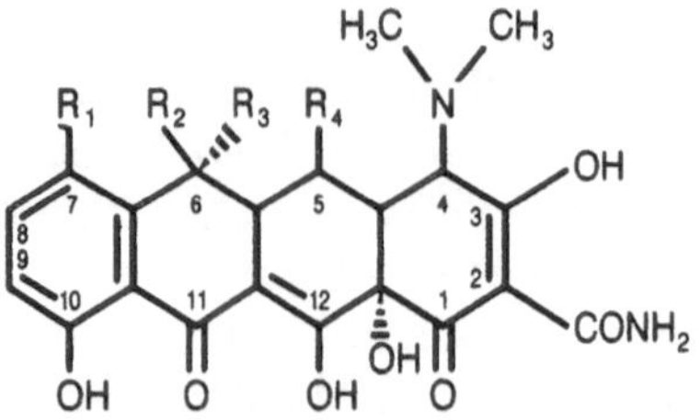

Joseph O'Sullivan

	R₁	R₂	R₃	R₄
	R_1	R_2	R_3	R_4
Tetracyline	H	CH_3	OH	H
Chlortetracycline	Cl	CH_3	OH	H
Oxytetracycline	H	CH_3	OH	OH
6-demethyltetracycline	H	H	OH	H
Minocycline	$N(CH_3)_2$	H	H	H
Doxycycline	H	CH_3	H	OH

Fig. 6. The structure of some tetracyclines.

aureofaciens (Behal *et al.*, 1979); the reaction requires NADPH and atmospheric oxygen and the specific activity of the enzyme was found to be sufficient to account for whole cell tetracycline production. The final step in the biosynthesis of tetracyclines is the reduction of the 5a—11a double bond; this is NADPH dependent and cell-free conversion of 7-chloro-5a, 11a-dehydrotetracycline to a bioactive product (chlortetracycline) which can be monitored by using *Staphylococcus aureus* (Miller & Nash, 1975b).

Tetracyclines exert their antibacterial activity by inhibiting protein synthesis at the ribosome level. Resistance to tetracyclines is widespread in clinical isolates and several classes of resistance have been described (Chopra & Howe, 1978; Foster, 1983; Levy, 1988). Classes A to E are found in Gram-negative bacteria and are associated with reduced accumulation (increased efflux) of the drug. Gram-positive bacteria also carry a group of tetracycline resistance determinants, K to O. Classes M and N would appear to be related to ribosomal protection and the exact mechanism has yet to be elucidated (Burdett, 1986). Recently, Speer & Salyers (1988, 1989) described a third resistance mechanism which involves chemical modification of the tetracycline to give a non-bioactive product. The gene for this resistance was first found on a plasmid from *Bacteroides* and when transferred to *E. coli* led to tetracycline resistance but only when the cells were grown aerobically. Whether the modification described by Speer & Salyers (1989) is the same as that described by Meyers & Smith (1962) is, as yet, unclear. They showed that *Xylaria digitata* was capable of degrading tetracycline, 7-chlortetracycline, 7-chlor-6-demethyltetracycline, 2-acetyl-2-decabroxamido-5-hydroxytetracycline and 12α-deoxytetracycline and while the degradation products were not identified, evidence suggested that neither the A or D rings were attacked. Non-degradative modification of a tetracycline by use of an organism which itself, does not produce tetracycline was demonstrated by Holmlund *et al.* (1959) who showed that *Curvularia lunata, C. pallescens* and *Botrytis cinera* were able to 12α-hydroxylate 12α-deoxytetracycline.

2.3 Aminocyclitol Glycosides (Aminoglycosides)

The aminocyclitol glycosides, also called aminoglycosides, constitute a large group of antibiotics containing aminosugar residues (Fig. 7). They can be divided according to structural similarities into four groups:

(1) those compounds that contain streptidine (streptomycin);
(2) those that contain 2-deoxystreptamine;
   (a) neomycin and butirosin, with adjacent (4, 5) hydroxyl substituents,
   (b) kanamycin and gentamycin with non-adjacent (4, 6) hydroxyl substituents,
   (c) monosubstituted compounds such as distomycin and hygromycin B,
(3) actinamine-containing compounds such as spectinomycin;
(4) monoaminocyclitol-containing antibiotics such as kasugamycin, valid-amycin and hygromycin A.

There are several other classification schemes and lists of naturally produced and semisynthetic aminocyclitols found in reviews by Davies & Yagisawa (1983), Okachi & Nara (1984) and Umezawa *et al.* (1986). Aminocyclitols are predominantly produced by *Streptomyces* species but also by *Micromonospora, Nocardia, Dactylosporangium, Stretoverticillium* and *Saccharopolyspora* and by *Bacillus* species.

Biosynthetically, this is an extensively studied group of compounds where the use of blocked mutants and mutasynthesis has helped identify some of the intermediates in the pathway (Pearce & Rinehart, 1981). Several biosynthetic questions remain, however, and the review by Davies & Yagisawa (1983) highlights some of these. The mechanism of action of aminocyclitols has been extensively studied and the primary target is the ribosome and protein synthesis (Cundliffe, 1981). Aminocyclitols are widely used for the treatment of Gram-negative infections but problems of ototoxicity and nephrotoxicity reduce their usage.

Resistance is also a problem in the use of aminocyclitols; the most important mechanisms of resistance are due to enzymatic inactivation by phosphorylation and adenylation of the hydroxyl groups and by acetylation of amino groups. Impermeability is also a mechanism of resistance (Bryan & Van den Elzen, 1977; Hancock, 1981) and ribosomal resistance is known but uncommon (Yamamoto *et al.*, 1982). Enzymes which modify the antibiotic are common in clinical settings but are also found in antibiotic-producing organisms (Shaw & Piwowarski, 1977; Yagisawa *et al.*, 1978). Table 1 is a compilation of modification mechanisms and the aminocyclitols which are substrates for the three classes of enzymes involved. The 3'-phosphotransferase (APH(3')) enzymes are widely distributed in resistant organisms. They are classified into

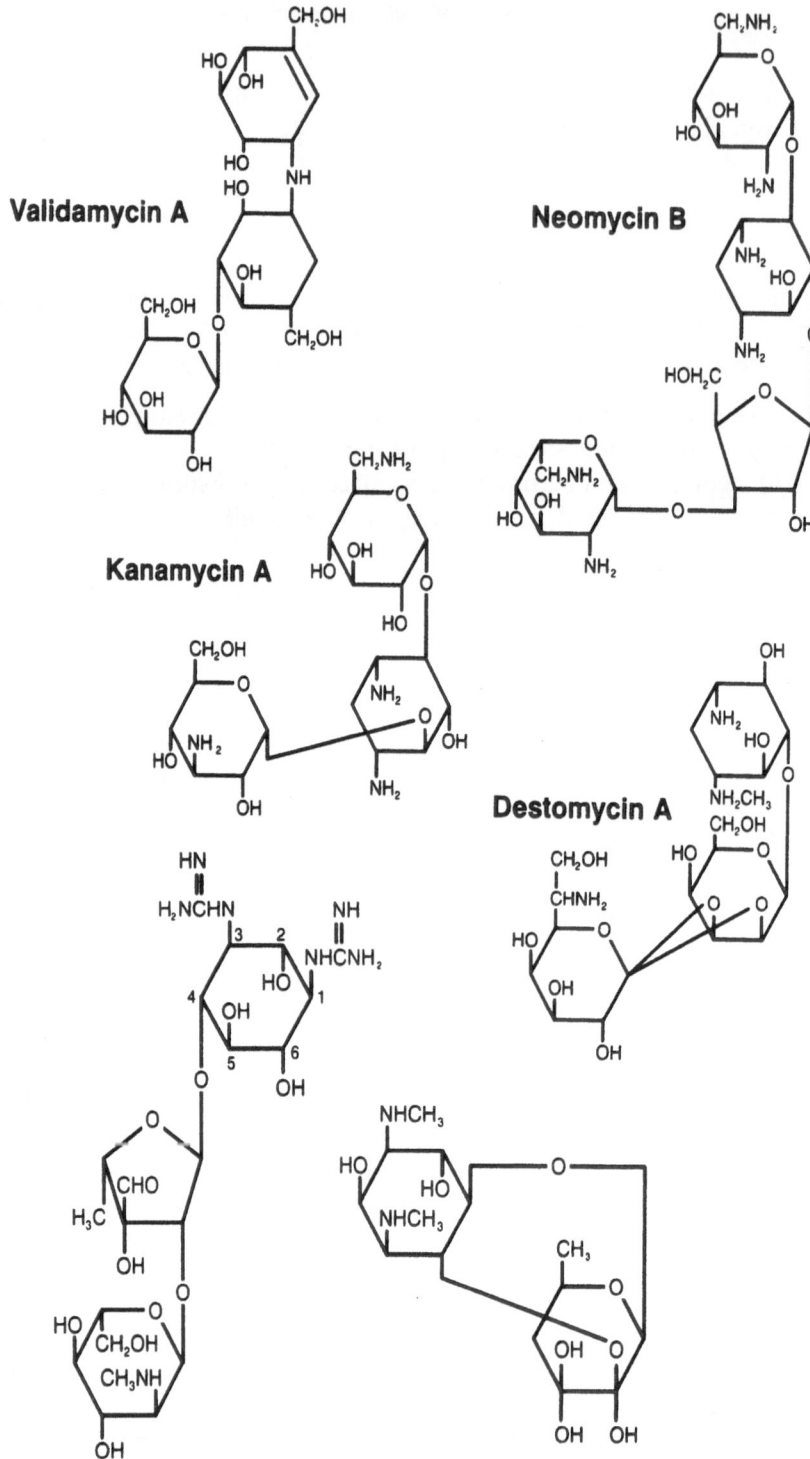

Fig. 7. The structure of selected aminocyclitol antibiotics.

Table 1

Enzymatic Resistance Mechanisms Among Aminocyclitols

Phosphorylation	Aminocyclitols
APH(3′)-I	Neomycin, kanamycin B, lividomycin
APH(3′)-II	Neomycin, kanamycin B, butirosin
APH(3′)-III	Neomycin, kanamycin B, lividomycin, butirosin
APH(2″)	Kanamycin B, tobramycin, gentamycin
APH(3″)	Streptomycin
APH(6)	Streptomycin
Acetylation	
AAC(6′)(I–IV)	Neomycin, kanamycin B, butirosin, tobramycin, gentamycin
AAC(3)(I–IV)	Kanamycin B, tobramycin, gentamycin, neomycin, lividomycin
AAC(2′)	Neomycin, kanamycin B, lividomycin, butirosin, tobramycin, gentamycin, sisomycin, paromomycin, dibekacin
Nucleotidylation	
AAD(3″)	Streptomycin, spectinomycin
AAD(6)	Streptomycin
ANT(2″)	Kanamycin B, tobramycin, gentamycin C
ANT(4′)	Neomycin, kanamycin B, lividomycin, butirosin, tobramycin, amikacin

APH = ATP-aminoglycoside phosphotransferase.
AAC = Acetyl-CoA aminoglycoside acetyltransferase.
AAD = ATP-aminoglycoside adenyltransferase.
ANT = NTP-aminoglycoside nucleotidyltransferase.

groups I, II and III depending on substrate specificity but extensive overlap exists (Table 1). APH(3′)-I is found in *E. coli* strains and in *Pseudomonas aeruginosa* T1–13 (Kondo *et al.*, 1972); APH(3′)-II was reported from *E. coli* JR66/W677 (Yagisawa *et al.*, 1972) and *P. aeruginosa* HA (Doi *et al.*, 1969) and APH(3′)-III was reported from *P. aeruginosa* 21–75 (Umezawa *et al.*, 1975) and *Staphylococcus aureus* 20240 (Davies & Yagisawa, 1983). APH(3′) enzymes do not exclusively phosphorylate the 3′ hydroxyl but may also act on the 5″ hydroxyl group of 3′,4′-dideoxyribostamycin as in the case of APH(3′)-I; the APH(3′)-III from *P. aeruginosa* 21–75 phosphorylates the 5″ hydroxyl of lividomycins as well as the 3′-hydroxyl of butirosins. APH(2″) activity was isolated from a *S. aureus* which was resistant to tobramycin, gentamycin and sisomycin (Le Goffic, 1977). APH(3″) activity has been reported from several sources: *S. aureus* B294, *P. aeruginosa* strains, *P. lachrymans* (Yano *et al.*, 1978) *Erwinia carotovora* (Fukusawa *et al.*, 1980), and from *E. coli* JR35 (Ozanne *et al.*, 1969). APH(6) was reported from streptomycin resistant *P. aeruginosa* (Kida *et al.*, 1975).

The acetyl transferases, catalyze the acetylaction of -6′, -2′ and -3 amino groups (Table 1), the acetyl donor is acetyl-coenzyme A. An enzyme that acetylates the -6' amino group of kanamycin was first described by Umezawa *et al.* (1967) from *E. coli* K12R5 which was resistant to kanamycin B but not to kanamycin C which lacks a -6' amino group. The enzyme has since been reported from other organisms. The AAC (-6') enzymes have been subdivided into four groups based on substrate specificity (Kawabe *et al.*, 1975). 3-Acetyltransferases have also been classified into four groups (Davies & Smith, 1978). Finally 2'-acetyltransferase which transfers the acetyl group of acetyl-CoA to the 2'-amino group of gentamycin, sisomycin etc. (Table 1) are found in *Proteus inconstance* strains (Chevereau *et al.*, 1974; Yamaguchi *et al.*, 1974).

The third class of enzyme associated with resistance to aminocyclitols is that leading to the adenylation or nucleotidylation of the -3″, -6, -2″ and -4' hydroxyl groups (Table 1). The 4'-nucleotidyltransferase (ANT-4) was found in *S. epidermidis* 109 (Santanam & Kayser, 1976) and in *S. aureus* ApO1 (LeGoffic *et al.*, 1976) and it adenylated the 4'-hydroxyl group of kanamycin, neomycin and others (Table 1) and also slowly acts on the -4″ hydroxyl group of dibekacin (which lacks a -4' hydroxyl group). 6-Adenyltransferase (AAD-6) was found in *S. aureus* MS27 (Suzuki *et al.*, 1975). The 3″-adenyltransferase (AAD(3″)) from *E. coli* K12 ML1629 acts on the 3″ hydroxyl group of streptomycin (Umezawa *et al.*, 1968) and on the 9-hydroxyl group of spectinomycin (Yamada *et al.*, 1968). A knowledge of the specificity of the enzymes responsible for aminocyclitol inactivation has been of value in designing chemical modification schemes. Some of these efforts have led to synthetic and semisynthetic products of clinical value (Umezawa *et al.*, 1986). The enzymes responsible for resistance are coded for by plasmid-borne genes and the clinical importance of these plasmids and their transfer has been reviewed (Davis, 1982).

2.4 Macrolides

The macrolides are a group of antibiotics containing a macrocyclic lactone ring to which amino or deoxysugars, or both, are attached (Woodward, 1957). They are classified into three major groups, 12-, 14- and 16-membered rings (Higashide, 1984), examples of which are shown in Fig. 8. In addition, atypical macrolides have been reported such as the 17-membered lankacidin (Gäumann *et al.*, 1960) bundlin B (Uramoto *et al.*, 1969) and the T-2636 series (Harada *et al.*, 1969; Higashide *et al.*, 1971). These compounds do not possess sugars but do appear to have a similar spectrum of activity and mode of action to the normal macrolides. Fused macrolides such as the avermectins (Burg *et al.*, 1979), nargenicins (Celmer *et al.*, 1980) and milbemycins (Takiguchi *et al.*, 1980) have been more recently reported and extend the structural diversity found within the macrolide class. Macrolides were first discovered in the 1950s with the identification of the 14-membered pikromycin (Brockmann & Henkel, 1951) and erythromycin (McGuire *et al.*, 1952); the 16-membered leukomycin (Hata *et al.*, 1953), the 12-membered methymycin (Donin *et al.*, 1953) and

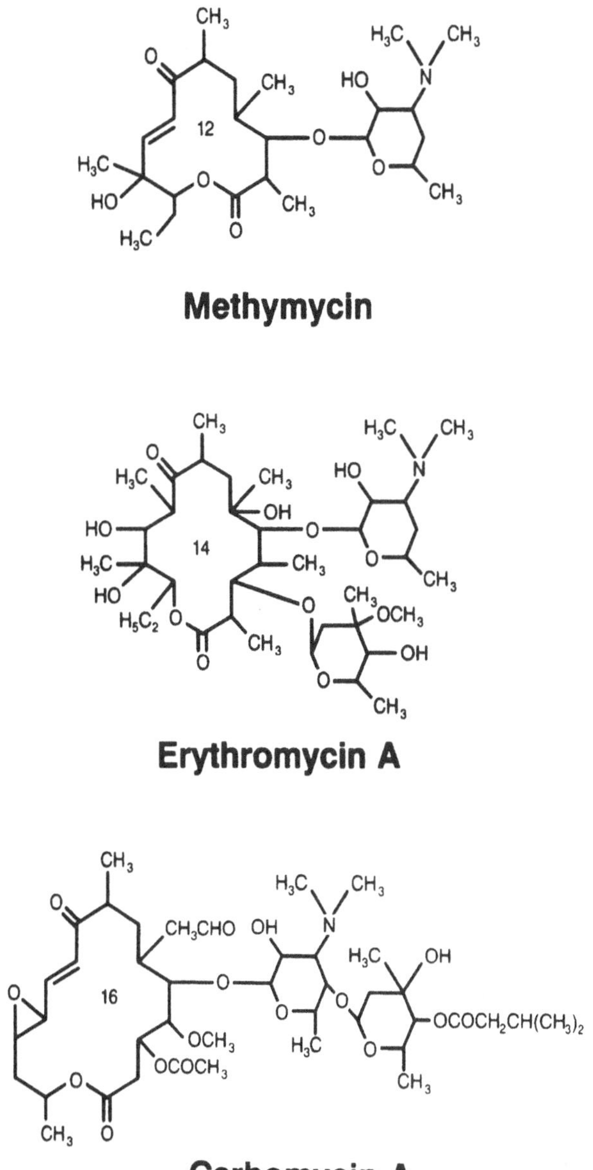

Methymycin

Erythromycin A

Carbomycin A

Fig. 8. The structure of a 12-, 14- and 16-membered macrolide.

numerous others (Omura & Tanaka, 1983; Higashide, 1984; Omura & Yoshitake, 1986). There are now well over fifty members of this class and this figure is likely to grow as natural product screening with new methodology proceeds. Macrolides are produced predominantly by *Streptomyces* species but *Micromonospora megalomicea* produces a 14-membered macrolide, megalomycin (Mallams *et al.*, 1969; Weinstein *et al.*, 1969) and other

Micromonospora species produce erythromycins and rosamycin (Wagman & Weinstein, 1980). Macrolides are typically active against Gram-positive bacteria and *Mycoplasma* and the activity is due to inhibition of protein synthesis mediated by binding to 50S ribosomes, thereby preventing peptidyl transfer. Resistance in clinical isolates has been reported and is due to a plasmid determined N^6,N^6-dimethylation of the ribosomal target (Gryczan *et al.* 1980; Horinouchi & Weisblum, 1980). Interestingly, the same mechanism of protection against macrolide inhibition is found in a producing organism (Yang Graham & Weisblum, 1979). Enzymatic modification of macrolides has been reported; crude extracts of *Streptomyces coelicolor* were shown to catalyze the 2'-O-phosphorylation of erthromycin, oleandomycin, tylosin, spiramycins and leucomycin A_3 (Wiley *et al.*, 1987). Oleandomycin-2'-O-phosphate and spiramycin-2'-O-phosphate had reduced antibacterial activity when compared with their unphosphorylated counterparts (Marshall *et al.*, 1989).

The biosynthesis of macrolides has been extensively studied. Kaneda *et al.* (1962) showed by labelling studies of erythromycin production by *S. erythraeus* that one propionate unit acts as a starter to which six additional propionates condense. Besides propionate, other short-chain fatty acids such as acetate, glycolate, *n*- and isobutyrate may be involved in the biosynthesis of various aglycones. A head to tail condensation of fatty acids similar to that for higher fatty acid biosynthesis is believed to be involved (Omura & Tanaka, 1983). The aglycone erythronolide B was first reported from *S. erythraeus* (Tardrew & Nyman, 1964) and later shown to be involved as a key intermediate in erythromycin biosynthesis by Hung *et al.* (1965) and Martin & Rosenbrook (1968). A detailed analysis of the enzymes involved in erythromycin biosynthesis has been presented by Corcoran (1981). The sugars involved in macrolide biosynthesis are derived from D-glucose. Because of the great diversity of macrolide structures and the wide range of producing organisms the number of conceivable biosynthetic permutations is enormous. Mutants blocked in various stages of the biosynthetic pathway have been valuable in elucidating the sequence of biosynthetic events (Martin & Rosenbrook, 1968; Martin & Goldstein, 1970) and in generating novel macrolide derivatives (LeMahieu *et al.*, 1976). Omura *et al.* (1980) successfully used the fatty acid synthesis condensing enzyme inhibitor cerulenin to block aglycone formation in the pikromycin (14-membered macrolide) producing organism and then incubating it with protylonide, the 16-membered aglycone tylosin, to give a 16-membered macrolide with desosamine as its sugar. This turned out to be a known natural product identified by Kinumaki *et al.*, (1977) but it illustrated the phenomenon of hybrid biosynthesis. Tylosin is a 16-membered branched lactone with three deoxy-sugar residues, mycaminose, mycarose and mycinose and is produced by *S. fradiae*. The biosynthesis of this commercially important antibiotic has been studied as has the genetics of the producing organism (Baltz & Seno, 1988). The biosynthesis of the aglycone, tylactone, was inhibited by cerulenin (Omura *et al.*, 1978) and labeling studies (Omura *et al.*, 1975) suggested a propionyl-CoA primer with condensations of acetyl-CoA. Several biosynthetic enzymes have been measured: macrocin *O*-methyltransferase catalyzed the conversion of macrocin to tylosin (Seno *et al.*, 1977). The specific activity of this enzyme peaked during maximal tylosin production and it was inhibited by its product (Seno & Baltz, 1981). Another tylosin-specific *o*-methyltransferase was also reported which converted demethylmacrocin to macrocin by methylation of the 2-hydroxyl group of the 6-deoxy-D-allose residue; this enzyme was also product inhibited (Seno & Baltz, 1981). A cell-free extract of the tylosin-producing strain *S. rimosus* catalyzed the conversion of TDP-D-glucose to TDP-mycarose in the presence of *S*-adenosyl-methionine; the intermediate TDP-4-keto-6-deoxy-D-glucose was generated by an oxidoreductase (Pape & Brillinger, 1973.) Extensive analysis of blocked mutants has helped unravel some of the complexities of tylosin production by *S. fradiae* and allowed the determination of preferred pathways and of some of the biosynthetic (mutasynthetic) possibilities of this organism with this class of compound (Baltz & Seno, 1988).

2.5 Chloramphenicol

Chloramphenicol (D-(−)-*threo*-1-*p*-nitrophenyl-2-dichloracetamido-1,3-propanediol) (Fig. 9) is a broad-spectrum bacteriostatic antibacterial agent produced by *Streptomyces venezuelae* and other actinomycetes (Malik, 1983). It is a potent inhibitor of protein synthesis by virtue of its binding to 50S ribosomes thereby preventing the peptidyltransferase reaction (Cundliffe & McQuillen, 1967).

Chloramphenicol is the agent of choice in the treatment of typhoid fever and other *Salmonella* infections and it is also very active against *Haemophilus influenzae*. Its use is restricted however, because of hematological toxicity. Biosynthesis of chloramphenicol proceeds from chorismic acid by amination to give *p*-aminophenylpyruvic acid; this step and the succeeding one (aminotransferase) have been studied by Jones *et al.* (1978). Hydroxylation leads to *p*-aminophenylserine which is then acetylated. Chlorination and oxidation of the *p*-amino group follow to give chloramphenicol (Vining & Westlake, 1984). Bacterial resistance to chloramphenicol is predominantly due to an *R* factor mediated acetyltransferase. This plasmid coded enzyme is constitutive in

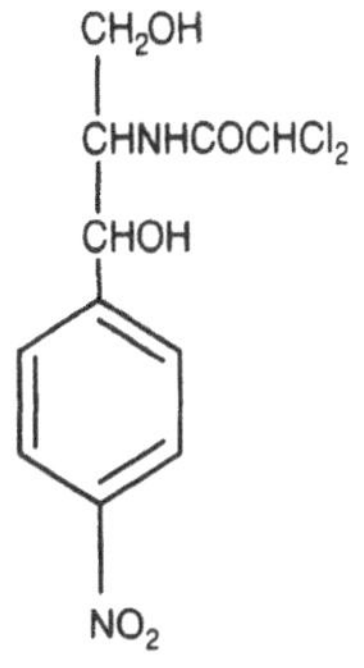

CH_2OH
|
$CHNHCOCHCl_2$
|
$CHOH$

Fig. 9. Chloramphenicol.

Gram-negative bacteria but is inducible in Gram-positives. Acetyltransferases can be classified on the basis of physical, kinetic and immunological criteria; thus, in enterics there are three groups and in staphylococci, five (Shaw, 1983). *S. venezuelae* and some bacteria produce an enzyme that hydrolyzes the amide bond to give the inactive 4-nitrophenylserinol which may be further modified by reduction of the nitro group (Lingrens *et al.*, 1966).

2.6 Rifamycin

Rifamycin belongs to a large class of compounds known as ansamycins (Prelog & Oppolzer, 1973). They were first isolated by Sensi *et al.* (1959) from *Nocardia mediterranea* (*Streptomyces mediterranea*). The fermentation produces several rifamycins but if sodium barbiturate is added to the medium, rifamycin B is predominantly produced (Margalith & Pagani, 1961). Rifamycin B can be chemically modified to give rifamycin S which is a major starting point for semisynthetic derivatization. When rifamycin S is converted to 3-formyl rifamycin SV, this compound can be reacted with 1-amino-4-methyl piperazine to give rifampicin (Fig. 10). Rifampicin is extensively used in the treatment of tuberculosis and other infectious diseases. The compounds act by inhibiting RNA polymerase in bacteria (Hartmann *et al.*, 1967). The biosynthesis of rifamycins is reminiscent of the macrolides (2.4) in the abundance of derivatives which can be contemplated. Genetic analysis of *N. mediterranea* has been carried out by Schupp *et al.* (1975) and through genetic recombination it has been possible to generate novel rifamycins (Schupp *et al.*, 1981). It was suggested that gene rearrangement may activate silent regions of the chromosome resulting in enhanced biosynthetic capacity (Ghisalba *et al.*, 1984).

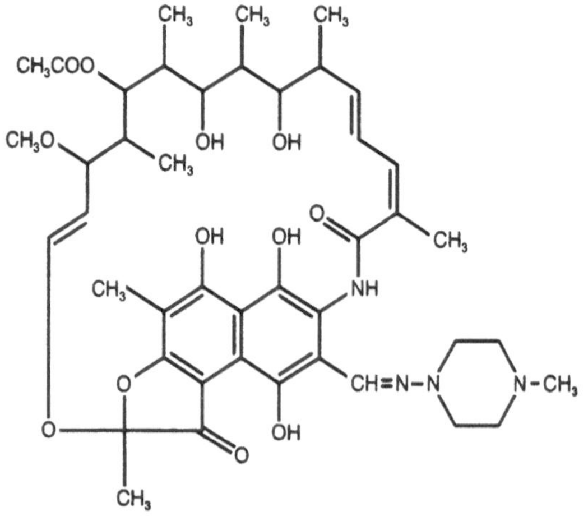

Fig. 10. The structure of rifampicin, a semisynthetic rifamycin.

3 STEROIDS

Steroids are a diverse class of compounds based on the cyclopentano-perhydrophenantrene structure (Fig. 11). They may be divided (Fig. 12) into hormonal steroids such as pregnanese (e.g. progesterone, pregnenolone), androstanes (e.g. testosterone, androsterone) and estranes (e.g. estrone, estradiol), bile acids (cholic acids, 5β cholanic acid), sterols (ergosterol, cholesterol), sapogenins (diosgenin), cardenolides (digitoxigenin), bufadieno-lides (bufalin) and steroidal alkaloids (tomatidine).

The hormonal steroids are of obvious importance for mammalian physiology and reproduction and annual sales of sex hormones have been estimated at about 1500 million dollars and corticosteroids at 1700 million dollars (Brown, 1984). The bile acids are important for digestion and absorption of lipid in mammals. Sterols are major structural components of cell membranes in eucaryotes; ergosterol is the major structural sterol in the membrane of yeast and cholesterol fulfills a similar structural role in mammals as well as being a precursor for bile acids and hormones. The cardinolides are used in the treatment of heart disease and the corticosteroids are extensively used as anti-inflammatory agents. Indeed, it was the discovery by Hench *et al.* (1950) of the powerful effect of cortisone in the treatment of rheumatoid arthritis that led to an explosion of interest in steroids, their bioactivity and biotransformation. Over the succeeding twenty years the collaborative interaction of chemists and microbiologists led to a prolific number of interesting and useful new steroids and to an appreciation of the resourcefulness of microbial enzymes in achieving complex steroidal transformations. This material has been extremely well reviewed in a number of books and chapters (Charney & Herzog, 1967; Iizuka & Naito, 1967; Kieslich, 1980; Sedlaczek, 1988).

The types of reactions which steroids undergo are listed in Table 2. Most of the type compounds in Fig. 12 can be extensively modified and the reviews by Charney & Herzog (1967), Kieslich (1980) and Iizuka & Naito (1967) contain extensive lists of reactions and of the microorganisms which effect these modifications. The prototype of many of these reactions is the classic case of

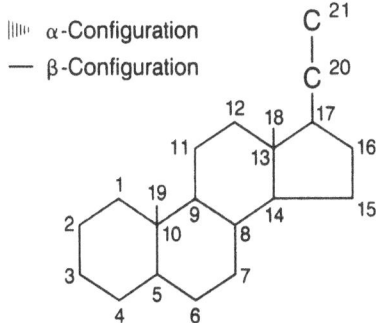

Fig. 11. The nuclear structure of steroids.

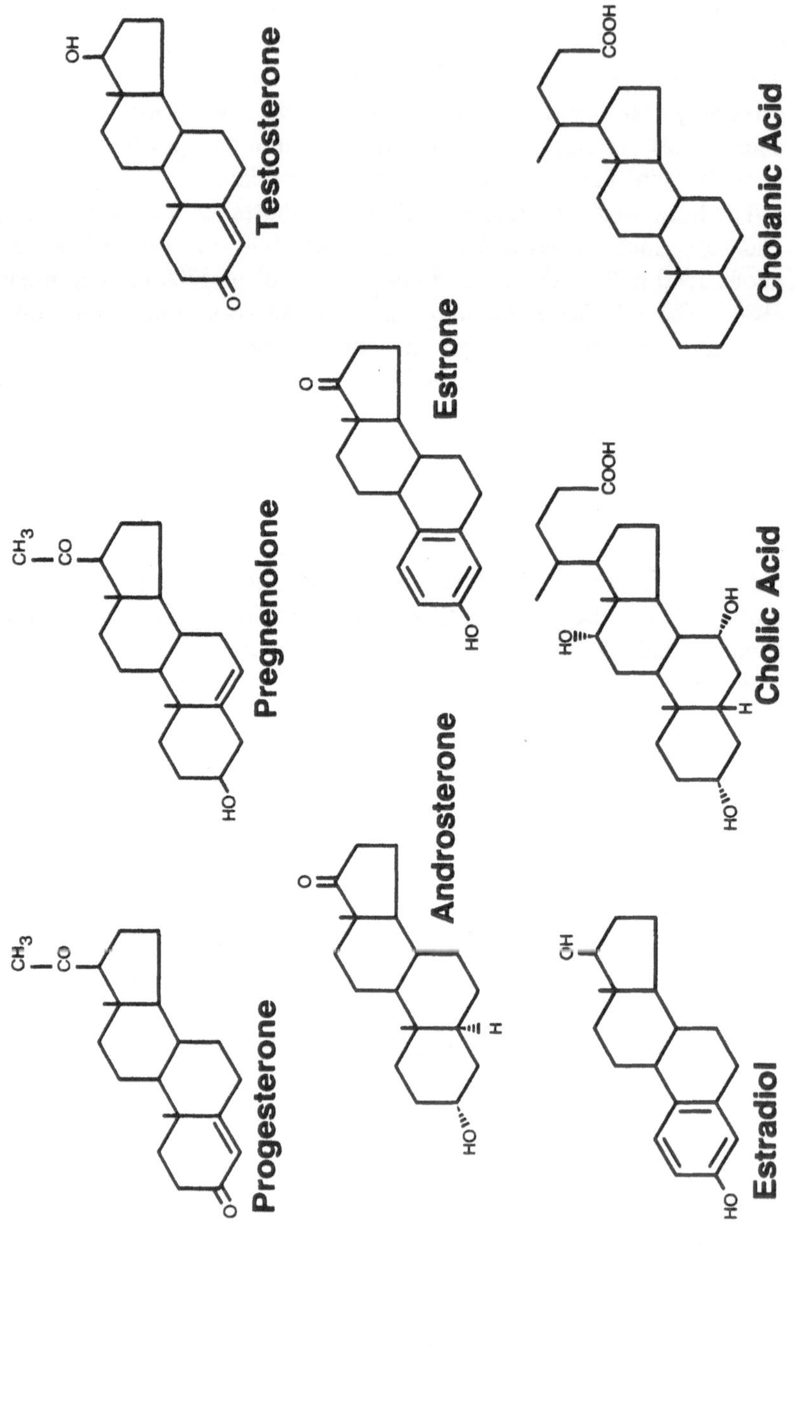

Testosterone

Cholanic Acid

Estrone

Pregnenolone

Cholic Acid

Androsterone

Progesterone

Estradiol

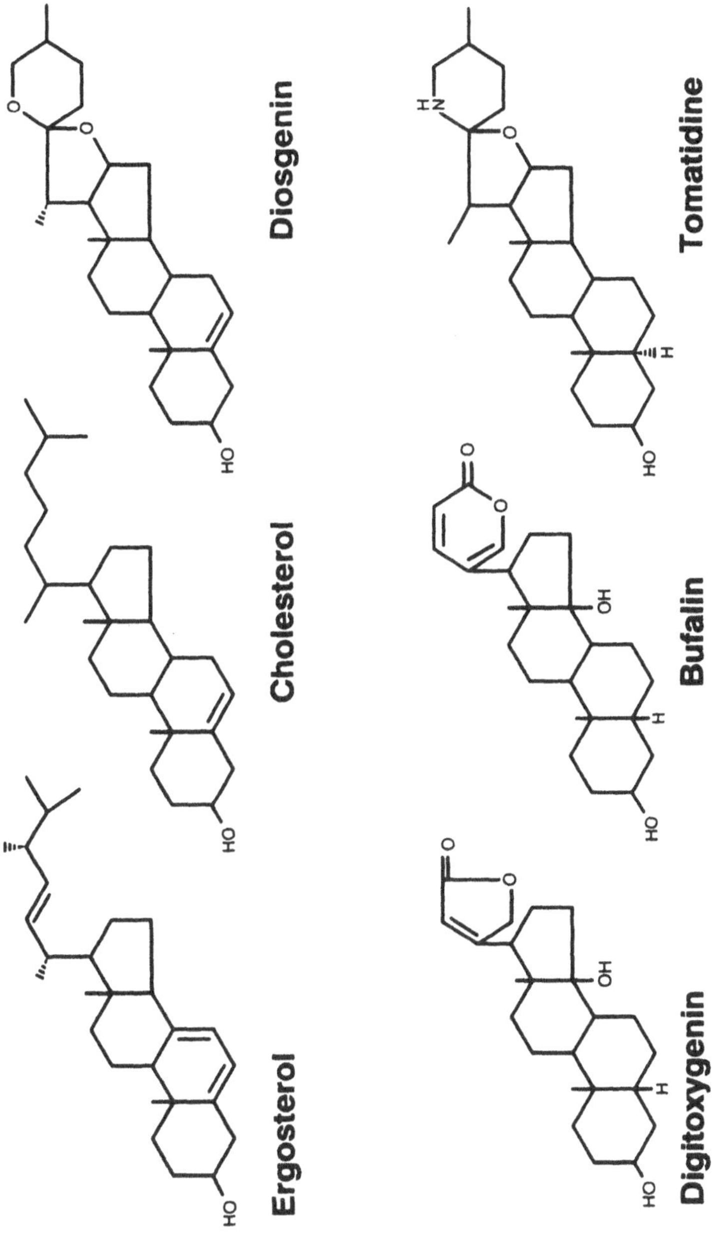

Fig. 12. Representative steroidal structures.

Table 2
Enzymatic Reactions in the Transformation of Steroids

Oxidation

Hydroxylation
Dehydrogenation
Epoxidation
Peroxidation
C—C cleavage
Heteroatom oxidation

Conjugations

Glycosidation
Acetylation
Phenol ether formation

Reduction

Ketones, acids, enols and aldehydes to alcohols
Double bonds
Bromide

Isomerization

Heteroatom introduction

Hydrolysis of ethers and esters

---

cortisone or hydrocortisone production. The effectiveness of cortisone in the treatment of rheumatoid arthritis and other diseases relies, in part, on the oxygen at C-11 and the chemical synthesis of cortisone which was described by Sarett (1946) required thirty two chemical steps to get from the starting compound deoxycholic acid to cortisone (Fig. 13). The cost of cortisone by this process was about $200/g. Peterson & Murray (1952) in a seminal publication, radically altered the production of sterols. They showed that 11α-hydroxylation of progesterone could be carried out in a one-step reaction using the fungus *Rhizopus* (Fig. 14) and yields of about 90% were achieved for this reaction. The 11α-hydroxyprogesterone can be readily converted to cortisone and hydrocortisone by chemical methods (Mancera *et al.*, 1952). Fried *et al.* (1952) independently showed that *Aspergillus niger* could accomplish the 11α

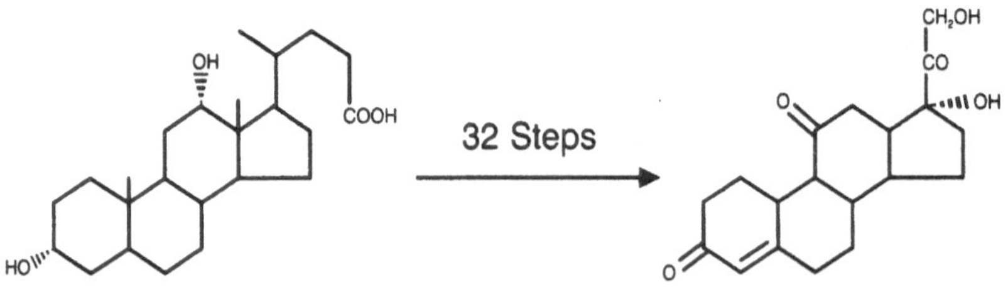

Fig. 13. The chemical conversion of deoxycholic acid to cortisone.

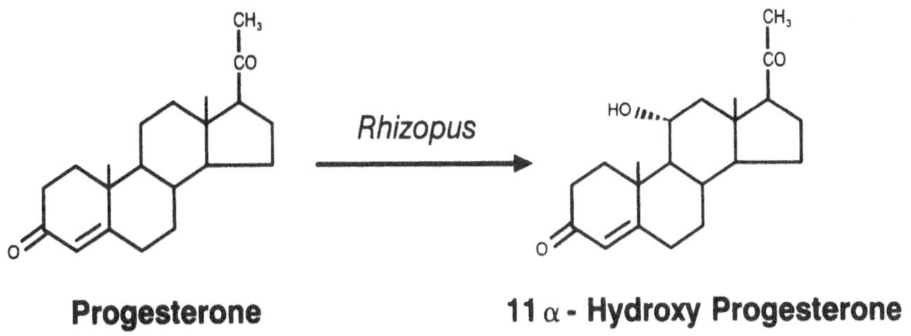

Progesterone   **11 α - Hydroxy Progesterone**

Fig. 14. The 11α-hydroxylation of progesterone by *Rhizopus*.

and 11β-hydroxylation of progesterone. Since then numerous investigators have identified other fungi, bacteria and actinomycetes which will accomplish these reactions. These studies led to significant reduction in the cost of corticosteroids (<$1/g) and stimulated an intense effort to find new transforming cultures. It became possible to achieve a vast range of steroid modifications but only three further reactions will be given here. 16α-Hydroxylation of progesterone using *Actinomyces*, was described by Perlman *et al.* (1952) and became important in the production of triamcinolone (Fig. 15) a potent anti-inflammatory substance; several other organisms were found to accomplish this process (Iizuka & Naito, 1967). The 1-dehydrogenation of steroids (Fig. 16) proved to be important because prednisone and prednisolone, the 1-dehydro products of α cortisone and hydrocortisone, proved not to lead to high salt retention. The dehydrogenation reaction was first described by Fried *et al.* (1953) and by Vischer & Wettstein (1953) along with other degradative events. Later *Arthrobacter simplex* was found to effect the 1-dehydrogenation without significant side reactions (Nobile *et al.* 1955; Herzog *et al.*, 1962). Fried & Sabo (1953) reported the synthesis of 9α-fluorinated sterols and this modification led to further advances in corticosteroid therapy. Triamcinolone (Fig. 15) has a fluorine at the 9α position, is 11β and 16α-hydroxylated and is dehydrogenated at the 1 position, thereby combining all the key modifications

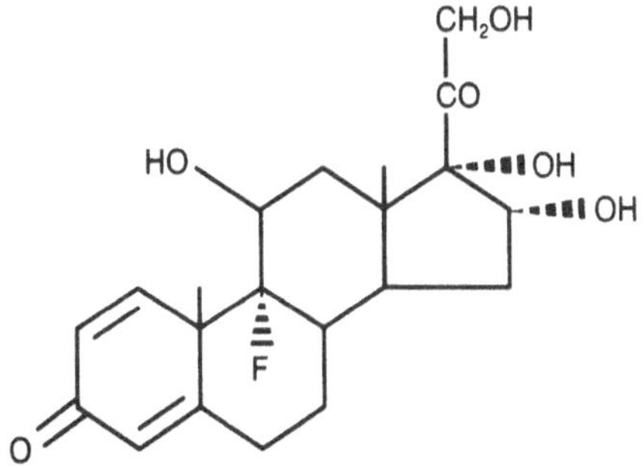

Fig. 15. Triamcinolone (9α-fluoro-11β,16α,17α,21-tetrahydroxypreg-4-ene-3,20-dione).

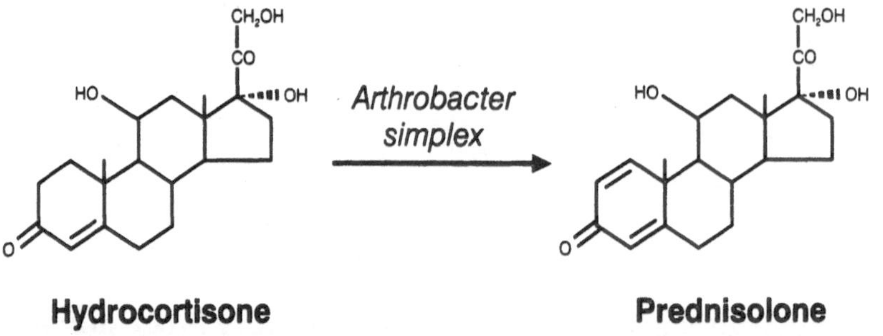

Hydrocortisone   **Prednisolone**

Fig. 16. The 1-dehydrogenation of hydrocortisone by *Arthrobacter simplex*.

described above. The success of triamcinolone led to further synthetic effort and today many steroids are commercially available. They fall into three classes: those based on the prednisolone structure such as medrol (9α-methylprednisolone), betamethasone (9α-fluoro-16β-methylprednisolone) and fluocinolone (6α,9α-difluoro-16α-hydroxyprednisolone); secondly, those based on 1-dehydrocorticosterone such as deoxymethasone (11β,21-dihydroxy-9α-fluoro-16α-methyl-1,4-pregnandien-3,20-dione) and thirdly, those based on 21-deoxyprednisolone such as oxylone (9α-fluoro-6α-methyl-21-deoxy-prednisolone).

Many of the key modifications to the steroid structure were determined between 1950 and 1965. What has occurred since then has been, for the most part, descriptions of new organisms with improved transformation yields, fewer side reactions, different growth kinetics, etc. In his excellent review of steroid transformation, Sedlaczek (1988) states that over the period 1967–78 when interest in microbiological steroid transformation was in decline, 1735 new bioconversion products formed by 180 species of microorganisms and plants, growing on 853 substrates were described. Some of the more significant developments have been the recognition of new steroid sources, new degradation processes, better fermentation technologies and the application of an improved understanding of enzyme regulation, gene–enzyme dosage effects and other technologies.

A wide variety of steroid sources are used for commercial purposes. Cholic acid and deoxycholic acid are derived from bile acids, cholesterol from mammalian sources such as sheep lanolin and conjugated estrogens can be recovered from equine urine. Sapogenins such as diosgenin are derived from the Mexican yam, *Dioscorea*, and *Yucca* and *Agave* are rich sources of hecogenin. Soybean has become an important source of stigmasterol and yeast and other fungi are good sources of ergosterol and other sterols.

Degradative pathways for cholesterol and other steroids have been studied; the cholesterol side-chain is cleaved by adrenal enzymes to yield pregnenolone and isocaproic aldehyde (Shimizu *et al.*, 1960) but also by microorganisms (Sih & Wang, 1965; Sih *et al.*, 1965). These workers showed that both the steroid nucleus and the hydrocarbon side-chain are attacked and the direction and extent of degradation could be controlled by selective inhibition. Thus, by use of selective modification using either microorganisms or cell free enzyme preparations, steroids from a variety of sources can be transformed to useful intermediates.

The enzymology of steroid transformation has largely been studied using whole cells either in growing cultures, resuspensions of induced cells in simple media or water, dried or powdered cells or spore suspensions. Cell-free studies have been carried out (reviewed by Sih & Whitlock, 1968) but not extensively used in production because of lability and cost. It is now possible to apply a wide range of technologies to steroid production and modification. Once the genes coding for enzymes of interest are identified they can be cloned and amplified on high copy plasmids or assembled with other genes on plasmids to combine several reactions in the one organism. Alternatively, undesirable reactions may be removed or reduced; or strains can be selected with other desirable properties such as the ability to grow on specific media, having increased growth rates, ease of harvesting or prolonged survival in immobilized systems.

REFERENCES

Abbot, B. J. (1976). *Advances in Applied Microbiology*, **20**, 203.

Abbot, B. J. & Fukuda, D. (1975a). *Applied Microbiology*, **30**, 413.

Abbot, B. J. & Fukuda, D. (1975b). *Antimicrobial Agents and Chemotherapy*, **8**, 282.

Abraham, E. P. & Chain, E. B. (1940). *Nature*, **146**, 837.

Albers-Schönberg, G., Arison, B. H., Hensens, O. D., Hirshfield, J., Hoogsten, K., Kaczka, E. A., Rhodes, R. E., Kahan, J. S., Kahan, F. M., Ratcliffe, R. W., Walton, E., Ruswinkel, L. J., Morin, R. B. & Christensen, B. G. (1978). *Journal of the American Chemical Society*, **100**, 6491.

Amyes, S. (1987). In *β-Lactamases: Current Perspectives*, ed. D. Livermore, Theracom, The Hague, p. 31.

Baltz, R. H. & Seno, E. T. (1988). *Annual Review of Microbiology*, **42**, 547.

Banyu Pharmaceutical Co. Ltd (1972). Japan Patent No. J5-2038-092.

Basker, M. J., Boon, R. J. & Hunter, D. A. (1980). *Journal of Antibiotics*, **33**, 878.

Batchelor, F. R., Doyle, F. P., Nayler, J. H. & Robinson, G. N. (1959). *Nature*, **183**, 257.

Behal, V., Hostalek, Z. & Vanek, Z. (1979). *Biotechnology Letters*, **1**, 177.

Behrens, O. K. (1948). In *The Chemistry of Penicillins*, eds H. T. Clarke, J. R. Johnson, & R. Robinson. Princeton University Press, Princeton, p. 657.

Blumberg, P. M. & Strominger, J. L. (1974). *Bacteriological Reviews*, **38**, 291.

Brockmann, H. & Henkel, W. (1951). *Chemische Berichte*, **84**, 284.

Brown, A. G., Butterworth, D., Cole, M., Hanscomb, G., Hood, J. D., Reading, C. & Robinson, G. N. (1976). *Journal of Antibiotics*, **29**, 668.

Brown, A. G., Corbett, D. F., Eglington, A. G. & Howarth, T. T. (1977). *Journal of the Chemical Society, Chemical Communications*, 523.

Brown, W. E. (1984). In *Annual Reports on Fermentation Processes*, ed. G. T. Tsao. Academic Press, New York, p. 135.

Bryan, L. E. & van den Elzen, H. M. (1977). *Antimicrobial Agents and Chemotherapy*, **12**, 163.

Burdett, V. (1986). *Journal of Bacteriology*, **165**, 564.
Burg, R. W., Miller, B. M., Baker, E. E., Birnbaum, J., Currie, S. A., Hartman, R., Kong, T., Monaghan, R. L., Olson, G., Putter, I., Tunal, J. B., Wallick, H., Stapley, E. O., Oiwa, R. & Omura, S. (1979). *Antimicrobial Agents and Chemotherapy*, **15**, 361.
Bush, K. (1988). *Reviews of Infectious Diseases*, **10**, 681.
Cassidy, P. J. (1981). *Developments in Industrial Microbiology*, **22**, 181.
Celmer, W. D., Chmurny, G. N., Moppett, C. E., Ware, R. S., Watts, P. C. & Wipple, E. B. (1980). *Journal of the American Chemical Society*, **102**, 4203.
Charnas, R. L. & Knowles, J. R. (1981). *Biochemistry*, **20**, 2732.
Charnas, R. L., Fisher, J. & Knowles, J. R. (1978). *Biochemistry*, **17**, 2185.
Charney, W. & Herzog, H. L. (1967). *Microbial Transformation of Steroids*. Academic Press, New York.
Chevereau, M., Daniels, P. J. L., Davies, J. & LeGoffic, F. (1974). *Biochemistry*, **13**, 598.
Chopra, I. & Howe, T. G. B. (1978). *Microbiological Reviews*, **42**, 707.
Corcoran, J. W. (1981). In *Antibiotics: Biosynthesis*, Vol. 4, ed. J. W. Corcoran. Springer-Verlag, New York, p. 132.
Cundliffe, E. (1981). In *The Molecular Basis of Antibiotic Action*, eds E. F. Gale, E. Cundliffe, P. E. Reynolds, M. H. Richmond & M. J. Waring. John Wiley, New York, p. 419.
Cundliffe, E. & McQuillen, K. (1967). *Journal of Molecular Biology*, **30**, 137.
Davies, J. E. & Smith, D. I. (1978). *Annual Review of Microbiology*, **32**, 469.
Davies, J. E. & Yagisawa, M. (1983). In *Biochemistry and Genetic Regulation of Commercially Important Antibiotics*, eds L. C. Vining. Addison-Welsey, London, p. 329.
Davis, B. D. (1982). *Reviews of Infectious Diseases*, **4**, 237.
Doi, O., Kondo, S., Tanaka, N. & Umezawa, H. (1969). *Journal of Antibiotics*, **22**, 273.
Donin, N. N., Pagano, J., Duitcher, J. D. & McKee, C. M. (1953). *Antibiotics Annual*, **1953/54**, 179.
Elander, R. P. (1983). In *Antibiotics Containing the β-Lactam Structure*, eds A. L. Demain & N. A. Solomon. Springer-Verlag, New York, p. 97.
Easton, C. J. & Knowles, J. R. (1982). *Biochemistry*, **21**, 2857.
Fawcett, P. A., Usher, J. J. & Abraham, E. P. (1975) *Biochemical Journal*, **151**, 741.
Fildes, R. A. Potts, J. R. & Farthing, J. E. (1974) US Patent Office 3,801,458.
Finlay, A. C., Hobby, G. L., Pan, S. Y., Regna, P. P., Routien, J. B., Seeley, D. B., Schull, G. M., Sobin, B. A., Solamons, I. A., Vinson, J. W. & Kane, J. H. (1950). *Science*, **III**, 85.
Fisher, J., Charnas, R. L. & Knowles, J. R. (1978). *Biochemistry*, **17**, 2180.
Fisher, J., Belasco, J. G., Khosla, S. & Knowles, J. R. (1980). *Biochemistry*, **19**, 2895.
Foster, T. J. (1983). *Microbiological Reviews*, **47**, 361.
Fried, J. & Sabo, E. F. (1953). *Journal of the American Chemical Society*, **75**, 2273.
Fried, J., Thoma, R. W., Gerke, J. R., Merz, J. E., Donin, M. N. & Perlman, D. (1952). *Journal of the American Chemical Society*, **74**, 3962.
Fried, J., Thoma, R. W. & Klingsberg, A. (1953). *Journal of the American Chemical Society*, **75**, 5764.
Fujisawa, Y., Shirafuji, N., Kida, M., Nara, K., Yoneda, M. & Kanzaki, T. (1973). *Nature*, **246**, 154.
Fukusawa, K., Sakurai, H., Shimizu, S., Naganawa, H., Kondo, S., Kawabe, H. & Mitsuhashi, S. (1980). *Journal of Antibiotics*, **33**, 122.
Gäumann, E., Hutter, R., Keller-Schierlein, W., Neipp, L., Prelog, V. & Zähnor, H. (1960). *Helvetica Chimica Acta*, **43**, 601.

Georgopapadakou, N. H. & Liu, R. Y. (1980). *Antimicrobial Agents and Chemotherapy*, **18**, 148.

Ghisalba, O., Auden, J. A. L., Schupp, R. & Nüesch, J. (1984). In *Biotechnology of Industrial Antibiotics*, ed. E. J. Vandamme. Marcel Dekker, Inc., New York, p. 281.

Ghuysen, J. M. (1988). *Reviews of Infectious Diseases*, **10**, 726.

Gryczan, T. J., Frandi, G., Hahn, J. & Dubnau, D. (1980). *Nucleic Acids Research*, **8**, 6081.

Hancock, R. E. W. (1981). *Journal of Antimicrobial Chemotherapy*, **8**, 429.

Harada, S., Higashide, E., Fugono, T. & Kishi, T. (1969). *Tetrahedron Letters*, **27**, 2239.

Hartmann, G., Honikel, K. O., Knüsel, F. & Nüesch, J. (1967). *Biochimica et Biophysica Acta*, **145**, 843.

Hata, T., Sano, Y., Ohki, N., Yokoyama, Y., Matsuma, A. & Ito, S. (1953). *Journal of Antibiotics*, Series A **6**, 87.

Hedges, R. W., Datta, N., Kontomichalou, P. & Smith, J. T. (1974). *Journal of Bacteriology*, **117**, 56.

Hench, P. S., Kendall, E. C., Slocumb, C. H. & Polley, H. F. (1950). *Archives of Internal Medicine*, **85**, 545.

Herzog, H. L., Payne, C. C., Hughes, M. T., Gentles, M. T., Herschberg, E. B., Nobile, A., Charney, W., Federbush, C., Sutter, D. & Perlman, P. L. (1962). *Tetrahedron*, **18**, 581.

Higashide, E. (1984). In *Biotechnology of Industrial Antibiotics*, ed. E. J. Vandamme. Marcel Dekker, Inc., New York, p. 451.

Higashide, E., Fugono, T., Hatano, T. & Schibata, M. (1971). *Journal of Antibiotics*, **24**, 1.

Holmlund, C. E., Anders, W. W. & Shay, A. J. (1959). *Journal of the American Chemical Society*, **81**, 4750.

Hood, J. D., Box, S. J. & Verall, M. S. (1979). *Journal of Antibiotics*, **32**, 295.

Hood, J. D., Brown, A. G., Elson, A. L., Gilpin, M. L. & Holmes, R. (1984). In *Recent Advances in the Chemistry of β-Lactam Antibiotics*, eds A. G. Brown & S. M. Roberts. The Royal Society of Chemistry, London, p. 52.

Hoover, J. R. (1983). In *Antibiotics Containing the β-Lactam Structure*, eds A. L. Demain & N. A. Soloman. Springer-Verlag, New York, p. 119.

Horinouchi, S. & Weisblum, B. (1980). *Proceedings of the National Academy of Sciences USA*, **77**, 7079.

Howarth, T. T., Brown, A. G. & King, T. J. (1976). *Journal of the Chemical Society, Chemical Communications*, 266.

Hung, P. P., Marks, C. L. & Tardrew, P. L. (1965). *Journal of Biological Chemistry*, **240**, 1322.

Ichikawa, W., Murai, Y., Yamamoto, S., Shibuya, Y., Fujii, T., Komatzu, K. & Kodaira, R. (1981). *Agricultural and Biological Chemistry*, **45**, 2225.

Iizuka, H. & Naito, A. (1967). *Microbial Transformation of Steroids and Alkaloids*. University of Tokyo Press, Tokyo.

Imada, A., Kitano, K., Kintaka, K., Muroi, M. & Asai, M. (1981). *Nature*, **289**, 590.

Jeffery, J. D'A., Abraham, E. P. & Newton, G. G. F. (1961). *Biochemical Journal*, **81**, 591.

Jensen, S. E., Westlake, D. W. S. & Wolfe, S. (1985). *Journal of Antibiotics*, **38**, 263.

Jones, A., Francis, M. M., Vining, L. C. & Westlake, D. W. S. (1978). *Canadian Journal of Microbiology*, **24**, 238.

Kaneda, T., Butte, J. C., Taubman, S. B. & Corcoran, J. W. (1962). *Journal of Biological Chemistry*, **237**, 322.

Kaufmann, W. & Bauer, K. (1960). *Naturwissenschaften*, **47**, 474.

Kawabe, H., Kondo, S., Umezawa, H. & Mitsuhashi, S. (1975). *Antimicrobial Agents and Chemotherapy*, **7**, 494.

Kawamori, M., Hashimoto, Y., Katsumata, R., Okachi, R. & Takayama, K. (1983). *Agricultural and Biological Chemistry*, **47**, 2503.

Kawate, S., Fukuo, T. & Kunito, K. (1978). *Technology Reports Kansas University*, **29**, 77.

Kida, M., Asako, T., Yoneda, M. & Mitsuhashi, S. (1975). In *Microbial Drug Resistance*, ed. S. Mitsuhashi. University of Tokyo Press, Tokyo, p. 441.

Kieslich, K. (1980). In *Economic Microbiology: Microbial Enzymes and Bio-conversions*, Vol. 5, ed. A. H. Rose. Academic Press, New York, p. 369.

Kim, I. H., Nam, D. H. & Ryu, D. D. Y. (1983). *Applied Biochemistry and Biotechnology*, **8**, 195.

Kintaka, K., Kitano, K., Nozaki, Y., Kawashima, F., Imada, A., Nakao, Y. & Yoneda, M. (1981). *Journal of Fermentation Technology*, **59**, 263.

Kinumaki, A., Harada, K., Suzuki, T., Suzuki, M. & Okuda, T. (1977). *Journal of Antibiotics*, **30**, 450.

Kitano, K. (1983). *Progress in Industrial Microbiology*, **17**, 37.

Klein, J. & Wagner, F. (1980). *Enzyme Engineering*, **5**, 335.

Kohsaka, M. & Demain, A. L. (1976). *Biochemical and Biophysical Research Communications*, **70**, 465.

Kondo, S., Yamamoto, H., Naganawa, H., Umezawa, H. & Mitsuhashi, S. (1972). *Journal of Antibiotics*, **25**, 483.

Konomi, T., Herschen, S., Baldwin, J. E., Yoshida, M., Hunt, N. A. & Demain, A. L. (1979). *Biochemical Journal*, **184**, 427.

Kropp, H., Sundelof, J. G., Hajdu, R. & Kahan, R. M. (1982). *Antimicrobial Agents and Chemotherapy*, **22**, 62.

Kubo, K., Ishikura, R. & Fukugawa, Y. (1984). *Journal of Antibiotics*, **37**, 1396.

Kushner, D. J. & Breuil, C. (1977). *Archives of Microbiology*, **112**, 219.

Labia, R. & Peduzzi, J. (1978). *Biochimica et Biophysica Acta*, **596**, 572.

LeGoffic, F. (1977). *Journal of Antibiotics*, **30**, S286.

LeGoffic, F., Martel, A., Capman, M. L., Baca, B., Goebel, P., Chardon, H., Soussy, C. J. Duval, J. & Bouanchaud, D. H. (1976). *Antimicrobial Agents and Chemotherapy*, **10**, 258.

LeMahieu, R. A., Ax, H. A., Blount, J. F., Carson, M., Despreaux, C., Pruess, D. L., Scannell, J. P., Weiss, F. & Kierstread, R. W. (1976). *Journal of Antibiotics*, **29**, 728.

Levy, S. B. (1988). *American Society of Microbiology*, **54**, 418.

Lingrens, F., Eberhardt, H. & Ottmanns, O. (1966). *Biochimica et Biophysica Acta*, **130**, 345.

Loder, P. B. (1972). *Progress in Hygiene and Experimental Medicine*, **26**, 493.

Loder, P. B., Newton, G. G. F. & Abraham, E. P. (1961). *Biochemical Journal*, **79**, 408.

Lowe, D. A. (1988). *Developments in Industrial Microbiology*, **30**, 121.

Malik, V. (1983). In *Biochemistry and Genetic Regulation of Commercially Important Antibiotics*, ed. L. C. Vining. Addison-Welsey, London, p. 293.

Mallams, A. K., Jaret, R. S. & Reimann, H. (1969). *Journal of the American Chemical Society* **91**, 7506.

Mancera, O., Zaffaroni, A., Rubin, B. A., Sondheimer, F., Rosenkranz, G. & Djerassi, C. (1952). *Journal of the American Chemical Society*, **74**, 3711.

Marconi, W., Bartoli, F., Cecere, F., Galli, G. & Morisi, F. (1975). *Agricultural and Biological Chemistry*, **39**, 277.

Margalith, P. & Pagani, H. (1961). *Applied Microbiology*, **9**, 325.

Marshall, V. P., Ciadella, J. I., Baczynskyj, L., Liggett, W. F. & Johnson, R. A. (1989). *Journal of Antibiotics*, **42**, 132.

Martin, J. R. & Goldstein, A. W. (1970). *Progress in Antimicrobial and Anticancer Chemotherapy*, **2**, 1112.

Martin, J. R. & Rosenbrook, W. (1968). *Biochemistry*, **7**, 1728.

Matsuda, A., Matsuyama, K., Yamamoto, K., Ichikawa, S. & Komatsu, K. (1987). *Journal of Bacteriology*, **169**, 5815.

Mayer, H., Collins, J. & Wagner, F. (1979). In *Plasmids of Medical, Environmental, and Commercial Importance*, eds K. N. Timmis & A. Puhle, Elsevier, Amsterdam, p. 459.

McCormick, J. R. D., Sjolander, N. O., Hirch, U., Jensen, E. R. & Doerschuk, A. P. (1957). *Journal of the American Chemical Society*, **79**, 4561.

McCormick, J. R. D., Joachim, U. H., Jensen, E. R., Johnson, S. & Sjolander, N. O. (1965). *Journal of the American Chemical Society*, **87**, 1798.

McGuire, J. M., Bunch, R. L., Anderson, R. C., Boaz, H. E., Flyan, E. H., Powell, H. M. & Smith, J. W. (1952). *Antibiotics and Chemotherapy*, **2**, 281.

Medeiros, A. A. & Jacoby, G. A. (1986). In *β-Lactam Antibiotics for Clinical Use*, eds S. F. Queener, J. A. Webber & S. W. Queener, Marcel Dekker, Inc., New York p. 49.

Mehta, R. J. & Nash, C. H. (1978). *Journal of Antibiotics*, **31**, 239.

Meyers, E. & Smith, D. A. (1962). *Journal of Bacteriology*, **84**, 797.

Miller, P. A. & Nash, J. H. (1975a). *Methods in Enzymology*, **43**, 603.

Miller, P. A. & Nash, J. H. (1975b). *Methods in Enzymology*, **43**, 606.

Miller, P. A., Saturnelli, A., Martin, J. H., Mitscher, L. A. & Bohonos, N. (1964). *Biochemical and Biophysical Research Communications*, **16**, 285.

Minieri, P. P., Firman, M. C., Histretta, A. G., Abbey, A., Brickner, C. E., Rigler, N. E. & Sokol, H. (1954). *Antibiotic Annual*, 81.

Nagarajan, R., Boeck, L. D., Gorman, M., Hamill, C. E., Hoehn, M. M., Stark, W. M. & Whitney, J. G. (1971). *Journal of the American Chemical Society*, **93**, 2308.

Nakagawa, J. I., Tamaki, S., Tomioka, S. & Matsuhashi, M. (1984). *Journal of Biological Chemistry*, **259**, 13937.

Nara, T., Misawa, M., Okachi, R. & Yamomoto, M. (1971). *Agricultural and Biological Chemistry*, **35**, 1676.

Newton, G. G. F. & Abraham, E. P. (1956). *Biochemical Journal*, **62**, 651.

Nicholas, R. A. & Strominger, J. L. (1988). *Reviews of Infectious Diseases*, **10**, 732.

Nobile, A., Charney, W., Perlman, P. L., Herzog, H. L., Payne, C. C., Tully, M. E. Jernik, M. A. & Herschberg, E. B. (1955). *Journal of the Americal Chemical Society*, **77**, 4184.

Norrby, S. R., Alestig, K., Bjornegard, B., Burman, L. A., Ferber, F., Huber, J. L., Jones, K. H., Kahan, F. M., Kahan, J. S., Kropp, H., Meisinger, M. A. P. & Sundeloff, J. G. (1983). *Antimicrobial Agents and Chemotherapy*, **23**, 300.

Nozaki, K. Y., Kitano, K. & Imada, A. (1984). *Agricultural and Biological Chemistry*, **48**, 37.

Okachi, R. & Nara, T. (1984). In *Biotechnology of Industrial Antibiotics*, ed. E. J. Vandamme. Marcel Dekker, Inc., New York, p. 329.

Okamura, K., Hirata, S., Okamura, Y., Fukugawa, Y., Shimanchi, Y., Kouno, K., Ishikura, R. & Lein, J. (1978). *Journal of Antibiotics*, **31**, 480.

Okamura, K., Sakamoto, M. & Ishikura, T. (1980). *Journal of Antibiotics*, **33**, 293.

Omura, S. & Tanaka, Y. (1983). In *Biochemistry and Genetic Regulation of Commercially Important Antibiotics*, ed. L. C. Vining. Addison-Welsey, London, p. 179.

Omura, S. & Yoshitake, T. (1986). In *Biotechnology*, Vol. 4, eds H. Pape & H. J. Rehm. VCH Verlagsgesellschaft, Weinheim, p. 359.

Omura, S., Nakagawa, A., Takeshima, H., Miyazawa, J. & Kitao, C. (1975). *Tetrahedron Letters* **50**, 4503.

Omura, S., Kitao, C., Miyazawa, J., Imai, H. & Takeshima, H. (1978). *Journal of Antibiotics,* **31,** 254.

Omura, S., Ikeda, H., Matsubara, H. & Sadakane, N. (1980). *Journal of Antibiotics,* **33,** 1570.

Osono, T., Watannabe, S., Saito, T., Gushima, H., Murakami, K., Takahashi, I., Yamaguchi, H., Sasaki, T., Susaki, K., Takamura, S., Miyoshi, T. & Oka, Y. (1980). *Journal of Antibiotics,* **33,** 1074.

O'Sullivan, J. & Abraham, E. P. (1980). *Biochemical Journal,* **186,** 613.

O'Sullivan, J. & Aklonis, C. A. (1984). *Journal of Antibiotics,* **37,** 804.

O'Sullivan, J. & Sykes, R. B. (1986). In *Biotechnology,* Vol. 4, eds H. Pape & H. J. Rehm. VCH Verlagsgesellschaft, Weinheim, p. 247.

O'Sullivan, J., Bleaney, R. C., Huddleston, J. A. & Abraham, E. P. (1979). *Biochemical Journal,* **184,** 421.

O'Sullivan, J., Gillum, A. M., Aklonis, C. A. Souser, M. L. & Sykes, R. B. (1982). *Antimicrobial Agents and Chemotherapy,* **21,** 558.

Ozanne, B., Benveniste, R., Tipper, D. & Davies, J. E. (1969). *Journal of Bacteriology,* **100,** 1144.

Pape, H. & Brillinger, G. U. (1973). *Archives of Microbiology,* **88,** 25.

Parker, W. L., Rathnum, M. L., Wells, J. S., Trejo, W. H., Principe, P. A. & Sykes, R. B. (1982). *Journal of Antibiotics,* **35,** 653.

Parker, W. L., O'Sullivan, J. & Sykes, R. B. (1986). *Advances in Applied Microbiology,* **31,** 181.

Pearce, C. J. & Rinehart, K. L. (1981). In *Antibiotics: Biosynthesis,* Vol. 4, ed. J. W. Corcoran. Springer-Verlag, New York, p. 74.

Perlman, D., Titus, E. & Fried J. (1952). *Journal of the American Chemical Society,* **74,** 2126.

Peterson, D. H. & Murray, H. C. (1952). *Journal of the American Chemical Society,* **74,** 1871.

Prelog, V. & Oppolzer, W. (1973). *Helvetica Chimica Acta,* **56,** 2279.

Richmond, M. H. & Sykes, R. B. (1973). *Advances in Microbial Physiology,* **9,** 31.

Rolinson, G. N., Batchelor, F. R., Butterworth, D., Cameron-Wood, J., Cole, M., Eustace, C. G., Hart, M. V., Richards, M. & Chain, E. B. (1960). *Nature,* **187,** 236.

Rosi, D., Drozd, M. L., Kuhot, M. F., Terminiello, L., Came, P. E. & Daum, S. J. (1981). *Journal of Antibiotics,* **34,** 341.

Samson, S. M., Belgaje, R., Blankenship, D. T., Chapman, J. C., Perry, D., Skatrud, P. L., VanFrank, R. M., Abraham, E. P., Baldwin, J. E., Queener, S. W. & Ingolia, T. D. (1985). *Nature,* **318,** 191.

Samson, S. M., Dotzlaf, J. E., Slisz, M. L., Becker, G. W., VanFrank, R. M., Veal, L. E., Yeh, N. K., Miller, J. R., Queener, S. W. & Ingolia, T. D. (1987). *Biotechnology,* **5,** 1207.

Sanders, C. C. (1987). *Annual Review of Microbiology,* **41,** 573.

Santanam, P. & Kayser, F. H. (1976). *Journal of Infectious Diseases,* **134,** 533.

Sarett, L. H. (1946). *Journal of Biological Chemistry,* **162,** 591.

Sato, T., Tosa, T. & Chibata, I. (1976). *European Journal of Applied Microbiology,* **2,** 153.

Savidge, T. A. (1984). In *Biotechnology of Industrial Antibiotics,* ed. E. J. Vandamme. Marcel Dekker Inc., New York, p. 171.

Scheidegger, A., Keunzi, M. T. & Nuesch, J. (1984). *Journal of Antibiotics,* **37,** 522.

Schumacher, G., Sigmann, D., Haug, H., Buckel, P. & Bock, A. (1986). *Nucleic Acids Research,* **14,** 5713.

Schupp, T., Hütter, R. & Auden, D. A. (1975). *Journal of Bacteriology,* **121,** 128.

Schupp, T., Traxler, P. & Hopwood, J. A. L. (1981). *Journal of Antibiotics,* **34,** 965.

Sedlaczek, L. (1988). *Critical Reviews in Biotechnology,* **7,** 187.

Seno, E. T. & Baltz, R. H. (1981). *Antimicrobial Agents and Chemotherapy*, **20**, 370.

Seno, E. T., Pieper, R. L. & Huber, F. M. (1977). *Antimicrobial Agents and Chemotherapy*, **11**, 455.

Sensi, P., Margalith, P. & Timbal, M. T. (1959). *Farmaco Edizione Scientifica*, **14**, 146.

Shaw, P. D. & Piwowarski, J. (1977). *Journal of Antibiotics*, **30**, 404.

Shaw, W. V. (1983). *CRC Critical Reviews in Biochemistry*, **14**, 1.

Shimizu, K., Hayano, M., Gut, M. & Dorfman, R. I. (1960). *Journal of Biological Chemistry*, **236**, 695.

Sih, C. J. & Wang, K. C. (1965). *Journal of the American Chemical Society*, **87**, 1387.

Sih, C. J. & Whitlock, H. W. (1968). *Annual Review of Biochemistry*, **37**, 682.

Sih, C. J., Lee, S. S., Tsong, Y. Y., Wang, K. C. & Chang, F. N. (1965). *Journal of the American Chemical Society*, **87**, 2765.

Singh, P. D., Ward, P. C., Wells, J. S. Ricca, C. M., Trejo, W. H., Principe, P. A. & Sykes, R. B. (1982). *Journal of Antibiotics*, **35**, 1397.

Singh, P. D., Young, M. G., Johnson, J. H., Cimarusti, C. M. & Sykes, R. B. (1984). *Journal of Antibiotics*, **37**, 773.

Slusarchyk, W. A., Dejneka, T., Gordon, E. M., Weaver, E. R. & Koster, W. H. (1984). *Heterocycles*, **21**, 191.

Speer, B. S. & Salyers, A. A. (1988). *Journal of Bacteriology*, **170**, 1623.

Speer, B. S. & Salyers, A. A. (1989). *Journal of Bacteriology*, **171**, 148.

Spencer, B. & Maung, C. (1970). *Biochemical Journal*, **118**, 29.

Spratt, B. G. (1975). *Proceedings of the National Academy of Sciences, USA*, **72**, 2999.

Spratt, B. G. (1977). *European Journal of Biochemistry*, **72**, 341.

Spratt, B. G. & Cromie, K. D. (1988). *Reviews of Infectious Diseases*, **10**, 699.

Spratt, B. G., Jobanputra, V. & Zimmermann, W. (1977). *Antimicrobial Agents and Chemotherapy*, **12**, 406.

Stapley, E. O., Jackson, M., Hernandez, S., Zimmerman, S. B., Currie, S. A., Mochales, S., Mata, J. M., Woodruff, H. B. & Hendlin, D. (1972). *Antimicrobial Agents and Chemotherapy* **2**, 122.

Suzuki, I., Takahashi, N., Shirota, S., Kawabe, H. & Mitsuhashi, S. (1975). In *Microbial Drug Resistance*, ed. S. Mitsuhashi. University of Tokyo Press, Tokyo, p. 463.

Sykes, R. B. & Matthew, M. (1976). *Journal of Antimicrobial Chemotherapy*, **2**, 115.

Sykes, R. B., Cimarusti, C. M., Bonner, D. P., Bush, K., Floyd, D. M., Georgopapadakou, N. H., Koster, W. H., Liu, W. C., Parker, W. L. Principe, P. A., Rathnum, M. L., Slusarchyk, W. A., Trejo, W. H. & Wells, J. S. (1981). *Nature*, **291**, 489.

Takiguchi, Y., Mishima, H., Okuda, M., Terao, M., Aoki, A. & Fukuda, R. (1980). *Journal of Antibiotics*, **33**, 1120.

Tardrew, P. L. & Nyman, M. A. (1964). US Patent 3,127,315.

Turner, M. K., Farthing, J. E. & Brewer, S. J. (1978). *Biochemical Journal*, **173**, 839.

Umezawa, H., Okaniski, M., Utahara, R., Maeda, K. & Kondo, S. (1967). *Journal of Antibiotics*, **A20**, 136.

Umezawa, H., Takasawa, S., Okanishi, M. & Utahara, R. (1968). *Journal of Antibiotics* **21**, 81.

Umezawa, Y., Yagisawa, M., Sawa, T., Takeuchi, T., Umezawa, H., Matsumoto, H. & Tazaki, T. (1975). *Journal of Antibiotics*, **28**, 845.

Umezawa, S., Kondo, S. & Ito, Y. (1986). In *Biotechnology*, Vol. 4, eds H. Pape & H. J. Rehm. VCH Verlagsgesellschaft, Weinheim, p. 309.

Uramoto, M., Otake, N., Ogawa, Y., Yonohara, H., Marumo, F. & Sato, Y. (1969). *Tetrahedron Letters*, 2249.

Vandamme, E. J. (1988). In *Bioreactor Immobilized Enzymes and Cells*, ed. M. Moo-Young. Elsevier, New York, p. 261.

Vining, L. C. & Westlake, D. W. S. (1984). In *Biotechnology of Industrial Antibiotics,* ed. E. J. Vandamme. Marcel Dekker Inc., New York, p. 387.

Vischer, E. & Wettstein, W. (1953). *Experentia,* **9,** 371.

Wagman, G. H. & Weinstein, M. J. (1980). *Annual Review of Microbiology,* **34,** 537.

Weinstein, M. J., Wagman, G. H., Marquez, J. A., Testa, R. T., Oden, E. & Waitz, J. A. (1969). *Journal of Antibiotics* **22,** 253.

Wells, J. S., Trejo, W. H., Principe, P. A., Bush, K., Georgopapadakou, N. H., Bonner, D. P. & Sykes, R. B. (1982*a*). *Journal of Antibiotics,* **35,** 184.

Wells, J. S., Trejo, W. H., Principe, P. A., Bush, K., Georgopapadakou, N. H., Bonner, D. P. & Sykes, R. B. (1982*b*). *Journal of Antibiotics,* **35,** 295.

Wiley, P. F., Baczynskyj, L., Dolak, L. A., Ciadella, J. I. & Marshall, V. P. (1987). *Journal of Antibiotics,* **40,** 195.

Williamson, J. M., Inamine, E., Wilson, K. E., Douglas, A. W., Liesch, J. M. & Albers-Schönberg, G. (1985). *Journal of Biological Chemistry,* **260,** 4637.

Woodward, R. B. (1957). *Angewandte Chemie,* **69,** 50.

Yagisawa, M., Yamamoto, H., Naganawa, H., Kondi, S., Takeuchi, T. & Umezawa, H. (1972). *Journal of Antibiotics,* **25,** 748.

Yagisawa, M., Huang, R. S. R. & Davies, J. E. (1978). *Journal of Antibiotics,* **31,** 809.

Yamada, T., Tipper, D. & Davies, J. (1968). *Nature,* **219,** 288.

Yamaguchi, A., Hirata, T. & Sawai, T. (1984). *Antimicrobial Agents and Chemotherapy,* **25,** 348.

Yamamoto, H., Hotta, K., Okami, Y. & Umezawa, H. (1982). *Journal of Antibiotics,* **35,** 1020.

Yang Graham, M. & Weisblum, B. (1979). *Journal of Bacteriology,* **137,** 1464.

Yano, H., Fujii, H., Mukoo, H., Shimura, M., Watanabe, T. & Sekizawa, Y. (1978). *Annals of the Phytopathological Society of Japan,* **44,** 413.

Chapter 10

REGULATION AND EXPLOITATION OF ENZYME BIOSYNTHESIS

Arnold L. Demain

Fermentation Microbiology Laboratory, Department of Biology, Massachusetts Institute of Technology, Cambridge, Massachusetts 02139, USA

CONTENTS

1 INTRODUCTION

Enzymes are valuable in manufacturing because of their rapid and efficient action at low concentrations under mild pH values and temperatures, their high degree of substrate specificity (which reduces side production formation), their low toxicity, and the ease of stopping their action by mild treatments.

A large number of enzymes are produced by microorganisms. Indeed the term 'enzyme,' first used by Kuhne in 1877, means 'in yeast'. When the field of biochemistry was born in 1897 due to the discovery by Buchner that cell-free extracts of yeast could carry out ethanol production from sugar, Buchner referred to the glycolytic enzyme complex as 'zymase', meaning the 'enzyme of yeast itself'. These enzymes make up from 30–65% of soluble proteins in yeast (Fraenkel, 1982). Over 1100 individual proteins have been observed in extracts of *Escherichia coli* during growth under one particular set of conditions (Neidhardt *et al.*, 1983). When grown under different conditions, a total of 1500 have been observed. The number of genes which have been mapped in *E. coli* is greater than 1000 (Neidhardt *et al.*, 1983) and that in *Bacillus subtilis* is greater than 420 (Workman, 1986). Some microbial strains produce very high concentrations of extracellular proteins. Wild strains of *Bacillus licheniformis* produce 5 g of protease per liter (Aunstrup, 1984) and commercial strains make 20 g per liter. High-yielding strains of *Aspergillus* spp. produce up to 20 g per liter of glucoamylase (Aunstrup, 1984; van Brunt, 1986). These figures attest to the ease with which enzyme levels can be increased by environmental and genetic manipulations. Thousand-fold increases have been recorded for catabolic enzymes, and biosynthetic enzymes have been increased by several hundred-fold. It is obvious that the higher the specific enzyme content of a culture, the simpler will be the job of enzyme isolation.

Additional reasons for using microbial cells as sources of enzymes are as follows: (a) enzyme fermentations are quite economical on a large scale due to short fermentation cycles and inexpensive media; (b) screening procedures are simple and thousands of cultures can be examined in a reasonably short time; and (c) different species produce somewhat different enzymes catalyzing the same reaction, allowing one flexibility with respect to operating conditions in the reactor. This versatility is illustrated by the fact that α-amylase from *Bacillus amyloliquefaciens*, a commercial enzyme used for years for hydrolysis of starch at a temperature as high as 90°C, was forced to compete in 1972 with a similar enzyme from *B. licheniformis* which could operate at 110°C (Aunstrup, 1979). The optimal temperatures for the *B. amyloliquefaciens* and the *B. licheniformis* α-amylases are 70°C and 92°C respectively (Stewart, 1987). To appreciate the vast spectrum of microbial enzyme activities, one can examine the extremes in properties of microbial glucose isomerases as shown in Table 1 (Ulmer, 1983).

In the last decade, microbial enzymes have increasingly been used for applications which traditionally employed plant and animal enzymes. These shifts include the partial replacement of (a) amylases of malted barley and wheat by amylases from *Bacillus* spp. and *Aspergillus* spp. in the beer, baking and textile industries; (b) plant and animal proteases by *Aspergillus* protease for chill-proofing beer and meat tenderization; (c) pancreatic proteases by *Aspergillus* and *Bacillus* proteases for leather bating and in detergent preparations; and (d) calf stomach rennet (chymosin) by *Mucor* rennins for cheese manufacture. Sales of *Mucor* rennins for the cheese industry have been

Table 1
Extremes in Properties of Microbial Glucose Isomerases (Ulmer, 1983)

Property	Units	Low extreme	High extreme
Turnover number	Glucose molecules per molecule per min	63	2 151
Molecular mass	Daltons	52 000	191 000
Temperature optimum	°C	50	90
Temperature stability	% loss in 10 min	100 at 60°C	0 at 70°C
pH optimum	pH units	←————— 3·5 —————→	
pH stability range	pH units	7–9	4–11

increasing steadily. In 1980, it only amounted to 5 million dollars (Aunstrup, 1980) out of a total rennet market of 120 million dollars. More recently, fungal rennin has replaced 50% of calf rennin for cheese making (Beppu, 1983). However the major application of microbial enzymes has been the use of glucose isomerase in conjunction with α-amylase and glucoamylase to convert starch to mixtures of glucose and fructose known as 'high fructose corn syrup'. The development of glucose isomerase permitted the corn wet milling industry to capture 30% of the sweetener business from the sugar industry. In the United States alone, high fructose corn syrup was produced at 1–2 billion (10^9) pounds (dry basis) in 1976, 4·6 billion pounds in 1980 and 10 billion pounds in 1985. Today, it outsells industrial sucrose in the USA (Poulsen, 1987).

The US and European enzyme market, mainly made up of protease, amylase, glucoamylase, microbial rennet and glucose isomerase, amounted to 300 million dollars in 1980, 500 million dollars in 1985 and 700 million dollars in 1987. Of the 700 million dollar market, microbial enzymes accounted for 400 million dollars. The annual rate of increase in the enzyme market recently has been 8%, although the microbial enzyme segment is increasing at an annual rate of 15% (Poulsen, 1987). In 1980, two companies dominated the market: Novo Industries at 42% and Gist-Brocades at 17% (O'Sullivan, 1981). In 1983, twelve enzymes exceeded worldwide sales of 10 million dollars and accounted for over 90% of the enzyme market (Ulmer, 1983). In 1986, the international enzyme market reached 1 billion dollars (Godtfredsen *et al.,* 1987). Recent additions to the list of commercial enzymes include thermostable lipase (e.g. *Mucor miehei* Lipozyme from Novo) for transesterification, ester synthesis and detergents, nitrilases (from Novo) for conversion of nitriles into amides or carboxylic acids, and cyanide-degrading enzymes for pollution abatement and detoxification of foods and feeds. ICI's cyclear degrades cyanide to formic amide while Novo's enzyme produces formic acid. The main problem limiting the further development of enzyme technology is the poor availability of enzymes for large scale application testing. Only about 10% of the 2000–3000 enzymes described in the literature are commercially available and over half of these sell for more than 1 million dollars per pound (Katchalski-Katzir, 1980). A review of enzyme production by microorganisms has been recently published (Frost & Moss, 1987).

2 IMPORTANT FACTORS IN ENZYME FERMENTATIONS

In devising an enzyme fermentation, it is very important to begin with the most active strain available. Often, known enzyme-producing strains can be obtained from workers in the field or from culture collections. In the absence of such cultures, a screening program of cultures from nature or culture collections must be set up. The major requirement here is simplicity so that rapid examination of a large number of strains is feasible. Before final selection of a strain, the requirements shown in Table 2 should be fulfilled to as great a degree as possible.

Once a good strain is obtained, fermentation parameters must be optimized to maximize growth and enzyme production. For example, the level of leucine biosynthetic enzymes in *E. coli* has been found to vary over a 2500-fold range merely by changing growth conditions (Davis & Calvo, 1977). Of importance here are temperature, pH and oxygen transfer. Also important is the nutrition of the microorganism especially with respect to sources of carbon, nitrogen, phosphorus, sulfur and mineral salts. Often, especially with extracellular enzymes, the addition of surfactants is important. For example, optimum production of ligninase by *Phanerochaete chrysosporium* in agitated submerged culture requires Tween 80 or Tween 20 (Jäger *et al.*, 1985). Non-ionic surfactants are generally preferred over anionic and cationic agents which are usually toxic to the microorganism.

Table 2
Some Important Factors for Enzyme Production by Microorganisms

1. Microbial strain
2. Temperature
3. pH
4. Oxygen transfer
5. Carbon dioxide level
6. Sources of energy, carbon, nitrogen, phosphorus, sulfur and other inorganic chemicals
7. Growth factors
8. Stage of batch growth cycle
9. Growth rate
10. Addition of surfactants
11. Induction
12. Feedback repression and attenuation
13. Nutrient repression
14. Gene dosage

If the fermentation is conducted as a batch culture, the stage of the growth cycle at which enzyme content is highest must be ascertained. In many cases, enzymes rapidly disappear after reaching their peak activity (Agathos & Demain, 1987). Even when enzyme activity does not disappear from the cells, its stability in a crude extract can vary widely depending upon the time of harvest. For example, the activity of SAM: macrocin-*O*-methyltransferase, the final enzyme of tylosin biosynthesis in *Streptomyces fradiae,* has a half-life of 38 min when harvested at 18 h but only 3 min when harvested at 66 h or later (Seno & Baltz, 1981). Such an observation is presumably due to the production of proteases at the later time. A chemostat can be quite useful for enzyme production since the formation of many enzymes is controlled by growth rate which can be kept constant in continuous culture.

3 REGULATION OF ENZYME PRODUCTION

In addition to the above environmentally manipulated factors, it is of extreme importance to exploit the genetic regulation of enzyme biosynthesis. Production of an enzyme can be increased by many different types of manipulation which affect regulatory mechanisms (Eveleigh & Montenecourt, 1979).

3.1 Enzyme Induction

The structural genes encoding many enzymes are normally inactive in the absence of the substrate, i.e. enzyme production is normally repressed. However, when a substrate is added, the structural gene is turned on and the enzyme is produced. Such a process is termed 'induction' or 'derepression' and the enzyme is considered to be 'inducible'.

The most thoroughly studied inducible enzyme system is that for lactose hydrolysis (*lac*) in *E. coli*, which provided the basis of a model for negative control of protein synthesis. Negative control means that the regulatory protein ('repressor') encoded by the regulator locus interferes with transcription. In the case of the *lac* operon in *E. coli*, about 10 molecules of repressor protein are made per regulator locus. The repressor binds to the operator locus and interferes with binding of RNA polymerase to the neighboring promoter locus; this interference is normally caused by the repressor protein overlapping into the promoter region when it is bound to the operator. The inducer (a β-galactoside in this case) binds to the repressor, bringing about an allosteric modification in the conformation of the protein so that it can no longer bind to the operator. The specificity of the repressor–operator interaction can be appreciated by the fact that of the three million base pairs of the *E. coli* chromosome, only 20 are bound by the *lac* repressor. This protein binds to those 20 DNA base pairs ten million times more tightly than it does to the rest of the organism's DNA and contacts its binding site at one thousand times the rate of simple diffusion (Takeda *et al.*, 1983).

Positive regulation of transcription by the regulator locus is another type of induction mechanism. Here the regulatory protein coded by the regulator is necessary for transcription to occur, i.e. it is an 'activator' rather than a repressor. Whereas induction in bacteria involves negative control (via repressor proteins) and positive control (via activator proteins), eukaryotes appear to use positive control only (Raibaud & Schwartz, 1984). In positive control, transcription initiation requires the binding of the activator protein to the promoter at a site different from the RNA polymerase site. Binding of the activator to the promoter allosterically increases the frequency of binding of RNA polymerase at its site. The activator is encoded by the regulatory locus (R) which can exist in two conformations, R_i (inactive) and R_a (active), which are in equilibrium. The inducer displaces the equilibrium to the R_a form. Promoters controlled by positive regulation have sequences which differ markedly from the consensus promoter sequence and thus RNA polymerase does not recognize them unless the activator is bound at its site.

Many catabolic enzymes fall into the inducible category. Typical inducers are substrates such as starch or dextrins for amylases, sucrose for invertase and urea for urease. Some inducers are extremely potent, e.g. certain galactosides increase β-galactosidase production in *E. coli* by a thousand-fold, yielding cells possessing several per cent of their cellular protein as this particular enzyme. Substrate analogues that are not attacked by the enzyme (i.e. gratuitous inducers) are often excellent inducers of enzyme synthesis. Although most inducers are substrates or substrate analogues, intermediates and products sometimes function as inducers. Maltodextrins induce amylase; xylobiose induces xylanase; fatty acids induce lipase; urocanic acid induces histidase; and galacturonic acid induces polygalacturonase. This is especially common in the induction of extracellular enzymes hydrolyzing large polymers. A low (basal) concentration of enzyme is produced even under conditions of repression. This basal level slowly acts on a few molecules of the polymer releasing lower molecular weight inducers which enter the cell and induce the enzyme. A probable mechanism for such a phenomenon has been revealed in *Cellulomonas fimi*. An endoglucanase (=cellulase) gene, *cenB*, has been shown to be expressed from two promoters, one constitutive and one inducible (Greenberg *et al.*, 1987). In the case of the renowned cellulase producer, *Trichoderma reesei*, the potent inducer, sophorose, as well as another suspected inducer, gentiobiose, are made during hydrolysis of cellulose or cellobiose. The reaction is a transglycosylation, thought to be catalyzed by β-glucosidase (Vaheri *et al.*, 1979). Especially when the substrate of an enzyme is an essential metabolite (e.g. an amino acid, or a purine) and the product is metabolically dispensible, induction is by the product rather than the substrate. Some coenzymes induce enzymes, as in thiamine induction of pyruvate decarboxylase.

Although induction has no role in controlling biosynthetic operons in *E. coli*, it does in other microorganisms. In *Pseudomonas putida*, tryptophan synthetase is induced by indoleglycerophosphate and the entire tryptophan branch is induced by chorismate in *B. subtilis*.

Regulatory mutations can be used to eliminate the dependence of enzyme formation on an inducer. In negative regulation, this occurs when the regulator locus is genetically modified to produce a nonfunctional repressor, or mutation in the operator locus eliminates the ability of repressor to bind. In positive control, the R gene mutation would result in an activator protein which does not require inducer to push the equilibrium toward the R_a (active) form. Such regulatory mutants, which produce a normally inducible enzyme in the absence of the inducer, are called 'constitutive' mutants.

Many procedures to select mutants of *E. coli* constitutive for β-galactosidase synthesis have been devised, based on the survival and growth of mutants under poor conditions of induction; parental cells, still requiring induction, are disadvantaged due to poor substrate utilization and become a smaller proportion of the population. Thus, maintaining *E. coli* in a chemostat with a low concentration of lactose selects for mutants that can efficiently use low levels of substrate, i.e. cells which are no longer repressed in the absence of inducing substrate (Horiuchi *et al.*, 1963). Similarly, growing mutagenized cells on a compound that is a good carbon source but a poor inducer or a noninducer (e.g. phenyl-β-galactoside or 2-nitrophenyl-α-L-arabinoside for constitutive β-galactosidase mutants) selects for those cells which use the substrate well and no longer require induction (Jacob & Monod, 1961; Jayaraman *et al.*, 1966).

The usual goal in obtaining constitutive mutants is to produce the parental level of enzyme without going to the expense of adding an inducer to the culture. However, in a number of cases, hyperproducing mutants are obtained which produce higher levels of enzyme in the absence of inducer than the parental strain does in the presence of inducer. This occurred while selecting constitutive maltase producers from *Saccharomyces italicus*; the noninducing substrate used was sucrose. In addition to constitutivity, the mutant obtained was a 2- to 3-fold hyperproducer (Schaeffer & Cooney, 1982).

Similar techniques have been successful in many systems. Constitutive acetamidase mutants were obtained when acrylamide was the sole nitrogen source (Hynes & Pateman, 1970). Use of D-serine as the limiting nitrogen source and inducer in a chemostat yielded constitutive 25-fold hyperproducers of D-serine deaminase (Sikyta & Kyslík, 1981). Constitutive producers of mandelate-catabolizing enzymes (Hegeman, 1966) and of ribitol dehydrogenase (Hartley, 1974) have similarly been obtained. In the latter case, xylitol cannot be used for growth of *Klebsiella aerogenes* unless the bacterium can produce high levels of ribitol dehydrogenase, an inducible enzyme. Since xylitol is not an inducer, only constitutive producers of ribitol dehydrogenase can grow on xylitol. Use of xylitol in a chemostat thus yields constitutive hyperproducers (Rigby *et al.*, 1974; Sikyta & Kyslík, 1981) some of which have increased copies of the ribitol dehydrogenase gene.

Constitutive tryptophanase producers were obtained (Pavlasová *et al.*, 1983) by plating tryptophan auxotrophs on glycerol–indole agar in the absence of an inducer of tryptophanase (tryptophanase would allow growth on indole acting in its reverse direction). One mutant (out of ten which developed on these plates) produced an amount of tryptophanase equivalent to the parental induced level. It was then grown in a nitrogen-limited chemostat using low tryptophan as a nitrogen source. A new mutant was isolated from the chemostat which produced five times the parental induced level. No higher production was observed upon addition of the inducer 5-methyltryptophan.

Means have been devised to detect constitutive clones after mutation and selection. For example, β-galactosidase constitutives are detected by plating on an agar medium containing a noninducing, nonenzyme substrate (e.g. glycerol) as carbon source. After growth, colonies are sprayed with o-nitrophenyl-β-galactoside, a colorless compound. Constitutive colonies turn yellow due to their ability to hydrolyze the compound to the yellow o-nitrophenol.

3.2 Feedback Repression and Attenuation

Feedback repression involves the 'turning off' of enzyme synthesis when a sufficient amount of the pathway product has been made and it starts to accumulate. The end product of the pathway acts as a corepressor. The aporepressor specified by the regulator locus is inactive in the absence of its corepressor and is unable to bind to the operator. However, in the presence of corepressor, an active repressor is formed which binds to the operator, overlaps the promoter, prevents binding and transcription by RNA polymerase and hence prevents enzyme synthesis.

Many of the amino acid biosynthetic pathways are regulated not by the amino acids themselves but by their charged tRNA molecules. Thus, whereas feedback repression is effected by the amino acid end products acting as corepressors and interfering with transcription initiation, another type of control called attenuation (= transcription termination control) involves charged tRNA and transcription termination. A significant number of the intracellular tRNA molecules for a particular amino acid must be in the uncharged state before the genes coding for that amino acid's biosynthetic enzymes can be efficiently transcribed; in the presence of an excess of charged tRNA, transcription is initiated but terminated before the first structural gene is transcribed (Kolter & Yanofsky, 1982). Attenuation is known to control six bacterial amino acid biosynthetic operons producing seven amino acids (threonine, isoleucine, valine, tryptophan, leucine, phenylalanine, histidine). Unlike the other operons, the *trp* operon is regulated by both repression and attenuation. Their combined action permits a degree of expression over a 600-fold range, repression being responsible for 80-fold and attenuation for a seven-fold range. Repression responds to the level of tryptophan in the cell and attenuation to the level of charged tRNA^{trp}. The two mechanisms act at different degrees of tryptophan deprivation. Repression acts first, i.e. when the tryptophan level drops to that of moderate starvation whereas attenuation acts in the moderate to severe tryptophan starvation range (Yanofsky *et al.*, 1984).

Whereas synthesis of catabolic enzymes is mainly controlled by induction, biosynthetic enzymes are controlled by repression and attenuation. Thus, by limiting the internal accumulation of end product corepressors, substantial increases in enzyme production can be realized. It goes without saying that the presence of the pathway end product as a constituent of the growth medium should be avoided.

Two major types of manipulations have been employed to relieve feedback repression. The first type prevents accumulation of repressive concentrations of end products in the cell, whereas the second alters the enzyme-forming system so that it is less sensitive or even totally insensitive to feedback repression.

3.2.1 Limiting Intracellular Accumulation of End Product Corepressors

One way to limit internal buildup of end products is to add an inhibitor of the pathway to the medium (Table 3). A more economical way is to obtain an auxotrophic (nutritional) mutant which fails to make the repressive product and requires it as a medium constituent for growth; limited feeding of the product partially starves the mutant of its end product requirement and thus repressive concentrations never occur intracellularly. As a result, high levels of biosynthetic enzymes are produced (Table 4).

Instead of limited feeding, one can use a slowly-utilized derivative of the required end product. Thus, growth of a uracil auxotroph on dihydroorotic acid derepressed aspartate transcarbamylase 1000-fold, yielding cells containing 7% of their protein as this enzyme (Sheperdson & Pardee, 1960).

Another means to derepress biosynthetic enzymes is to grow a bradytroph (an auxotroph which can grow in minimal medium slowly but is stimulated by its growth factor; also called a partial or 'leaky' auxotroph) in the total absence of its stimulatory end product. Thus, a 500-fold increase in aspartic transcarbamylase was obtained by growing a pyrimidine bradytroph in minimal medium (Moyed, 1961).

3.2.2 Feedback Resistance Mutations

One of the most common ways to eliminate feedback repression is to obtain mutants that are resistant to toxic structural analogues (i.e. antimetabolites) of

Table 3
Derepression of Biosynthetic Enzyme Synthesis by Addition of Pathway Inhibitors

Inhibitor added	Derepressed enzyme	Increase
2-Thiazolealanine	Ten enzymes of histidine biosynthesis	Up to 30-fold
Psicofuranine	IMP dehydrogenase and XMP aminase	5-fold
Adenine	Four enzymes of thiamine biosynthesis	5- to 10-fold
β-Chloroalanine	Enzymes of valine and isoleucine biosynthesis	2- to 7-fold
Glycyl-L-leucine	Threonine deaminase; acetohydroxyacid synthetase	10-fold; 3-fold

Table 4
Derepression of Biosynthetic Enzyme Synthesis by Limited
Feeding of Requirement to Auxotrophs

Auxotrophic requirement	Derepressed enzyme	Increase
Histidine	Ten enzymes of hisidine biosynthesis	25-fold
Leucine	Acetohydroxyacid synthetase	40-fold
Thiamine	Four enzymes of thiamine biosynthesis	Up to 1500-fold
Biotin	7-Oxo-8-amino-pelargonate amino-transferase	400-fold
Guanine	IMP dehydrogenase	45-fold
Guanine	XMP aminase	25-fold
Tryptophan	Anthranilate synthetase	7-fold
Methionine[a]	ATP sulfurylase	140-fold
Leucine[a]	α-IMP synthetase; β-IPM dehydrogenase	60-fold 47-fold

[a] Chemostat culture used.

the end product of the pathway. Mutants resistant to a particular antimetabolite are often altered at the level of enzyme synthesis, constitutively producing the appropriate biosynthetic enzymes. Like mutants which no longer require inducer, these are also called 'constitutive mutants' and are thought to be genetically altered at the regulatory locus so that a modified aporepressor is produced which combines only poorly or not at all with corepressor. Alternatively, they may be mutated at the operator locus so that binding of the repressor does not occur or occurs only poorly.

Such constitutive mutants, which produce enzyme in the presence of normally-repressing levels of end products, often are hyperproducers of the enzyme. For example, ethionine-resistant mutants have been isolated which show 120-fold increased content of cystathionine synthetase, an enzyme of methionine biosynthesis (Holloway *et al.*, 1970). A constitutive mutant of *Salmonella typhimurium* was obtained which formed twice as much tryptophan biosynthetic enzymes in the presence of exogenous tryptophan, than its parent did in the absence of the corepressor (Balbinder *et al.*, 1970). Other constitutive mutants of *S. typhimurium*, produce 25 times more aspartic transcarbamylase than their parent even in the absence of the exogeneously added corepressor uracil (O'Donovan & Neuhard, 1970). A canavanine-resistant mutant produced two to three times more ornithine transcarbamylase in the presence of arginine (Bechet *et al.*, 1970). *S. typhimurium* mutated at the regulatory *cys*B gene (a positive regulatory gene necessary for production of six enzymes of cysteine biosynthesis) overproduced *O*-acetylserine sulfhydrylase by three-fold (Borum & Monty, 1976).

Selection of mutants resistant to end product antimetabolites can also yield overproducers due to mutation in RNA polymerase. Selection of mutants of *S. typhimurium* resistant to 5-fluorouracil plus 5-fluorouridine led to a constitutive mutant in the RNA polymerase gene producing derepressed levels of aspartate transcarbamylase (40-fold), orotate phosphoribosyltransferase (7-fold), carbamoyl-phosphate synthetase (3-fold) and orotidine 5'-monophosphate decarboxylase (3-fold) (Jensen *et al.*, 1982).

Occasionally selection for resistance to antimetabolites results in gene amplification. The *his* operon was found to be amplified in certain 3-amino-1,2,4-triazole-alanine-resistant mutants of *S. typhimurium* (Anderson, *et al.*, 1976). In this case, production of enzymes of the *his* operon was doubled.

If analogues fail to inhibit growth (i.e. fail to act as antimetabolites,) one can often achieve inhibition by changing the carbon or nitrogen source, or by adding a detergent to the medium. Shift to a poorer carbon or nitrogen source can decrease interference with antimetabolite transport and use of a detergent can also increase its uptake.

3.3 Carbon Source Repression

Carbon source repression, also known as carbon catabolite repression, is another conservative mechanism which safeguards against waste of a cell's protein-synthesizing machinery; it operates when more than one utilizable carbon substrate is present in the environment. The cell produces enzymes to catabolize the most rapidly assimilated carbon source; synthesis of enzymes utilizing other substrates is repressed until the primary substrate is exhausted. The enzymes involved are usually inducible. Carbon source repression is a phenomenon usually caused by glucose, but in different organisms other rapidly metabolized carbon sources can cause repression and, indeed, sometimes even repress catabolism of glucose. An example of this occurs in *Pseudomonas aeruginosa*, where citrate is the preferred carbon source over glucose (Clarke & Lilly, 1969).

In enteric bacteria, carbon source repression is mediated by insufficient intracellular levels of cyclic 3',5'-adenosine monophosphate (cAMP). In *E. coli*, cAMP is necessary for synthesis of all inducible enzymes. cAMP binds to the promoter region after attachment to a specific binding protein (cAMP–receptor protein or CRP), a dimer of identical subunits and two separate domains. The N-terminal of CRP attaches to cAMP and the C-terminal to DNA. Each CRP subunit binds one cAMP molecule. After binding cAMP, the CRP undergoes an allosteric transition to an active state in which it binds to a specific portion of the promoter, increasing the affinity of RNA polymerase to its site on that particular promoter and thus the frequency of transcription increases. This is an example of positive regulation.

In bacteria, positive regulation is useful for two general purposes: (1) controlling specific metabolic pathways (e.g. for metabolism of L-arabinose, maltose or L-rhamnose in *E. coli*, xylene in *Pseudomonas putida*); (2) setting priorities between pathways that serve the same final purpose, e.g. CRP in carbon source regulation; the *ntrC* gene product, an activator which turns on ammonia-repressible genes in enterobacteria; the *phoM* and *phoB* gene products, activators of phosphate regulation (see below).

In carbon source repression, the *E. coli* promoter has a sequence quite different from the consensus for *E. coli* promoters. The consensus sequence to which the cAMP–CRP complex binds [aa-TGTGA(N_7)CACa-t] occurs at a variety of locations in different promoters relative to the start site for transcription (Gottesman, 1984).

cAMP reverses carbon source repression of many enzymes in enteric bacteria. Since promoters of different operons have different affinities for the cAMP–CRP complex (Piovant & Lazdunski, 1975), not all promoters are binding the complex and undergoing transcription at the same time. During glucose assimilation, the intracellular concentration of cAMP is depressed 1000-fold, whereas metabolism of a nonrepressive carbon source such as acetate has little effect on cAMP levels.

Carbon source repression in *E. coli* is caused not by a specific catabolite of the rapidly assimilated carbon source but by uptake of the latter compound (Peterkofsky, 1976). In permeation of glucose by the phosphoenolpyruvate (PEP)–phosphotransferase (PTF) system, there is a small protein, enzyme III–Glc (molecular weight of 22 000), which is phosphorylated by protein Hpr and in turn transfers the phosphate to glucose. The glucose phosphate is then transported into the cell through a high affinity enzyme II membrane protein. The gene for enzyme III–Glc is called *crr* because mutants lacking this gene are resistant to carbon source repression. Enzyme III–Glc when unphosphory-lated (i.e. glucose present) mediates inducer exclusion and fails to activate adenylate cyclase, the enzyme converting ATP to cAMP. In its phosphorylated form (glucose absent), Enzyme III–Glc has no ability to exclude inducers and activates adenylate cyclase (Danchin, 1985). Other transport systems which lead to inhibition of adenylate cyclase are proton symport (used for lactose uptake) and facilitated diffusion (used for glycerol transport) (Botsford, 1981).

Mutants of *E. coli* that cannot make effective CRP or adenylate cyclase fail to grow or grow poorly on lactose, glycerol and other carbon sources, whereas mutants lacking cAMP phosphodiesterase (which degrades cAMP to AMP) are insensitive to catabolite repression (Monard *et al.*, 1969). Mutants in the *crp* gene, known as CRP*, make CRP independent of cAMP for binding to the promoter. In these mutants, it appears that the conformation of CRP necessary for binding to the promoter exists even in the absence of cAMP (Harman & Dobrogosz, 1983).

Outside the realm of enteric bacteria, cAMP does not appear to play a major role in carbon source repression although some exceptions have been reported. In *P. putida* and *P. aeruginosa*, where cAMP is found, it has no effect on succinate- or lactate-repression of histidase (Phillips & Mulfinger, 1981). In other organisms where carbon source repression occurs (e.g. some *Bacillus* species, *P. aeruginosa*, *Arthrobacter crystallopoietes*, *Rhizobium meliloti* and anaerobic bacteria such as *Bacteroides fragilis*), cAMP has not been detected (Siegel *et al.*, 1977) nor has it been shown to have an effect on carbon source repression of enzymes when added (Botsford, 1981).

The situation in *Bacillus* is unclear at the moment. Many bacilli (e.g. *Bacillus megaterium, Bacillus cereus*) make no cAMP. Although *B. subtilis* had been considered to be incapable of making cAMP, the regulatory nucleotide is made under oxygen limitation along with adenylate cyclase and cAMP phosphodiesterase (Mach *et al.*, 1984). One strain of *Bacillus circulans* forms cAMP only in media rich in glucose (Esteban *et al.*, 1984). Furthermore cAMP, upon addition to *B. circulans*, repressed xylanase, 1,3-β-D-glucanase and 1,6-β-D-glucanase. Other *B. circulans* strains failed to produce cAMP. Carbon source repression of sporulation in *B. subtilis* can be decreased by mutation in the β and β' subunits of RNA polymerase (Sun & Takahashi, 1984). Such *crs* mutations can be suppressed by rifamycin resistance mutations mapping in the same region.

Glucose repression of cellulase in the actinomycete, *Thermomonospora curvata*, might involve cAMP. A hyperproducing mutant had a higher cAMP level than its parent and showed an increase in the level as growth slowed as opposed to a decrease in the parent when its growth rate decreased (Fennington *et al.*, 1983).

In molds, cAMP appears to have no role in carbon source repression (Arst, 1981). At one time, cAMP was considered important in carbon repression in yeasts. For example, the cAMP content of *Saccharomyces fragilis* cells grown in lactate was higher than those grown in glucose (Sy & Richter, 1972) and the cAMP content of *Saccharomyces cerevisiae* cells grown in galactose was 70% higher than those grown in glucose (van Wijk & Konijn, 1971). However it was later shown that glucose repression of galactokinase did not involve cAMP as determined with *S. cerevisiae* mutants able to take up cAMP as well as mutants unable to synthesize cAMP (Matsumoto *et al.*, 1983). On the other hand, cAMP does stimulate galactokinase synthesis in the absence of glucose (Matsumoto *et al.*, 1982). It thus plays a role in yeast enzyme formation but this role is distinct from carbon source repression. The ability of cAMP to reverse glucose repression of respiratory adaptation in *S. cerevisiae* protoplasts was later shown to be duplicated by AMP and ATP (Fang & Burow, 1970). Often in yeasts, cAMP levels are higher under conditions of carbon source repression than under derepressed conditions (Eraso & Gancedo, 1984).

Yeasts contain a number of sugar phosphorylating enzymes such as glucokinase, hexokinase PI and hexokinase PII. Hexokinase PII appears to be the repressor protein of carbon source repression, in addition to its action as an enzyme phosphorylating the repressive sugars glucose, fructose and mannose (Entian & Fröhlich, 1984). Many mutations in this enzyme lead to derepressed synthesis of invertase, maltase, malate dehydrogenase and respiratory enzymes. In *Saccharomyces carlsbergensis*, mutants in hexokinase PII are resistant to carbon source repression of α-glucosidase and invertase. High glucose conditions lead to increased hexokinase PII whereas glucose limitation leads to decreased hexokinase PII. Addition of xylose to high glucose led to

98% inactivation of hexokinase PII and derepression of invertase (Fernández et al., 1984).

Since cAMP occurs in yeasts, what is its role? It binds to cAMP-dependent protein kinase and activates it, resulting in phosphorylation of other enzymes such as trehalase, phosphorylase, glycogen synthase and fructose-1,6-bisphosphatase (Pall, 1984). cAMP stimulates glycolysis, glycogenolysis and trehalase degradation via protein phosphorylation (phosphorylated phosphofructokinase is the activated form) and inhibits gluconeogenesis (phosphorylated fructose 1,6-bisphosphatase is the inhibited form and the form more labile to proteolysis) (Pohlig & Holzer, 1985). Carbon source inactivation of gluconeogenic enzymes (fructose-1,6-bisphosphatase, cytoplasmic malate dehydrogenase, PEP carboxykinase) by glucose involves cAMP. It is thought that cAMP activates protein kinase which phosphorylates the proteins prior to proteolysis (Torotora et al., 1983).

Carbon source repression is very important in commercial practice since many enzymes of current or potential industrial importance are subject to this type of regulation. Some examples are shown in Table 5. As can be seen, glucose is not always the repressing carbon source. Indeed, inducible alkaline

Table 5
Carbon Source Repression of Enzymes of Commercial Interest

Enzyme	Microorganism	Repressing carbon source
Acid protease	Aspergillus niger	Glucose, fructose, xylose
α-Amylase	Bacillus licheniformis	Glucose, fructose, maltose, glycerol, acetate, succinate
β-Amylase	Bacillus megaterium	Glucose, maltose
Cellulase	Trichoderma reesei	Glucose, glycerol, starch
Glucoamylase	Aspergillus niger	Glycerol, sorbitol, xylose, fructose, pyruvate, α-ketoglutarate
β-1,3 Glucanase	Streptomyces sp.	Glucose, glycerol
β-Glucosidase	Trichoderma reesei	Glucose
α-Galactosidase	Monascus sp.	Glucose, sucrose, glycerol
β-Galactosidase	Escherichia coli	Glucose
Invertase	Saccharomyces cerevisiae	Glucose, fructose, mannose
Pectinase	Acrocylindrium sp.	Glucose, fructose, mannose, xylose, glycerol, malate, α-ketoglutarate, pyruvate, succinate, fumarate, acetate
Glucose isomerase	Streptomyces phaeochromogenes	Glucose
Protease	Bacillus megaterium	Glucose
Acid protease	Aspergillus niger	Glucose, fructose, xylose
Aspartase	Escherichia coli	Glucose
Penicillin acylase	Escherichia coli	Glucose
Tryptophanase	Escherichia coli	Glucose

phosphatase in bacilli is repressed by lactate, pyruvate and succinate and derepressed by glucose (Hydrean *et al.*, 1977).

Carbon source repression can be avoided by using a nonrepressive carbon source (if utilizable nonrepressive carbon sources are available and costs remain competitive), or by limiting the rate of growth by controlled feeding of the repressive carbon source. Production of cellulase by *Pseudomonas fluorescens* var. *cellulosa* is suppressed by glucose and galactose; growing this organism on mannose results in cells producing over 1500 times as much cellulase. Similar relief from carbon source represssion of cellulase production in this organism can be achieved by slow feeding of glucose, which increases cellulase production by almost 200-fold.

Addition of an analogue of the repressing carbon source has been known to stimulate production. Induction of invertase and α-glucosidase in *S. cerevisiae* and that of β-galactosidase, α-galactosidase and β-glucosidase in *Cryptococcus laurentii* is repressed by glucose. Addition of 3-*O*-methyl-D-glucose stimulated enzyme production many-fold (Bhanot & Brown, 1980).

Occasionally the peculiar situation is encountered in which rapid catabolism of an inducer causes carbon source repression. This has caused problems in the production of cellulase, invertase, dextranase and β-galactosidase. In such cases, enzyme production may be increased by using a gratuitous inducer, by feeding the substrate-inducer slowly, or by using a slowly-utilizable derivative of the inducer such as an ester. As an example, use of sucrose monopalmitate instead of sucrose led to an 80-fold increase in invertase production (Reese *et al.*, 1969).

Mutants can be selected that are resistant to carbon source repression. Many such mutants are apparently modified in their glucose catabolic pathway, which causes slower glucose utilization and derepression of a number of usually repressed enzymes. Mutation to resistance to carbon source repression of specific enzymes has also been observed. For example, screening for *B. subtilis* mutants able to hydrolyze starch in the presence of glucose yielded a glucose-resistant mutant (Nicholson & Chambliss, 1985). The mutation mapped very close to the α-amylase structural gene. The mutation was specific for α-amylase; it did not relieve glucose repression of sporulation, extracellular proteases or RNAase. Since enzyme production in the mutants was still delayed until the end of exponential growth, temporal regulation and carbon source repression are separate phenomena.

One way to select for derepressed mutants is by making this type of mutation obligatory for growth. This occurs, for example, when histidine is offered as the sole nitrogen source for growth of *K. aerogenes* on media containing glucose. Glucose usually represses production of histidine-degrading enzymes. However, growth cannot occur unless histidine-degrading enzymes are produced because they liberate nitrogen from histidine for cell growth. Hence only derepressed mutants will grow under these conditions (Neidhardt, 1960). Glucose–proline agar selects for such mutants in *S. typhimurium* (Newell & Brill, 1972), since proline oxidase is normally repressed by glucose. Only mutants resistant to carbon source repression are able to oxidize proline and to obtain nitrogen from the amino acid. Up to five times more proline oxidase and Δ'-pyrroline carboxylic acid dehydrogenase was formed by a mutant than by its parent even when glucose was absent.

Aspartase in *E. coli* is repressed by glucose. Glucose-resistant mutants were selected by chemostat growth with glucose as carbon source and aspartate as nitrogen source. Derepression was 25-fold in a medium containing glucose plus aspartate and 5-fold in a medium containing aspartate as sole carbon source. Also derepressed were β-galactosidase, tryptophanase and three TCA cycle enzymes (Nishimura & Kisumi, 1984).

A pleiotropic mutant of *B. subtilis* resistant to glucose repression of several enzymes (aconitase, α-glucosidase, histidase) was selected by growth in glucose plus histidine after mutation to glutamate auxotrophy (glutamate synthase deficient) (Fisher & Magasanik, 1984). Selection was possible since glutamate dehydrogenase does not supply significant glutamate in *B. subtilis* and the conversion of histidine to glutamate is repressed by glucose. Other mutants were obtained which were specific for histidase but the pleiotropic mutant was of greater interest. It appears to have a mutation in the glycolytic pathway. The intracellular concentration of PEP was increased whereas pyruvate, α-ketoglutarate and oxaloacetate were decreased.

By forcing *E. coli* to depend upon its penicillin G amidase for growth on a nonutilizable amide as sole nitrogen source and to do this under conditions of glucose repression, strains were isolated which were hyperproducers of the amidase, constitutive, and resistant to glucose repression (Daumy *et al.*, 1985). The mutant produced 670-fold more enzyme than the parent under noninducing, nonrepressing conditions, 13-fold more under inducing nonrepressing conditions and 460-fold more under inducing, repressing conditions.

In *P. aeruginosa*, amidase production is repressed by succinate. Growth on lactamide (a good inducer but a poor substrate) plus succinate does not normally occur since amidase is repressed and no nitrogen can be obtained for growth. Thus, repression-resistant mutants can be selected on this type of medium (Clarke & Lilly, 1969). Such mutants produced as much as 10% of their cell protein as amidase (Betz *et al.*, 1974). In *Aspergillus nidulans*, mutants obtained by increased growth on acetamide as sole nitrogen source in the presence of glucose, were found to be partially resistant to glucose repression of acetamidase and other enzymes (Kelly & Hynes, 1977).

Another method relies on the fact that synthesis of normally repressed enzymes takes a particular period of time after the organisms have been transferred from a repressing to a nonrepressing carbon source. Such a lag in initiation of growth occurs when *E. coli* is transferred from glucose to a medium containing lactose, maltose, acetate, or succinate (Hsie & Rickenberg, 1967). By alternating serial transfers between media containing glucose or lactose, β-galactosidase derepressed mutants were selected.

Mutants resistant to carbon source repression can also be isolated by selection in the presence of substrate plus 2-deoxyglucose (DOG), a glucose antimetabolite. This powerful technique has been successful in many microorganisms in the cases of invertase, α-glucosidase, malate dehydrogenase, glucoamylase, α-amylase, cellulase and β-glucosidase. A striking example of this technique was the isolation of a glucose-resistant mutant of *Candida wickerhamii* producing β-glucosidase (Leclerc *et al.*, 1985). The mutant was a hyperproducer, producing six- to ten-times more enzyme than the parent on all carbon sources tested. A mutation program which included screening for larger zones of cellulose digestion in the presence of glucose and lactose and selection for resistance to DOG led to a superior partially glucose-derepressed *T. reesei* producer of cellulase and cellobiase. The culture, CL847, is now being used at the factory level in France (Durand *et al.*, 1988).

D-Glucosamine has been used instead of DOG to select for derepressed mutants of yeast. The D-glucosamine can be phosphorylated but not metabolized further. Mutants isolated in the presence of maltose plus D-glucosamine were derepressed for maltase, galactokinase, α-galactosidase, NADH-cytochrome C reductase and cytochrome C oxidase (Michels & Romanowski, 1980).

Although not as efficient as selection methods, screening can be used to obtain mutants resistant to carbon source repression. A deregulated strain of *T. reesei* was obtained by mutagenesis, growth on glucose and testing colonies for clear zone formation on insoluble cellulose. The mutant produced various cellulolytic enzymes and β-glucosidase in glucose, unlike the parent strain (Mishra *et al.*, 1982). In a similar fashion, invertase mutants, lacking carbon source repression, were obtained by growing cultures on the repressing carbon source, fructose, and spraying colonies with sucrose plus the glucostat reagent. Repression-resistant colonies produced glucose which was detected by the reagent (Montenecourt *et al.*, 1973).

Mutation to carbon source derepression by the DOG technique failed in the case of *Kluyveromyces fragilis* inulinase. However, derepressed mutants were obtained by plating survivors on a complex agar medium containing insoluble inulin and glucose (Tsang & GrootWassink, 1985). Clear zones were produced by glucose-resistant colonies. The best mutant produced 170 units/ml in excess glucose (8%) and in limiting glucose. The parent produced 12 units/ml and 40 units/ml, respectively under the same conditions.

Mutagenesis and screening for mutants which clear crystalline cellulose on plates in the presence of glucose yielded a culture of *Cellulomonas* which produced an eight-fold increase in extracellular cellulase and a three-fold increase in cell-bound cellulase (Choi *et al.*, 1978).

Sequential repetition of the technique eventually yielded a hyperproducing 'regulation reversal' mutant with respect to cell-bound β-glucosidase, i.e. the parent was repressed by glucose whereas the hyperproducing mutant was stimulated (Haggett *et al.*, 1978). Thus the mutant produced in the presence of glucose ten-fold more enzyme than the parent produced in its absence.

3.4 Nitrogen Source Repression

Certain enzymes including catabolic and transport enzymes are repressed by NH_4^+ or rapidly-utilized amino acids; the enzymes usually are those acting on nitrogenous substrates. Such enzymes include proteases, urease, nitrate reductase, nitrite reductase, ribonuclease, amidases and amino acid-degrading enzymes. Somehow involved in this control are enzymes of nitrogen assimilation such as NADP-glutamate dehydrogenase, glutamine synthetase, glutamate synthase and alanine dehydrogenase but the mechanisms are still not completely clear. At one time, it was felt that one or more of these enzymes might act as a regulatory (repressor or activator) protein in addition to its catalytic role but this does not appear to be the case. More likely is the possibility that a high or low pool size of one or more substrates and/or products of these enzymes, e.g. glutamine, glutamate or alanine, brings about repression or derepression.

A regulatory gene (*glnG*) in *E. coli* and other enteric bacteria codes for NR_I, a dimer protein with a subunit M_r of 54 000. NR_I (nitrogen regulator I) is produced at a high level (90 molecules/cell) under nitrogen limitation and at a low level (5 molecules/cell) under nitrogen excess. It activates transcription of glutamine synthetase (the *glnA* gene product). NR_I binds to DNA at or near the promoter of *glnA* and is thought to regulate production of all nitrogen-regulated systems in enteric bacteria (Reitzer & Magasanik, 1983). In addition to NR_I, the protein products of two other genes are necessary to activate transcription of the glutamine synthetase regulatory genes: *ntrA* and *ntrC*. The product of *ntrA* in *Salmonella* appears to be a sigma factor different from σ-70 (Hirschman *et al.*, 1985).

In *P. aeruginosa*, glutamine appears to be the negative effector of nitrogen source repression (Janessen *et al.*, 1981). In a glutamine synthetase-negative mutant, urease and histidase were derepressed when growth was limited for glutamine; addition of ammonia or glutamate had no effect. Addition of glutamine caused repression of these two enzymes as well as NADP-glutamate dehydrogenase.

A gene controlling nitrogen source repression (*areA*) has been identified in *Aspergillus* (Marzluf, 1981). The gene codes for a regulatory protein exerting positive control on transcription. The regulatory protein appears to be active under conditions of derepression (e.g. low ammonium supply) and inactive under repressive conditions (e.g. high ammonium supply). The intracellular effector appears to be glutamine which is thought to bind to the regulatory protein and decrease its affinity to the nitrogen recognition sites adjacent to each nitrogen controlled structural gene. Upon NH_4^+ limitation, glutamine concentration would drop, the regulatory protein would assume an active conformation and bind at the recognition sites of the structural genes, turning on transcription.

Mutations in the *areA* gene are of the *areA'* type in which a large variety of nitrogen sources can no longer be utilized for growth and the enzymes catalyzing their usage cannot be derepressed, or of the $areA^d$ type in which the enzymes cannot be repressed by ammonia but all still require inducer. Some $areA^d$ type mutants are hyperproducers.

In *Neurospora crassa*, the gene analogous to *areA* is called *nit-2*. Growth on NH_4^+, glutamine or glutamate represses a large number of enzymes acting on nitrogenous substrates. Ammonium ions and glutamate presumably act via glutamine formation. These compounds repress the formation of the *nit-2* gene product which acts as a positive control agent for use of poor nitrogen sources. Glutamine does not appear to act directly to repress *nit-2*. Instead another gene, *nmr-1*, is thought to produce a protein which when combined with glutamine represses *nit-2*. When glutamine is low, the *nmr-1* gene product is an inactive repressor thus allowing the *nit-2* gene product to derepress synthesis of nitrogen repressible enzymes. Thus mutations of *nmr-1* allow production of these enzymes in the presence of glutamine, NH_4^+ or glutamate (Debusk & Ogilvie, 1984*a*). One of the nitrogen repressible enzymes in *N. crassa* is an extracellular L-amino acid deaminase. Its expression requires inducer (one of many amino acids), lifting of nitrogen source repression and the presence of the *nit-2* gene product (Debusk & Ogilvie, 1984*b*). Nitrogen (ammonium) repression of uricase in *N. crassa* is clearly mediated by glutamine (Wang & Marzluf, 1979). In a glutamine synthetase mutant, glutamine (but not NH_4^+ or glutamate) represses enzyme synthesis.

In *Chlamydomonas reinhardii*, the corepressor of nitrate reductase appears to be ammonium rather than glutamine (Franco *et al.*, 1984).

Since nitrogen repression is so widespread, it is obvious that limitation of NH_4^+ and certain amino acids will increase the production of many enzymes. Of particular commercial importance are the proteases.

Mutants resistant to nitrogen source repression can be selected by growth in the presence of the ammonium antimetabolite, methylammonium (Arst & Cove, 1969). Screening can be done by detection of clear zones around colonies growing on milk agar containing NH_4^+ (Cohen, 1972).

3.5 Other Forms of Nutrient Repression

3.5.1 Phosphorus Source Repression

Phosphate represses many nucleases and phosphatases in microorganisms. Limitation of phosphate in *E. coli* dramatically derepressed alkaline phosphatase production to a point where this enzyme represented 5% of the cell protein (Garen & Otsuji, 1964). Constitutive mutants can be selected by growth in the presence of phosphate on a substrate requiring action of the particular phosphate-repressible enzyme. For example, phosphatase constitutive mutants are selected when a mutagenized population is plated on β-glycerol phosphate as the sole carbon source in the presence of a high concentration of inorganic phosphate (Torriani & Rothman, 1961).

Derepressed mutants of *S. cerevisiae* have been isolated which produce alkaline and acid phosphatase in the presence of inorganic phosphate (Schurr

& Yagil, 1971). They form four to seven times more enzyme than their parent does in the absence of phosphate.

Phospholipase C is a heat-labile extracellular phosphate-regulated enzyme which is a hemolysin of *P. aeruginosa* and contributes to its virulence. A phosphate-regulatory mutant was obtained by screening of colonies for hydrolysis of *p*-nitrophenyl-phosphorylcholine while growing in the presence of 10 mM phosphate (Gray *et al.*, 1982). The mutant was a hyperproducer of phospholipase C. Production of alkaline phosphatase and other phosphate-repressible extracellular proteins was also derepressed. The mutant was found to be deficient in phosphate uptake.

In *E. coli*, over 40 proteins can be phosphorylated including RNA polymerase and two protein kinases have been isolated (Enami & Ishihama, 1984). Phosphate control could involve phosphorylation–dephosphorylation control and thus changes in enzyme activity rather than enzyme formation.

3.5.2 Sulfur Source Repression

Protease formation by certain organisms is sensitive to repression by sulfur compounds and optimum protease yields in these cases are obtained under conditions of sulfate limitation. Arylsulfatase is subject to sulfur repression by growth in sulfate or cysteine. When a *Pseudomonas* strain was grown on nonrepressive sulfur sources such as methionine or choline-*O*-sulfate, arylsulfatase production increased one thousand-fold (Fitzgerald & Payne, 1972). Heparinase production by *Flavobacterium heparinum* is repressed markedly by sulfate (Galliher *et al.*, 1982). Heparin is a carbohydrate polymer containing sulfur.

Enzyme biosynthesis is often controlled by more than one effector. For example, some proteases and RNases are derepressed by limitation of carbon, nitrogen or sulfur. In each case, the same enzyme is made and hence the same gene must have been activated (Lindberg & Drucker, 1984). In the case of *N. crassa*, both protease and RNAase formation also requires the presence of a protein as an 'inducer'.

3.6 Additional Control Mechanisms

The rate of growth has a major effect on production of microbial enzymes. Growth rate is known to control expression of operons by a mechanism(s) independent of induction and cAMP in the case of the *lac* operon (Wanner *et al.*, 1977) and independent of repression and attenuation in the case of the *trp* operon (Botsford & Shimizu, 1983).

The stage of growth (and/or growth rate) can have an effect on the type of regulation exerted. For example, ornithine transcarbamylase (an enzyme involved in biosynthesis of arginine and, in some organisms, in degradation of arginine) is repressed in *B. subtilis* by arginine during the exponential phase and induced by arginine during the stationary phase. The repressed and induced enzymes are the same (Neway & Switzer, 1983).

Some growth rate-related processes are controlled by increased RNA stability at low growth rates (Nilsson *et al.*, 1984).

Other types of regulation of possible importance in enzyme technology include 'stringent' [or guanosine tetraphosphate (ppGpp)] control and regulatory inactivation. Control by ppGpp involves the effect of amino acid deficiency on a large number of physiological activities in bacteria (Gallant, 1979). Deficiency of any amino acid leads to production of ppGpp from GTP. This intracellular effector redirects the cell's activities to correction of the amino acid deficiency by activating some enzymes and inhibiting or repressing others. For example, ppGpp causes repression of aspartate transcarbamylase, thus cutting off synthesis of pyrimidine nucleotides (Turnbough, 1983). ppGpp shuts off stable RNA synthesis generally, i.e. rRNA and tRNA, and mRNA synthesis specifically for ribosomal proteins. The rate of total mRNA synthesis drops only modestly however. The effect on RNA synthesis appears to be at the level of transcription with ppGpp interfering with binding of RNA polymerase to promoters of stringently-controlled operons (Lamond & Travers, 1985). The significance of ppGpp in eucaryotes is doubtful.

Regulatory inactivation refers to the selective inactivation of enzymes by two different mechanisms (Switzer, 1977). *In modification inactivation,* the enzyme remains intact but its physical state is changed or it is covalently modified. Covalent modifications include phosphorylation of a specific serine or threonine residue, nucleotidylation of a specific tyrosine residue, ADP-ribosylation of an arginine residue, methylation of a glutamate or aspartate carboxyl group, acetylation of an ε-amino group of a lysine residue or tyrosinolation of a protein terminal carboxyl group (Chock *et al.*, 1980). *In degradative inactivation,* at least one peptide bond is broken; it may represent the first step in protein turnover. Degradation is carried out by proteases which are restricted from acting nonselectively by confinement in vacuoles or by protease inhibitors. Regulatory inactivation usually occurs after the exponential phase of growth, especially after exhaustion of a source of carbon or nitrogen. This inactivation serves to avoid futile cycles of metabolism, to destroy enzymes no longer needed, or to divert branch point metabolism from one branch to another.

Certain controls ('global') affect a large number of operons. When this occurs, the operons controlled by a common regulatory protein, if they represent genes in more than one metabolic pathway, constitute a global regulon. Some of these regulons involve heat-shock, carbon source repression, the stringent response, aerobic/anaerobic shifts, phosphorus source repression and nitrogen source repression (Gottesman, 1984). An operon which is part of a global regulon can still have its own individual regulation. An operon can be a member of more than one global regulon.

The phosphate deficiency global regulon leads to induction of a high affinity scavenging system for phosphate (Gottesman, 1984). It is very complicated, involving some 18 genes in *E. coli*, all of which are transcriptionally activated by phosphate limitation. Some of these are *phoA* (alkaline phosphatase), *phoS* and *phoE* (phosphate transport systems) and *ugpA* and *ugpB* (glycerol phosphate transport systems). The intracellular signal is not known; it may be the level of intracellular phosphate or the rate of phosphate transport. At least three regulatory genes are involved: *phoB*, *phoR* and *phoM*; *phoR* and *phoM* appear to regulate *phoB* expression. The *phoB* product is the positive regulator of all the *pho* global genes.

In organisms capable of differentiating into spores, appearance of certain enzymes occurs as the cells leave the exponential phase. Since this usually is the time when the carbon source is exhausted and also the time when growth rate is falling, it is of interest whether enzyme synthesis is controlled by a single mechanism involving both growth rate (temporal control) and nutrient repression or by independent controls. At least in the case of α-amylase synthesis in *B. subtilis*, carbon source repression and temporal repression are two separate control mechanisms (Nicholson & Chambliss, 1986).

3.7 Additional Means to Increase Enzyme Production

Enzyme hyperproduction can be achieved by isolation of mutants resistant to the toxic effects of inhibitors of the enzyme. For example, mutants of *P. putida* resistant to pentafluoromandelate are hyperproducers of mandelate dehydrogenase, an enzyme catabolizing the conversion of mandelate to benzoate (Hegeman & Root, 1976).

In sporulating organisms, elimination of sporulation can have a beneficial effect on enzyme production. In *B. licheniformis,* an asporogenous mutant was found to produce higher levels of protease (Aunstrup, 1980). A similar type of mutant in *B. cereus*, obtained by selection for rifamycin resistance, produced seven-fold more β-amylase than its parent (Shinke *et al.*, 1981). Reversion to sporogeny was accompanied by a return in production to the parental level. Glucose-resistant *B. subtilis* mutants which can form spores in the presence of the sugar were found to be hyperproducers of protease (Dobrzhanskaya *et al.*, 1978).

A change in colony morphology can accompany changes in enzyme production. Morphological mutants of *T. reesei* which have increased hyphal branching produced increased amounts of β-glucosidase (2- to 3-fold), chitinase (17-fold) and laminarinase (2-fold) (Labudova *et al.*, 1981).

Hydrolysis of cellulose by colonies on agar resulting in clear zones was the screening procedure used in obtaining a mutant of *T. reesei* which was 2·5-fold improved in carboxymethylcellulase, 3-fold improved in hydrolysis of crystalline cellulose and 6-fold improved in β-glucosidase (Nevalainen *et al.*, 1980).

Application of flow cytometry to screening for increased enzyme production was attempted with *Rhizopus arrhizus* (Betz *et al.*, 1984). As a result, lipase production was increased 6-fold.

Genetic recombination is a powerful tool for obtaining increases in enzyme production. Individual mutations in α-amylase production were obtained in *B. subtilis*, each of which increased enzyme from 2- to 7-fold. By genetic transformation, all five mutations ($amyR^3$, $amyS$, $papS$, tmr and papM118) were combined in a single strain. As a result, the multiple transformant produced 250-fold more α-amylase than its parent.

4 ENZYME IMMOBILIZATION

Enzyme engineering has advanced greatly since its beginning in the early 1970s resulting in novel ways to immobilize enzymes (and cells) and novel types of enzyme reactors. Today virtually any new application of an enzyme involves immobilization since it offers the following potential benefits: increased stability, shifted pH optima in either direction, purer products, minimized effluent problems, reduced substrate inhibition, reduced product inhibition and facilitated recovery and reuse. However, the cost of the immobilization procedure and the loss of activity during immobilization are negative factors which must be considered.

Despite the potential of immobilized enzymes, most of the applications have been in the area of hydrolytic and isomerizing enzymes i.e. enzymes that do not require cofactors such as ATP or the pyridine nucleotides, NAD, NADH, NADP and NADPH. The reluctance to use cofactor-requiring enzymes derives from the cost of cofactor regeneration. However, considerable progress has been made on this problem as described below.

The glycolysis and kinase systems of yeast can be released from cells and used for continuous ATP regeneration (Asada *et al.*, 1978). Alternately, acetone-dried yeast cells can be immobilized and used to regenerate ATP from adenosine (Asada *et al.*, 1979). Other means include the regeneration of ATP from AMP by adding acetylphosphate in the presence of immobilized acetate kinase and adenylate kinase (Archer *et al.*, 1976) and the use of polyphosphate (Okamoto *et al.*, 1986) or pyrophosphate (Wood, 1977) rather than ATP for those enzymes that can use such forms of phosphate as ATP replacements.

5 RECOMBINANT DNA TECHNOLOGY

5.1 Gene Amplification

Over the years, it was occasionally found that certain regulatory mutants produced extremely high levels of enzyme, much higher than could be accounted for by mere blockage of repressor formation or action. In some of these cases, it was noted that the mutants contained multiple copies of the structural gene coding for the enzyme. The *E. coli* mutants of Horiuchi and co-workers (Horiuchi *et al.*, 1963) isolated in a lactose-limited chemostat and producing up to 25% of their cellular protein as β-galactosidase were not merely constitutives; they also contained up to four copies of the *lacZ* gene. Similarly *K. aerogenes* mutants which had been selected for growth on xylitol,

354 Arnold L. Demain

and which overproduced ribitol dehydrogenase were found to have amplified
their ribitol dehydrogenase gene (Hartley, 1974; Rigby *et al.*, 1974).

Since increasing the number of copies of a gene would normally be expected
to increase production of its specific enzyme, it was deemed desirable to be
able to do this intentionally. During the 1960s and early 1970s, it was
discovered that an increase in gene copies of *E. coli* and *Proteus mirabilis* by
genetic manipulation could be accomplished by transferring extrachromosomal
DNA segments (plasmids) or by the use of transducing phage. Production of
β-galactosidase, penicillinase, chloramphenicol transacetylase and aspartate
transcarbamylase was increased by transfer of plasmids containing the respective structural genes into recipient cultures especially when the plasmid or the
phage replicated faster than the host chromosome.

With the development of recombinant DNA technology (Cohen, 1975), it
became possible to transfer genes via plasmids or phage from one species to a
completely unrelated one. This gene cloning approach became important in
efforts to make enzymes available for large scale application testing. The
cloning and expression in microorganisms of genes from higher plants and
animals, such genes coding for unavailable or very expensive enzymes, is
playing an important role in solving this problem.

The tools which are used in recombinant DNA technology are DNA
sequencing, chemical DNA synthesis, DNA enzymology and cloning. Thus the
DNA to be introduced can be isolated from a chromosome, chemically
synthesized, or enzymatically synthesized from messenger RNA by use of the
enzyme reverse transcriptase. Recombinant DNA is DNA which is covalently
ligated in juxtaposition to heterologous DNA sequences with the purpose of
propagating it in an appropriate host. Molecular cloning is the isolation and
propagation of these molecules within a single colony. For good expression of
foreign genes, the engineered DNA sequence should have an efficiently
regulated promoter, a strong ribosome binding site and a high but controllable
plasmid copy number.

The power of cloning can be appreciated by the fact that although wild-type
E. coli, when induced, can produce 1000–2000 times the amount of β-
galactosidase it produces in glucose, amounting to 2% of total protein, it can
do much better when the *lacZ* gene is cloned on a 'high copy' plasmid, i.e.
greater than 50% of the protein synthetic machinery is shifted into manufacture of β-galactosidase (Gelfand *et al.*, 1978).

The production of many enzymes has been increased by recombinant DNA
technology. Some of these are shown in Table 6. Production of mammalian
enzymes in microorganisms is another active area. One example is the cloning
of the calf rennet gene (Beppu, 1983).

After construction of recombinant strains, conventional genetic methods are
useful for strain improvement. A recombinant *Aspergillus awamori* strain
containing the calf rennin gene was improved five-fold by conventional
mutagenesis and screening for increased proteolytic zones on agar plates
(Lamsa & Bloebaum, 1988). Classical parasexual techniques successfully

improved glucoamylase production in an *Aspergillus niger* strain possessing an integrated glycoamylase gene constructed by recombinant DNA technology (Crawford *et al.*, 1988).

5.2 Transcription

Of great importance in expression of genes are the promoters which are used. These are DNA sequences which direct binding of RNA polymerase and initiation of transcription. Natural promoters may be strong or weak. The strong promoters bind RNA polymerase very effectively which starts the transcription process. A weak promoter binds poorly and RNA polymerase may fall off the promoter before initiating transcription. Strong promoters are evidently needed for producing the major enzymes of the cell while weak promoters are for enzymes needed in only small amounts.

Strong promoters have certain related sequences in the -10 and -35 nucleotide regions, counting upstream from the position of the start of transcription. These sequences signal the beginning of the transcription process. The consensus sequences are:

$$(-35) \quad \text{TTGACA}$$
$$(-10) \quad \text{TATAAT}$$

The importance of the -35 and -10 regions of the promoter and the space between them was shown using the penicillinase gene *ampC* of *E. coli*. The sequences of the wild-type *ampC* promoter are TTGTCA (-35) and TACAAT (-10) separated by 16bp. They differ by one base in each region from the consensus sequences and from the 17bp consensus spacing. Mutations changing these differences to the correct consensus sequences and spacing (up-promoter mutations) showed 7- to 21-fold increases in *ampC*-specific mRNA and a 16-fold increase in enzyme production (Jaurin & Cohen, 1984).

Certain promoters are extremely useful in increasing expression of heterologous genes. Inducible promoters are important for cloning when expression of the foreign gene interferes with growth. The culture can be grown in the repressed state and then induced. The P_L promoter of phage lambda is often used because it has a temperature-sensitive repressor. Thus, mRNA and protein are made only when temperature is raised. The *lac* UV5 promoter is turned on by addition of a β-galactoside and the *trp* promoter by the exhaustion of tryptophan from the medium. Use of the *trp* promoter system has led to the production of human, mammalian and viral products at 10–15% of total *E. coli* protein (Johnson & Somerville, 1985).

The *trp* promoter is 2–3 times stronger than the *lac* UV5 promoter (Windass *et al.*, 1982). The *tac* promoter combines the -10 region of the *lac* UV5 promoter and the -35 region of the *trp* promoter. It is 5–10 times more efficient than the *lac* UV5 promoter. Using it under optimal conditions, lambda repressor was made at 30% of *E. coli* protein (Amann *et al.*, 1983). The *E. coli* lipoprotein promoter is 30 times as active as *lac* UV5 (Nakamura &

Table 6
Increasing Enzyme Production by Recombinant DNA Technology

Protein	Gene donor	Gene recipient	Type vector	Increase (fold)	% of total protein	Reference
Aspartate-β-semialdehyde dehydrogenase	*Escherichia coli*	*E. coli*	Plasmid	65	18–25	Preiss *et al.*, 1982
Benzylpenicillin acylase	*E. coli*	*E. coli*	Plasmid	8	—	Mayer *et al.*, 1980
α-Amylase	*Bacillus amylolique-faciens*	*Bacillus subtilis*	Plasmid	10	—	Sibakov *et al.*, 1983
α-Amylase	*Bacillus stearo-thermophilus*	*B. stearo-thermophilus*	Plasmid	5	—	Aiba *et al.*, 1983
α-Amylase	*B. stearo-thermophilus*	*Bacillus brevis*	Plasmid	100	—	Tsukagoshi, *et al.*, 1985
Glucoamylase	*Saccharomyces diastaticus*	*Saccharomyces cerevisiae*	Plasmid	5–10	—	Yamashuta & Fukui, 1983
Cellulase (exoglucanase)	*Cellulomonas fimi*	*E. coli*	Plasmid	—	>20	O'Neill *et al.*, 1986
Cellulase (endoglucanase)	*Clostridium thermocellum*	*E. coli*	Plasmid	—	10–15	Schwarz *et al.*, 1987
Cellulase (endoglucanase)	*C. thermo-cellum*	*B. subtilis*	Plasmid	3–5	50 (of extra-cellular protein)	Soutschek-Bauer & Stauden-bauer, 1987
Neutral protease	*B. stearo-thermophilus*	*B. subtilis*	Plasmid	15	—	Fujii *et al.*, 1983
Glutamate dehydrogenase	*Salmonella typhimurium*	*E. coli*	Plasmid	>100	—	Miller & Brenchley, 1984
δ-Aminolevulinic acid synthase	*S. cerevisiae*	*S. cerevisiae*	Cosmid	16	—	Urban-Grimal *et al.*, 1984
EcoRI restriction endonuclease	*E. coli*	*E. coli*	Plasmid	50–100	—	Cheng *et al.*, 1984
Modification methylase	*E. coli*	*E. coli*	Plasmid	50–100	—	Cheng *et al.*, 1984
Aspartase	*E. coli*	*E. coli*	Plasmid	30	—	Komatsu-bara *et al.*, 1986
Phosphatidyl-serine synthase	*E. coli*	*E. coli*	Plasmid	15	—	Raetz *et al.*, 1977
Tryptophan synthase	*E. coli*	*E. coli*	Plasmid	16	—	Nagahari *et al.*, 1977
γ-Glutamyl phosphate reductase	*E. coli*	*E. coli*	Plasmid	17	—	Hayzer & Leisinger, 1980
β-Isopropyl malate dehydro-genase	*E. coli*	*E. coli*	Plasmid	20	10	Davis & Calvo, 1977
DNA poly-merase I	*E. coli*	*E. coli*	Phage	100	—	Kelley *et al.*, 1987

Table 6—*contd.*

Protein	Gene donor	Gene recipient	Type vector	Increase (fold)	% of total protein	Reference
DNA ligase	*E. coli*	*E. coli*	Phage	500	5	Panasenico *et al.*, 1977
δ'-Pyrroline-5 carboxylate -reductase	*E. coli*	*E. coli*	Plasmid	190	5 (of soluble protein)	Deutch *et al.*, 1982
Glucose (xylose) isomerase	*E. coli*	*E. coli*	Plasmid	20	—	Stevis & Ho, 1985

Inoyue, 1982). 'Runaway' replication vectors are available which replicate over 2000 plasmid copies upon increase in temperature (Uhlin *et al.*, 1979).

Often the choice of the right vector is important in consideration of regulatory controls. Cloning of the *E. coli* D-xylose isomerase (=glucose isomerase) gene fused with the *tac* or *lac* promoter on high copy number plasmids resulted in 20-fold overproduction of the enzyme in *E. coli*. Similar manipulations with the natural promoter did not yield overproduction due to tight control of the natural promoter (Stevis & Ho, 1985). Cloning of the *Bacillus pumilus* xylose-induced xylanase and β-xylosidase into *E. coli* and *B. subtilis* via a plasmid vector resulted in constitutive synthesis (Panbangred *et al.*, 1985).

Regulatory genes have been cloned. Such a gene of *Bacillus natto* responsible for overproduction of extracellular alkaline protease, neutral protease and levansucrase has been cloned and sequenced (Nagami & Tanaka, 1986). It codes for a protein of 60 amino acid residues and does not affect production of other extracellular enzymes such as α-amylase, RNase and alkaline phosphatase.

Transcription is terminated by special DNA structures, called terminators, located next to the final nucleotide transcribed into mRNA. There is no consensus structure but these transcription terminating structures contain complementary sequences which form hairpin loops.

In cloned systems, when degradation of mRNA is a problem, nuclease-deficient mutants are useful (Hautala *et al.*, 1979).

5.3 Translation

Between the operator and structural gene of an operon is the ribosomal binding site (RBS). This DNA sequence is transcribed so that the mRNA can bind to the ribosome and thus initiate translation. The sequence of bases determines the strength of the RBS and the efficiency of translation, since a weak RBS will allow the mRNA to fall off the ribosome before translation begins. In *E. coli*, most RBSs contain the minimal sequence ... AGGA ... A consensus RBS devised from many RBSs is ... TAAGGAGGTG ..., the

so-called Shine-Delgarno sequence. These sequences have base-pairing potential with the 3'-end of 16S ribosomal RNA. In addition to a RBS, translation
initiation requires an initiation codon at the start of the open reading frame.
All eukaryotic and most prokaryotic genes use the translation initiation codon
AUG, which codes for methionine. This N-terminal methionine is sometimes
cleaved off enzymatically and if so, is not present in the final protein. Nine
known genes use GUG and three use UUG. The adenylate cyclase gene is one
of those using UUG. Using oligonucleotide-directed mutagenesis, UUG was
changed to GUG and AUG. GUG doubled gene expression and AUG
increased expression three- to six-fold (Reddy *et al.*, 1985). Stop signals for
translation are UAA, UAG and UGA.

5.4 Secretion

Proteins in bacteria are made in the cytoplasm but many are 'secreted' (or
'exported') to extracytoplasmic locations in the envelope (periplasm and outer
membrane) or 'excreted' into the extracellular medium. These proteins usually
are produced as larger precursors containing, at the amino terminal, a 'signal'
sequence of 20–40 amino acids which is required for secretion or excretion
(Oliver, 1985). The signal sequence is removed by 'signal peptidases'. Since
signal sequences from eukaryotes work in bacteria (and vice versa), the
sequences have been conserved both structurally and functionally. The amino
terminus of signal peptides contains one to three positively charged amino
acids such as arginine or lysine. Next to this is a stretch of 14–20 neutral,
primarily hydrophobic, amino acids (the 'hydrophobic core') which tend to
form an α-helix. Certain amino acids are usually found adjacent to the signal
peptidase cleavage site with a conserved sequence of A–X–B. B is immediately next to the site of processing and is alanine, glycine or serine. A contains
alanine, glycine, serine, leucine, valine or isoleucine. The ultimate location of
the protein is not determined by the signal sequence itself but probably by the
sequence of the mature protein.

   E. coli has two signal peptidases located in the plasma membrane: the leader
peptidase (LP) and lipoprotein signal peptidase (LSP). LP has a wider activity
spectrum. LSP acts only on precursors having a glyceride–fatty acid–modified
cysteine residue right after the cleavage site. The cleaved signal peptide is
degraded by membrane peptidase IV in *E. coli*. Plasma membrane proteins in
E. coli are generally synthesized without signal sequences (Pugsley &
Schwartz, 1985).

   Excretion into the extracellular medium is desirable for economical isolation
of proteins but *E. coli* is a poor excretor of proteins. Even with signal
sequences, proteins in *E. coli* usually get no further than the periplasm. On the
other hand, members of the genus *Bacillus* often excrete proteins to the
outside of the cell. Cloning of the *Bacillus coagulans* extracellular α-amylase
gene with its own signal sequence in *E. coli* resulted in its secretion into the
periplasm of the Gram-negative host (Cornelis *et al.*, 1982). In contrast, *E. coli*

β-lactamase and human interferon were secreted into the medium when cloned in *B. subtilis* using the *B. subtilis* α-amylase signal sequence (Palva *et al.*, 1983).

The leader sequence of the extracellular yeast sexual pheromone, α-Factor, is being used for excretion of foreign proteins in yeast. α-Factor is a 13 amino acid peptide, excreted by yeast of the α-mating type that promotes conjugation with the α-mating type. It is synthesized as a precursor containing 165 residues which contains an 83-residue leader and four α-Factor coding regions, each of which is separated by a short spacer peptide. The leader and spacers contain signals for proteolytic processing and excretion. Attachment of the α-Factor leader region to the yeast invertase gene rendered extracellular this normally cell wall-bound enzyme (Emr *et al.*, 1983). *Pichia pastoris*, a methanol-utilizing yeast, has been genetically engineered to excrete *S. cerevisiae* invertase into the extracellular medium at a concentration of 100 mg/liter (van Brunt, 1986). The system employs the invertase gene signal sequence and the alcohol oxidase promoter which is turned on by growth in methanol.

Streptomyces lividans excretes β-galactosidase, an enzyme which is intracellular in most microorganisms. The *S. lividans* enzyme has an unusually long signal sequence (56 residues) which is very high in arginine residues. When this signal sequence is attached to the *E. coli lacZ* gene, the *E. coli* β-galactosidase is excreted into the medium by *S. lividans*. Such excretion has never been obtained with non-streptomycetes (Eckhardt *et al.*, 1986).

A completely different approach is the use of leaky mutants of *E. coli* which release periplasmic enzymes into the culture fluid. Such mutants have been used as cloning hosts to yield high extracellular levels of the normally periplasmic alkaline phosphatase (Lazzaroni & Protalier, 1982).

5.5 The Protease Problem

Although members of the genus *Bacillus* are looked on favorably for production of extracellular proteins, a major problem is the production of extracellular proteases by bacilli which leads to degradation of the valuable product of cloning. Thus, attempts have been made to eliminate these proteases. For example, a *B. subtilis* mutant completely deficient in its extracellular protease [alkaline serine protease (=subtilisin)] and neutral metal protease has been used to increase recombinant staphyloccal protein A production from one up to three g/liter (Fahnestock & Fisher, 1987). Using another *B. subtilis* mutant, a foreign β-lactamase was excreted at high levels and was not degraded by proteases (Wong *et al.*, 1986). This system utilized a plasmid containing the β-lactamase gene, the signal peptidase cleavage site and glucose-resistant promoter of the subtilisin gene, catabolite repression of sporulation, a double protease-deficient mutant (extracellular neutral and alkaline proteases) which produced only low levels of glucose-repressible extracellular esterase, and a rich medium containing 3% glucose which supported a high cell density and production of high levels of the recombinant gene product starting in the log phase and remaining stable for at least 100 h. However, even with this system, heterologous mammalian gene products are still destroyed by an unknown mechanism after their production (Doi, personal communication).

5.6 Plasmid Copy Number and Stability

Plasmid copy number is partly controlled by the particular DNA replication origin on the plasmid. A very active DNA replication origin allows the plasmid to be replicated very often; a weak one allows only one or a few replications during any one cell division cycle. A sequence yielding temperature-sensitive control of plasmid copy number is very useful, often increasing the normal copy number ten to fifteen times when temperature is elevated. Such manipulations can result in production of one million molecules of a foreign protein per *E. coli* cell.

An important activity in the exploitation of cloning is the stability of engineered plasmids. Plasmid instability is of two types: (1) segregational, in which the plasmid can be lost by unequal distribution among daughter cells; and (2) structural, in which part of the plasmid is lost or inactivated by deletion or rearrangement respectively. In both segregational and structural instability, the major problem is the ability of the resulting strain (either lacking the plasmid or possessing the deleted or rearranged plasmid) to outgrow the parental strain. Solutions to segregational plasmid instability include (a) the addition of antibiotics to the medium to maintain plasmids containing antibiotic-resistance genes, (b) elimination from a chemically defined medium of growth factors to which the organism is auxotrophic with regard to chromosomal genes but prototrophic with respect to plasmid genes, and (c) the insertion of the *par* gene into the plasmid, which effects efficient segregation and partitioning of replicated plasmid molecules to daughter cells upon division (Meacock & Cohen, 1980). The need for expensive chemically-defined media to stabilize plasmids in an auxotrophic strain has been bypassed by making a double mutant which cannot grow even if the growth factor is present in the medium (Loison *et al.*, 1986). The original host was a uracil auxotroph blocked in URA3, the gene encoding OMP decarboxylase, an enzyme involved in *de novo* uridine monophosphate (UMP) synthesis. This block is bypassed normally by taking up uracil from the medium and converting it to UMP by uracil phosphoribosyl-transferase, the product of the FUR1 gene. By converting the *ura3* host to a double-mutant *ura3, fur1,* it cannot grow unless a plasmid is present containing at least one of these genes. Use of a plasmid containing URA3 and the gene for human 1-antitrypsin allowed stable production of the foreign protein in complex industrial media.

Another means of eliminating the need for a chemically-defined medium involves placing the alanine racemase gene on the plasmid and using a host lacking its chromosomal alanine racemase gene. Loss of the plasmid results in death of the cell since it cannot form D-alanine and thus cannot produce cell walls. Since D-alanine does not occur in complex media, industrial-type media can be used (Ferrari *et al.*, 1985).

Structural plasmid instability is more difficult to correct than segregational instability. However, recent results indicate that both media modification and eliminating certain parts of the plasmid vector can stabilize cultures containing structurally-unstable plasmids in *B. subtilis* (Shoham, 1987).

6 MODIFICATION OF THE PROPERTIES OF ENZYMES

During years of using mutation to increase the production of enzymes, it had been noted that the properties of the enzymes were occasionally changed. It thus became obvious that by mutagenesis plus selection or screening, one could improve the properties of an enzyme. Mutation has yielded enzymes desensitized to feedback inhibition (higher K_i), with altered pH optima, substrate specificities, V_{max} or K_m values (Mortlock, 1982; Langridge, 1969). A few examples are as follows.

A mutation in the *lac* permease gene was detected by the requirement of an *E. coli* mutant for high (100 mM instead of 5 mM) lactose concentrations for growth. This mutation resulted in a change of substrate specificity (Markgraf *et al.*, 1985) allowing the mutant to grow on maltose in the absence of a maltose transport system. Accompanying this increased transport of maltose was a diminished transport of lactose. Codon 266 was found to have changed from ACA (threonine) to ATA (isoleucine). A mutant of *T. reesei*, whose β-glucosidase is less subject to glucose inhibition, was obtained by plating mutagenized populations on cellulose agar and screening for increased glucose production around the colony by a chromogenic method (Cuskey *et al.*, 1980).

Modification of allosteric inhibition has been obtained by mutation and selection for resistance to end product antimetabolites, or by a double mutation technique, the first to auxotrophy and the second (a suppressor mutation) back to prototrophy. Such double mutations have resulted in enzymes modified in heat stability, heat activation, substrate affinity, activity and allosteric inhibition (Fincham, 1973). Second-site suppression is known in the case of the α-subunit of tryptophan synthetase. Activity was lost when glycine 211 was replaced by glutamate and restored when tyrosine 175 was replaced by cysteine (Helinski & Yanofsky, 1963).

A mutant of *Diplococcus pneumoniae* resistant to the folate analogue, amethopterin, and exhibiting a 100-fold higher activity of dihydrofolate reductase, was found to produce a modified enzyme resistant to inhibition (Shaw, 1987).

Occasionally, the genetic change that results is a 'regulation-reversal' mutation. For example, a phenylalanine-inhibited prephenate dehydratase was changed by mutation into a phenylalanine-activated enzyme (Coats & Nester, 1967).

Cloning of a selectable or screenable mesophilic gene into the thermophile, *B. stearothermophilus,* is a useful technique for isolating thermostable mutant enzymes (Liao *et al.,* 1986). With this technique, the kanamycin resistance (KNTase) gene from *Staphylococcus aureus* whose product has a 17 min half-life at 50°C was converted by a spontaneous mutation into a protein with a half-life of 17–21 min at 60°C via replacement of aspartic 80 by tyrosine. Further selection at a higher temperature resulted in replacement of threonine 130 by lysine and an enzyme with a half-life of 12–15 min at 65°C. In an independent study using mutagenesis, almost identical results were obtained (Matsumura & Aiba, 1985).

After the development of recombinant DNA technology, its techniques were used to change the structure and properties of enzymes in a more rational way. Thus in-vitro mutagenesis (= site-directed mutagenesis) has developed into the important and broad field of 'protein engineering'. Small insertions, deletions and substitutions at single specific sites can be made in cloned genes. A small oligonucleotide primer containing the desired nucleotide modification is hybridized to the appropriate site and then DNA polymerase is used to replicate the (unmodified) rest of the gene. Much basic information on protein structure and function has been learned from protein engineering studies (Raibaud & Schwartz, 1984). In the following paragraphs, examples are given to demonstrate the broad applicability of this technology to the improvement of enzymes.

By site-directed mutagenesis, the substrate specificity of β-lactamase was changed so that it had enhanced activity towards cephalosporins (Hall & Knowles, 1976). In another study, alanine was substituted for glycine at positions 216 and 226 in the binding cavity of trypsin, resulting in three trypsin mutants. Trypsin usually attacks proteins at a lysine or arginine residue; the mutant trypsins had modified substrate specificities with respect to arginine and lysine peptides (Craik *et al.,* 1985). Site-directed mutagenesis has been used to modify substrate specificity of subtilisin, tyrosyl-tRNA synthetase, carboxypeptidase, alcohol dehydrogenase and aspartate aminotransferase (Carter & Wells, 1987).

Many studies have been done on increasing thermostability of enzymes. Insertion of a disulfide bridge in RT4 lysozyme increased its half-life at 67°C and pH 8 from 11 min to over 6 h (Perry & Wetzel, 1986). Since deamidation of asparagine residues is one mechanism of irreversible enzyme inactivation, protein engineering of yeast triosephosphate isomerase was used to convert asparagine 14 to threonine and asparagine 78 to isoleucine, doubling the half-life of the enzyme at 100°C and pH 6 (Ahern *et al.,* 1987).

Oxidative inactivation is a major problem with some enzymes. Instability is usually due to the presence of methionine or cysteine residues in or around the active sites, the sulfur groups being subject to oxidation. Subtilisin, a serine protease of *B. amyloliquefaciens* is inactivated by H_2O_2 via conversion of methionine 222 to methionine sulfoxide. Methionine 222 is adjacent to serine 221 which is part of the catalytic site. Site-directed mutagenesis changing the methionine to cysteine improved stability to H_2O_2 but only partially. Substitution with serine, alanine or leucine rendered the enzyme resistant to inactivation. The cysteine enzyme had a relative specific activity of 138%; alanine enzyme, 53%; serine enzyme, 35%; and leucine enzyme, 12% (Estell *et al.*, 1985). Protein engineering of *B. amyloliquefaciens* subtilisin has also shown that replacement of individual amino acids can lead to marked changes in substrate specificity, activity (K_{cat}) and pH profile (Estell, 1986).

Substitution of leucine for methionine 358 in α-antitrypsin prevented oxidative inactivation yet activity was retained at a level of 82%. Substitution with phenylalanine at 358 changed the substrate specificity of the enzyme eliminating activity on elastase but retaining cathepsin G as a substrate. The arginine 358 protein failed to act on elastase but was broadened in its substrate specificity to include thrombin, kallikrein and factor XII (McCormick, 1986).

Lowering of K_m was accomplished with site-directed mutagenesis of tyrosinyl-tRNA synthetase by converting an active-site cysteine to a serine. As predicted, the K_m for ATP was decreased (Winter *et al.*, 1982). An increase in enzyme activity (K_{cat}/K_m) of 25-fold was produced with tyrosinyl-tRNA synthetase by changing an active site threonine to a proline (Wilkinson *et al.*, 1984).

REFERENCES

Agathos, S. N. & Demain, A. L. (1987). In *Horizons of Biochemical Engineering*, ed. S. Aiba. University of Tokyo Press, Tokyo, p. 147.

Ahern, T. J., Casal, J. I., Petsko, G. & Klibanov, A. M. (1987). *Proceedings of the National Academy of Science USA*, **84**, 675.

Aiba, S., Kitai, K. & Imanaka, T. (1983). *Applied and Environmental Microbiology*, **46**, 1059.

Amann, E., Brosius, J. & Ptashne, M. (1983). *Gene*, **25**, 167.

Anderson, R. P., Miller, C. G. & Rogher, J. R. (1976). *Journal of Molecular Biology*, **105**, 201.

Archer, M. C., Colton, C. K., Cooney, C. L., Demain, A. L., Wang, D. I. C., & Whitesides, G. M. (1976). In *Enzyme Technology*, ed. E. K. Pye. The University of Pennsylvania, Pennsylvania, p. 1.

Arst, H. N., Jr. (1981). *Symposium of the Society of General Microbiology*, **31**, 131.

Arst, H. N., Jr. & Cove, D. J. (1969). *Journal of Bacteriology*, **98**, 1284.

Asada, M., Nakanishi, K., Matsuno, R., Kariya, Y., Kimura, A. & Kamikubo, T. (1978). *Agricultural and Biological Chemistry*, **42**, 1533.

Asada, M., Morimoto, K., Nakanishi, K., Matsuno, R., Tanaka, A., Kimura, A. & Kamikubo, T. (1979). *Agricultural and Biological Chemistry*, **43**, 1773.

Aunstrup, K. (1979). *Applied Biochemistry and Bioengineering*, **2**, 27.

Aunstrup, K. (1980). In *Economic Microbiology 5. Microbial Enzymes and Bioconversions*, ed. A. H. Rose. Academic Press, New York, p. 49.

Aunstrup, K. (1984). *Proceedings of the Third European Congress on Biotechnology*, **4**, 143.

Balbinder, E., Callahan, R., McCann, P. P., Cordaro, J. C., Weber, A. R., Smith, A. M. & Angelosanto, F. (1970). *Genetics*, **66**, 31.

Bechet, J., Grenson, M. & Wiame, J. M. (1970). *European Journal of Biochemistry*, **12**, 31.

Beppu, T. (1983). *Trends in Biotechnology*, **1**, 85.

Betz, J. L., Brown, P. R., Smyth, M. J. & Clarke, P. H. (1974). *Nature*, **247**, 261.

Betz, J. W., Aretz, W. & Härtel, W. (1984). *Cytometry*, **5**, 145.

Bhanot, P. & Brown, R. G. (1980). *Canadian Journal of Microbiology*, **26**, 1289.

Borum, P. R. & Monty, K. J. (1976). *Journal of Bacteriology*, **125**, 94.

Botsford, J. L. (1981). *Microbiology Reviews*, **45**, 620.

Botsford, J. L. & Shimizu, R. W. (1983). *FEMS Microbiology Letters*, **17**, 19.

Carter, P. & Wells, J. A. (1987). *Science*, **237**, 394.

Cheng, S.-C., Kim, R., King, K., Kim, S.-H. & Modrich, P. (1984). *Journal of Biological Chemistry*, **259**, 11571.

Chock, P. B., Rhee, S. G. & Stadtman, E. R. (1980). *Annual Review of Biochemistry*, **49**, 813.

Choi, W. Y., Haggett, K. D. & Dunn, N. W. (1978). *Australian Journal of Biological Science*, **31**, 553.

Clarke, P. H. & Lilly, M. D. (1969). *Symposium of the Society of General Microbiology*, **19**, 113.

Coats, J. H. & Nester, E. W. (1967). *Journal of Biological Chemistry*, **242**, 4948.

Cohen, B. L. (1972). *Journal of General Microbiology*, **71**, 293.

Cohen, S. N. (1975). *Scientific American*, **233**, 24.

Cornelis, P., Digneffe, C. & Willemot, K. (1982). *Molecular and General Genetics*, **186**, 507.

Craik, C. S., Largman, C., Fletcher, T., Roczniak, S., Barr, P. J., Fletterick, R. & Rutter, W. J. (1985). *Science*, **228**, 291.

Crawford, M. S., Soliday, C. L., Cronk, J. & Finkelstein, D. B. (1988). Abstract S-76, Annual Meeting of the Society for Industrial Microbiology, Chicago.

Cuskey, S. M., Schamhart, D. H. J., Chase, Jr, T., Montenecourt, B. S. & Eveleigh, D. E. (1980). *Developments in Industrial Microbiology*, **21**, 471.

Danchin, A. (1985). *Therapeutic Agents Produced by Genetic Engineering. 'Quo Vadis?' Symposium*. Sanofi Group, Toulouse-Labége, France, p. 37.

Daumy, G. O., Danley, D., McColl, A. S., Apostolakos, D. & Vinick, F. J. (1985). *Journal of Bacteriology*, **163**, 925.

Davis, M. G. & Calvo, J. M. (1977). *Journal of Bacteriology*, **131**, 997.

Debusk, R. M. & Ogilvie, S. (1984a). *Journal of Bacteriology*, **160**, 656.

Debusk, R. M. & Ogilvie, S. (1984b). *Journal of Bacteriology*, **160**, 493.

Deutch, A. H., Smith, C. J., Rushlow, K. E. & Kretschmer, P. J. (1982). *Nucleic Acids Research*, **10**, 7701.

Dobrzhanskaya, E. O., Erokhina, L. I. & Bol'Shakova, T. N. (1978). *Genetika*, **14**, 1175.

Durand, H., Clanet, M. & Tiraby, G. (1988). *Enzyme and Microbial Technology*, **10**, 341.

Eckhardt, T., Rosenberg, M. & Fare, L. (1986). *Abstracts of the First Annual ASM Conference on Biotechnology* **218**, 26.

Emr, S. D., Schekman, R., Flessel, M. C. & Thorner, J. (1983). *Proceedings of the National Academy of Science USA*, **80**, 7080.

Enami, M. & Ishihama, A. (1984). *Journal of Biological Chemistry*, **259**, 526.

Entian, K. D. & Fröhlich, K. U. (1984). *Journal of Bacteriology*, **158**, 29.

Eraso, P. & Gancedo, J. M. (1984). *European Journal of Biochemistry*, **141**, 195.

Esteban, R., Nebreda, A. R., Villanueva, J. R. & Villa, T. G. (1984). *FEMS Microbiology Letters*, **23**, 91.

Estell, D. A., Graycar, T. P. & Wells, J. A. (1985). *Journal of Biological Chemistry*, **260**, 6518.

Estell, D. A. (1986). *Chimicaoggi*, **6** (3), 29.

Eveleigh, D. E. & Montenecourt, B. S. (1979). In: *Advances in Applied Microbiology*, Vol. 25, ed. D. Perlman. Academic Press, New York, p. 58.

Fahnestock, S. R. & Fisher, K. (1987). *Applied and Environmental Microbiology*, **53,** 379.

Fang, M. & Burow, R. A. (1970). *Biochemical and Biophysical Research Communications*, **41,** 1579.

Fennington, G., Neubauer, D. & Stutzenberger, F. (1983). *Biotechnology and Bioengineering*, **25,** 2271.

Fernández, R., Herrero, P., Gascon, S. & Moreno, F. (1984). *Archives of Microbiology*, **139,** 139.

Ferrari, E., Henner, D. J. & Yang, M. Y. (1985). *Bio/Technology*, **3,** 1003.

Fincham, J. R. S. (1973). In *Genetics of Industrial Microorganisms: Actinomyces and Fungi*, eds Z. Vanek, Z. Hostalek & J. Cudlin. Academia, Prague, p. 97.

Fisher, S. H. & Magasanik, B. (1984). *Journal of Bacteriology*, **157,** 942.

Fitzgerald, J. W. & Payne, W. J. (1972). *Microbios*, **6,** 147.

Fraenkel, D. G. (1982). In *Molecular Biology of the Yeast* Saccharomyces, eds J. N. Strathern, E. W. Jones & J. R. Broach. Cold Spring Harbor Laboratory, New York, p. 1.

Franco, A. R., Cárdenas, J. & Fernández, E. (1984). *FEBS Letters*, **176,** 453.

Frost, G. M. & Moss, A. D. (1987). In *Biotechnology*, Vol. 7A, eds H. J. Rehm & G. Reed. VCH Verlagsgesellschaft, Weinheim, Federal Republic of Germany, p. 65.

Fujii, M., Takagi, M., Imanaka, T. & Aiba, S. (1983). *Journal of Bacteriology*, **154,** 831.

Gallant, J. A. (1979). *Annual Review of Genetics*, **13,** 393.

Galliher, P. M., Linhardt, R. J., Conway, L. J., Langer, R. & Cooney, C. L. (1982). *European Journal of Applied Microbiology and Biotechnology*, **15,** 252.

Garen, A. & Otsuji, N. (1964). *Journal of Molecular Biology*, **8,** 841.

Gelfand, D. H., Shepard, H. M., O'Farrell, P. H. & Polisky, B. (1978). *Proceedings of the National Academy of Science USA*, **75,** 5869.

Godtfredsen, S. E., Ingvorsen, K., Elgtved, P. & Hansen, T. T. (1987). In *Proceedings of the 4th European Congress on Biotechnology 1987*, Vol. 4, eds. O. M. Neijssel, R. R. van der Meer & K. C. A. M. Luyben. Elsevier, Amsterdam, p. 13.

Gottesman, S. (1984). *Annual Review of Genetics*, **18,** 415.

Gray, G. G., Berka, R. M. & Vasil, M. L. (1982). *Journal of Bacteriology*, **150,** 1221.

Greenberg, N. M., Warren, R. A. J., Kilburn, D. G. & Miller, Jr, R. C. (1987). *Journal of Bacteriology*, **169,** 4674.

Haggett, K. D., Choi, W. Y. & Dunn, N. W. (1978). *European Journal of Applied Microbiology and Biotechnology*, **6,** 189.

Hall, A. & Knowles, J. R. (1976). *Nature*, **264,** 803.

Harman, J. G. & Dobrogosz, W. J. (1983). *Journal of Bacteriology*, **153,** 191.

Hartley, B. S. (1974). *Symposium of the Society of General Microbiology*, **24,** 151.

Hautala, J. A., Bassett, C. L., Giles, N. H. & Kushner, S. R. (1979). *Proceedings of the National Academy of Science USA*, **76,** 5774.

Hayzer, D. J. & Leisinger, T. (1980). *Journal of General Microbiology*, **118,** 287.

Hegeman, G. D. (1966). *Journal of Bacteriology*, **91,** 1161.

Hegeman, G. D. & Root, R. T. (1976). *Archives of Microbiology*, **110,** 19.

Helinski, D. R. & Yanofsky, C. (1963). *Journal of Biological Chemistry*, **238,** 1043.

Hirschman, J., Wong, P.-K., Sei, K., Keener, J. & Kustu, S. (1985). *Proceedings of the National Academy of Science USA*, **82,** 7525.

Holloway, C. T., Greene, R. C. & Su, C.-H. (1970). *Journal of Bacteriology*, **104,** 734.

Horiuchi, T., Horiuchi, S. & Novick, A. (1963). *Genetics*, **48,** 157.

Hsie, A. W. & Rickenberg, H. V. (1967). *Biochemical and Biophysical Research Communications*, **29,** 303.

Hydrean, C., Ghosh, A., Nallin, M. & Ghosh, B. K. (1977). *Journal of Biological Chemistry*, **252,** 6806.

Hynes, M. J. & Pateman, J. A. (1970). *Molecular and General Genetics*, **108,** 97.

Jacob, F. & Monod, J. (1961). *Journal of Molecular Biology,* **3,** 318.
Jäger, A., Croan, S. & Kirk, T. K. (1985). *Applied and Environmental Microbiology,* **50,** 1274.
Janessen, D. B., Herst, P. M., Joosten, H. M. L. J. & van der Drift, C. (1981). *Archives of Microbiology,* **128,** 398.
Jaurin, B. & Cohen, S. N. (1984). *Gene,* **28,** 83.
Jayaraman, K., Muller-Hill, B. & Rickenberg, H. V. (1966). *Journal of Molecular Biology,* **18,** 339.
Jensen, K. F., Neuhard, J. & Schack, L. (1982). *EMBO Journal,* **1,** 69.
Johnson, D. I. & Somerville, R. L. (1985). *Developments in Industrial Microbiology,* **26,** 87.
Katchalski-Katzir, E. (1980). In *Enzyme Engineering,* Vol. 5, eds H. H. Weetall & G. P. Royer. Plenum Press, New York, p. 3.
Kelley, W. S., Chalmers, K. & Murray, N. E. (1987). *Proceedings of the National Academy of Science USA,* **74,** 5632.
Kelly, J. M. & Hynes, M. J. (1977). *Molecular and General Genetics,* **156,** 87.
Kolter, R. & Yanofsky, C. (1982). *Annual Review of Genetics,* **16,** 113.
Komatsubara, S., Taniguchi, T. & Kisumi, M. (1986). *Journal of Biotechnology,* **3,** 281.
Labudova, I., Farkas, V., Bauer, S., Kolarova, N. & Branyik, A. (1981). *European Journal of Applied Microbiology and Biotechnology,* **12,** 16.
Lamond, A. I. & Travers, A. A. (1985). *Cell,* **41,** 6.
Lamsa, M. & Bloebaum, P. (1988). Abstract S-77, Annual Meeting of the Society for Industrial Microbiology, Chicago.
Langridge, J. (1969). *Molecular and General Genetics,* **105,** 74.
Lazzaroni, J. C. & Protalier, R. (1982). *European Journal of Applied Microbiology and Biotechnology,* **16,** 146.
Leclerc, M., Blondin, B., Ratomahenina, R., Arnaud, A. & Galzy, P. (1985). *FEMS Microbiology Letters,* **30,** 389.
Liao, H., McKenzie, T. & Hageman, R. (1986). *Proceedings of the National Academy of Science USA,* **83,** 576.
Lindberg, R. A. & Drucker, H. (1984). *Journal of Bacteriology,* **157,** 380.
Loison, G., Nguyen-Juilleret, M., Alouani, S. & Marquet, M. (1986). *Bio/Technology,* **4,** 433.
Mach, H., Hecker, M. & Mach, F. (1984). *FEMS Microbiology Letters,* **22,** 27.
Markgraf, M., Bocklage, H. & Müller-Hill, B. (1985). *Molecular and General Genetics,* **198,** 473.
Marzluf, G. A. (1981). *Microbiological Reviews,* **45,** 437.
Matsumoto, K., Uno, I., Toh-e, A., Ishikawa, T. & Oshima, Y. (1982). *Journal of Bacteriology,* **150,** 277.
Matsumoto, K., Uno, I., Toh-e, A., Ishikawa, T. & Oshima, Y. (1983). *Journal of Bacteriology,* **156,** 898.
Matsumura, M. & Aiba, S. (1985). *Journal of Biological Chemistry,* **260,** 15298.
Mayer, H., Collins, J. & Wagner, F. (1980). In *Enzyme Engineering,* Vol. 5, eds H. H. Weetall & G. P. Royer. Plenum Press, New York, p. 61.
McCormick, D. (1986). *Bio/Technology,* **4,** 698.
Meacock, P. A. & Cohen, S. N. (1980). *Cell,* **20,** 529.
Michels, C. A. & Romanowski, A. (1980). *Journal of Bacteriology,* **143,** 674.
Miller, E. S. & Brenchley, J. E. (1984). *Journal of Bacteriology,* **157,** 171.
Mishra, S., Gopalkrishnan, K. S. & Ghose, T. K. (1982). *Biotechnology and Bioengineering,* **24,** 251.
Monard, D., Janacek, J. & Rickenberg, H. V. (1969). *Biochemical and Biophysical Research Communications,* **35,** 584.
Montenecourt, B. S., Kuo, S.-C. & Lampen, J. O. (1973). *Journal of Bacteriology,* **114,** 233.

Mortlock, R. P. (1982). *Annual Review of Microbiology,* **36,** 259.

Moyed, H. S. (1961). *Cold Spring Harbor Symposium on Quantitative Biology,* **26,** 323.

Nagahari, K., Tanaka, T., Hishinuma, F., Kuroda, M. & Sakaguchi, K. (1977). *Gene,* **1,** 141.

Nagami, Y. & Tanaka, T. (1986). *Journal of Bacteriology,* **166,** 20.

Nakamura, K. & Inoyue, M. (1982). *EMBO Journal,* **1,** 771.

Neidhardt, F. C. (1960). *Journal of Bacteriology,* **80,** 536.

Neidhardt, F. C., Vaughn, V., Phillips, T. A. & Bloch, P. L. (1983). *Microbiological Reviews,* **47,** 231.

Nevalainen, K. M. H., Palva, E. T. & Bailey, M. J. (1980). *Enzyme and Microbial Technology,* **2,** 59.

Neway, J. O. & Switzer, R. L. (1983). *Journal of Bacteriology,* **155,** 512.

Newell, S. L. & Brill, W. J. (1972). *Journal of Bacteriology,* **111,** 375.

Nicholson, W. L. & Chambliss, G. H. (1985). *Journal of Bacteriology,* **161,** 875.

Nicholson, W. L. & Chambliss, G. H. (1986). *Journal of Bacteriology,* **165,** 663.

Nilsson, G., Belasco, J. G., Cohen, S. N. & von Gabain, A. (1984). *Nature,* **312,** 75.

Nishimura, N. & Kisumi, M. (1984). *Applied and Environmental Microbiology,* **48,** 1072.

O'Donovan, G. A. & Neuhard, J. (1970). *Bacteriological Reviews,* **34,** 278.

Okamoto, N., Tei, H., Murata, K. & Kimura, A. (1986). *Journal of General Microbiology,* **132,** 1519.

Oliver, D. (1985). *Annual Review of Microbiology,* **39,** 615.

O'Neill, G. P., Kilburn, D. G., Warren, A. J. & Miller, R. C. Jr (1986). *Applied and Environmental Microbiology,* **52,** 737.

O'Sullivan, D. (1981). *Chemical Engineering News,* **59**(2), 37.

Pall, M. L. (1984). *Molecular and Cellular Biochemistry,* **58,** 187.

Palva, I., Lehtovaara, P., Sarvas, M., Takkinen, K., Kalkkinen, N., Pettersson, R. & Kääriäinen, L. (1983). In *Genetics of Industrial Microorganisms 1982,* eds Y. Ikeda & T. Beppu. Kodansha, Tokyo, p. 287.

Panasenico, S. M., Cameron, J. R., Davis, R. W. & Lehman, I. R. (1977). *Science,* **196,** 188.

Panbangred, W., Fukasaki, E., Epifanio, E. C., Shinmyo, A. & Okada, H. (1985). *Applied Microbiology and Biotechnology,* **22,** 259.

Pavlasová, E., Stejskalová, E. & Sikyta, B. (1983). *Biotechnology Letters,* **5,** 223.

Perry, L. J. & Wetzel, R. (1986). *Biochemistry,* **25,** 733.

Peterkofsky, A. (1976). *Advances in Cyclic Nucleotide Research,* **7,** 1.

Phillips, A. T. & Mulfinger, L. M. (1981). *Journal of Bacteriology,* **145,** 1286.

Piovant, M. & Lazdunski, C. (1975). *Biochemistry,* **14,** 1821.

Pohlig, G. & Holzer, H. (1985). *Journal of Biological Chemistry,* **260,** 13818.

Poulsen, P. B. (1987). Presented at the Meeting of European Producers of Antibiotics, Stockholm, Sweden.

Preiss, J., Mazelis, M. & Greeberg, E. (1982). *Current Microbiology,* **7,** 263.

Pugsley, A. P. & Schwartz, M. (1985). *FEMS Microbiology Reviews,* **32,** 3.

Raetz, C. R. H., Larson, T. J. & Dowhan, W. (1977). *Proceedings of the National Academy of Science USA,* **74,** 1412.

Raibaud, O. & Schwartz, M. (1984). *Annual Review of Genetics,* **18,** 173.

Reddy, P., Peterkofsky, A. & McKenney, K. (1985). *Proceedings of the National Academy of Science USA,* **82,** 5656.

Reese, E. T., Lola, J. E. & Parrish, F. W. (1969). *Journal of Bacteriology,* **100,** 1151.

Reitzer, L. J. & Magasanik, B. (1983). *Proceedings of the National Academy of Science USA,* **80,** 5554.

Rigby, P. W. J., Burleigh, B. C. & Hartley, B. S. (1974). *Nature,* **251,** 200.

Schaeffer, E. J. & Cooney, C. L. (1982). *Applied and Environmental Microbiology,* **43,** 75.

Schurr, A. & Yagil, E. (1971). *Journal of General Microbiology*, **65**, 291.

Schwarz, W. H., Schimming, S. & Staudenbauer, W. L. (1987). *Applied Microbiology and Biotechnology*, **27**, 50.

Seno, E. T. & Baltz, R. H. (1981). *Antimicrobial Agents and Chemotherapy*, **20**, 370.

Shaw, W. V. (1987). *Biochemical Journal*, **246**, 1.

Sheperdson, M. & Pardee, A. B. (1960). *Journal of Biological Chemistry*, **235**, 3233.

Shinke, R., Aoki, K., Nishiva, H. & Yuki, S. (1981). In: *Advances in Biotechnology 3. Fermentation Products*, eds C. Vezina & K. Singh, Pergamon Press, Toronto, p. 307.

Shoham, Y. (1987). PhD Thesis, Massachusetts Institute of Technology, Cambridge, MA.

Sibakov, M., Sarvas, M. & Palva, I. (1983). *FEMS Microbiology Letters*, **17**, 81.

Siegel, L. S., Hyleman, P. B. & Phibbs, P. V., Jr (1977). *Journal of Bacteriology*, **129**, 87.

Sikyta, B. & Kyslík, P. (1981). In *Advances in Biotechnology. Scientific and Engineering Principles*. eds M. Moo-Young, C. W. Robinson & C. Vezina. Pergamon Press, Toronto, p. 215.

Soutschek-Bauer, E. & Staudenbauer, W. L. (1987). *Molecular and General Genetics*, **208**, 537.

Stevis, P. E. & Ho, N. W. Y. (1985). *Enzyme and Microbial Technology*, **7**, 592.

Stewart, G. G. (1987). *CRC Critical Reviews of Biotechnology*, **5**, 89.

Sun, D. & Takahashi, I. (1984). *Canadian Journal of Microbiology*, **30**, 423.

Switzer, R. L. (1977). *Annual Review of Microbiology*, **31**, 135.

Sy, J. & Richter, D. (1972). *Biochemistry*, **11**, 2788.

Takeda, Y., Ohlendorf, D. H., Anderson, W. F. & Mathews, B. W. (1983). *Science*, **221**, 1020.

Torotora, P., Burlini, N. & Leoni, F. (1983). *FEBS Letters*, **155**, 39.

Torriani, A. & Rothman, F. (1961). *Journal of Bacteriology*, **81**, 835.

Tsang, E. W. T. & GrootWassink, J. W. D. (1985). *Biotechnology Letters*, **7**, 179.

Tsukagoshi, N., Iritani, S., Sasaki, T., Takemura, T., Ihara, H., Idota, Y., Yamagata, H. & Udaka, S. (1985). *Journal of Bacteriology*, **164**, 1182.

Turnbough, Jr, C. L. (1983). *Journal of Bacteriology*, **153**, 998.

Uhlin, B. E., Molin, S., Gustafsson, P. & Nordstrom, K. (1979). *Gene*, **6**, 91.

Ulmer, K. M. (1983). *Science*, **219**, 666.

Urban-Grimal, D., Ribes, V. & Labbe-Bois, R. (1984). *Current Genetics*, **8**, 327.

Vaheri, M., Leisola, M. & Kauppinen, V. (1979). *Biotechnology Letters*, **1**, 41.

van Brunt, J. (1986). *Bio/Technology*, **4**, 1057.

van Wijk, R. & Konijn, T. M. (1971). *FEBS Letters*, **13**, 184.

Wang, L. C. & Marzluf, G. A. (1979). *Molecular and General Genetics*, **176**, 385.

Wanner, B. L., Kodaira, R. & Neidhardt, F. C. (1977). *Journal of Bacteriology*, **130**, 212.

Wilkinson, A. J., Fersht, A. R., Blow, D. M., Carter, P. & Winter, G. (1984). *Nature*, **307**, 187.

Windass, J. D., Newton, C. R., DeMeyer-Guignard, J., Moore, V. E., Markham, A. F. & Edge, M. D. (1982). *Nucleic Acids Research*, **10**, 6639.

Winter, G., Fersht, A. R., Wilkinson, A. J., Zoller, M. & Smith, M. (1982). *Nature*, **299**, 756.

Wong, S. L., Kawamura, F. & Doi, R. H. (1986). *Journal of Bacteriology*, **168**, 1005.

Wood, H. G. (1977). *Federation Proceedings*, **36**, 2197.

Workman, W. E. (1986). *CRC Critical Reviews of Biotechnology*, **3**, 234.

Yamashita, I. & Fukui, S. (1983). *Agricultural and Biological Chemistry*, **47**, 2689.

Yanofsky, C., Kelley, R. L. & Horn, V. (1984). *Journal of Bacteriology*, **158**, 1018.

Chapter 11

IMMOBILIZED BIOCATALYST TECHNOLOGY

Lawson W. Powell

*Beecham Pharmaceuticals, Clarendon Road, Worthing,
West Sussex BN14 8QH, UK*

CONTENTS

1 INTRODUCTION

1.1 Aim of the Review

It would be impossible within a review of this size to cover in depth all of the numerous facets of immobilized biocatalyst technology. Indeed, whole books are devoted to specific topics such as the analytical uses of immobilized enzymes (Guilbault, 1984) or the principles and applications of immobilized cells (Tampion & Tampion, 1987). It is the intention of this review to try and indicate the breadth of topics within the umbrella of immobilized biocatalyst technology and to provide references allowing anyone interested to research a particular area further.

1.2 Definition of Terms

The term biocatalyst has been used because it is a general term referring to biological entities exhibiting enzyme activity. Thus it covers a single enzyme at one end of the spectrum and a viable cell (microbial, plant or animal) with its complex array of enzymes at the other. An immobilized biocatalyst is one that has been modified such that it is easily separable from the medium in which it is acting. This property enables its use in successive reactions or for continuous processes and thus the useful life of the catalyst can be extended.

1.3 The Importance of Biocatalysts

Man has employed the action of biocatalysts since prehistory. Vinegar, a product of the microbial oxidation of ethanol, was recorded in the Bible, and the trickling process for vinegar manufacture, devised around 1815, was an early use of immobilized cells. Acetic acid bacteria adhering to beech or birch shavings oxidized the alcohol as it was trickled over them. Other ancient processes include alcoholic beverage production and flax retting. Today, enzymic processes exist for the production of a wide range of products including antibiotics, organic and amino acids, steroids, dairy products, glucose, high-fructose syrups. In some instances a single enzyme may be used (e.g. hydrolysis of penicillin G to produce 6-aminopenicillanic acid, the precursor of a number of semi-synthetic penicillins) and in others whole cells are grown in fermentations to provide a product involving a complex biosynthetic pathway (e.g. penicillin G production). However, over the last thirty years there has been an increasing interest in the application of enzymes for a wide variety of chemical transformations. The growth of interest can be judged from the fact that biotransformations are the subject of the government's sixth LINK programme with industry. Enzymes have two major advantages over chemical catalysts. They often have high substrate specificity and this selectivity, particularly for enantiomers, is becoming increasingly important. In the pharmaceutical industry for example there is a growing need to produce only one enantiomeric form of a compound. The second advantage of biocatalysts is their ability to act under mild conditions of pH, temperature and pressure. They are also more efficient catalysts. Potential applications of biocatalysts are reviewed by Rosazza (1982), Whitesides (1985), Soda & Yonaha (1987) and Yamada & Shimizu (1988). There is a growing availability of enzymes. In the 1989 catalogue published by the Sigma Chemical Company, some 313 enzymes are included, covering the six major types, oxidoreductases (88), transferases (50), hydrolases (127), lyases (34), isomerases (9) and ligases

(5). The International Union of Biochemistry (1979) catalogued some 2100 enzymes and this number must have reached 3000 by now. Cheetham (1987) in a review of screening techniques and their applications in providing novel biocatalysts for food, chemical, pharmaceutical and waste disposal processes, emphasizes the need for more goal-orientated screening for new enzyme systems. He notes that many experts estimate that less than one percent of the world's microorganisms have been properly studied for their metabolic properties. As well as looking for new biocatalysts the properties of those we have need to be studied more extensively. Alberti & Klibanov (1982) found that glucose oxidase could be used to oxidize benzoquinone to hydroquinone; while highly specific for glucose it was non-specific in respect of its electron acceptor. Genetic recombination techniques are enabling the manipulation of the genes associated with biocatalytic activity, thus allowing modifications of enzyme structure and activity (Fersht, 1985; Smith, 1985). Wilkinson *et al.* (1984) engineered a single point mutation in the tRNA synthetase of *Bacillus stearothermophilus* which increased its K_m by a factor of 100. Estell *et al.* (1985) increased the resistance of subtilisin to oxidation by substituting serine, alanine or leucine for a methionine at a particular site (Met 222) on the enzyme. Enzyme production can be linked to strong promoters in order to increase productivity. It is also possible to remove enzyme production from the control of a repressor. All of these factors are increasing the range of enzyme systems available and their versatility.

Having located the required biocatalytic activity the next stage is to develop a reactor system and it is at this stage that immobilization technology is needed.

1.4 Why Immobilize Biocatalysts?

The immobilization procedure primarily makes the biocatalyst separable from the process stream. By attaching it to or entrapping it inside a particulate support, the biocatalyst can easily be recovered by filtration or centrifugation. It can also be retained in a reactor system by the use of an appropriate outlet filter. This ensures clean process streams for extraction of product thereby improving yield and quality. By enabling reuse of the enzyme, costs can be minimized. Immobilization also enables biocatalysts to be used in analytical and biomedical devices thereby extending their application range and increasing their versatility.

Immobilization often modifies the properties of enzymes. Stability can be enhanced, particularly where the conformation is retained by multi-point covalent linkages to a support. By introducing diffusional limitations product inhibition may be reduced. There may be pH and temperature shifts for optimum activity and stability. However, it would be a naive person who would attempt to increase an enzyme's stability solely by immobilization. It is essential to understand the factors involved in the inactivation of an enzyme and to operate the immobilized enzyme under the conditions where denaturation of the free enzyme is minimal.

In the next section the various procedures used for biocatalyst immobilization will be described.

2 IMMOBILIZATION METHODOLOGIES

There are three basic techniques employed for biocatalyst immobilization. In the first the biocatalyst is adsorbed to a retaining support medium. For the second method the biocatalyst is chemically linked (covalently attached) to the support. The third system relies on entrapment of the biocatalyst in a supporting medium or in a microcapsule or within a reactor by the use of an appropriate filter. These methods are not mutually exclusive and many hybrid systems have been developed.

2.1 Enzyme Immobilization

For intracellular enzyme systems there is obviously a need to choose between using the extracted enzyme or the cell-bound enzyme. In general, the latter is favoured for systems requiring regeneration of a cofactor or where a sequence of several enzymes (a multi-step reaction) is operating. In these cases it is often easier to use the whole cell. The cell would also be favoured for enzymes which are difficult to stabilize when they are released from the cell. In every other case the use of a cell-free enzyme would be favoured. The specific activity of immobilized enzymes are usually higher than for whole cell systems and the possibility of unwanted side-reactions is minimized (it can be avoided altogether if a purified enzyme is used).

The extent to which an enzyme is purified before immobilization will depend on the specific activity required and this, in turn, will depend upon the activity required in the reactor. It will also be related to the stability required (i.e. number of reuses) if the stability of an enzyme is reduced on purification a compromise may have to be reached where some activity improvement is traded off for a better operational stability. For large-scale industrial processes it is important to minimize the number of purification steps required in order to improve biocatalyst yield and minimize overall costs. Thus it is obviously advantageous to increase the productivity of the enzyme source to maximize the specific activity of the enzyme released from the cells and minimize the extent of further purification. There are many reviews that have dealt with enzyme immobilization (Zaborsky, 1973; Goldstein & Manecke, 1976; Chibata, 1978; Powell, 1984; Rosevear, 1984; Kennedy & Cabral, 1987; Hartmeier, 1988). A very useful step-by-step guide to the subject is that of Rosevear *et al.* (1987). A highly readable introduction to enzyme immobilization is given by Trevan (1980).

2.1.1 Adsorption

This is the easiest procedure for immobilizing an enzyme and must be the first to be tested by anyone developing a commercial process with an immobilized enzyme. A wide range of supports has been used including activated charcoal, alumina, glass, ion-exchange resins, chitin, chitosan, celluloses, agaroses, bentonite and Sephadexes. A recent addition was sandy alumina (Fadda *et al.*, 1989). The earliest immobilized enzyme was invertase adsorbed to charcoal (Nelson & Griffin, 1916). To maximize protein loading it is necessary to have a large surface area for adsorption. This can be achieved by using a porous support and if it is particulate by having a small particle size. The enzyme is held in place on the support by a number of forces. These range from the relatively weak van der Waal's forces and hydrogen bonding to the stronger hydrophobic and ionic links. All of these bonds can be broken by changes in pH, ionic strength, temperature and solvent and herein lies the weakness of this method of immobilization. For maximum operational stability it is essential that the optimum conditions for enzyme adsorption are the same as the conditions for the enzyme reaction. Acid phosphatase, active at pH 5·0, was stable to 2 M acetate buffer at this pH when adsorbed to CM-cellulose (Vaidya *et al.*, 1987). In the case of porous supports desorption of protein is impeded by the pores and this appears to be minimal if the pore size is roughly twice the major axis or spin diameter of the enzyme (Messing, 1975).

Adsorption of a protein to a porous support does not result in a uniform distribution within the particles. Pederson *et al.* (1985) showed that even after 10 h the majority of the β-galactosidase adsorbed to Duolite S-761, a phenol-formaldehyde resin, was confined to the outer half of the support. In practice it is preferable to have enzyme near enough to the surface of the support where it can be active but sufficiently deep to be protected from inimical agencies.

Adsorption can be improved either by modification of the support or the enzyme. Krill chitin treated with CS_2 showed stronger binding powers than untreated material (Synowiecki *et al.*, 1987). This was explained by conversion of 30% of the amino groups on the support to dithiocarbamino groups which, being negatively charged, bound to amino, guanidium, imidazole, indole and thiol groups on the protein. The untreated amino groups bound only to the protein carboxyl groups. Thus a wider range of linkages was possible with the treated chitin. A range of agaroses with affinity ligands is commercially available. Lectins can bind glycoproteins and Woodward (1985) reported a stable cellobiase preparation on Con-A-Sepharose. Hutchinson & Collier (1987) immobilized bacteriophage T4 polynucleotide kinase on phenyl Sepharose. Viera *et al.* (1988) adsorbed soybean β-amylase to phenylboronate-agarose. Tannin bound to aminohexylcellulose was able to specifically bind proteins (Chibata *et al.*, 1986).

Hydrophobic binding can be improved by introducing hydrophobic side-chains into the enzyme molecule. Ampon & Means (1988) improved the binding of enzymes to polymethacrylate supports (Amberlite XAD-7 and

XAD-8) by reaction of the proteins with a hydrophobic imidoester, methyl 4-phenyl butyrimidate. Perfluoroalkylated proteins have been found to bind very strongly to fluorocarbon supports (Kobos *et al.*, 1989). In this instance the nature of the binding is unknown—it may be hydrophobic or fluorophilic. To retain activity on immobilization it is necessary to have a spacer arm between the perfluoroalkylating agent and the enzyme. This methodology is the basis of Perflex® affinity supports developed by E. I. du Pont de Nemours and Co. Inc.

Inevitably, when enzymes are adsorbed to supports, occlusion of active sites occurs. Fusek *et al.* (1988) were able to reduce the activity loss on adsorption of chymotrypsin by using a support to which was immobilized a polyclonal antibody which bound the enzyme with minimal occlusion or distortion of the active site.

The stability of adsorbed enzymes can be improved by cross-linking and the most popular reagent for this purpose is glutaraldehyde. Rucka & Turkiewicz (1989) stabilized a lipase on PTFE membrane by cross-linking with glutaraldehyde. It is cheap, readily available and has low toxicity. Glyoxal and formaldehyde have also been used. The anti-microbial activity of these reagents is useful in sanitizing the enzyme preparation. Numerous other bifunctional reagents are available including carbodi-imides and bi-functional imidoesters, the latter being useful for retaining the overall charge of the protein. Rosevear (1975) precipitated enzyme inside the pores of a range of materials including titania spheroids, glass beads, wood chips, and cross-linked the enzyme with glutaraldehyde, formaldehyde, glyoxal, bis-diazonium salts or diethyl pyrocarbonate.

If the adsorbent has free amino groups and the enzyme is cross-linked with glutaraldehyde then inevitably chemical bonds are formed between the enzyme and the support and the enzyme is now covalently bonded as well as being physically bonded to the support.

2.1.2 Covalent Attachment

Covalent attachment methodologies are generally favoured over purely adsorbed systems because of the greater stability of the bond between the enzyme and the support. It is questionable whether the same is true where the adsorbed enzyme is stabilized by cross-linking. Indeed, cross-linking can be used to further stabilize covalently bound enzymes. The stronger binding requires more harsh conditions (e.g. acidic/alkaline hydrolysis, pyrolysis) to remove the protein if it is essential to reuse the support. If reuse of the support is a requirement then ones resistant enough to withstand such treatments e.g. inorganic supports or synthetic polymers, would be favoured. An alternative approach would be to use a reversible covalent bond to immobilize the enzyme. Thiol-containing supports have been used for this purpose (Royer *et al.*, 1977). Thiol groups can be introduced into enzymes by treatment with reagents such as *N*-acetyl homocysteine thiolactone or methyl-3-mercaptoproprionimidate.

Covalent attachment would seem essential for the soluble-immobilized enzyme preparations. In these enzyme is linked to a polymer which is soluble under the reaction conditions for the enzyme but can be precipitated by change of pH to allow recovery of the enzyme–polymer by filtration. Poly-L-glutamic acid was used to immobilize cellulase by Takeuchi & Makino (1987). It was soluble at pH 4·5 and could be recovered at pH 3·0. Fujimura *et al.* (1987) used a methacrylic acid–methacrylate–methylmethacrylate co-polymer for immobilizing a number of proteases. It was soluble above pH 5·8 and insoluble below pH 4·8. Immobilized papain was more active against haemoglobin and casein than enzyme covalently bonded to CM-Sepharose; no difference could be found with a small molecular weight substrate. As well as better activity against macrosubstrates, a soluble immobilized enzyme would be more active on insoluble substrates and could be separated easily from an insoluble product. The enzyme has to be tolerant to pH changes to enable this system to be used. Enzymes immobilized on soluble polymers can be recovered by ultrafiltration and this has formed the basis of one of the reactor systems to be discussed later.

Smith (1976) formed an amphipathic enzyme–polymer conjugate which could be recovered by flotation in a water-immiscible solvent (*n*-decanol).

The range of materials used as immobilization supports has been very wide and even poultry bone residues has been reported as a promising matrix (Findlay *et al.*, 1986). An equally impressive array of immobilization processes has been developed. In general the support is activated in the absence of the enzyme, thus allowing the use of reagents and conditions that would denature the protein. The matrix is then washed free of reactants and enzyme is coupled under mild conditions. One of the simplest methods is the activation of an amine-bearing support by glutaraldehyde. Braun *et al.* (1989) used this technique to attach penicillin acylase to chitosan. Nylon needs to have its surface hydrolysed first to provide free amine groups for glutaraldehyde linking (Jain & Wilkins, 1987). Daka *et al.* (1988), after hydrolysis of the surface of a nylon disc, converted the carboxyl groups to acid chloride groups by reaction with thionyl chloride and then coupled these with 3,3'-diamobenzidine. This provided extra amino groups on the support surface to which lactate dehydrogenase could be immobilized by glutaraldehyde. Nylon can also be activated by alkylation, coupled with a spacer arm such as polyethylene imine to which enzyme can be linked by glutaraldehyde (Lozano *et al.*, 1988). Aromatic amines can be activated by diazotization and this method was used by Simionescu *et al.* (1987) to couple invertase to 4-aminobenzyl cellulose. Mansfeld & Schellenberger (1987) activated a porous polystyrene anion exchange resin using either glutaraldehyde or benzoquinone for the immobilization of invertase. Both amine and carboxyl groups can be activated by carbodi-imides. Hence these reagents can be used to couple enzymes via their amino groups to supports having free carboxyls or to couple enzymes via their carboxyl groups to supports with amine groupings. Takeuchi & Makino (1987) coupled cellulase to poly-L-glutamic acid using dicyclohexyl carbodi-imide and

Domínguez *et al.* (1988) used 1-ethyl-3-(3-dimethylaminopropyl) carbodi-imide to immobilize β-galactosidase to alginate beads.

A simple technique applicable to supports having free hydroxyl groups was used by Jin & Toda (1988) to immobilize papain to wood chips and to cotton. Both materials were treated with sodium periodate to produce aldehyde groups with which enzyme was reacted. The immobilized enzyme was very stable when used to treat beer to prevent haze formation. A more complicated system used by Makino *et al.* (1988) involved tosylating glycidylmethacrylate gels in pyridine, aminating with hexamethylenediamine and linking enzyme using glutaraldehyde.

Sørensen & Emborg (1989) covalently bound lactase and glucose isomerase to Luxopor, a naturally occurring SiO_2 material. It was acid treated to remove $CaCO_3$ and immobilization was effected by silanization with 3-aminopropyl triethoxysilane followed by glutaraldehyde treatment. Manjón *et al.* (1985) oxidized Glycophase G ™ controlled pore glass with periodate and coupled an arylamine which was diazotized to provide a linkage point for enzyme. Crumbliss *et al.* (1988) coupled carbonic anhydrase to an aniline derivative of controlled pore glass via glutaraldehyde.

Carbon is an important support because of its use in biosensors. Bourdillon *et al.* (1988) coupled glucose oxidase to carbon felt. Refluxing of the felt in nitric acid oxidized the surface to produce carboxyl groups which were activated with carbodi-imide. Crumbliss *et al.* (1988) oxidized the surface of graphite rods in an rf oxygen plasma reactor and formed an activated ester with *N*-hydroxysuccinimide and carbodi-imide to which enzyme could be coupled.

A simple technique which has been applied with variations to a range of supports is the transition metal chelation method (Kennedy & Cabral, 1985). With inorganic supports it is important to dry the support–transition metal complex in order to produce a stable immobilized enzyme. As with other methods spacer molecules can be inserted between the enzyme and the support.

The growing interest in biotransformations has led to an increasing supply of enzyme supports, activating agents and cross-linking agents being available commercially. Some of the supports available for covalent immobilization are shown in Table 1. The advantage in having these materials available lies in the removal of the need to carry out the activation process which can necessitate the use of highly toxic chemicals. A good example of this is cyanogen bromide-activated Sepharose which has been one of the most popular supports for many years.

A detailed description of the very large number of covalent attachment systems available and discussion of their relative merits and disadvantages can be found in the reviews cited earlier. It is noticeable that the flow of new methodologies for enzymes has slowed considerably. Most publications now concentrate on cell immobilization for which entrapment systems have been particularly popular. However they have also been applied to enzymes.

Table 1
Commercially-Available Supports for Covalent Immobilization of Enzymes

Supplier	Support
E.C.C. International Ltd, Household Products Division, John Keay House, St. Austell, Cornwall PL25 4DJ, UK.	Kaolinite (Biofix[R]) activated by glutaraldehyde
ICN Biochemicals, PO Box 28050, Cleveland, Ohio 44128, USA.	Polyvinyl chloride silica membrane (Protrans[R] Affinity Disks) activated by glutaraldehyde
Koch-Light Ltd, Hollands Road, Haverhill Suffolk CB9 8PU, UK.	Derivatized polyacrylamide (Enzacryls) and coated inorganics, nylon, cellulose
Life Science Laboratories Ltd, Sedgewick Road, Luton, Bedfordshire LU4 9DT, UK.	Controlled pore glass, Glycophase[R] glass, derivatized polystyrene
Sterling Organics, Dudley, Cramlington, Northumberland NE23 7QG, UK.	Kieselguhr (Macrosorbs) activated by glutaraldehyde, cyanogen bromide, Tresyl chloride
Pharmacia Ltd, Pharmacia LKB Biotechnology Division, Midsummer Boulevard, Central Milton Keynes, Buckinghamshire MK9 3HP, UK.	Derivatized agaroses (Sepharose)
Röhm Pharma GmbH, Postfach 4347, D-6100 Darmstadt-1, FRG.	Derivatized acrylic beads (Eupergit[R])

2.1.3 Entrapment

Immobilization of enzymes by entrapment can be achieved by a diversity of methodologies and some are opening up exciting new technologies.

An enzyme can be immobilized within a reactor by employing a membrane to prevent enzyme loss from the system while allowing product to pass out. Ultrafiltration cells and hollow fibre devices have been used in this way. In the latter the enzyme is usually retained on the shell side of the reactor and substrate is pumped through the fibres. Details of commercially available ultrafiltration systems including large-scale devices are given by Cooper (1984). Stabilization of the enzyme can be effected by chemical modification prior to introduction into the reactor. Hydrophilization has been very effective with trypsin and chymotrypsin (Melik-Nubarov *et al.*, 1987; Mozhaev *et al.*, 1988).

Alternatively the enzyme could be linked to a soluble polymer by multiple attachment to rigidify the structure and impede unfolding with concomitant inactivation. The use of a polymer-bound enzyme would allow a more open membrane to be used.

Where liquor is being passed directly through a membrane a layer of protein builds up at the membrane surface. This can act as a protection for enzymes entrapped within it. This can be increased by the addition of a soluble polymer to the reactor (Greco *et al.*, 1983) which builds up and forms a polymeric network around the enzyme. If polyelectrolytes are used the polymer network produces microenvironmental effects as a result of the charged groupings surrounding the enzyme. Thus, the pH optimum can apparently move either to a more acid or a more alkaline value and this could be used to advantage to develop a reactor with enzymes of differing pH optima which could be fed with one process stream at an intermediate pH value (Gianfreda *et al.*, 1989). Concentration of enzyme at the membrane also improves reaction rate and thus increases the process intensity.

Membrane reactors are probably the most practical system for cofactor-requiring enzymes because these materials can be coupled to soluble polymers such as polyethylene glycol or dextran to effect retention in the reactor (Wandrey *et al.*, 1982).

An alternative entrapment system is microencapsulation which entails forming a semi-permeable membrane around an aqueous microdroplet of enzyme. Microencapsulated enzymes have been used for enzyme replacement therapy (Chang, 1977). Two processes are used for their manufacture (Chang, 1976). In one an emulsion of enzyme is made in a water-immiscible solvent containing the polymer (e.g. cellulose nitrate). A second solvent is added to precipitate the polymer and this occurs at the boundary between the microdroplet and the solvent. As well as nitrocellulose, ethylcellulose, polystyrene and polyvinyl acetate have been used. In the second process the enzyme emulsion is made in a water-immiscible solvent containing a hydrophilic monomer. The second hydrophobic monomer is then added and polymerization occurs at the microdroplet/solvent interface. This process has been used to produce polyurethane, polyurea, polyester and nylon microcapsules. Haemoglobin is usually encapsulated with the enzyme to protect it from denaturation. The semi-permeable membrane, while allowing ready ingress of substrate (low molecular weight) and egress of products, protects the enzyme from the action of other enzymes or antibodies. Enzyme can be stabilized by cross-linking prior to microencapsulation. Cofactors can be encapsulated and will be retained if linked to a soluble polymer. Enzymes can be co-immobilized in the same microcapsule. Gu & Chang (1988) prepared nylon-polyethylene imine microcapsules containing L-glutamic dehydrogenase, yeast alcohol dehydrogenase, urease and dextran-NAD$^+$. The system was used to produce L-glutamic acid from urea and α-ketoglutarate. Ethanol, a co-substrate, was converted to acetaldehyde to regenerate NADH. Over four days of continuous operation the activity fell to 38% of the original level.

Poncelet *et al.* (1989) analysed the factors which were important in producing uniformly sized microcapsules of collodion. These included type of agitator, emulsification time, presence of a surfactant and the ratio of emulsified aqueous to organic phase.

A large-scale application of microencapsulation is the use of fibre-entrapped enzymes (Dinelli *et al.*, 1976). In these, enzyme is immobilized in aqueous droplets within cellulose acetate fibres as they are spun. Fibre-entrapped penicillin acylase has been used for the industrial-scale production of 6-aminopenicillanic acid.

Liposomes have also been used for enzyme entrapment (Gregoriadis, 1976). Their selective permeability particularly to ions has been used to protect β-D-glucosidase from the action of copper (Sada *et al.*, 1988).

Although not coming strictly within our definition of immobilized it can be considered that another form of entrapment is exemplified by the use of enzymes in organic solvents. In these systems the enzyme is acting in an aqueous environment, which can vary from being a discrete phase to less than a monomolecular layer around the enzyme, within an organic solvent. Interest in the use of organic solvents stems from a need to be able to use enzymes to transform substrates that are insoluble in water. In addition it has been found that in organic solvents hydrolases can be used for synthetic reactions (Cassells & Halling, 1989; Kawamoto *et al.*, 1987; Tai *et al.*, 1989). In solvents enzymes may be protected from substrate and product inhibition and side-reactions can be minimized. The enzyme may be in a discrete aqueous phase dispersed as an emulsion within the organic medium. Alternatively, the water may be provided by a support with which enzyme is associated and no discrete aqueous phase is present. Halling (1987) has reviewed the use of these systems. Enzyme denaturation can be minimized by selection of a suitable (usually non-polar) solvent that does not distort the hydration shell of the enzyme and it can also be reduced by immobilizing the enzyme to prevent unfolding of the protein structure (Khmelnitsky *et al.*, 1988; Abramowicz & Keese, 1989). Semenov *et al.* (1987) showed that product yield in a biphasic system can in principle be higher than in water or solvent alone because the equilibrium is moved completely towards the product.

Rather than have a discrete enzyme phase in the reactor the enzyme may be immobilized in reversed micelles, transparent microdroplets of water embedded in water-immiscible solvent and stabilized by a surfactant. Chloroperoxidase entrapped in reversed micelles formed from buffer, cetyltrimethyl ammonium bromide, pentanol and octane halogenated 2-monochlorodimedon and 1,3-dihydroxy benzene at reaction rates double those in aqueous media (Franssen *et al.*, 1988). Reversed micelles have the advantages of very large interfacial areas ($\sim100\,m^2/ml$) and minimal diffusion limitation. Applications of reversed micelle systems in organic synthesis, analysis and therapy are discussed by Martinek *et al.* (1987). Recovery of enzymes entrapped in reversed micelles is difficult. However, Fadnavis & Luisi (1989) overcame this problem by preparing reversed micelles containing trypsin and chymotrypsin entrapped in polyacrylamide gel. The supported enzyme was easily recovered.

In another approach enzymes have been linked to polyethylene glycol, which allows solubilization of the complex in solvent systems such as benzene, toluene and chlorinated hydrocarbons (Inada *et al.*, 1986). Yoshimoto *et al.* (1984) reused a polyethylene glycol-linked lipase in benzene for ester synthesis, recovering it by precipitating with *n*-hexane or petroleum ether. The enzyme retained 50% of its initial activity after three months storage in benzene at room temperature.

Interestingly enzyme immobilized by adsorption to a support may be more stable in an organic system than in water because the bonds between enzyme and support are less easily broken (Cassells & Halling, 1989). Enzymes can work in anhydrous solvents requiring only 'essential' water which may be less than a unimolecular layer around the protein (Klibanov, 1989). Under these conditions, stability can be enhanced, substrate specificity can be altered and novel reactions can be carried out. It is thus possible to 'solvent engineer' an enzyme and this is going to be a useful facility to use in addition to protein engineering (Klibanov, 1989).

The aqueous biphasic systems based on polyethylene glycol (PEG) in combination with dextran or salt, which have been developed for enzyme purification (Hustedt, 1986) would seem to offer potential for entrapping an enzyme. The ideal system would be where enzyme and substrate partitioned in one phase with product in a second phase. Anderrson *et al.* (1984) partitioned (immobilized) penicillin acylase into the lower phase of a PEG/phosphate system and reused the enzyme five times for converting penicillin G to 6-aminopenicillanic acid. They acknowledged the difficulty in finding an optimum phase system. Larsson *et al.* (1989) described a reactor for converting starch to glucose where the substrate partitioned into the lower (dextran) phase of a PEG/dextran mixture and glucose was recovered via an ultrafiltration unit from the top phase; both phases being recirculated.

The final entrapment method is where a three-dimensional gel is formed around the enzyme. To retain enzyme the pore size needs to be very small and these systems tend to very very diffusion-limited. On the whole this technique has been more successful with cells and for this reason it will be considered under cell immobilization.

2.2 Cell Immobilization

A wide variety of cells has been immobilized and the methods used have been derived from those developed for enzyme immobilization. The whole area of cell immobilization was comprehensively reviewed by Tampion & Tampion (1987).

2.2.1 *Animal Cells*
Materials derived from animal cells include viral vaccines, monoclonal antibodies, cell surface antigens, enzymes such as plasminogen activators, growth factors, hormones, tumour antigens, immunoregulators, viral bioinsecticides and reconstituted skin (Mizrahi, 1986). Animal cells are either

anchorage-dependent and require a surface for growth and proliferation or anchorage-independent and can be grown quite happily in suspension culture. Many of the products listed above are produced by the former. The most widely used method for culturing anchorage-dependent cells is on small solid or semi-solid beads termed microcarriers (Hirtenstein & Clark, 1983). The cells grow on the beads which are kept suspended in nutrient medium by gentle agitation to avoid shear. This is important with animal cells because of their fragility. Microcarriers can be fabricated from cross-linked dextran such as the Cytodex carriers developed by Pharmacia and used by Smiley *et al.* (1989) for growing Chinese hamster ovary cells for the production of human interferon. Other materials that are in use include the gelatin-based CultiSpher-G carriers developed by Percell Biolytica of Sweden and the glass Siran Microcarriers developed by Schott Glaswerke of West Germany.

Anchorage-dependent cells produce an extracellular matrix which 'cements' the cells together and to the surface on which they are growing (Varani *et al.*, 1989).

Hollow fibre culture systems have also been developed (Hopkinson, 1983; Evans & Miller, 1988) and can be used for both anchorage-dependent and anchorage-independent cells. Microencapsulation has also been used. Boag & Sefton (1987) encapsulated human fibroblasts in microcapsules formed from Eudragit RL, a commercially-available polyacrylate. Gin *et al.* (1987) used agarose for encapsulating Islets of Langerhans. King *et al.* (1988) encapsulated insect cells in multiple membrane alginate/polylysine microcapsules.

2.2.2 Algae

Interest in immobilizing algae emanates from their use as biocatalysts for carrying out biotransformations and biosyntheses, in the production of energy (hydrogen and electricity), for providing oxygen or reduced NADP for coimmobilized heterotrophs and for waste water treatment.

Immobilization has usually been accomplished by entrapment within either natural polymers such as agar, κ-carrageenan and alginate or synthetic polymers such as polyurethane. Entrapment has either been by forming the polymer around the cells (active entrapment) or by allowing the cells to grow within the polymer matrix (passive entrapment). Further details of these techniques are given in Section 2.2.4.

The immobilization of algae and their potential applications are well reviewed by Robinson *et al.* (1986) and Trevan & Mak (1988).

2.2.3 Plant Cells

The immobilization of plant cells has been reviewed by Brodelius (1983, 1985, 1988) and Rosevear & Lambe (1985). The importance of plants can be gauged from the fact that over 20 000 known natural materials are derived from them (Brodelius, 1988). Shikonin, a red dye and an anti-inflammatory agent, is produced commercially by plant cell culture of *Lithospermum erythrorhizon* by Mitsui Petrochemical Industries Ltd. For immobilization of plant cells, entrapment processes have been favoured, particularly the use of alginate gels which are made under mild conditions and give high cell yields. Immobilization obviously allows a higher concentration of cells within a reactor and the cell–cell contact induces differentiation to some extent mimicking what takes place during tissue development in the growing plant. Immobilization also protects the cells which are more fragile than microbial cells. Archambault *et al.* (1989) immobilized *Catharanthus* cells to man-made fibres by surface adhesion. Electron microscopy showed a secretion between the cells that could have been acting as a gluing agent. Cells have also been surface-grown on beads of gelatin (Kargi & Freidel, 1988) and calcium alginate (Kargi, 1988). These biofilm cultures have been used in packed bed reactors (Kargi, 1988). Hollow fibre devices have also been used.

A major limitation of plant cell cultures is that secondary metabolites are often stored within the cells in vacuoles and the cells need to be permeabilized to allow release of the desired product. While this is feasible, difficulty can be experienced in maintaining cell viability and other methods need to be evaluated. Pu *et al.* (1989) for example have used an iontophoretic device to apply a current of 1–2 mA to *Carantharus* cells to increase the release of alkaloids. The problem of product release and the unpredictability of growth and metabolism may limit immobilized plant cell usage to high-cost products or those difficult or impossible to make by other routes.

2.2.4 Microorganisms

The immobilization of microorganisms has been the subject of many reviews including Kolot (1980, 1981*a, b*), Messing (1980), Bucke (1983), Brodelius & Vandamme (1987), Rosevear *et al.* (1987), Tampion & Tampion (1987) and Hartmeier (1988). A wide range of bacterial, yeast and fungal genera have been immobilized. In its simplest form the microbial cell or hypha acts as a support for a single enzyme and its viability is unimportant. Permeabilization of the cell may be required for full expression of the enzyme activity and this can be achieved by chemical treatment with a surface active agent or solvent or by physical disruption. The disadvantage of a cell-bound enzyme is the lower specific activity that can be obtained on immobilization compared with the extracted enzyme. However, the enzyme may be unstable after release from the cell or the cost of extraction may be uneconomic. At the other extreme is the immobilized viable cell which may undergo division while associated with the support system. The full biosynthetic capacity of the cell is required for the production of a metabolite and the system is akin to a fermentation with the advantage that the process can be run continuously and at a higher cell concentration than may be possible in a conventional fermentation system. Cell growth rate can be uncoupled from metabolite production rate with an immobilized cell system (Dalili & Chau, 1988); very large inocula can be used which is valuable for slow-growing organisms (Champagne *et al.*, 1989) and process streams can be much cleaner thus aiding extraction of product. A disadvantage of using cells is the presence of unwanted side-reactions and it may be necessary to suppress these either by using inhibitors or by genetic engineering techniques. In general, entrapment systems have been favoured and are much more amenable to cells than enzymes because of their larger size. With enzymes pore sizes needed for efficient entrapment can produce severe diffusional limitation particularly with large molecular weight substrates. With cells a more open matrix is possible. Living cells require a mild entrapment procedure and the most favoured has been ionotropic gelation in calcium alginate. Its popularity stems from its simplicity; cells suspended in sodium alginate (preferably alginate with a high L-guluronic acid content according to Martinsen *et al.*, 1989) are dropped into calcium chloride solution whereon the cells immobilized in precipitated calcium alginate are formed as beads. The concentration of the reactants and the reaction time are important in determining bead strength (Cheetham *et al.*, 1979). Gel strength can be improved by a factor of two by a wash with aluminium nitrate (Rochefort *et al.*, 1986). This treatment also stabilizes the gel against solubilization in high phosphate media. Incorporation of silica gel within the alginate also increases gel strength (Fukushima *et al.*, 1988). Solubilization can be precluded by incorporating calcium chloride into the reaction medium or by preparing barium alginate. However, the toxicity of barium limits its usefulness with living cells. As well as beads, alginate can be formed into fibres and papers (Kobayashi *et al.*, 1986). The porosity of alginate can be reduced by forming a polyelectrolyte complex of potassium poly (vinyl alcohol) sulphate and trimethylammonium glycol chitosan iodide within the alginate (Kokufuta *et al.*, 1988*a, b*). Strict anaerobes can be immobilized within alginate by incorporating sodium sulphide and sodium tetrathionate in the gelling mixture (Joubert & Britz, 1988).

Other entrapment media include chitosan which is polymerized by ionotropic gelation in a polyanion such as sodium tri-poly phosphate and is resistant to phosphate solubilization (Vorlop & Klein, 1981), κ-carrageenan and the cheaper genucarrageenan polymerized by adding to potassium chloride solution (Malaníková *et al.*, 1988), the thermohardening gels such as agar and agarose (Birnbaum *et al.*, 1988), polyacrylamide, prepolymerized polyacrylamide hydrazide cross-linked with glyoxal (Freeman, 1984; Silberger & Freeman, 1987) and silicone foam (Oriel, 1988). The last system has an open structure and allows greater diffusion of oxygen and CO_2, which is advantageous for living systems. A particularly attractive entrapment matrix is polyurethane. The microorganism can either be immobilized directly by forming the gel with the cells present (Mulchandani *et al.*, 1989) or it can be allowed to grow within the foam (Kutney *et al.*, 1988) and the foams can be formed with a wide range of pore sizes. Kokufuta *et al.* (1988*a, b*) noted increased diffusion of *ortho*-nitrophenyl-β-D-galactoside into alginate beads containing a lectin and there may be other materials that can act by facilitating diffusion of substrate molecules to entrapped enzyme systems. Entrapment within a gel protected lactic streptococci from bacteriophage (Steenson *et al.*, 1987) although any free cells released from the matrix were susceptible.

Champagne *et al.* (1988) postulated that while 'phage was excluded by the matrix a virolysin might be able to penetrate into the cells. Entrapment has been reported to enhance plasmid stability of recombinant cultures of *Escherichia coli* (de Taxis du Pöet *et al.*, 1986).

Of growing interest have been adsorbed cell systems. Many microorganisms are capable of adhering to surfaces and this property is relied upon for the trickling filter systems used for waste water treatment. Materials that have been used as supports include ultraporous fired bricks (Opara & Mann, 1987), porous ceramics (Messing, 1982) and needle-punched polyester cloth (Duff, 1988). The ability of some microorganisms to form flocs has been exploited as a simple immobilization system. Ongcharit *et al.* (1989) immobilized *Thiobacillus denitrificans* in a macroscopic floc by coculturing it with floc-forming heterotrophs from activated sludge and used the preparation to remove hydrogen sulphide from a gas stream.

2.2.5 Protoplasts and Organelles

The diffusional constraints of the cell wall can be avoided by using protoplasts. Because of their poor mechanical stability immobilization within a protective matrix is necessary. It is also necessary to maintain an osmotically stable environment. Deshpande *et al.* (1987) immobilized *Sclerotium rolfsii* protoplasts in calcium alginate and found 36% retention of endogluconase and 26% retention of β-glucosidase activity after five reuses. The reduction in activity was possibly due to extensive mycelium formation having taken place. While useful on the small scale it is difficult to see large-scale application of protoplasts.

There is interest in using photosynthetic systems for the conversion of solar energy to chemical energy and for the production of hydrogen by biophotolysis of water (Hallenbeck, 1983; Rao & Hall, 1984). The use of isolated chloroplasts and chloroplast membranes has been limited by poor stability. This can be improved by immobilization. Thomasset *et al.* (1988) found a serum albumin–glutaraldehyde matrix to be particularly suitable for chloroplast membrane immobilization. A requirement of any support is to be permeable to light.

Sofer (1979) reviewed the potential value of microsomal enzymes particularly their role as carcinogen activators and detoxifying agents. An obvious way to use these enzymes is to immobilize the microsome. Camp & Sofer (1987) used calcium alginate to entrap liver microsomes. Ahern *et al.* (1983) reported the synthesis of a prostaglandin from arachidonic acid using ram seminal microsomes entrapped in a photocrosslinkable acrylate gel.

Godbole *et al.* (1983) entrapped rat liver mitochondria in polyacrylamide to provide an immobilized fumarase preparation. No activity loss was noted over 12 reuses.

The immobilization of organelles was reviewed by Tanaka & Fukui (1983).

2.3 Cofactor Immobilization

The activity of 30–40% of enzymes is dependent upon a cofactor. Where the cofactor is bound to the enzyme (it is termed a prosthetic group) and is regenerated *in situ* no problem is encountered on immobilization because the cofactor accompanies the enzyme. This is the case for FAD and pyridoxal phosphate. However, others including NAD(P), NAD(P)H, ATP and coenzyme A are not tightly associated with enzymes. They are freely mobile in the cell and move between enzymes acting as redox, phosphate and acyl carriers respectively. Regeneration occurs at each alternate move. The action of cofactors is discussed by Dixon & Webb (1979). Thus in order to develop a biotransformation process with an enzyme requiring one of these cofactors it is necessary to have methodologies for recycling and regenerating the cofactor because cost precludes its addition to each reaction. The cofactor has to be immobilized in an active configuration (Lowe, 1981) because it has to be able to fit into a complementary part of the active site. Lowe (1983) has further discussed the problems associated with using cofactor-requiring enzymes.

Chenault & Whitesides (1987) have reviewed the systems used to regenerate nicotinamide cofactors. These have included chemical, electrochemical, photochemical, biological and enzymic methodologies. They concluded that any regeneration system should recycle the cofactor from 10^2 to over 10^5 times and that it should be highly selective (99·3%) in forming active cofactor. They concluded that this latter requirement was most effectively met by enzymic regeneration. Enzymic methods were also favoured for ATP regeneration by Langer *et al.* (1976).

The regeneration system of choice for NADH appears to be formate and formate dehydrogenase. This system was used by Wandrey *et al.* (1982) in their membrane reactor. The cofactor was retained in the reactor by covalent bonding to polyethylene glycol 10 000. The reactor produced L-leucine from α-ketoisocaproic acid using L-leucine dehydrogenase. The number of cofactor recycles exceeded 100 000 and a reactor producing 1 kg of L-*tert*-leucine per day has been reported to be operated by Degussa (Schmidt *et al.*, 1986). Formate is cheap, harmless, acts as a mild disinfectant and produces only carbon dioxide and water. However, Carrea *et al.* (1988) noted the low activity of commercial formate dehydrogenase preparations. Ethanol/alcohol dehydrogenase can be used for regenerating NADH and NADPH although acetaldehyde can be a problem. It can be swept out by a nitrogen purge or converted to acetate using yeast aldehyde dehydrogenase. Regeneration of NAD and NADP can be achieved using α-ketoglutarate and glutamate dehydrogenase. Several other enzyme systems are available and their relative merits and problems are discussed by Chenault & Whitesides (1987).

As previously mentioned the membrane reactor with polymer-bound cofactors has been the most useful system to date for cofactor-requiring enzymes. A dextran-linked NAD is commercially available from Sigma Chemical Company. Concentration polarization needs to be avoided and a tangential flow of substrate across the membrane rather than a straight through flow is better. The operation of such reactors is discussed by Wandrey & Wichmann (1985).

Acetate kinase with acetyl phosphate was the system of choice for regenerating ATP from ADP in a membrane reactor system in which glucose was converted to glucose-6-phosphate by hexokinase (Berke *et al.*, 1988). The cofactor was bound to polyethylene glycol 20 000 via an aminohexyl spacer arm and was recycled 20 000 times.

While the enzyme systems have given promising results, there are problems due to stability, enzyme availability and activity, unwanted products and it is not too surprising that whole cells have been used for biotransformations requiring cofactor regeneration.

3 REACTOR TECHNOLOGIES

A number of different reactor systems have been developed for immobilized biocatalysts some of which have already been mentioned. The stirred tank reactor, developed primarily for fermentations, has been used. It is necessary to minimize attrition and low shear agitator systems have been developed. If the pH has to be maintained by titrant addition it is necessary to balance the need for good mixing to prevent denaturation of the enzyme with the requirement to minimize biocatalyst break-up. A system which achieves both is the recirculation reactor where the enzymic conversion is carried out in one reactor through which substrate is circulated and from which it then passes into a well stirred tank reactor where the titrant is added. A system of this type was used by Lagerlöf *et al.* (1976) Where there is little pH change during the reaction or where it is economically and technically feasible to use a buffered system, a packed bed reactor can be used. By adjusting the flow rate it is possible to achieve total conversion of substrate to product within the reactor. Where there are gaseous products or particulate matter in the process stream it may be preferable to flow substrate upwards through the bed of biocatalyst and fluidize the particles. Gist Brocades operate a fluidized bed reactor for the anaerobic treatment of effluent from yeast and penicillin fermentations (Gorris *et al.*, 1989). The trickling bed filter used for waste water treatment is a type of packed bed reactor, impurities being degraded by the microorganisms in the biofilm on the filter particles.

Shama (1988) has reviewed the reactor systems developed for ethanol production using flocculated or support-associated yeast cells.

Membrane bioreactors have been comprehensively reviewed by Belfort (1989). Using selectively permeable membranes the reactor system can have an integral extraction system for product removal and concentration (Matson & Quinn, 1986).

Analytical devices using immobilized biocatalysts are also forms of bioreactor. Guilbault (1984) reviewed the analytical use of immobilized enzymes and described the various devices. He listed the commercially-available soluble and immobilized enzymes and the commercially-available analytical instruments using immobilized enzymes which are available for measuring glucose, urea, uric acid, creatinine, lactose, sucrose, maltose, galactose, amino acids, pesticides and cholesterol. Ugarova & Lebedeva (1987) reviewed the use of immobilized luciferase for bioluminescent analyses. Biosensors, where an enzyme system (enzyme, whole cell, tissue slice, organelle) is immobilized on to a sensing device (pH, ion-selective, oxygen probe) have been the growth area. These are reviewed by Karube (1984), Lowe (1984), Guilbault & de Olivera Neto (1985) and McCann (1987). Baker (1988) discussed the use of field effect transistors (FETs) and ion-selective FETs (ISFETs) in biosensors. Guilbault (1989) described an immunobiosensor for detecting the salmonella in foods based on an antibody for *Salmonella typhimurium* immobilized on a piezo-electric crystal.

Immobilized biocatalysts are also being incorporated into extracorporeal and intracorporeal devices for treating medical disorders. The former are based on haemodialysis units where blood is circulated through the reactor system. One example is where blood is detoxified by urease acting on blood urea. Obviously special needs have to be met for such devices such as sterility, compatibility with blood, non-toxicity, non-antigeniaty, non clot-forming. Klein & Lange (1986) have reviewed the current state of development of these treatments. Progress has been slow because of the stringent requirements for any device particularly intracorporeal enzyme preparations.

The range of methodologies for biocatalysts is extensive but it is being matched by the range of devices in which these materials operate. It is probably true to say that for any application a range of methodologies exist for preparing the biocatalyst in a suitable form. The early years when immobilized enzymes were interesting curiosities are over and emphasis is now well on their application.

4 IMMOBILIZED BIOCATALYST APPLICATIONS

Apart from their applications in the analytical and medical fields, immobilized biocatalysts have been used on an industrial scale. Immobilized cells have long been an integral part of wastewater treatment plants and processes are being developed for treating effluents containing materials such as phenol (Wisecarver & Fan, 1989), pear-canning industry effluent and high strength cheese whey (Bermúdez *et al.*, 1988), papermill sludge (Gijzen *et al.*, 1988), cyanide (Thompson *et al.*, 1988) and landfill leachate (Gourdon *et al.*, 1989). In addition there are systems for the anaerobic processing of waste to produce biogas (methane and CO_2) using methanogenic microorganisms.

An interesting use of immobilized biomass is the recovery of minerals by biosorption. Tsezos *et al.* (1988) utilized immobilized *Rhizopus arrhizus* for recovering uranium.

Other industrial processes include L-amino acid production using an aminoacylase (Tanabe Seiyaku) or an L-amino acid dehydrogenase (Degussa) or aspartase (Tanabe Seiyaku). 6-Aminopenicillanic acid, the precursor of a range of semi-synthetic penicillins is produced from penicillin G or penicillin V using the appropriate penicillin acylase. The largest industrial application has been the production of fructose syrups with glucose isomerase. β-Galactosidase has been used to remove lactose from milk and to convert lactose in whey to glucose and galactose (Maugh, 1984). Lipases have shown useful potential for hydrolysis and transesterification of oils and fats, being able to provide products that are sometimes unobtainable by chemical techniques (Macrae, 1983; Yamane, 1987).

There are many more potential applications for immobilized biocatalysts which are currently being investigated.

5 CONCLUDING REMARKS

It has not been possible to do more than indicate the breadth of immobilized biocatalyst technology and even then some aspects remain untouched. Immobilization enables biocatalysts to be developed as usable and reusable catalytic agents. In addition, in the case of soluble enzymes it also converts a homogeneous catalyst into a heterogeneous catalyst subject to diffusional restrictions on the movement of substrate and product. This can be beneficial if an inhibitor is restricted from the active site of the enzyme but it limits the usefulness of enzymes for acting on macrosubstrates. Often immobilization enhances the stability of an enzyme and this may allow it to act at a higher temperature but this is not so for all enzymes and destabilization has been noted. A useful review of factors affecting enzyme stability is given by O'Fagain *et al.* (1988). The support can profoundly affect the kinetics of an enzyme by reason of its charge and microenvironmental pH within a support may differ from that of the bulk liquor causing apparent changes in pH optimum. Charge-mediated attraction and repulsion of substrate and product can profoundly alter the kinetic properties of an enzyme. These effects are more fully discussed in the other reviews cited. Oxygen limitation is a problem with immobilized living cells sometimes restricting growth to the outer area of a bead. Oxygen carriers such as perfluorocarbons (Cho & Wang, 1988) are helping to overcome this. Membrane systems for enriching the oxygen content of air are also being developed but their use would appear to be limited to high value products.

There are undoubtedly other aspects that have been omitted from this review but it must be apparent that immobilized biocatalyst technology is a rapidly growing field and many new process developments must be anticipated.

REFERENCES

Abramowicz, D. A. & Keese, C. R. (1989). *Biotechnology and Bioengineering*, **33**, 149.

Ahern, T. J., Katoh, S. & Sada, E. (1983). *Biotechnology and Bioengineering*, **25**, 881.

Alberti, B. N. & Klibanov, A. M. (1982). *Enzyme and Microbial Technology*, **4**, 47.

Ampon, K. & Means, G. E. (1988). *Biotechnology and Bioengineering*, **32**, 689.

Anderrson, E., Mattiasson, B. & Hahn-Hägerdal, B. (1984). *Enzyme and Microbial Technology*, **6**, 301.

Archambault, J., Volesky, B. & Kurz, W. G. W. (1989). *Biotechnology and Bioengineering*, **33**, 293.

Baker, C. J. S. (1988). *Laboratory Practice*, **37**(3), 13.

Belfort, G. (1989). *Biotechnology and Bioengineering*, **33**, 1047.

Berke, W., Schüz, H.-J., Wandrey, C., Morr, M., Denda, G. & Kula, M.-R. (1988). *Biotechnology and Bioengineering*, **32**, 130.

Bermúdez, J. J., Jimeno, A., Canovas-Diaz, M., Manjon, A. & Iborra, J. L. (1988). *Process Biochemistry*, **23**, 178.

Birnbaum, S., Buelow, L., Hardy, K. & Mosbach, K. (1988). *Enzyme and Microbial Technology*, **10**, 601.

Boag, A. H. & Sefton, M. V. (1987). *Biotechnology and Bioengineering*, **30**, 954.

Bourdillon, C., Lortie, R. & Laval, J. M. (1988). *Biotechnology and Bioengineering*, **31**, 553.

Braun, J., Le Chanu, P. & Le Goffic, F. (1989). *Biotechnology and Bioengineering*, **33**, 242.

Brodelius, P. (1983). In *Immobilized Cells and Organelles* Vol. 1, ed. B. Mattiasson. CRC Press, Boca Raton, Florida, p. 27.

Brodelius, P. (1985). In *Immobilized Cells and Enzymes a Practical Approach*, ed. J. Woodward. IRL Press Ltd, Oxford and Washington, DC, p. 127.

Brodelius, P. (1988). In *Bioreactor Immobilized Enzymes and Cells. Fundamentals and Applications*, ed. M. Moo-Young. Elsevier Applied Science, London and New York, p. 167.

Brodelius, P. & Vandamme, E. J. (1987). In *Biotechnology*, Vol. 7a, ed. J. F. Kennedy. VCH Verlagsgesellschaft mbH, D-6940, Weinheim, FRG, p. 405.

Bucke, C. (1983). *Philosophical Transactions of the Royal Society of London* **B 300**, 369.

Camp, C. E. & Sofer, S. S. (1987). *Enzyme and Microbial Technology*, **9**, 685.

Carrea, G., Riva, S., Bovara, R. & Pasta, P. (1988). *Enzyme and Microbial Technology*, **10**, 333.

Cassells, J. M. & Halling, P. J. (1989). *Biotechnology and Bioengineering*, **33**, 1489.

Champagne, C. P., Girard, F. & Marin, N. (1988). *Biotechnology Letters*, **10**, 463.

Champagne, C. P., Baillargeon-Côte, C. & Goulet, J. (1989). *Journal of Applied Bacteriology*, **66**, 175.

Chang, T. M. S. (1976). In *Methods in Enzymology*, Vol. 44, ed. K. Mosbach. Academic Press, New York, San Francisco and London, p. 201.

Chang, T. M. S. (1977). *Biomedical Applications of Immobilized Enzymes and Proteins*, Vols 1 & 2. Plenum Press, New York and London.

Cheetham, P. S. J. (1987). *Enzyme and Microbial Technology*, **9**, 194.

Cheetham, P. S. J., Blunt, K. W. & Bucke, C. (1979). *Biotechnology and Bioengineering*, **21**, 2155.

Chenault, H. K. & Whitesides, G. M. (1987). *Applied Biochemistry and Biotechnology*, **14**, 147.

Chibata, I. (1978). *Immobilized Enzymes*. John Wiley and Sons, New York, London, Sydney and Toronto.

Chibata, I., Tosa, T., Mori, T., Watanabe, T. & Sakata, N. (1986). *Enzyme and Microbial Technology*, **8**, 130.

Cho, M. H. & Wang, S. S. (1988). *Biotechnology Letters*, **10**, 855.

Cooper, A. C. (1984). *Chemistry in Britain*, **20**, 815.

Crumbliss, A. L., McLachian, K. L., O'Daly, J. P. & Henkins, R. W. (1988). *Biotechnology and Bioengineering*, **31**, 796.

Daka, J. N., Laidler, K. J., Sipehia, R. & Chang, T. M. S. (1988). *Biotechnology and Bioengineering*, **32**, 213.

Dalili, M. & Chau, P. (1988). *Biotechnology Letters*, **10**, 331.

Deshpande, M. V., Balkrishnan, H., Ranjekar, P. K. & Shankar, V. (1987). *Biotechnology Letters*, **9**, 49.

De Taxis du Pöet, P., Dhulster, P., Barbotin, J.-N. & Thomas, D. (1986). *Journal of Bacteriology*, **165**, 871.

Dinelli, D., Marconi, W. & Morisi, F. (1976). In *Methods in Enzymology*, Vol. 44, ed. K. Mosbach. Academic Press, New York, San Francisco and London, p. 227.

Dixon, M. & Webb, E. C. (1979). *Enzymes*, 3rd edn. Longman Group Ltd, London, p. 468.

Domínguez, E., Nilsson, M. & Hahn-Hägerdal, B. (1988). *Enzyme and Microbial Technology*, **10**, 606.

Duff, S. J. B. (1988). *Biotechnology and Bioengineering*, **31**, 345.

Estell, D. A., Graycar, T. P. & Wells, J. A. (1985). *Journal of Biological Chemistry*, **260**, 6518.

Evans, T. L. & Miller, R. A. (1988). *Biotechniques*, **6**, 762.

Fadda, M. B., Dessi, M. R., Rinaldi, A. & Satta, G. (1989). *Biotechnology and Bioengineering*, **33**, 777.

Fadnavis, N. W. & Luisi, P. L. (1989). *Biotechnology and Bioengineering*, **33**, 1277.

Fersht, A. R. (1985). *Enzyme Structure and Mechanism*, 2nd edn. W. H. Freeman and Company, New York, p. 369.

Findlay, C. J., Parking, K. L. & Yada, R. Y. (1986). *Biotechnology Letters*, **8**, 649.

Franssen, M. C. R., Weijnen, J. G. J., Vincken, J. P., Laane, C. & van der Plas, H. C. (1988). *Biocatalysis*, **1**, 205.

Freeman, A. (1984). *Annals of the New York Academy of Sciences*, **434**, 418.

Fujimura, M., Moro, T. & Tosa, T. (1987). *Biotechnology and Bioengineering*, **29**, 747.

Fukushima, Y., Okamura, K., Imai, K. & Motai, H. (1988). *Biotechnology and Bioengineering*, **32**, 584.

Fusek, M., Turkova, J., Stovickova, J. & Franek, F. (1988). *Biotechnology Letters*, **10**, 85.

Gianfreda, L., Pirozzi, D. & Greco, G. Jr (1989). *Biotechnology and Bioengineering*, **33**, 1067.

Gijzen, H. J., Schoenmakers, T. J. M., Caerteling, C. G. M. & Vogels, G. D. (1988). *Biotechnology Letters*, **10**, 61.

Gin, H., Dupuy, B., Baquey, C., Ducassou, D. & Aubertin, J. (1987). *Journal of Microencapsulation*, **4**, 239.

Godbole, S. S., Kaul, R., D'Souza, S. F. & Nadkarni, G. B. (1983). *Biotechnology and Bioengineering*, **25**, 217.

Goldstein, L. & Manecke, G. (1976). In *Applied Biochemistry and Bioengineering*, Vol. 1, eds L. Wingard, E. Katchalski-Katzir & L. Goldstein. Academic Press, New York, San Francisco and London, p. 23.

Gorris, L. G. M., van Deursen, J. M. A., van der Drift, C. & Vogels, G. D. (1989). *Biotechnology and Bioengineering*, **33**, 687.

Gourdon, R., Comel, C., Vermande, P. & Véron, J. (1989). *Biotechnology and Bioengineering*, **33**, 1167.

Creco, G. Jr, Veronese, F., Largajolli, R. & Gianfreda, L. (1983). *European Journal of Applied Microbiology and Biotechnology*, **18**, 333.

Gregoriadis, G. (1976). In *Methods in Enzymology*, Vol. 44, ed. K. Mosbach. Academic Press, New York, San Francisco and London, p. 218.

Gu, K. F. & Chang, T. M. S. (1988). In *Bioreactor Immobilized Enzymes and Cells. Fundamentals and Applications*, ed. M. Moo-Young. Elsevier Applied Science, London and New York, p. 59.

Guilbault G. G. (1984). *Analytical Uses of Immobilized Enzymes.* Marcel Dekker Inc., New York and Basel.

Guilbault, G. G. (1989). *Biotechnology*, **7**, 349.

Guilbault, G. G. & de Olivera Neto, G. (1985). In *Immobilized Cells and Enzymes a Practical Approach*, ed. J. Woodward. IRL Press Ltd, Oxford and Washington, DC, p. 55.

Hallenbeck, P. C. (1983). *Enzyme and Microbial Technology*, **5**, 171.

Halling, P. J. (1987). *Biotechnology Advances*, **5**, 47.

Hartmeier, W. (1988). *Immobilized Biocatalysts—an Introduction.* Springer-Verlag, Berlin, Heidelberg.

Hirtenstein, M. & Clark, J. (1983). In *Immobilized Cells and Organelles*, Vol. 1, ed. B. Mattiasson. CRC Press, Boca Raton, Florida, p. 58.

Hopkinson, J. (1983). In *Immobilized Cells and Organelles*, Vol. 1, ed. B. Mattiasson. CRC Press, Boca Raton, Florida, p. 89.

Hustedt, H. (1986). *Biotechnology Letters*, **8**, 791.

Hutchinson, D. W. & Collier, R. (1987). *Biotechnology and Bioengineering*, **29**, 793.

Inada, Y., Takahashi, K., Yoshimoto, T., Ajima, A., Matsushima, A. & Saito, Y. (1986). *Trends in Biotechnology*, **4**, 190.

International Union of Biochemistry (1979). *Enzyme Nomenclature.* Academic Press Inc., New York.

Jain, P. & Wilkins, E. S. (1987). *Biotechnology and Bioengineering*, **30**, 1057.

Jin, F. & Toda, K. (1988). *Biotechnology Letters*, **10**, 221.

Joubert, W. A. & Britz, T. J. (1988). *Biotechnology Letters*, **10**, 49.

Kargi, F. (1988). *Biotechnology Letters*, **10**, 181.

Kargi, F. & Freidel, I. (1988). *Biotechnology Letters*, **10**, 409.

Karube, I. (1984). In *Biotechnology and Genetic Engineering Reviews*, Vol, 2, ed. G. E. Russell. Intercept Ltd, Newcastle-upon-Tyne, p. 313.

Kawamoto, T., Sonomoto, K. & Tanaka, A. (1987). *Biocatalysis*, **1**, 137.

Kennedy, J. F. & Cabral, J. M. S. (1985). In *Immobilized Cells and Enzymes a Practical Approach*, ed. J. Woodward. IRL Press Ltd, Oxford and Washington, DC, p. 19.

Kennedy, J. F. & Cabral, J. M. S. (1987). In *Biotechnology*, Vol. 7, ed. J. F. Kennedy. VCH Verlagsgesellschaft mbH, D-6940 Weinheim, FRG, p. 347.

Khmelnitsky, Y. L., Levashov, A. V., Klyachko, N. L. & Martinek, K. (1988). *Enzyme and Microbial Technology*, **10**, 710.

King, G. A., Dauglis, A. J., Faulkner, P., Bayly, D. & Goosen, M. F. A. (1988). *Biotechnology Letters*, **10**, 683.

Klein, M. D. & Lange, R. (1986). *Trends in Biotechnology*, **4**, 179.

Klibanov, A. M. (1989). *Trends in Biochemical Sciences*, **14**, 141.

Kobayashi, Y., Matsuo, R., Ohya, T. & Yokoi, N. (1986). *Biotechnology and Bioengineering*, **30**, 451.

Kobos, R. K., Eveleigh, J. W. & Arentzen, R. (1989). *Trends in Biotechnology*, **7**(4), 101.

Kokufuta, E., Shimizu, N., Tanaka, H. & Nakamura, I. (1988*a*). *Biotechnology and Bioengineering*, **32**, 756.

Kokufuta, E., Yamaya, Y., Shimada, A. & Nakamura, I. (1988*b*). *Biotechnology Letters*, **10**, 301.

Kolot, F. B. (1980). *Process Biochemistry*, **15**(7), 2.

Kolot, F. B. (1981*a*). *Process Biochemistry*, **16**(5), 2.

Kolot, F. B. (1981*b*). *Process Biochemistry*, **16**(6), 30.
Kutney, J. P., Berset, J. D., Heewitt, G. M. & Singh, M. M. (1988). *Applied and Environmental Microbiology*, **54**, 1015.
Lagerlöf, E., Nathorst-Westfelt, L., Ekström, B. & Sjoberg, B. (1976). In *Methods in Enzymology*, Vol. 44, ed. K. Mosbach. Academic Press, New York, San Francisco and London, p. 759.
Langer, R. S., Hamilton, B. K., Gardner, C. R., Archer, M. C. & Colton, C. K. (1976). *AICHE Journal*, **22**, 1079.
Larsson, M., Arasaratnam, V. & Mattiasson, B. (1989). *Biotechnology and Bioengineering*, **33**, 758.
Lowe, C. R. (1981). In *Topics in Enzyme and Fermentation Biotechnology*, Vol. 5, ed. A. Wiseman. Ellis Horwood Ltd, Chichester, p. 13.
Lowe, C. R. (1983). *Philosophical Transactions of the Royal Society of London*, **B 300**, 335.
Lowe, C. R. (1984). *Trends in Biotechnology*, **2**, 59.
Lozano, P., Manjón, A., Romojaro, F. & Iborra, J. L. (1988). *Process Biochemistry*, **23**, 75.
Macrae, A. R. (1983). *Journal of the American Oil Chemists Society*, **60**, 291.
Makino, K., Maruo, S.-I., Morita, Y. & Takeuchi, T. (1988). *Biotechnology and Bioengineering*, **31**, 617.
Malaníková, M., Malaník, V., Pšenička, I. & Marek, M. (1988). *Biotechnology Letters*, **10**, 579.
Manjón, A., Llorca, F., Bonete, M. J., Bastida, J. & Iborra, J. L. (1985). *Process Biochemistry*, **20**, 17.
Mansfeld, J. & Schellenberger, A. (1987). *Biotechnology and Bioengineering*, **29**, 72.
Martinek, K., Berezin, I. V., Khmelnitski, Y. L., Klyachko, N. L. & Levashov, A. V. (1987). *Biocatalysis*, **1**, 9.
Martinsen, A., Skjåk-Braek, G. & Smidsrød, O. (1989). *Biotechnology and Bioengineering*, **33**, 79.
Matson, S. L. & Quinn, J. A. (1986). *Annals of the New York Academy of Sciences*, **469**, 152.
Maugh, T. H. II (1984). *Science*, **223**, 474.
McCann, J. (1987). *Laboratory Practice*, **36**(5), 17.
Melik-Nubarov, N. S., Mozhaev, V. V., Siksnis, S. & Martinek, K. (1987). *Biotechnology Letters* **9**, 725.
Messing, R. A. (1975). *Journal of Non-crystalline Solids*, **19**, 277.
Messing, R. A. (1980). *Annual Reports on Fermentation Processes*, **4**, 105.
Messing, R. A. (1982). *Biotechnology and Bioengineering*, **24**, 1115.
Mizrahi, A. (1986). *Process Biochemistry*, **21**, 108.
Mozhaev, V. V., Siksnis, V. A., Melik-Nubarov, N. S., Galkantaite, N. Z., Denis, G. J., Butkus, E. P., Zaslavsky, B. Y., Mestechkina, N. M. & Martinek, K. (1988). *European Journal of Biochemistry*, **173**, 147.
Mulchandani, A., Luong, J. H. T. & Le Duy, A. (1989). *Biotechnology and Bioengineering*, **33**, 306.
Nelson, J. M. & Griffin, E. G. (1916). *Journal of the American Chemical Society*, **38**, 1109.
O'Fagain, C., Sheehan, H., O'Kennedy, R. & Kilty, C. (1988). *Process Biochemistry*, **23**, 166.
Ongcharit, C., Dauben, P. & Sublette, K. C. (1989). *Biotechnology and Bioengineering*, **33**, 1077.
Opara, C. C. & Mann, J. (1987). *Biotechnology and Bioengineering*, **31**, 470.
Oriel, P. (1988). *Biotechnology Letters*, **10**, 113.
Pederson, H., Furler, L., Venkatsubramanian, K., Prenosil, J. & Stuker, E. (1985). *Biotechnology and Bioengineering*, **27**, 961.

Poncelet, B., DeSmet, D., Poncelet, D. & Neufeld, R. J. (1989). *Enzyme and Microbial Technology*, **11**, 29.

Powell, L. W. (1984). In *Biotechnology and Genetic Engineering Reviews*, Vol. 2, ed. G. R. Russell. Intercept Ltd, Newcastle-upon-Tyne, p. 409.

Pu, H. T., Yang, R. Y. K. & Saus, F. L. (1989). *Biotechnology Letters*, **11**, 83.

Rao, K. K. & Hall, D. O. (1984). *Trends in Biotechnology*, **2**, 124.

Robinson, P. K., Mak, A. L. & Trevan, M. D. (1986). *Process Biochemistry*, **21**, 122.

Rochefort, W. E., Rehg, T. & Chau, P. C. (1986). *Biotechnology Letters*, **8**, 115.

Rosazza, J. P. (1982). *Microbial Transformations of Bioactive Compounds*, Vols 1 & 2. CRC Press, Boca Raton, Florida.

Rosevear, A. (1975). British Patent 1,514,707.

Rosevear, A. (1984). *Journal of Chemical Technology and Biotechnology*, **34B**, 127.

Rosevear, A. & Lambe, C. A. (1985). *Advances in Biochemical Engineering*, **31**, 37.

Rosevear, A., Kennedy, J. F. & Cabral, J. M. S. (1987). *Immobilized Enzymes and Cells*. Adam Hilger, Bristol and Philadelphia.

Royer, G. P., Ikeda, S. & Aso, K. (1977). *FEBS Letters*, **80**, 89.

Rucka, M. & Turkiewicz, B. (1989). *Biotechnology Letters*, **11**, 167.

Sada, E., Katoh, S., Terashima, M. & Tsukiyama, K.-I. (1988). *Biotechnology and Bioengineering*, **32**, 826.

Schmidt, E., Bossow, B., Wichmann, R. & Wandrey, C. (1986). *Kemija u Industriji*, **35**, 71.

Semenov, A. N., Khelmnitsky, Y. L., Berezin, I. V. & Martinek, K. (1987). *Biocatalysis*, **1**, 3.

Shama, G. (1988). *Process Biochemistry*, **23**, 138.

Silberger, E. & Freeman, A. (1987). *Biotechnology and Bioengineering*, **30**, 675.

Simionescu, C., Popa, M. I. & Dumitriu, S. (1987). *Biotechnology and Bioengineering*, **29**, 361.

Smiley, A. L., Hu, W.-S. & Wang, D. I. C. (1989). *Biotechnology and Bioengineering*, **33**, 1182.

Smith, M. (1985). *Annual Review of Genetics*, **19**, 423.

Smith, R. A. G. (1976). *Nature*, **262**, 519.

Soda, K. & Yonaha, K. (1987). In *Biotechnology*, Vol. 7a, ed. J. F. Kennedy. VCH Verlagsgesellschaft mbH, D-6940, Weinheim, FRG p. 605.

Sofer, S. S. (1979). *Enzyme and Microbial Technology*, **1**, 3.

Sørensen, J. E. & Emborg, C. (1989). *Enzyme and Microbial Technology*, **11**, 26.

Steenson, L. R., Klaenhammer, T. R. & Swaisgood, H. E. (1987). *Journal of Dairy Science*, **70**, 1121.

Synowiecki, J., Sikorska-Siondalska, A. & El-Bedaway, A. El-Fath (1987). *Biotechnology and Bioengineering*, **29**, 352.

Tai, D.-F., Fu, S.-L., Chuang, S.-F. & Tsai, H. (1989). *Biotechnology Letters*, **11**, 173.

Takeuchi, T. & Makino, K. (1987). *Biotechnology and Bioengineering*, **29**, 160.

Tampion, J. & Tampion, M. D. (1987). *Immobilized Cells: Principles and Applications*. Cambridge University Press, Cambridge.

Tanaka, A. & Fukui, S. (1983). In *Immobilized Cells and Organelles*, Vol. 1, ed. B. Mattiasson. CRC Press, Boca Raton, Florida, p. 102.

Thomasset, B., Thomas, D. & Lortie, R. (1988). *Biotechnology and Bioengineering*, **32**, 764.

Thompson, L. A., Knowles, C. J., Linton, E. A. & Wyatt, J. M. (1988). *Chemistry in Britain*, **24**, 900.

Trevan, M. D. (1980). *Immobilized Enzymes. An Introduction and Applications in Biotechnology*. John Wiley & Sons, Chichester, New York, Brisbane and Toronto.

Trevan, M. D. & Mak, A. L. (1988). *Trends in Biotechnology*, **6**, 68.

Tsezos, M., Noh, S. H. & Baird, M. H. I. (1988). *Biotechnology and Bioengineering*, **32**, 545.

Ugarova, N. N. & Lebedeva, O. V. (1987). *Applied biochemistry and Biotechnology*, **15**, 35.

Vaidya, S., Srivastasa, R. & Gupta, M. N. (1987). *Biotechnology and Bioengineering*, **29**, 1040.

Varani, J., Fligiel, S. E. G., Inman, D. R., Helmreich, D. L., Bendelow, M. J. & Hillegas, W. (1989). *Biotechnology and Bioengineering*, **33**, 1235.

Viera, F. B., Barragan, B. B. & Busto, B. L. (1988). *Biotechnology and Bioengineering*, **31**, 711.

Vorlop, K.-D. & Klein, J. (1981). *Biotechnology Letters*, **3**, 9.

Wandrey, C. & Wichmann, R. (1985). In *Enzymes and Immobilized Cells in Biotechnology*, Vol. 8, Biotechnology series 5, ed. A. I. Laskin. Benjamin/Cummings Publishing Company, Mendo Park, California, p. 177.

Wandrey, C., Wichmann, R. & Jandel, A.-S. (1982). *Enzyme Engineering*, **6**, 61.

Whitesides, G. M. (1985). In *Enzymes in Organic Synthesis*, eds R. Porter & S. Clark. Pitman, London, p. 76.

Wilkinson, A. J., Fersht, A. R., Blow, D. M., Carter, P. & Winter, G. (1984). *Nature* **307**, 187.

Wisecarver, K. D. & Fan, L.-S. (1989). *Biotechnology and Bioengineering*, **33**, 1029.

Woodward, J. (1985). In *Immobilized Cells and Enzymes a Practical Approach*, ed. J. Woodward. IRL Press Ltd, Oxford and Washington, DC, p. 3.

Yamada, H. & Shimizu, S. (1988). *Angewandte Chemie—International Edition in English*, **27**, 622.

Yamane, T. (1987). *Journal of the American Oil Chemists' Society*, **64**, 1657.

Yoshimoto, T., Takahishi, K., Nishimura, H., Ajima, A., Tamaura, Y. & Inada, Y. (1984). *Biotechnology Letters*, **6**, 337.

Zaborsky, O. R. (1973). *Immobilized Enzymes*. CRC Press, Boca Raton, Florida.

Chapter 12

ENZYMES IN ORGANIC SYNTHESIS

S. M. ROBERTS

Department of Chemistry, University of Exeter, Stocker Road, Exeter EX4 4QD, UK

CONTENTS

1 INTRODUCTION

The use of enzymes as catalysts for key steps in the synthesis of complex organic molecules is gaining popularity at a great rate. There are several reasons for this trend. First enzymes operate under mild conditions of temperature (c. 30°C) and pH (c. 7·0) and hence sensitive non-participating functional groups are not affected. Secondly enzymes can accomplish reactions which are difficult to emulate using more conventional chemical methods. For example, using oxygenase enzymes highly regioselective functionalisation of un-activated carbon centres can be achieved. Thirdly enzymes, being chiral catalysts, are able to effect kinetic resolution of racemates, and are also able to distinguish enantiotopic functional groups in prochiral systems, to generate optically active synthons. The preparation of chiral intermediates is of interest to a wide range of synthetic chemists particularly those scientists concerned with the preparation of new pharmaceuticals, agrochemicals, fragrances and flavours (Davies *et al.*, 1990).

The types of reactions catalysed by enzymes that are of interest to synthetic organic chemists are as follows:

Hydrolysis reactions
    Hydrolysis of carboxylic acid esters
    Hydrolysis of phosphate esters
    Hydrolysis of amides
    Hydrolysis of nitriles
    Other hydrolysis reactions
Esterification reactions, transesterification reactions, and the synthesis of amides
    Esterification and transesterification reactions
    Synthesis of amides
Reduction reactions
    Reduction of alkanones and alkanediones
    Reduction of oxoalkanoates
    Reduction of $\alpha\beta$-unsaturated carbonyl compounds and related species
Oxidation reactions
    Oxidation of alcohols to carbonyl compounds
    Formation of alkanols by the oxidation of aromatic and aliphatic compounds
    Hydroxylation of alicyclic and aliphatic compounds
    Oxidation of sulphides to sulphoxides
    Oxidation of ketones to esters (lactones)
Carbon–carbon bond-forming reactions
    The aldol reaction
    Cyanohydrin formation
    The acyloin reaction
Other biotransformations

When considering the use of biotransformations in organic synthesis it is sometimes necessary to consider the respective advantages and disadvantages of using a microorganism as opposed to an isolated enzyme. The 'pros and cons' of using a whole cell system on the one hand, or an isolated partially purified enzyme on the other, have been discussed elsewhere (Butt & Roberts, 1987).

Many other enzyme-catalysed reactions have been investigated to a greater or lesser extent and some have been used in synthetic organic chemistry. However transformations falling under the above headings account for ≫95% of the reported, synthetically useful reactions. A comprehensive survey of all the reactions from the above classes is impossible in the limited space of this review; a more expansive test should be consulted for a broader view (Davies *et al.*, 1989*b*). In this text a selection of reactions will be taken from each category so as to demonstrate the real and potential utility of the processes. Relevant examples reported during the period 1987–mid-1989 are quoted whenever possible to ensure an up-to-date perspective on the field.

Hydrolysis and esterification reactions are generally carried out using isolated enzymes (e.g. using commercially available esterases, lipases, acylases, kinases and phosphatases). Similarly the hydrolysis and preparation of amides involve processes that are efficiently catalysed by enzymes. The reduction of ketones by secondary alcohols can be accomplished using a microorganism (e.g. *Saccharomyces cerevisiae*) or a dehydrogenase enzyme. The reduction and oxidation of alkenes, the oxidation of aromatic compounds, and the acyloin reaction are processes generally effected using a whole-cell system, while the oxidation of alcohols to carbonyl compounds, aldol reactions and the formation of cyanohydrins are generally achieved using enzymes. Examples of all these processes are described below.

2 HYDROLYSIS REACTIONS

2.1 Hydrolysis of Carboxylic Acid Esters

Enzyme-catalysed hydrolysis of a carboxylic acid ester is often employed as a means of access to optically active materials. Simple racemic esters can be resolved using biotransformations. Thus (±)-3-acetoxyoct-1-yne is hydrolysed using lyophilised yeast (Glänzer *et al.*, 1987) or *Mucor miehei* lipase (Chan *et al.*, 1988) to give 3(*S*)-oct-1-yn-3-ol (**1**) (80–100% e.e.) and recovered optically

(1)   (2)   (3)

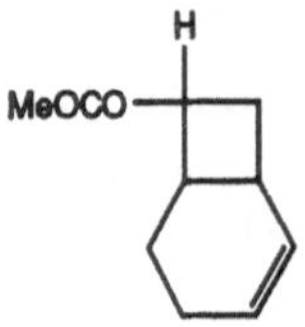

(±) - (4) 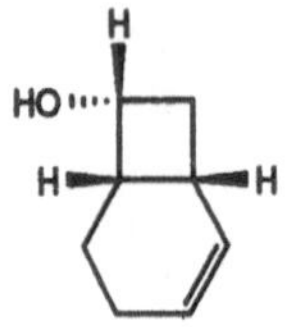   (5) 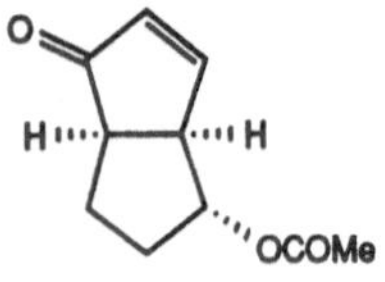   (6)

active 3(R)-acetate (2). The alcohol (1) has been used as a synthon for 13(S)-hydroxyoctadecadienoic acid (3) (Chan et al., 1989), a compound produced in mammals and purported to have a pronounced effect on the cardiovascular system.

More complex molecules can be resolved in a similar manner. Thus the racemic ester (4) is enantioselectively hydrolysed using *M. miehei* lipase or porcine pancreatic lipase to give optically pure alcohol (5) (and optically pure recovered 1(S), 6(R), 7(S)-acetate) (Cotterill et al., 1988a). The alcohol (5) was used to prepare useful bicyclo[3.3.0]octenones (6) (Cotterill et al., 1988b) by a series of stereocontrolled reactions.

The diacetate (7) was hydrolysed using cholesterol esterase from bovine pancreas to give the mono-ol (8) and this material was used as the starting point in an efficient synthesis of optically active D-myoinositol-1,4,5-triphosphate (9) (Liu & Chen, 1989). Enzyme-catalysed hydrolyses of 1,2- and 1,3-diacyloxycycloalkanes using pig liver esterase have also been reported (Crout et al., 1986; Baudin et al., 1988).

One of the most useful strategies for the employment of hydrolase enzymes in organic synthesis is the partial hydrolysis of prochiral diesters to give optically active compounds, inasmuch as the product could theoretically be formed in optically pure form in quantitative yield. While this optimum transformation is rarely achieved, nevertheless optically enriched compounds are often obtained. For example the diester (10) was hydrolysed using *Candida cylindracea* lipase to give the mono-ol (11) (97·5% e.e.) (Honig et al., 1989) *en route* to deoxyinositols.

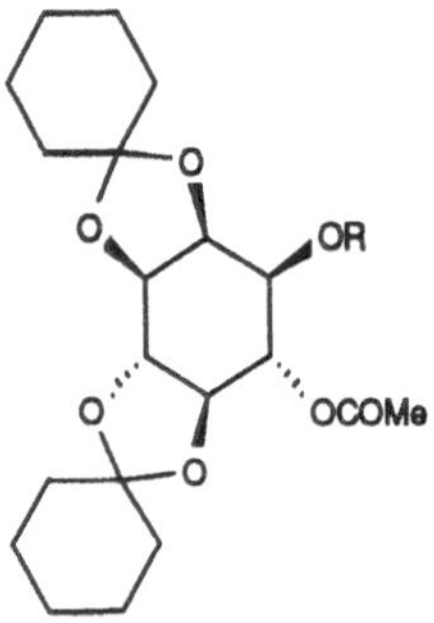

(7) R = COMe

(8) R = H

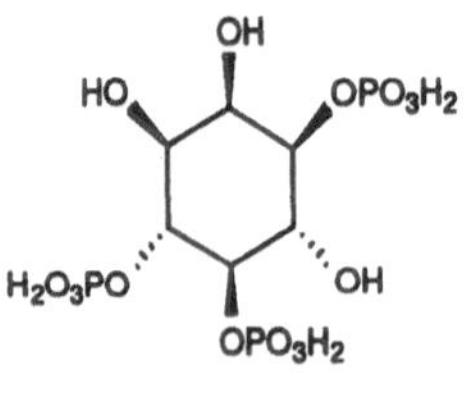

(9)

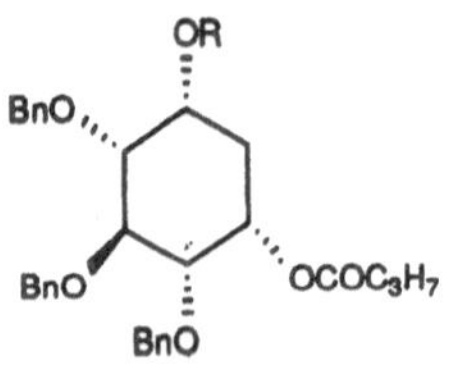

(10) R = COC$_3$H$_7$

(11) R = H

Up to this point the hydrolysis of esters carrying chiral centres in the alcohol moiety has been discussed. An equal body of work has been published describing hydrolysis reactions on materials in which the chirality is present in the acid portion of the molecule. The 2-benzyloxyesters (12) were hydrolysed using *Corynebacterium equi* and the recovered starting material (38–42% yield) was found to be the optically active (S)-form (99% e.e.). The acid was not recovered from this whole cell biotransformation (Kato *et al.*, 1987). Epoxy esters of the type (13) are hydrolysed by esterases to give the (S)-acid and by proteases to give the (R)-acid. Compounds of this type are well-established building blocks in organic synthesis.

(12) R = small alkyl   (13)   (14)

The enzyme-catalysed hydrolysis of cyclic *meso*-diesters having the structure (14) has been described (Sabbioni & Jones, 1987). Certain monoesters produced in this work have been used by Ohno's group to prepare thienamycin and 1β-methylpenems (Kaga *et al.*, 1989). Ohno's other noteworthy work in this area includes the conversion of the diester (15) into the carboxylic acid (16) (*c.* 80% e.e.) using pig liver esterase as a key step in the synthesis of aristeromycin (17) (Ohno, 1985). More recently it has been shown that the diester (18) can be hydrolysed using pig liver esterase to give the monoester (19) (92% yield, 96% e.e.) in work aimed at the synthesis of other carbocyclic ribonucleosides (Zemlicka *et al.*, 1989).

Acyclic diesters can also be hydrolysed with high selectivity; malonyl diesters, such as the oxaziridine derivative (20) (Bucciarelli *et al.*, 1989) and glutaric acid diesters have been popular substrates. In the latter series the acid (21) has been prepared by pig liver esterase-catalysed hydrolysis of the corresponding diester and subsequently used in the synthesis of the chiral diene (22) (a sexual pheromone of yellow scale) (Alvarez *et al.*, 1988) and (+)-faranal (23) (a component of the trail pheromone of Pharaoh's ant) (Poppe *et al.*, 1988). The use of hydrolase enzymes in organic synthesis is considerably enhanced by the availability of models for the active sites of the following enzymes, pig liver esterase (Mohr *et al.*, 1984), α-chymotrypsin (Cohen, 1969), *Candida cylindracea* lipase (Oberhauser *et al.*, 1989), *Pseudomonas fluorescens* lipase (Xie *et al.*, 1988).

2.2 Hydrolysis of Phosphate Esters

Acid-catalysed hydrolysis of polyprenyl pyrophosphates is difficult due to the acid lability of the resulting alcohols. Potato phosphatase hydrolyses these phosphates readily under relatively mild conditions (Fujii *et al.*, 1982).

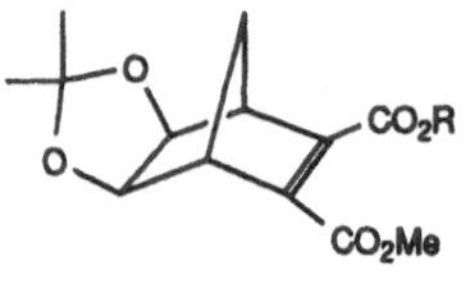

(15) R = Me
(16) R = H

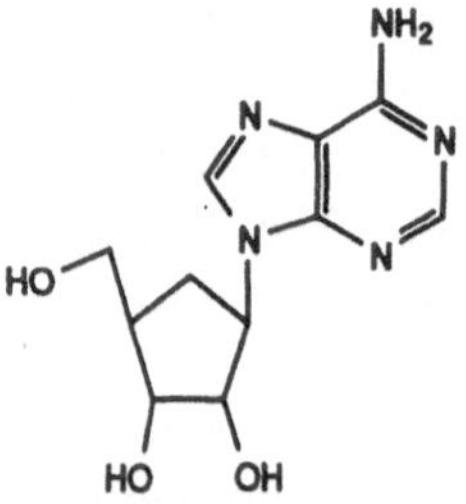

(17)

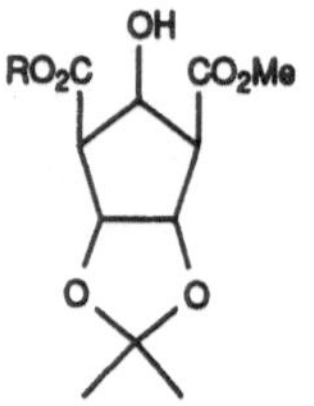

(18) R = Me
(19) R = H

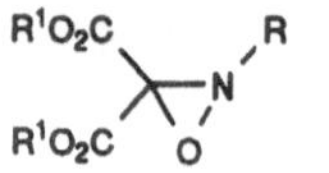

(20) R = ^{i}Pr, ^{t}Bu
R^1 = Me, Et

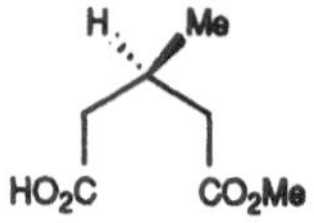

(21)

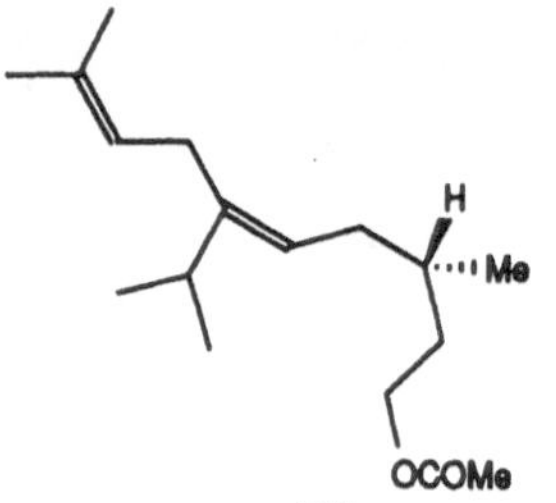

(22)

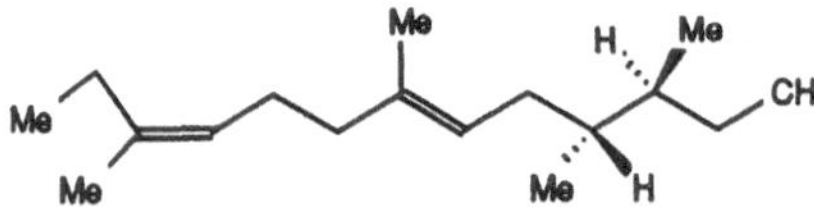

(23)

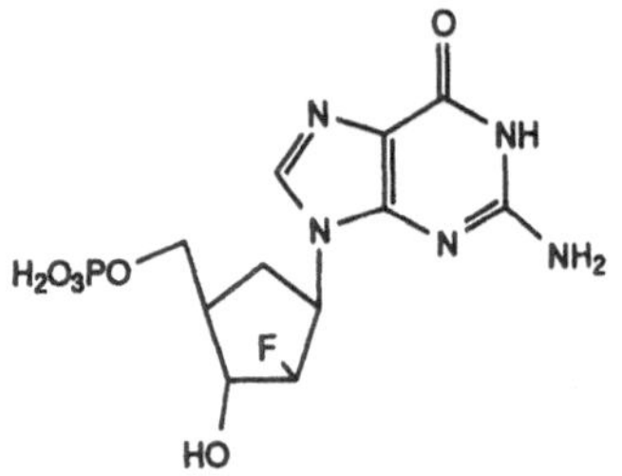

(24)

The enantioselective hydrolysis of a carbocyclic nucleoside 5'-phosphate has been achieved. Thus the racemic monoalkyl phosphate (24) was hydrolysed using 5'-nucleotidase (EC 3.1.3.5) from *Crotalus atrox* venom to give a nucleoside analogue with very potent anti-herpes activity (Borthwick *et al.*, 1988).

2.3 Hydrolysis of Amides

The enzyme-catalysed hydrolysis of the amide linkage has assumed particular importance in the synthesis of semi-synthetic penicillins and cephalosporins. The conversion of compounds such as penicillin-G (25) into 6-aminopenicillanic acid (26) can be accomplished on a very large scale: immobilization of the penicillin acylase enzyme on a solid support such as Sephadex has been achieved (Lagerlöf *et al.*, 1976).

Carboxypeptidases have been used for the resolution of a wide variety of

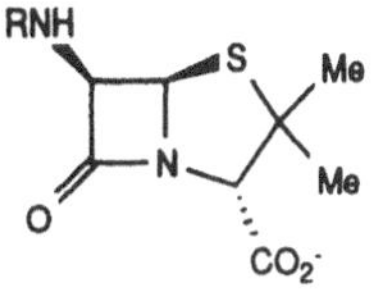

(25) R = PhCH₂CO

(26) R = H

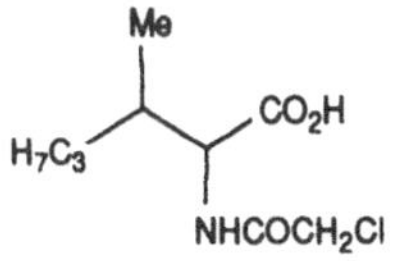

(27)

amino acids (Jones *et al.*, 1976; Dirlam *et al.*, 1987). Aminoacylases have also been employed for the production of L-amino acids such as alanine, methionine, tryptophane, and valine from the corresponding racemic *N*-acetyl precursors, often on a very large scale (Wandrey, 1986). The racemic *N*-chloroacetylamino acid (**27**) was resolved using an acylase from *Aspergillus* sp. Both the amino acid and the recovered starting material were obtained in a highly optically pure state and both compounds were used in syntheses of *threo*-4-methylheptan-3-ols, pheromone components of the destructive elm-bark beetle (Mori & Iwasawa, 1980).

Penicillin acylase from *E. coli* (EC 3.5.1.11) has also been recommended for the removal of the N-terminal protecting group in *N*-phenylacetyl-dipeptide esters, since peptide and ester bonds are unaffected (Waldmann, 1988).

2.4 Hydrolysis of Nitriles

The ease of hydrolysis of alkyl and aryl nitriles using enzymes is attractive to a synthetic organic chemist since the alternative chemical procedures require quite stringent reaction conditions. Thus resting cells of *Rhodococcus rhodochrous* (previously grown on benzonitrile) catalyse the conversion of 1,3-dicyanobenzene into 3-cyanobenzoic acid in 95% yield (Bengis-Garber & Gutman, 1988; Kobayashi *et al.*, 1988). Similarly, 3-cyanopyridine and 2-cyanopyridine can be converted into nicotinic acid and isonicotinamide respectively in quantitative yield (Mauger *et al.*, 1988) and a nitrile hydratase from *Pseudomonas chlorophis* B23 has been used to prepare acrylamide from acrylonitrile on a substantial scale (Endo & Watanabe, 1989).

The microorganism *Brevibacterium* sp. R312 possesses a nitrile hydratase and an amidase with wide substrate selectivities. Thus highly functionalised acids, such as compound (**28**) can be prepared from the corresponding nitrile in good yield (Vo-Quang *et al.*, 1987).

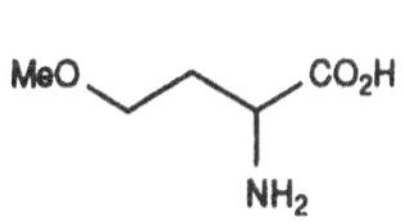

(28) 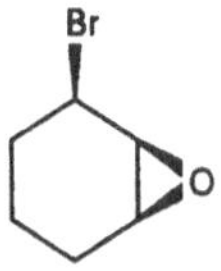   (29) 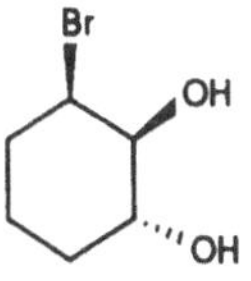   (30)

2.5 Other Hydrolysis Reactions

The resolution of γ-, δ- and ε-lactones has been effected using porcine pancreatic lipase, horse liver esterase and pig liver esterase (Blanco *et al.*, 1988) while the regioselective ring opening of α-substituted cyclic acid anhydrides has been accomplished using a lipase (Hiratake *et al.*, 1989).

The hydrolysis of various cyclohexene epoxides by rabbit liver microsomes has been studied. The utility of the method is limited to small scale reactions. A recent paper describing the epoxide hydrolase-catalysed hydrolysis of the bromo compound (**29**) noted several interesting effects. First, one enantiomer of the epoxide competitively inhibited the hydrolysis of the other enantiomer. Secondly, the preferred pattern of ring opening utilised the conformation showing (3,4 M) helicity to afford the diol (**30**) (60% e.e. at 30% conversion) (Bellucci *et al.*, 1988).

3 ESTERIFICATION REACTIONS, TRANSESTERIFICATION REACTIONS AND THE SYNTHESIS OF AMIDES

3.1 Esterification and Transesterification Reactions

While the ability of enzymes to effect esterification reactions has been known for ninety years the seminal work of Klibanov showed that the process could be used to resolve a racemic carboxylic acid. Thus 2-bromopropanoic acid was esterified with high enantioselectivity, using yeast lipase to catalyse the reaction between the acid and an alcohol (e.g. *n*-butanol) in an inert solvent (e.g. hexane) containing a trace quantity of water (Kirchner *et al.*, 1985). Over the past five years such uses of lipases to effect the condensation of alcohols and carboxylic acids in organic solvents has become very popular (Hennen *et al.*, 1988). One of the major advantages of this type of biotransformation is that the product(s) (and starting material(s)) are recovered easily. The enzymes are often used in an immobilised form to further simplify the work-up procedure.

The work of Sonnet (1987) suggested that chiral secondary alcohols can be acylated regioselectivity using an immobilised form of *Mucor miehei* lipase (Lipozyme®). ($\pm$)-Bicyclo[3.2.0]hept-2-en-6-*endo*-ol is acylated using cyclohexane carboxylic acid in hexane, employing Lipozyme as catalyst, to give the ester (**31**) (94% e.e.) and recovered optically active alcohol. The optically active components can be separated and used in enantiocomplementary syntheses of prostaglandins, for example prostaglandin-F$_2\alpha$ (**32**) (Newton & Roberts, 1980). Some advantage is gained by utilising a transesterification procedure involving the same enzyme, solvent and acylating agent but using ($\pm$)-6-acetoxybicyclo[3.2.0]hept-2-ene as the substrate. In this case the cyclohexanoate (**31**) was obtained in 97·2% e.e.

(31)   (32)   (33)

The esterification of a racemic sample of the pheromone sulcatol using 2,2,2-trifluoroethyl dodecanoate and dried porcine pancreatic lipase in ether gave biologically active (*S*)-sulcatol (**33**) (>97% e.e.) and the (*R*)-ester (*c.* 90% e.e.) (Stokes & Oehlschlager, 1987). Similarly, lipase-catalysed transesterifications between trichloroethyl butyrate and C-6-blocked monosaccharides gave C-2 and C-3-acylated products; protease-catalysed regioselective esterification of sugars and related compounds can be effected in anhydrous dimethylformamide (Riva *et al.*, 1988). Similarly 1,4-anhydro-5-*O*-hexadecyl-D-arabinitol is acylated at the 2-position using *Pseudomonas fluorescens* lipase as the catalyst and at the 3-position using *Rhizopus japonicus* lipase as the catalyst (Nicotra *et al.*, 1989).

The value of employing an enol ester in an ester interchange process is that the alcohol released in the biotransformation instantly tautomerises to the more stable 'keto' form thus ensuring that the reaction is irreversible (Degueil-Castaing *et al.*, 1987). The strategy has been used for the enantio-/regio-selective esterification of a range of alcohols (Wang *et al.*, 1988); for example acylation of 2-*O*-benzylglycerol using *Pseudomonas* lipase and 2-propenyl acetate gave the acetate (**34**) in optically pure form (Wang & Wong, 1988).

(34)   (35)   (36)   (37)

In a contemporaneous paper a variety of secondary alcohols of the type (aryl)CHOH(alkyl) were shown to be enantioselectively acylated using vinyl acetate in methyl *tert*-butyl ether using Lipase SAM II as the catalyst. Observed enantiomeric excesses of products were often >95% (Laumen *et al.*, 1988). Similar enantioselective esterifications can be accomplished using *Pseudomonas fluorescens* lipase (adsorped onto Celite) and acetic, propionic or butyric *anhydrides* as the acylating agents (Bianchi *et al.*, 1988). The same enzyme can be used to resolve racemic hydroperoxides (Baba *et al.*, 1988).

Intramolecular esterification has been achieved to provide δ-lactones such as the compound (**35**). In other cases (such as δ-valerolactone) intermolecular oligomerisation was observed (Gutman *et al.*, 1987). The synthesis of macrocyclic lactones such as compound (**36**) has been studied using long-chain alkanedioic acids, alkanediols, and a lipase catalyst (Zhi-Wei & Sih, 1988).

The process of enzyme-catalysed transesterification of fats has been explored for many years by scientists in the food industry. Triglycerides enriched with ω-3 polyunsaturated fatty acids are obtained on *Mucor miehei* lipase-catalysed interesterification (Haraldsson *et al.*, 1989).

3.2 Synthesis of Amides

The reactions of a variety of amines such as aniline and *n*-butylamine with ($\pm$)-ethyl chloropropionate under catalysis by *Candida cylindracea* lipase are enantioselective and optically active amides (**37**) are produced (Gotor *et al.*, 1988).

The steady increase in the interest of peptides in the biological properties has prompted a considerable amount of work towards the enzyme-catalysed synthesis of such compounds (Kullmann, 1987). The use of proteases has been explored as exemplified by Barbas & Wong (1988) who reported a one-pot synthesis of a tripeptide from three single amino-acid derivatives using papain as catalyst. The formation of a D-Ala–D-Ala containing tripeptide in aqueous solution has been accomplished using muramoylpentapeptide carboxypeptidase (EC 3.4.17.8) as the catalyst (Ekberg *et al.*, 1989).

The synthesis of aspartyl containing dipeptides has assumed a high profile following the commercial success of the sweetener Aspartame. Thermolysin and α-chymotrypsin have been used to couple Z–Asp to Phe–OMe or Phe–NH_2 (Adisson *et al.*, 1988). The use of *N*-formyl amino acid esters in enzyme-catalysed peptide synthesis had been advocated (Flörsheimer & Kula, 1988).

Lipsases have also been employed for the synthesis of di- and tripeptides. The advantage of using a lipase for this process lies in the fact that the enzyme has less of a tendency to promote the undesirable hydrolysis reaction (Margolin *et al.*, 1987).

4 REDUCTION OF KETONES TO SECONDARY ALCOHOLS

4.1 Reduction of Alkanones and Alkanediones

The reduction of ketones to secondary alcohols using microorganisms or isolated enzymes (with the appropriate cofactor(s)) is a well-established practice in synthetic organic chemistry. Simple alcohols such as (*S*)-sulcatol (**33**) can be prepared from the corresponding ketone (**38**) using yeast (80% yield, 94% e.e.) or growing cells of *Thermoanaerobium brockii* (100% yield, >99% e.e.). ((*R*)-Sulcatol can be obtained from 6-methylhept-5-en-2-one using *Aspergillus niger* (Belan *et al.*, 1987).) Reduction of the ketone (**39**)

(38)   (39)   (40)   (41)

with *A. niger* or *Geotrichum candidum* affords the (*R*)-alcohol (e.e. 90%) while bakers' yeast can be employed to produce the (*S*)-alcohol (Bernardi *et al.*, 1987).

The use of enzymes to reduce relatively simple ketones is illustrated by the conversion of the α-keto acid (40) into (*S*)-2-hydroxybutanoic acid (99% yield, >99% e.e.) using lactate dehydrogenase and reduced nicotinamide adenine dinucleotide (NADH) as the cofactor. The hydroxyacid was converted into the epoxide (41) in three simple chemical steps (Kim & Whitesides, 1988).

Microorganisms have also been employed to reduce ketones with pre-existing chiral centres. For example, the reduction of the ketone moiety in the β-lactam (42) was studied using a variety of microorganisms in an effort to

(42)   (43)   (44)   (45)

prepare a synthon for thienamycin and analogues (Hirai & Naito, 1989). The bicyclic ketone (±)-(43) was reduced by the fungus *Mortierella ramanniana* to give the diastereoisomeric alcohols (44) (80% e.e.) and (45) (>95% e.e.). The two diastereoisomeric alcohols are easily separated and the alcohol (44) was converted into eldanolide (46), a pheromone of the West African sugar cane borer (Butt *et al.*, 1987). The racemic ketone (43) can also be resolved using *Thermoanaerobium brockii* dehydrogenase in the presence of NADH in enantioselective fashion to give optically pure 6(*S*)-alcohol (44) and recovered optically active ketone. *Both* these chiral compounds have been used to prepare the chemotactic agent leukotriene-B$_4$ (47) (Davies *et al.*, 1985).

The reduction of (±)-tricarbonyl(2-methoxy benzaldehyde) chromium with bakers' yeast allows the ready preparation of chiral alcohol (48) (66% e.e.) and aldehyde complexes in optically active form (Top *et al.*, 1988).

The reduction of 1,2-diketones using whole cells and enzymes has been studied. For example the diketone (49) was reduced using bakers' yeast to afford the ketone (50) which was converted into L-digitoxose a component of digitalis (Fujisawa *et al.*, 1985). Similarly, benzil is converted into the

(46)   (47)   (48)

(R)-hydroxyketone using *Xanthomonas campestris* pv. *oryzae* (Konishi *et al.*, 1985).

Pentan-2,4-dione is reduced to (S)-4-hydroxypentan-2-one (90% yield, 99% e.e.) using yeast and to (R)-4-hydroxypentan-2-one (75% yield, 95% e.e.) using A. *niger* (Fauve & Veschambre, 1988). Similarly hexan-2,5-dione is reduced by bakers' yeast to give 2(S), 5(S)-hexanediol, a useful intermediate for the production of 2,5-dimethyl pyrrolidine (Short *et al.*, 1989). The latter compound is used as a chiral ligand and reactions of the derived amides have been investigated. 4-Methylheptane-3,5-dione is reduced by *Geotrichum candidum* to give the hydroxyketone (51) (70% e.e.) under aerobic conditions and the diastereoisomer (52) (100% e.e.) (the natural pheromone sitophilure) under anaerobic conditions (Fauve & Veschambre, 1987).

(49)   (50)   (51) R^1 = Me; R^2 = H

                                                            (52) R^1 = H; R^2 = Me

(53)   (54)   (55)

The yeast reduction of 2,2-disubstituted 1,3-diones in five- and six-membered ring systems has been explored. For cyclopentane derivatives the 2(S), 3(S)-compound is the major product: for cyclohexane derivatives the predominant product is the 2(R), 3(S)-isomer (Brooks *et al.*, 1987a). There are numerous examples of the use in synthesis of cyclic hydroxyketones that have been prepared by biotransformation. The cyclohexane derivative (53) has been used in the synthesis of (+)-baiyunol, (54) the diterpene aglycone of the

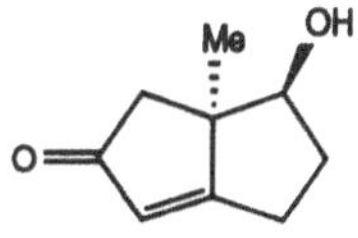

(56) 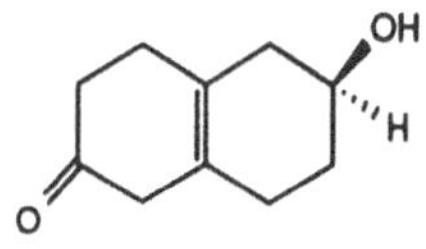 (57) 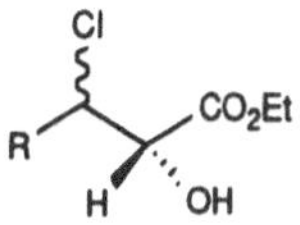 (58)

sweet glycoside baiyuroside (Mori & Komatsu, 1987). The hydroxycyclopen-tanone (55) has been converted into the bicyclo-[3.3.0]octanone (56) en route to the complex natural product coriolin (Brooks *et al.*, 1985).

Bicyclo[4.4.0]dec-1-en-3,8-dione is reduced by horse liver alcohol de-hydrogenase (in the presence of the cofactor, NADH) to give the hydroxy-ketone (57) (64% yield, >98% e.e.) (Dodds & Jones, 1988).

4.2 Reduction of Oxoalkanoates

Treatment of ethyl 3-chloro-2-oxoalkanoates with yeast yields diastereo-isomeric mixtures of products (58) which all possess exclusively the (S)-stereochemistry at the 2-position. Esters of this type can be converted into chiral 2,3-epoxyalcohols to provide an alternative to the Sharpless oxidation (see Section 5.4) (Tsuboi *et al.*, 1987).

The bio-reduction of β-ketoesters has been widely studied. The reduction of alkyl 3-oxobutanoates can lead to either 3(S)- or 3(R)-hydroxyesters depend-ing on the microorganism used and the nature of the alcohol moiety. Bakers' yeast reduction of ethyl 3-oxobutanoate furnishes the corresponding 3(S)-hydroxyester in c. 70% yield and >90% e.e. (Ehrler *et al.*, 1986). Similarly octyl 3-oxopentanoate is reduced to give octyl 3(S)-hydroxypentanoate (59) and this compound has been used to produce the cyclic acetal (60) a pheromone of the male swift moth *Hepialus hecta* L. (De Shong *et al.*, 1986).

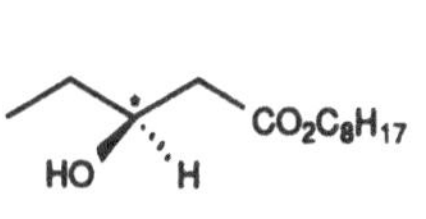

(59) 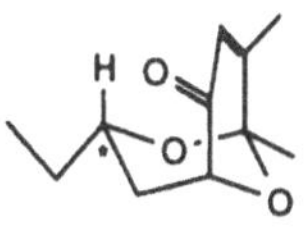 (60) 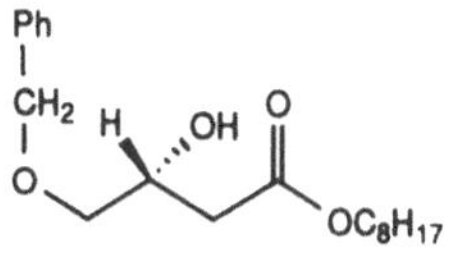 (61)

Substitution at the 4-position of the ethyl butanoates leads to yeast reduction producing the 3(R)-compound in excess (Butt & Roberts, 1986), but the balance can be restored to the production of 3(S)-products by employing long chain *n*-alkyl esters (Brooks *et al.*, 1987b). Thus the yeast-catalysed reduction of octyl 4-benzyloxy-3-oxobutanoate gave the alcohol (61) (96% e.e.) a synthon for unnatural (S)-lipoic acid.

The reduction of simple 3-oxobutanoates with other organisms can produce increased quantities of the product with the (R)-configuration (Seebach *et al.*, 1984) and the microorganism *Aspergillus niger* is particularly useful in this regard (Bernardi *et al.*, 1984). The best method for obtaining large quantities of ethyl 3(*R*)-hydroxybutanoate (**62**) is by ethanolysis of poly-3-hydroxy-butyrate produced by *Zoogloea ramigera* (Mori & Watanabe, 1981). The latter compound has been used as the starting material in a number of synthesis of natural products including β-lactams (Oguni & Ohkawa, 1988) and grandisol (**63**) (Mori & Miyake, 1987).

(62)   (63)   (64)   (65)

The bio-reduction of cyclic β-ketoesters is also well researched. Non-fermenting bakers' yeast reduction of β-ketoester units in cyclopentane, cyclohexane, piperidone and tetralone systems have been studied (Seebach *et al.*, 1987). The ketoester (**64**) was reduced using bakers' yeast in the presence of sucrose and Triton-X at pH 8 to give the hydroxyester (**65**) which was used as a synthetic building block in convergent syntheses of talaromycins A and B (Mori & Ikunaka, 1987). In another example, Mori & Tsuji (1988) have shown that the ketoester (**66**) is transformed by bakers' yeast into the hydroxyester (**67**) (35% yield, 100% e.e.) and the process has been used in a preparation of the natural product (−)-pentalenolactone E methyl ester.

2-Substituted 3-oxobutanoates can often be reduced by yeasts to give a mixture of diastereoisomers with compounds possessing the 3(*S*)-configuration predominant. Thus ethyl 2-(prop-2-enyl)-butanoate is reduced to the 2(*S*), 3(*S*)- and 2(*R*), 3(*S*)-isomers (**68**) and these compounds have been used to prepare a lyngbyatoxin A intermediate (Kosikowski & Cheng, 1987). A similar biotransformation of the more complex 2,4-substituted butanoate gave the optically active hydroxyester (**69**) with good selectivity (Buisson *et al.*, 1987).

(±) - (66)   (67)   (68)

The stereochemical course of yeast mediated reductions of 3- and 4-oxoesters has also been studied (Manzocchi *et al.*, 1987).

4.3 Reduction of αβ-Unsaturated Carbonyl Compounds and Related Species

The reduction of α,β-unsaturated aldehydes, ketones, and esters using bakers' yeast is well documented. For example the aldehyde (70) provides the alcohol (71) in optically active form (Gramatica *et al.*, 1985). Similarly bakers' yeast reduction of the formyl ester (72) provides access to the alcohol (73), a compound which was converted into 7(R),11(R)-phytol (Gramatica *et al.*, 1987). Incubation of the aldehyde (74) with bakers' yeast for a period of three weeks furnished the acid (75) in optically pure form (Sato *et al.*, 1988).

The reduction of various cyclic enones can be achieved using yeast or by inducing an enone reductase in *Beauvaria sulfurescens* (Fauve *et al.*, 1987).

The requisite allylic alcohol can be used as the substrate for reductions of the type described above. Thus 2-methylpenta-2,4-dienol is converted into 2(S)-methylpent-4-enol, presumably through the intermediacy of the dienal (Gramatica *et al.*, 1988). Similarly allenic alcohols of the type (76) are reduced to alken-3-ols (77).

A variety of unsaturated nitro compounds of the type (78) are reduced by yeast to give optically active saturated nitro compounds (e.e. values 66–98%) (Ohta *et al.*, 1989).

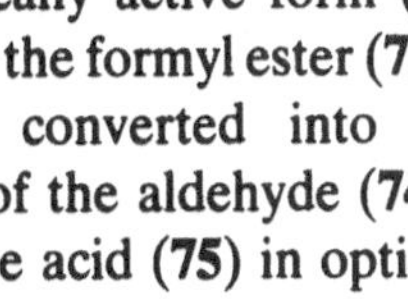

(69)

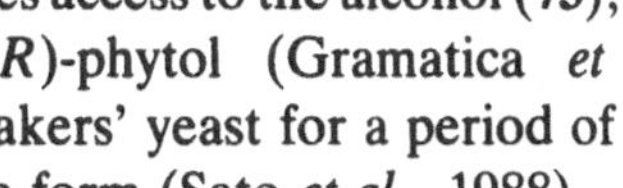

(70)

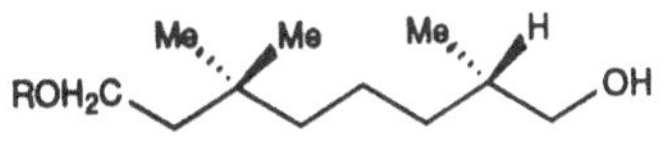

(71)

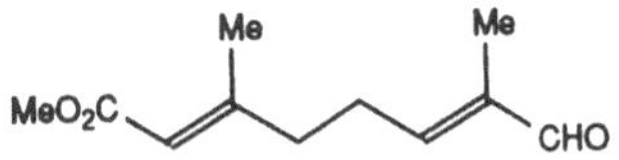

(72)

(73)

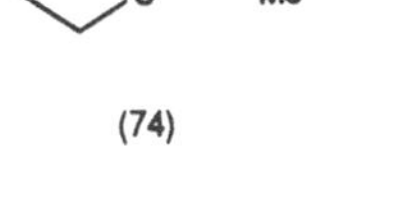

(74)

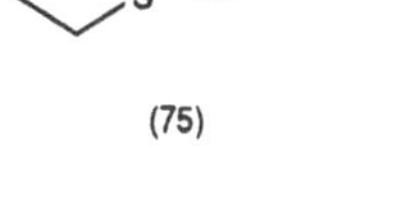

(75)

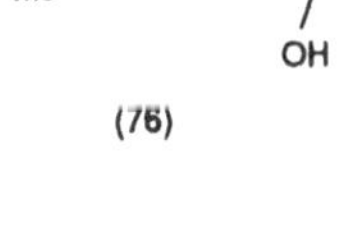

(76)

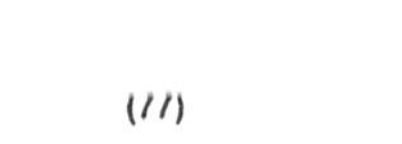

(77)

(78)

5 OXIDATION REACTIONS

5.1 Oxidation of Alcohols to Carbonyl Compounds

The oxidation of a secondary alcohol to a ketone can be accomplished using a whole-cell system or a partially purified enzyme, such as horse liver alcohol dehydrogenase (HLAD). However, the strategy is not commonly used in organic synthesis. The major reason is that, in the majority of cases, a useful chiral centre is lost during the transformation. Unless this loss is compensated by other stereochemical benefits the potential use of the biotransformation will be limited, bearing in mind that there is a plethora of mild oxidising agents available for this conversion.

In an example of a whole-cell oxidation the alcohol (79) was incubated with *Corynebacterium equi* at pH 8 to give the corresponding ketone and recovered starting material (50% yield, 95% e.e.) (Ohta *et al.*, 1987).

The oxidation of alcohols such as cyclopentanols of type (80) using HLAD has been explored by Lemière *et al.* (1985). The cofactor NAD^+ was recycled by coupling the required oxidation reaction to the reduction of acetaldehyde to isopropanol.

(79)   (±) - (80)   (81)

(82)   (83)   (84)   (85)

The oxidation of *meso*-diols to hydroxyaldehydes is the most important reaction in this section. The diol (81) was converted into the lactone (82) using HLAD (the intermediate hydroxy aldehyde exists as the tautomeric lactol which is oxidised further under the reaction conditions). The cofactor NAD^+ was recycled using flavin mononucleotide FMN (Jones *et al.*, 1982). The recycling of $NAD(P)^+$ is somewhat more problematic than the regeneration of NADPH despite the fact that much attention has been focussed on the former issue (Drakesmith & Gibson, 1988; Komoschinski & Steckhan, 1988).

5.2 Formation of Alkanols by the Oxidation of Aromatic and Aliphatic Compounds

Microbial and enzyme (e.g. horseradish peroxidase)-catalysed oxidation of aromatic compounds to furnish phenols is known, but the process has not been

CO$_2$H
...OH
''OH
R

(86)

RO$_2$C,,, OH
OH
R^1
F
R^2

(87) R^1 = H; R^2 = F
(88) R^1 = F; R^2 = H

R
OH

(89) R = H
(90) R = OH

used to a great extent by synthetic organic chemists. The oxidation of benzene (and derivatives) by *Pseudomonas* spp. to give (3-substituted) cyclohexadiene-1,2-diols has received more attention recently. Thus *cis*-cyclohexa-3,5-diene-1,2-diol has been converted into the glycosidase inhibitor conduritol A (**83**) by two routes (Sutbeyaz *et al.*, 1988; Carless & Oak, 1989). The same diol has been converted into inositol triphosphate (**84**) by a multi-step sequence of reactions (Ley & Sternfeld, 1988).

3-Methylcyclohexa-3,5-diene-1,2-diol (derived from toluene by microbial oxidation) has been used in a synthesis of prostaglandins (Hudlicky *et al.*, 1988) and employed as a starting material to make some interesting novel organometallic compounds (**85**) (Howard *et al.*, 1988). The dioxygenase enzyme(s) in the microorganism can accommodate disubstituted benzene systems and the cyclohexadienediols (**86**) have been made available (Taylor *et al.*, 1987). The absolute stereochemistry of this group of compounds has been carefully defined. Fluorobenzoate esters are oxidised by *Pseudomonas putida* in an exceptional manner producing compounds of the type (**87**) (**88**) (Rossiter *et al.*, 1987).

5.3 Hydroxylation of Alicyclic and Aliphatic Compounds

Dihydronaphthalene is hydroxylated in the benzylic position by *P. putida* and the first formed compound (**89**) is further oxidised to give the diol (**90**) (Boyd *et al.*, 1989). Detailed studies have been made of fungal hydroxylation of side chains of methyl and ethyl substituted aromatic compounds (Holland *et al.*, 1988*a*). Biohydroxylation of α-cedrene and cedrol provides rapid access to components of cedarwood oil (Lamare *et al.*, 1987). The hydroxylation of amide derivatives of norbornane and camphane have been investigated by the same research group (Fourneron *et al.*, 1987).

The regioselective microbial hydroxylation of complex natural products has been investigated and assessed as a means of access to new compounds which would be difficult to obtain in other ways. Obviously the hydroxylation of the steroid nucleus has been investigated in detail and the large scale preparation of selected 11-hydroxysteroids was made possible by this methodology. The hydroxylation of 1,9-dideoxyforskolin (**91**) and 7-deacetyl-1,9-dideoxyforskolin has been carried out to give various mono- and di-hydroxylated derivatives (Khandelwal *et al.*, 1988). Incubation of deoxyvulgarin (**92**) with *Rhizopus nigricans* afforded erivarin (**93**) in 14% yield (Arias *et al.*, 1987). Similarly

(91)   (92)   (93)

(94)   (95)   (96)

eudesmane (94) is hydroxylated at positions C-12 (22%) and C-13 (20%) using *Curvularia lunata* (Arias *et al.*, 1988).

O- and N-debenzylation reactions have been noted during fermentations involving *Mortierella isabellina* and *Helminthospora* sp. (Holland *et al.*, 1988*b*).

5.4 Oxidation of Alkenes

The oxidation of allylic alcohols by titanium-catalysed hydroperoxide reactions in the presence of dialkyl tartrates (the Sharpless oxidation) is a first-class method for the production of optically active epoxyalkanols. The stereoselective oxidation of simple alkenes using chemical methods is more difficult and some microorganisms have been found to be useful in promoting such epoxidation reactions. For example, 2-methylhept-1-ene is oxidised to the epoxide (95) (56% yield, 88% e.e.) using *Nocardia corallina* B-276 (Takahashi *et al.*, 1989).

5.5 Oxidation of Sulphides to Sulphoxides

Microbiological oxidation of sulphides to produce optically active sulphoxides is well known. For example, incubation of allyl aryl sulphides and alkyl aryl sulphides with *Corynebacterium equi* gave sulphoxides with high enantiomeric excesses (Butt & Roberts, 1986). Over-oxidation of the sulphoxide to the sulphone can be a problem. Chloroperoxidase oxidation of sulphides Me-S-Ar produces the (R)-sulphoxide in 25–92% e.e. (Colonna *et al.*, 1988). (Note that an N(S)-ethylated flavinophane acted as an asymmetric autorecycling catalyst for mono-oxidation of the same type of sulphides using H_2O_2 as the ultimate oxidant (Shinkai *et al.*, 1988).)

The oxidation of dithianes of type (**96**) is of considerable interest in terms of synthetic potential. The maximum stereo-preference for the *pro-(R)* sulphur atom in (**96**) (66%) was shown by *Mortierella isabellina* which may be compared with the oxidation of the *pro-(S)*-sulphur atom by *Helminthosporium* sp. (Auret *et al.*, 1988).

5.6 Oxidation of Ketones to Esters (Lactones)

The Baeyer–Villiger oxidation of a ketone to an ester, or cyclic ketone to a lactone is often performed using a peracid such as peracetic acid. Baeyer–Villiger oxidations can also be performed using monooxygenase enzymes (such as cyclohexanone oxygenase) contained in various microorganisms. The opportunities offered by oxygenase-catalysed reactions have been reviewed by Walsh & Chen (1988) and a number of synthetically useful transformations have been recorded. For example the hydroxyketone (**97**) is initially trans-formed into the lactone (**98**) using *Curvularia lunata* (Butt & Roberts, 1986; Ouazzani-Chahdi *et al.*, 1987).

A recent report by Taschner & Black (1988) outlines the oxidation of ketones of the type (**99**) with the cyclohexanone oxygenase from *Acinetobacter* NC1B 9871 to provide the lactone (**100**) (88% yield, >98% e.e.). A very recent disclosure reports that the racemic ketone (**101**) is oxidised to the lactone (**102**) (a prostaglandin intermediate) and the isomer (**103**) using a cyclooxygenase enzyme (Alphand *et al.*, 1989). Finally racemic 5-bromo-7-fluoro-norbornan-2-one is oxidised on incubation with whole cells of *Acinetobacter* 9871 to give recovered ketone with the absolute configuration depicted in structure (**104**) and the lactone (**105**). The ketone (**104**) was

(97)   (98)   (99)   (100)

(101)   (102)   (103)   (104)

(105)   (106)   (107)   (108)

oxidised chemically using *meta*-chloroperoxybenzoic acid to give the lactone
(**106**) along with a small amount of the regioisomeric 2-oxabicyclooctanone,
and the lactone was converted, in seven steps, into the analogue (**107**) of the
anti-HIV agent azidothymidine.

6 CARBON–CARBON BOND-FORMING REACTIONS

6.1 The Aldol Reaction

The potential value of aldolases in organic synthesis has been recognised;
rabbit muscle aldolase has been investigated in some detail. The
enzyme catalyses the reaction of dihydroxyacetone phosphate,
$HOCH_2COCH_2OPO_3H_2$ (DHAP), and a wide range of aldehydes (RCHO) to
give aldols of the type (**108**) (Bednarski *et al.*, 1989). The enzyme is highly
selective in terms of one component of the reaction (i.e. DHAP—only minor

(109)   (110)   (111)

(112)   (113)   (114)

changes can be made to this substrate (Drueckhammer *et al.*, 1989)) but a
variety of aldehydes are accepted as substrates. It is noteworthy that a
convenient preparation of dihydroxyacetone phosphate has been published
(Effenberger & Straub, 1987). The aldehyde (**109**) was reacted with DHAP
under catalysis by fructose diphosphate aldolase to give the triols (**110**) (after
dephosphorylation using a phosphatase enzyme) and the latter compounds
were converted into 1-deoxymannojirimycin and 1-deoxynojirimycin (Pederson
et al., 1988; Ziegler *et al.*, 1988).

6.2 Cyanohydrin Formation

Mandelonitrile lyase (EC 4.1.2.10) immobilised on cellulose, is effective in catalysing the addition of HCN to a wide range of aldehydes in organic solvents such as ethyl acetate to afford cyanohydrins (111) possessing the (*R*)-configuration (Effenberger *et al.*, 1987). For example, enzyme-catalysed addition of HCN to benzaldehyde gives the adduct (111, R = Ph) in 65% yield (99% e.e.) and this compound can be converted into the acyloins (112) (e.g. >92%) in two simple chemical steps (Brussee *et al.*, 1988).

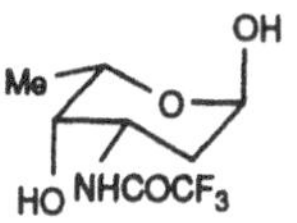

(115)

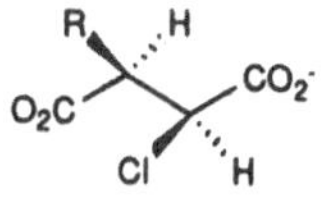

(116) R = OH
(117) R = NH$_2$

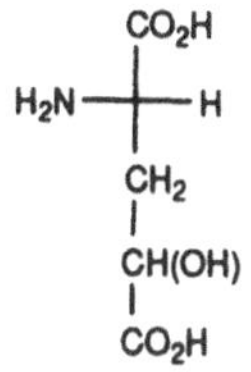

(118)

An alternative strategy for the production of cyanohydrins, namely the enantio-selective hydrolysis of cyanohydrin acetates using *Pichia miso*, has been reported (Ohta *et al.*, 1988).

6.3 The Acyloin Reaction

Fuganti *et al.* (1984) have shown that $\alpha\beta$-unsaturated aromatic aldehydes of the type (113) are converted into the diols (114) using bakers' yeast fermenting on D-glucose, in a yield of *c.* 25%. The diols have been used to make amino-sugars such as the daunosamine derivative (115).

7 OTHER BIOTRANSFORMATIONS

Pig heart fumarase [EC 4.2.1.2] has been used to transform 2-chlorofumaric acid into L-*threo*-chloromalic acid (116), a compound employed in the synthesis of 2-deoxy-D-ribose and D-*erythro*-sphingosine (Findeis & Whitesides, 1987). Similarly 3-methylaspartate ammonia lyase [EC 4.3.1.2] can be used to prepare the amino acid (117) from chlorofumaric acid (Akhtar *et al.*, 1987).

Immobilised glutamic oxaloacetic aminotransferase catalyses the trans-amination of α-hydroxy-α-ketoglutaric acid and cysteine sulphinic acid, providing a convenient route to α-hydroxy-L-glutamic acids (118) (Passerat & Bolte, 1987). Reductive amination of 3-fluoroketoglutarate using glutamate dehydrogenase furnishes 2(*R*), 3(*R*)- and 2(*R*), 3(*S*)-3-fluoroglutamic acids (Vidal-Cros *et al.*, 1989).

Bakers' yeast catalyses the addition of water across the α,β-double bond of compounds such as 4-benzyloxybut-2-enal, with concomitant reduction of the aldehyde moiety, to give the diol (119) (25% yield) (Fronza *et al.*, 1988).

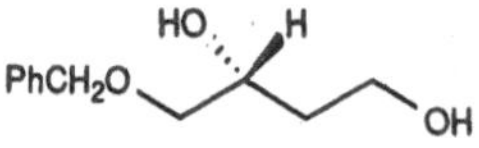

(119) 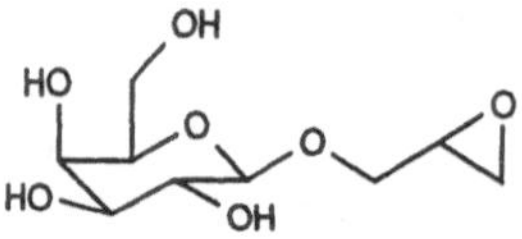   (120)

The haloperoxidase from *Caldariomyces fumago* converts thymine into thymine bromohydrin (Itoh *et al.*, 1987), while chloroperoxidase and phosphate buffer containing potassium bromide has been used to convert uracil into 5-bromouracil.

The enzyme-catalysed synthesis and modification of sugars is an area of increasing importance. The developing role of aldolases in organic synthesis, and in particular the preparation of unnatural sugars, has been briefly mentioned above. The preparation of oligosaccharides (for example using glycosyl transferases) and polysaccharides are areas of equal importance and are areas where enzyme-catalysed reactions will be most useful (Toone *et al.*, 1989). β-Galactosidase has been used to prepare the epoxide (**120**) by coupling *o*-nitrophenyl-β-D-galactopyranoside and allyl epoxide. The product (**120**) was used to make monoacylgalactoglycerides (Bjorkling & Godtfredsen, 1988). Shinoyama & Yashui (1988) have studied several glycosidases regarding their hydrolytic and transglycosylating activity in alcoholic media. Enzymes were classified as follows:

(i)   no transglycosylating activity e.g. yeast α-glucosidase,
(ii) weak transglycosylating activity e.g. *Aspergillus niger* α-glucosidase,
(iii) modest transglycosylating activity with selected alcohols (e.g. *A. oryzae* β-galactoxidase),
(iv) strong transglycosylating activity in the presence of a variety of alcohols (e.g. *A. niger* β-xylosidase).

The field concerned with the synthesis and modification of sugars will be one which will feature procedures involving multiple enzyme-catalysed reactions working in a consecutive manner. The reaction cascade set up by Reimer *et al.* (1986) is one elegant example of the number of sequential reactions that can be set up in one 'pot' (Scheme 1).

The ability of enzymes to promote phosphorylation reactions is shown in Scheme 1. Another important example involves the generation of cytidine-5'-triphosphate from the monophosphate using adenylate kinase (EC 2.7.4.3), pyruvate kinase (EC 2.7.1.40) and phosphoenol pyruvate on a gram scale (Simon *et al.*, 1988).

Potentially interesting anti-viral compounds have been obtained by transdideoxyribosylation using 2',3'-dideoxyuridine (chemically derived from uridine) and adenine to give 2',3'-dideoxyadenosine. *Escherichia coli* AJ 2595 was found to be a useful organism for this transformation (Shirae *et al.*, 1989).

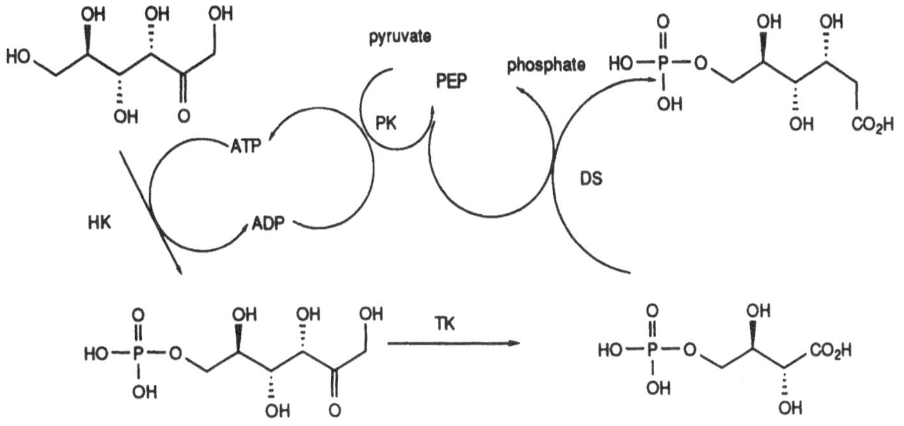

HK =hexokinase; PK = pyruvate kinase; TK = transketalase; DS = DAHP synthetase;

PEP = phosphoenol pyruvate

Scheme 1

8 CONCLUSIONS AND OBVIOUS DIRECTIONS FOR FUTURE WORK

The use of esterases, acylases, and lipases will become more widespread in laboratories specialising in organic synthesis. Enantioselective hydrolysis reactions will become more popular and the employment of immobilised enzymes and non-aqueous systems will be necessary when esterification reactions offer advantages in terms of simpler work-up procedures, etc. Reduction reactions involving simple-to-use microorganisms (such as bakers' yeast) or commercially available enzymes will also be used increasingly, and profitably, by chemists interested in producing novel fine chemicals.

The possibilities offered to the non-specialist of using enzymes and whole cells for oxidation reactions is more limited. The employment of enzyme-catalysed Baeyer–Villiger reactions in organic synthesis will increase slowly. Obviously, the functionalisation of non-activated carbon centres using enzymes is a *very* attractive proposition for the future but much more research needs to be undertaken before the regio- and stereoselectivity of the oxidation reactions can be predicted.

Enzyme-catalysed reactions will feature prominently in the chemistry and biochemistry laboratories engaged in work with sugars and glycoproteins.

While research work involving the use of enzymes in organic synthesis has increased in volume almost exponentially over the past ten years, and a number of noteworthy breakthroughs have been recorded, it must be remembered that new and valuable chemical procedures have also been discovered in the past decade, and indeed, the synthesis of many complex organic molecules has been accomplished in the laboratory without the aid of enzymes (Corey & Cheng, 1989).

The Sharpless asymmetric epoxidation (Carlier et al., 1988) and bis-hydroxylation reactions (Wai et al., 1989) have proved to be a tremendous boon to synthetic organic chemists. Reactions catalysed by transition-metal reagents (e.g. palladium compounds) have been investigated in depth in recent years (Heck, 1985). The use of asymmetrical reagents in organic synthesis (ApSimon & Collier, 1986) and the employment of cheap naturally-occurring chiral starting materials for the construction of target molecules (Hanessian, 1983) have been popular approaches with organic chemists and this situation will remain the same well into the next millenium.

Man-made chiral catalysts will complement the range of natural catalysts that are available; both sets of agents will prove useful to the organic chemist. Some of the well-known unnatural catalysts such as the asymmetric rhodium complexes capable of reducing alkenes (Morimoto et al., 1988), and the related catalysts discovered by Noyori et al. (1987) for the asymmetric reduction of ketones, are powerful weapons in the armoury of the organic chemist.

At the interface between natural and unnatural catalysts a number of research groups have embarked on research programmes aimed at mimicking the exquisite selectivity displayed by enzymes. Considerable efforts have been focussed on constructing mimics of cytochrome P-450s. Model systems such as iron-containing porphyrins (haemins) and oxygen atom transfer agents such as the iodosyl benzenes parallel P-450s in much of their chemistry (e.g. alkene epoxidation, alkane hydroxylation): the regiospecificity of attack is often similar (Dolphin et al., 1989). A Mn(III) porphyrin has also been used as an epoxidising agent for alkenes (Lee et al., 1989). Non-porphyrin complexes acting as P-450 mimics are less well documented. Current interest in this area coincides with reports on methane mono-oxygenase enzymes which have tentatively been assigned as having a μ-oxo-di-iron non-porphyrin active site (Ericson et al., 1988). Certain optically active Ni(II)-cyclan complexes act as catalysts for the PhIO oxidation of alkenes (Kinneary et al., 1988). Manganese clusters $Mn_{3-4}O_{1-2}(RCO_2^-)$ (pyridine) are catalysts that convert C—H bonds to C—OH bonds in C_2, C_3 and cyclic C_6 hydrocarbons in the presence of tert-butyl-hydroperoxide as the mono-oxygen transfer agent (Fish et al., 1988).

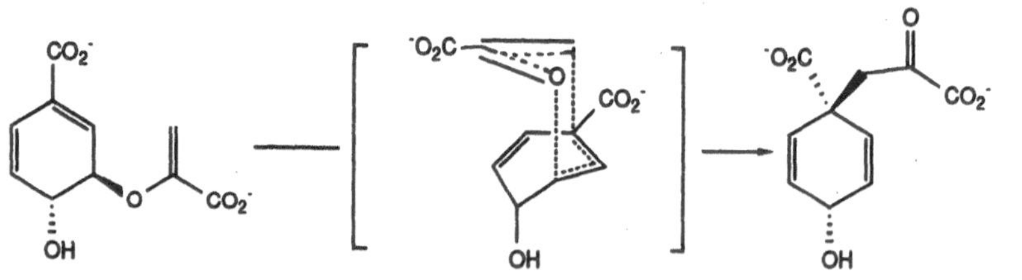

Scheme 2

The oxidation of steroids (for example the conversion of cholest-4-en-3-one to progesterone) has been achieved using the iron cluster $Fe_3O(OAc)_6$ (pyridine)$_{3.5}$ as catalyst, acetic acid/pyridine as solvent, zinc, and air as the oxidant (Barton *et al.*, 1989).

One of the most intriguing areas of research for the future, and one which links the studies involving the use of enzymes and those involving unnatural asymmetric catalysts in organic synthesis, is concerned with the preparation and use of abzymes. In a recent example of this phenomenon Hilvert & Nared (1988) and, independently Jackson *et al.* (1988) raised antibodies to a diacid designed as a mimic of the transition state of the chorismate to prephenate transformation (Scheme 2). In very exciting findings, the antibody/enzymes accelerated the [3,3]-rearrangement by a factor $< 1 \times 10^4$ at 10°C, pH 7.

REFERENCES

Adisson, L., Bolte, J., Denuynck, C. & Mari, J.-C. (1988). *Tetrahedron*, **44**, 2185.

Akhtar, M., Botting, N. P., Cohen, M. A. & Gani, D. (1987). *Tetrahedron*, **43**, 5899.

Alphand, V., Archelas, A. & Furstoss, R. (1989). *Tetrahedron Letters*, **28**, 3663.

Alvarez, E., Cuvigny, T., Hervé du Penhoat, C. & Julia, M. (1988). *Tetrahedron*, **44**, 119.

ApSimon, J. W. & Collier, T. L. (1986). *Tetrahedron*, **42**, 5157.

Arias, J. M., Breton, J. L., Gavin, J. A., Garcia-Granados, A., Martinez, A. & Onorato, M. E. (1987). *J. Chem. Soc. Perkin Transactions* **1**, 471.

Arias, J. M., Garcia-Grandos, A., Martinez, A., Onorato, M. E. & Rivas, F. (1988). *Tetrahedron Letters*, **29**, 4471.

Auret, B. J., Boyd, D. R., Dunlop, R. & Drake, A. F. (1988). *J. Chem. Soc. Perkin Transactions* **1**, 2827.

Baba, N., Mimura, M., Hiratake, J., Uchida, K. & Oda, J. (1988). *Agricultural and Biological Chemistry*, **52**, 2685.

Barbas, C. F. & Wong, C.-H. (1988). *Tetrahedron Letters*, **29**, 2907.

Barton, D. H. R., Boivin, J. & Lelandais, P. (1989). *J. Chem. Soc. Perkin Transactions* **1**, 463.

Baudin, G., Glanzer, B. I., Swaminathan, K. S. & Vasella, A. (1988). *Helvetica Chimica Acta*, **70**, 1367.

Bednarski, M. D., Simon, E. S., Bischofberger, N., Fessner, W.-D., Kim, M.-J., Lees, W., Saito, T., Waldmann, H. & Whitesides, G. M. (1989). *Journal of the American Chemical Society*, **111**, 627.

Belan, A., Bolte, J., Fauve, A., Gowry, J. G. & Veschambre, H. (1987). *Journal of Organic Chemistry*, **52**, 256.

Bellucci, G., Ferretti, M., Lippi, A. & Marioni, F. (1988). *J. Chem. Soc. Perkin Transactions* **1**, 2715.

Bengis-Garber, C. & Gutman, A. L. (1988). *Tetrahedron Letters*, **29**, 2589.

Bernardi, R., Cardillo, R. & Ghiringhelli, D. (1984). *Journal of the Chemical Society, Chemical Communications*, 460.

Bernardi, R., Cardillo, R., Ghiringhelli, D. & Vajna de Pava, O. (1987). *J. Chem. Soc., Perkin Transactions* **1**, 1607.

Bianchi, D., Cesti, P. & Battistel, E. (1988). *Journal of Organic Chemistry*, **53**, 5531.

Bjorkling, F. & Godtfredsen, S. E. (1988). *Tetrahedron*, **44**, 2957.

Blanco, L., Guibé-Jampel, E. & Rousseau, G. (1988). *Tetrahedron Letters*, **29**, 1915.

Borthwick, A. D., Butt, S., Biggadike, K., Exall, A. M., Roberts, S. M., Youds, P. Kirk, B. E., Booth, B. R., Cameron, J. M., Cox, S. W., Marr, C. L. P. & Shill, M. D. (1988). *J. Chem. Soc., Chemical Communications*, 656.

Boyd, D. R., McMordie, R. A. S., Sharma, N. D., Dalton, H., Williams, P. & Jenkins, R. O. (1989). *J. Chem. Soc., Chemical Communications*, 339.

Brooks, D. W., Mazdiyasni, H. & Sallay, P. (1985). *Journal of Organic Chemistry*, **50**, 3411.

Brooks, D. W., Mazdiyasni, H. & Grothaus, P. G. (1987a). *Journal of Organic Chemistry*, **52**, 3223.

Brooks, D. W., Kellogg, R. P. & Cooper, C. S. (1987b). *Journal of Organic Chemistry*, **52**, 192.

Brussee, J., Roos, E. C. & Van der Gen, A. (1988). *Tetrahedron Letters*, **29**, 4485.

Bucciarelli, M., Forni, A., Moretti, I. & Prati, F. (1989). *Tetrahedron Letters*, **30**.

Buisson, D., Henrot, S., Larcheveque, M. & Azerad, R. (1987). *Tetrahedron Letters*, **28**, 5033.

Butt, S. & Roberts, S. M. (1986). *Natural Product Reports*, **3**, 495.

Butt, S. & Roberts, S. M. (1987). *Chemistry in Britain*, 127.

Butt, S., Davies, H. G., Dawson, M. J., Lawrence, G. C., Leaver, J., Roberts, S. M., Turner, M. K., Wakefield, B. J., Wall, W. F. & Winders, J. A. (1987). *J. Chem. Soc., Perkin Transactions* 1, 903.

Carless, H. A. J. & Oak, O. Z. (1989). *Tetrahedron Letters*, **30**, 1719.

Carlier, P. R., Mungall, W. S., Schröder, G. & Sharpless, K. B. (1988). *Journal of the American Chemical Society*, **110**, 2978.

Chan, C., Cox, P. B. & Roberts, S. M. (1988). *J. Chem. Soc., Chemical Communications*, 971.

Chan, C., Cox, P. B. & Roberts, S. M. (1990). *Biocatalysis*, **3**, 111.

Cohen, S. G. (1969). *Transactions of the New York Academy of Science*, **31**, 705.

Colonna, S., Gaggero, N., Manfredi, A., Casella, L. & Gullotti, M. (1988). *J. Chem. Soc., Chemical Communications*, 1451.

Corey, E. J. & Cheng, X. M. (1989). *The Logic of Chemical Synthesis*. J. Wiley and Sons, Chichester.

Cotterill, I. C., Macfarlane, E. L. A. & Roberts, S. M. (1988a). *J. Chem. Soc., Perkin Transactions* 1, 3387.

Cotterill, I. C., Finch, H., Reynolds, D. P., Roberts, S. M., Rzepa, H. S., Short, K. M., Slawin, A. M. Z., Wallis, C. J. & Williams, D. J. (1988b). *J. Chem. Soc., Chemical Communications*, 470.

Crout, D. H. G., Gaudet, V. S. B., Laumen, K. & Schneider, M. P. (1986). *J. Chem. Soc., Chemical Communications*, 808.

Davies, H. G., Green, R. H., Kelly, D. R. & Roberts, S. M. (1989). *Biotransformations in Preparative Organic Chemistry: The Use of Enzymes and Whole Cell Systems in Synthesis*. Academic Press, London.

Davies, H. G., Green, R. H., Kelly, D. R. & Roberts, S. M. (1990). *Recent Advances in the Generation of Chiral Intermediates Using Enzymes*. CRC Press Inc., Boca Raton.

Davies, H. G., Roberts, S. M., Wakefield, B. J. & Winders, J. A. (1985). *J. Chem. Soc., Chemical Communications*, 1166.

Degueil-Castaing, M., De Jeso, B., Drouillard, S. & Maillard, B. (1987). *Tetrahedron Letters*, **28**, 953.

De Shong, P., Lin, M.-T. & Perez, J. J. (1986). *Tetrahedron Letters*, **27**, 2091.

Dirlam, N. C., Moore, B. S. & Urban, F. J. (1987). *Journal of Organic Chemistry*, **52**, 3587.

Dodds, D. R. & Jones, J. B. (1988). *Journal of the American Chemical Society*, **110**, 577.

Dolphin, D., Matsumoto, A. & Shortman, C. (1989). *Journal of the American Chemical Society*, **111**, 411.

Drakesmith, F. G. & Gibson, B. (1988). *J. Chemical Society, Chemical Communications*, 1493.

Drueckhammer, D. G., Durrwachter, J. R., Pederson, R. L., Crans, D. C., Daniels, L. & Wong, C.-H. (1989). *Journal of Organic Chemistry*, **54**, 70.

Effenberger, F. & Straub, A. (1987). *Tetrahedron Letters*, **28**, 1641.

Effenberger, F., Ziegler, T. & Forster, S. (1987). *Angewandte Chemie, International Edition*, **26**, 458.

Ehrler, J., Giovannini, F., Lamatsch, B. & Seebach, D. (1986). *Chimia*, **40**, 172.

Ekberg, B., Lindbladh, C., Kemp, M. & Mosbach, K. (1989). *Tetrahedron Letters*, **30**, 583.

Endo, T. & Watanabe, I. (1989). *FEBS Letters*, **243**, 61.

Ericson, A., Hedman, B., Hodgson, K. O., Green, J., Dalton, H., Bentsen, J. C., Beer, R. H. & Lippard, S. J. (1988). *Journal of the American Chemical Society*, **110**, 2330.

Fauve, A. & Veschambre, H. (1987). *Tetrahedron Letters*, **28**, 5037.

Fauve, A. & Veschambre, H. (1988). *Journal of Organic Chemistry*, **53**, 5215.

Fauve, A., Renard, M. F. & Veschambre, H. (1987). *Journal of Organic Chemistry*, **52**, 4893.

Findeis, M. A. & Whitesides, G. M. (1987). *Journal of Organic Chemistry*, **52**, 2838.

Fish, R. H., Fong, R. H., Vincent, J. B. & Cristou, G. (1988). *J. Chemical Society, Chemical Communications*, 1504.

Flörsheimer, A. & Kula, M. R. (1988). *Monatsh. Chemie*, **119**, 1323.

Fourneron, J.-D., Archelas, A., Vigne, B. & Furstoss, R. (1987). *Tetrahedron*, **43**, 2273.

Fronza, G., Fuganti, C., Grasselli, P., Poli, G. & Servi, S. (1988). *Journal of Organic Chemistry*, **53**, 6153.

Fuganti, C., Graselli, P., Servi, S., Spreafico, F. & Zirotti, C. (1984). *Journal of Organic Chemistry*, **49**, 4087.

Fujii, H., Koyama, T. & Ogura, K. (1982). *Biochimica et Biophysica Acta*, **712**, 716.

Fujisawa, T., Kojima, E., Itoh, T. & Sato, T. (1985). *Tetrahedron Letters*, **26**, 6089.

Glänzer, B. I., Faber, K. & Griengl, H. (1987). *Tetrahedron*, **43**, 5791.

Gotor, V., Brieva, R. & Rebelledo, F. (1988). *Tetrahedron Letters*, **29**, 6973.

Gramatica, P., Manitto, P. & Poli, L. (1985). *Journal of Organic Chemistry*, **50**, 4625.

Gramatica, P., Manitto, P., Monti, D. & Speranza, G. (1987). *Tetrahedron*, **43**, 4481.

Gramatica, P., Manitto, P., Monti, D. & Speranza, G. (1988). *Tetrahedron*, **44**, 1299.

Gutman, A. L., Oren, D., Boltanski, A. & Bravdo, T. (1987). *Tetrahedron Letters*, **28**, 5367.

Haraldsson, G. G., Hoskuldsson, P. A., Sigurdsson, S. T., Thorsteinsson, F. & Gudbjarnason, S. Y. (1989). *Tetrahedron Letters*, **30**, 1671.

Hanessian, S. (1983). *Total Synthesis of Natural Products: The Chiron Approach*. Pergamon Press, Elmsford, NY.

Heck, R. F. (1985). In *Palladium Reagents in Organic Syntheses*. eds A. R. Katritzky, O. Meth-Cohn & C. W. Rees. Academic Press, Orlando, p. 204.

Henner, W. J., Sweers, H. M., Wang, Y.-F. & Wong, C.-H. (1988). *Journal of Organic Chemistry*, **53**, 4939.

Hilvert, D. & Nared, K. D. Y. (1988). *Journal of the American Chemical Society*, **110**, 5593.

Hirai, K. & Naito, A. (1989). *Tetrahedron Letters*, **30**, 1107.

Hiratake, J., Yamamoto, K., Yamamoto, Y. & Oda, J. (1989). *Tetrahedron Letters*, **30**, 1555.

Holland, H. L., Brown, F. M., Munoz, B. & Ninniss, R. W. (1988a). *J. Chemical Society, Perkin Transactions II*, **29**, 1557.

Holland, H. L., Conn, M., Chenchaiah, P. C. & Brown, F. M. (1988*b*). *Tetrahedron Letters*, **29**, 6393.

Hönig, H., Seufer-Wasserthal, P., Stütz, A. E. & Zenz, E. (1989). *Tetrahedron Letters*, **30**, 811.

Howard, P. W., Stephenson, G. R. & Taylor, S. C. (1988). *J. Chemical Society, Chemical Communications*, 1603.

Hudlicky, T., Luna, H., Barbieri, G. & Kwart, L. D. (1988). *Journal of the American Chemical Society*, **110**, 4735.

Itoh, N., Izumi, Y. & Yamada, H. (1987). *Biochemistry*, **26**, 282.

Jackson, D. Y., Jacobs, J. W., Sugasawara, R., Reich, S. H., Bartlett, P. A. & Schultz, P. G. (1988). *Journal of the American Chemical Society*, **110**, 4841.

Jones, J. B., Sih, C. J. & Perlman, D. (1976). *Tech. Chem.*, **10**, 107.

Jones, J. B., Finch, M. A. W. & Jakovac, I. J. (1982). *Canadian Journal of Chemistry*, **60**, 2007.

Kaga, H., Kobayashi, S. & Ohno, M. (1989). *Tetrahedron Letters*, **30**, 113.

Kato, Y., Ohta, H. & Tsuchihashi, G. (1987). *Tetrahedron Letters*, **28**, 1303.

Khandelwal, Y., Inamder, P. K., de Souza, N. J., Rupp, R. H., Chatterjee, S. & Ganguli, B. N. (1988). *Tetrahedron*, **44**, 1661.

Kim, M.-J. & Whitesides, G. M. (1988). *Journal of the American Chemical Society*, **110**, 2959.

Kinneary, J. F., Wagler, T. R. & Burrows, C. J. (1988). *Tetrahedron Letters*, **29**, 877.

Kirchner, G., Scollar, M. P. & Klibanov, A. M. (1985). *Journal of the American Chemical Society*, **107**, 7072.

Kobayashi, M., Nagasawa, T. & Yamada, H. (1988). *Applied Microbiol. Biotechnology*, **29**, 231.

Komoschinski, J. & Steckhan, E. (1988). *Tetrahedron Letters*, **29**, 3299.

Konishi, J., Ohta, H. & Tsuchihashi, G. (1985). *Chemistry Letters*, 111.

Kosikowski, A. P. & Cheng, X.-M. (1987). *Tetrahedron Letters*, **28**, 3189.

Kullmann, W. (1987). *Enzymatic Peptide Synthesis*. CRC Press, Boca Raton, Florida.

Lagerlöf, E., Nathorst-Westfelt, L., Ekström, B. & Sjoberg, B. (1976). *Methods in Enzymology*, **44**, 759.

Lamare, V., Fourneron, J. D., Furstoss, R., Ehret, C. & Carbier, B. (1987). *Tetrahedron Letters*, **28**, 6269.

Laumen, K., Breitgoff, D. & Schneider, M. P. (1988). *J. Chem. Soc., Chemical Communications*, 1459.

Liu, Y.-C. & Chen, C.-S. (1989). *Tetrahedron Letters*, **30**, 1617.

Lee, R. W., Nakagaki, P. C. & Bruice, T. C. (1989). *Journal of the American Chemical Society*, **111**, 1368.

Lemière, G. L., Lepoivre, J. A. & Alderweireldt, F. C. (1985). *Tetrahedron Letters*, **26**, 4527.

Ley, S. V. & Sternfeld, F. (1988). *Tetrahedron Letters*, **29**, 5305.

Manzocchi, A., Cosati, R., Fiecchi, A. & Santaniello, E. (1987). *J. Chem. Soc., Perkin Transactions 1*, 2753.

Margolin, A. M., Tai, D. & Klibanov, A. M. (1987). *Journal of the American Chemical Society*, **109**, 7885.

Mauger, J., Nagasawa, T. & Yamada, H. (1988). *Biotechnology*, **8**, 87.

Mohr, P., Waespe-Sarcevic, N., Tamm, C., Gawronska, K. & Gawronski, J. K. (1984). *Helvetica Chimica Acta*, **66**, 2501.

Mori, K. & Ikunaka, M. (1987). *Tetrahedron*, **43**, 39.

Mori, K. & Iwasawa, H. (1980). *Tetrahedron*, **36**, 2209.

Mori, K. & Komatsu, M. (1987). *Tetrahedron*, **43**, 3409.

Mori, K. & Miyake, M. (1987). *Tetrahedron*, **43**, 2229.

Mori, K. & Tsuji, M. (1988). *Tetrahedron*, **44**, 2835.

Mori, K. & Watanabe, H. (1981). *Tetrahedron*, **37**, 1341.

Morimoto, T., Chiba, M. & Achiwa, K. (1988). *Tetrahedron Letters*, **29**, 4755.

Newton, R. F. & Roberts, S. M. (1980). *Tetrahedron*, **36**, 2163.

Nicotra, F., Riva, S., Secundo, F. & Zucchelli, L. (1989). *Tetrahedron Letters*, **30**, 1703.

Noyori, R., Ohkuma, T., Kitamura, M., Takaya, H., Oayo, N., Kumobayashi, H. & Akutagawa, S. (1987). *Journal of the American Chemical Society*, **109**, 5856.

Oberhauser, T., Faber, K. & Griengl, H. (1989). *Tetrahedron*, **45**, 1679.

Oguni, N. & Ohkawa, Y. (1988). *J. Chem. Soc., Chemical Communications*, 1377.

Ohno, M. (1985). In *Enzymes in Organic Synthesis. Ciba Foundation Symposium III*. eds R. Parker & S. Clark. Pitman, Bath, p. 171.

Ohta, H., Kato, Y. & Tsuchihashi, G. (1987). *Journal of Organic Chemistry*, **52**, 2735.

Ohta, H., Kimura, Y. & Sugamo, Y. (1988). *Tetrahedron Letters*, **29**, 6957.

Ohta, H., Kobayashi, N. & Ozaki, K. (1989). *Journal of Organic Chemistry*, **54**, 1802.

Ouazzani-Chahdi, J., Buisson, D. & Azerad, R. (1987). *Tetrahedron Letters*, **28**, 1109.

Passerat, N. & Bolte, J. (1987). *Tetrahedron Letters*, **28**, 1277.

Pederson, R. L., Kim, M.-J. & Wong, C.-H. (1988). *Tetrahedron Letters*, **29**, 4645.

Poppe, L., Novák, L., Kolonits, P., Bata, A. & Szántay, C. (1988). *Tetrahedron*, **44**, 1477.

Reimer, L. M., Conley, D. L., Pompliano, D. L. & Frost, J. W. (1986). *Journal of the American Chemical Society*, **108**, 8010.

Riva, S., Chopineau, J., Kieboom, A. P. G. & Klibanov, A. M. (1988). *Journal of the American Chemical Society*, **110**, 584.

Rossiter, J. T., Williams, S. R., Cass, A. E. G. & Ribbons, D. W. (1987). *Tetrahedron Letters*, **28**, 5173.

Sabbioni, G. & Jones, J. B. (1987). *Journal of Organic Chemistry*, **52**, 4565.

Sato, T., Hanayama, K. & Fujisawa, T. (1988). *Tetrahedron Letters*, **29**, 2197.

Seebach, D., Züger, M. F., Giovannini, F., Sonnleitner, B. & Fiechter, A. (1984). *Angewandte Chemie International Edition*, **23**, 151.

Seebach, D., Roggo, S., Maetzke, T., Braunschweiger, H., Cercus, J. & Krieger, M. (1987). *Helvetica Chimica Acta*, **69**, 1605.

Shinkai, S., Yamaguchi, T., Manabe, O. & Toda, F. (1988). *J. Chemical Society, Chemical Communications*, 1399.

Shirae, H., Kobayashi, K., Shiragami, H., Irie, Y., Yasuda, N. & Yokozeki, R. (1989). *App. Env. Microbiology*, **55**, 419.

Short, R. P., Kennedy, R. M. & Masamune, S. (1989). *Journal of Organic Chemistry*, **54**, 1755.

Simon, E. S., Bednarski, M. D. & Whitesides, G. M. (1988). *Tetrahedron Letters*, **29**, 1123.

Sonnet, P. E. (1987). *Journal of Organic Chemistry*, **52**, 3477.

Stokes, T. M. & Oehlschlager, A. C. (1987). *Tetrahedron Letters*, **28**, 2091.

Sutbeyaz, Y., Seçen, H. & Balci, M. (1988). *J. Chemical Society, Chemical Communications*, 1330.

Takahashi, O., Umezaka, J., Furuhashi, K. & Takagi, M. (1989). *Tetrahedron Letters*, **30**, 1583.

Taschner, M. J. & Black, D. J. (1988). *Journal of the American Chemical Society*, **110**, 6892.

Taylor, S. J. C., Ribbons, D. W., Slawin, A. M. Z., Widdowson, D. A. & Williams, D. J. (1987). *Tetrahedron Letters*, **28**, 6391.

Toone, E. J., Simon, E. S., Bednarski, M. D. & Whitesides, G. M. (1989). *Tetrahedron*, **45**, 5365.

Top, S., Jaouen, G., Gillois, J., Baldoli, C. & Maiorana, S. (1988). *J. Chem. Soc., Chemical Communications*, 1284.

Tsuboi, S., Furutani, H., Utaka, M. & Takeda, A. (1987). *Tetrahedron Letters*, **28**, 2709.

Vidal-Cros, A., Gaudry, M. & Marquet, A. (1989). *Journal of Organic Chemistry,* **54,** 498.

Vo-Quang, Y., Marais, D., Vo-Quang, L., Le Goffic, F., Thiery, A., Maestracci, M., Arnaud, A. & Galzy, P. (1987). *Tetrahedron Letters,* **28,** 4057.

Wai, J. S. M., Marko, I., Svendsen, J. S., Finn, M. G., Jacobsen, E. N. & Sharpless, K. B. (1989). *Journal of the American Chemical Society,* **111,** 1123.

Waldmann, H. (1988). *Tetrahedron Letters,* **29,** 1131.

Walsh, C. T. & Chen, Y.-C. J. (1988). *Angewandte Chemie, International Edition,* **27,** 333.

Wandrey, C. (1986). In *Enzymes as Catalysts in Organic Synthesis,* ed. M. P. Schneider, D. Reidel, Dordrecht, Holland, p. 263.

Wang, Y.-F. & Wong, C. H. (1988). *Journal of Organic Chemistry,* **53,** 3127.

Wang, Y.-F., Lalonde, J. J., Momongan, M., Bergbreiter, D. E. & Wong, C.-H. (1988). *Journal of the American Chemical Society,* **110,** 7200.

Xie, Z. F., Nakamura, I., Suemune, H. & Sakai, K. (1988). *J. Chem. Soc., Chemical Communications,* 966.

Zemlicka, J., Craine, L. E., Heeg, M.-J. & Oliver, J. P. (1989). *Journal of Organic Chemistry,* **54,** 937.

Zhi-Wei, G. & Sih, C. J. (1988). *Journal of the American Chemical Society,* **110,** 1999.

Ziegler, T., Straub, A. & Effenberger, F. (1988). *Angewandte Chemie, International Edition,* **27,** 216.

Chapter 13

MICROBIOSENSORS AND IMMUNOSENSORS

Isao Karube, Atsushi Seki & Koji Sode

*Research Center for Advanced Science and Technology, University of Tokyo,
4-6-1 Komaba, Meguro-ku, Tokyo 153, Japan*

CONTENTS

1 INTRODUCTION

Chemical reactions in organisms proceed smoothly because of the action of catalysts such as enzymes. These are highly specific for chemical substances with which they form complexes to promote their reactions. By using enzymatic analyses, therefore, many substances may be measured. The products of enzyme reactions are usually quantified by spectrophotometry or spectrofluorometry. But these methods need complicated procedures and are time consuming. If, however, the substrate or products can be detected by electric devices, these biosensors with good selectivity might be constructed using a combination of enzymes. In this chapter, conventional biosensors, microbiosensor development and novel biosensors using a new transducer are discussed. Emphasis is placed on the practical preparation of such devices.

2 CONVENTIONAL BIOSENSORS

Biosensors, composed of a bio-functional membrane and transducer, have been developed and applied in analytical fields, clinical analysis, the food industry and environmental science measurements (Turner *et al.*, 1987). Immobilized-enzymes, immobilized-microorganisms and immobilized-antibody membranes are used as molecular recognition materials (Fig. 1). As transducers, oxygen electrodes and hydrogen peroxide electrodes are most widely used. Various enzymes have been used as molecular recognition elements. Enzyme electrodes are composed of enzyme-immobilized membranes and electrodes. The principle of an enzyme electrode is based on the detection of electroactive compounds produced or consumed by the enzyme reaction. For example, glucose oxidase oxidizes glucose with consumption of oxygen and produces gluconolactone and hydrogen peroxide. By measuring the consumption of oxygen with an oxygen electrode or production of hydrogen peroxide with a hydrogen peroxide electrode, the concentration of glucose can be determined. A glucose sensor of this type has been commercialized and is used in the diagnosis of diabetes. Various kinds of biosensor using the same

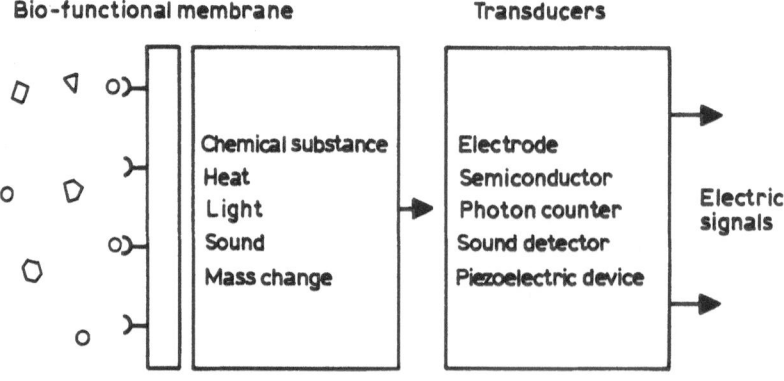

Fig. 1. Biosensor principles.

principle and devices, are being developed and used in the fields of clinical analysis and measurement of foodstuffs.

Microorganisms have also been utilized as molecular recognition elements. A microbial sensor consists of a microorganism-immobilized membrane and an electrode. Various kinds of microbial sensors have been developed and applied to the measurement of organic compounds. The principle of a microbial sensor is based on measurement of either the change in respiration or the amount of metabolites produced as a result of assimilation of a substrate. By using auxotrophic mutants, many kinds of substances can be selectively determined. For example, a vitamin B12 sensor was constructed by using immobilized *Escherichia coli* 215 (Karube *et al.*, 1987). *E. coli* 215 requires vitamin B12 for growth. In this sensor, a linear relationship was obtained between 5×10^{-9}–25×10^{-9} g/ml. During 25 days the decrease in the response of the sensor was about 8%.

Recently, microbial sensors using thermophilic bacteria have been developed. The advantages of using thermophilic bacteria include reduction in contamination by other microorganisms and/or to achievement of long term stability. For example, BOD (biological oxygen demand) and CO_2 sensors were constructed using thermophilic bacteria, isolated from a hot spring. A good linear correlation was observed with this sensor in the range of 1–10 mg/liter BOD (JIS) at 50°C (Suzuki *et al.*, 1988*b*). The sensor signal was stable and reproducible for more than 40 days. With the CO_2 sensor, a linear relationship was obtained between 1 mM and 8 mM at 50°C, with $NaHCO_3$ and the sensor's response time was 5–10 min (Karube *et al.*, 1989). A linear relationship was also observed with CO_2 concentrations from 3–8%.

Attention is also currently being focused on miniaturization and integration. Microbiosensors may be implanted in the human body and may thus be suitable for in-vivo measurements. In addition, such microbiosensors may be integrated in one chip and may be used to measure various substances simultaneously in a small amount of sample solution. Since semiconductor fabrication technology is used to make these microbiosensors, it may be possible to develop disposable transducers for biosensors by mass production. Recently, novel biosensors using piezoelectric devices, surface acoustic wave (SAW) devices, image sensor and optical electrodes (optrode) have been also reported. In the following sections, current trends in microbiosensor development and novel biosensors are discussed.

3 MICROBIOSENSOR DEVELOPMENT

3.1 Microbiosensors Based on ISFET

An ion-sensitive field effect transistor (ISFET) was first reported by Bergveld (1970). Matsuo & Wise (1974) improved the ISFET by using silicon nitride as the gate insulator to construct a micro-pH sensitive device. The advantages of ISFET include rapid response, low power consumption, low noise and no

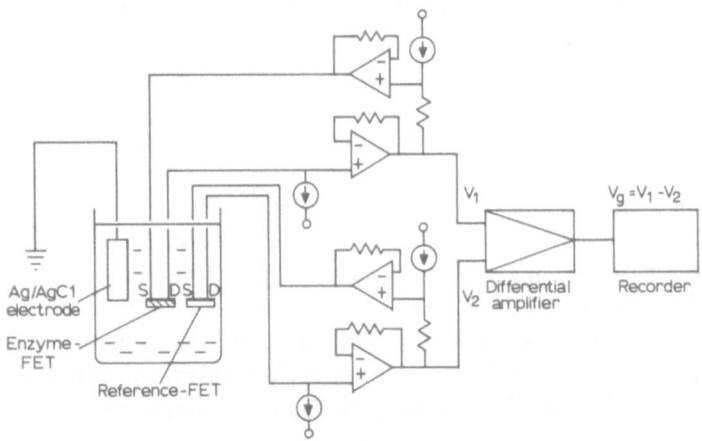

Fig. 2. Circuit diagram of the measuring system.

requirement for a high impedance amplifier. A general circuit diagram for measuring the gate output voltage is shown in Fig. 2. In this circuit, the voltage between the source and drain is held constant, and the current between source and drain is also held constant. The Ag/AgCl electrode is placed in the same solution as the ISFET. The surface potential on the silicon nitride insulator of the ISFET is affected by the pH of the solution with concomitant change in the gate voltage proportional to the change in surface potential. Therefore, the surface potential change on the ISFET, caused by the change of pH, can be measured as a change in the gate output voltage. ISFET is fabricated using semiconductor technology. Hence, it is easy to miniaturize and integrate ISFETs on one tip. ISFET is used as a potentiometric transducer and therefore it is suitable for use with enzymes which cause pH changes in their reactions, such as, urease and glucose oxidase. Since the first report of the penicillin sensitive penicillinase-immobilized ISFET by Caras & Janata (1980), many reports about the enzyme-modified ISFET have been published. ISFETs have been used as transducers for many microbiosensors.

In the following section, ISFET-biosensors for the measurement of urea, ATP, acetylcholine and alcohol are described. In addition amorphous silicon ISFET and its application to biosensors is discussed.

3.1.1 Urea Sensor (Karube et al., 1986)

A urea sensor consisting of a urease immobilized-membrane and an electrode such as a pH electrode, an ammonium electrode and a pCO_2 electrode have been reported. Urease-catalyzed reactions cause a pH change, and so ISFET may be used as a transducer. A micro-urease sensor is fabricated as follows. An ISFET is set inside a vacuum chamber and γ-aminopropyltriethoxysilane (γ-APTES) is vaporized at 80°C, 0·5 Torr for 30 min and then glutaraldehyde is vaporized under the same conditions. The chemically modified ISFET is covered with a cellulose acetate membrane containing 1,8-diamino-4-aminomethyloctane and glutaraldehyde (GA) and then immersed in a urease solution. This sensor gives a linear relationship between the initial rate of the output gate voltage and the logarithm value of urea concentration in the range 16·7–167 mM. It can be used for 20 days with only slight loss of enzyme activity.

3.1.2 ATP Sensor (Gotoh et al., 1986)

An ATP sensor based on H^+-ATPase immobilized via a polyvinylbutyral resin on a pH-sensitive field effect transistor has been reported. The enzyme H^+-ATPase catalyzes the hydrolysis of ATP to ADP and orthophosphate. Therefore, a micro-ATP sensor can be made by using an ISFET and H^+-ATPase immobilized on a membrane. The ATP sensor was constructed as follows: 0·1 g of polyvinylbutyral resin and 1 ml of 1,8-diamino-4-aminomethyloctane were dissolved in 10 ml of dichloromethane. After stirring for 30 min, this polymer solution was cast over the gate insulator of an ISFET. The ISFET was then immersed in 5%(v/v) glutaraldehyde solution at room temperature for 24 h. The H^+-ATPase was immobilized on the membrane by immersing the tip into a 5 mg/ml H^+-ATPase solution at 4°C for 24 h. The response of this sensor to ATP showed an initial rapid increase, followed by a gradual decay. It is considered that the initial rapid increase is due to the diffusion of the hydrogen ions generated by the enzyme reaction through the membrane on the ISFET. The gradual decrease of the output is attributed to the following phenomena: a pH change in the vicinity of the enzyme immobilized membrane causes a decrease of enzyme activity and substrate concentration surrounding the immobilized—enzyme membrane gradually decreases because of diffusion limitations in an unstirred solution. This sensor shows a linear relationship between the initial rate of change of the differential gate output voltage and the logarithm of the ATP concentration in the range 0·2–1·0 mM ATP. The system exhibited a 90% decrease in response to 1 mM ATP after 18 days.

3.1.3 Acetylcholine Sensor (Gotoh et al., 1987)

By using receptors as molecular recognition elements, biosensors can be constructed. In this section, an acetylcholine receptor and its application in a biosensor is discussed. The acetylcholine receptor consists of five kinds of subunit named α, α_2, B, γ, δ. The purified receptor was reconstituted in lipid vesicles and in planar bilayers. The amino acid sequences of all four kinds of protein has been elucidated. It was demonstrated that the subunit structure of the receptor contains both the agonist binding sites and the cation channel that is regulated by the agonist binding sites. The α-subunit has one acetylcholine-binding site. In the absence of acetylcholine, the channel is in a closed state. When acetylcholine binds to the receptor, the channel opens and sodium ions rush through the channel. Sodium ion influx can be detected by an ISFET. A novel biosensor for the measurement of acetylcholine and sodium ion flux consisting of the ISFET and the immobilized acetylcholine receptor membrane with a lipid membrane has been reported. Acetylcholine receptor was immobilized by using lipid as follows: 0·5 mg/ml acetylcholine receptor was dissolved in a solution containing 3 mg/ml lecithin, 150 mM KCl and 10 mM tris–HCl buffer (pH 7·0) using an ultrasonic disruptor at 30 W for 5 min. Then, the gate area of the ISFET covered with a polyvinylbutyral membrane was immersed in this solution at 4°C for 2 h. The acetylcholine receptor immobilized on the ISFET with lipid and the Ag/AgCl reference electrode were immersed in tris–HCl buffer (pH 7·0) containing 150 mM KCl and 10 mM NaCl. Then an acetylcholine sample solution was injected into the system. The surface potential change was recorded. When 10 μM acetylcholine solution was injected, this system provided about 2 mV of gate voltage. The difference is caused by influx of sodium ions through the acetylcholine receptor channel. With the lipid membrane, the gradient of sodium ions is set between both sides of the acetylcholine receptor channel. When the acetylcholine links to the binding site of its receptor, the channel opens, increasing sodium ion influx through the channel. This method can be applied to measuring the function of other receptors and channels.

3.1.4 Alcohol Sensor (Tamiya et al., 1988a)

The study of an alcohol-sensitive microbiosensor using an ISFET with the enzyme system existing in the cell membrane has been reported. The cell membrane of acetic acid-producing bacteria has a complex enzyme system which oxidizes ethanol to acetic acid via acetaldehyde. This system consists of membrane-bound alcohol dehydrogenase (ADH), aldehyde dehydrogenase (ALDH) and an electron transfer system. This complex enzyme system could, therefore, be used with an ISFET.

Gluconobacter suboxydans cells were passed through a French press and centrifuged to remove intact cells. The supernatent was then ultracentrifuged at 100 000 g, and the cell debris used as the cell membrane fraction. The cell membrane was immobilized using calcium alginate gel coated with nitrocellulose as follows. The mixture of cell membrane and sodium alginate and pyrroloquinolinequinone (PQQ) is cast on the gate surface of the ISFET, which is then dipped in $CaCl_2$ solution in order to form a calcium alginate gel layer containing the cell membrane. This ISFET is dipped in a solution of nitrocellulose in acetone and immediately dried in air, resulting in the formation of a nitrocellulose coating layer.

The differential mode circuit described previously is used and the differential gate output is displayed on a recorder. The output of the sensor (V_g) changes immediately after the injection of the substrate and reaches a steady state after approximately 10 min. A linear relationship is observed between the differential gate output (V_g) and the ethanol concentration up to 20 mg/liter.

3.1.5 Biosensors Using Amorphous Silicon ISFET

The ISFET device can only be manufactured by using a silicon wafer. In recent years, devices made with amorphous silicon have been receiving widespread attention because of their great potential in various applications. Various materials such as glass, plastics, etc. can be used for preparing amorphous silicon, and transistors can be fabricated with a number of different structures such as the needle of a syringe. In this section, the construction of an amorphous silicon ISFET (a-Si:H ISFET) is described.

3.1.5.1 CONSTRUCTION OF a-Si:H ISFET DEVICE (GOTOH *et al.*, 1989)

The device used in this series of experiments is mainly made by radio frequency plasma discharge. The type of glow-discharge apparatus is the capacitively-coupled discharge deposition system. A $0.05\,\mu m$ n^+ layer (3000 ppm PH_3 in silane) is deposited over an evaporated aluminum layer on glass (Corning 7059) in order to ensure ohmic contact between the a-Si:H and aluminum. After etching, the deposition of an amorphous silicon layer and an amorphous silicon nitride layer is performed successively in the same capacitively-coupled glow-discharge deposition system operating at 13·56 MHz. The amorphous silicon layer is grown from a mixture of silane and hydrogen, and the amorphous silicon nitride layer from silane and ammonia. All three layers are deposited at 300°C. Finally, a silicon oxide layer is evaporated over the amorphous silicon nitride layer. Figure 3 shows the structure of the a-Si:H ISFET.

The surface potential on the silicon oxide insulator of the a-Si:H ISFET is affected by the pH of the solution with a concomitant change in the gate voltage proportional to the change in surface potential. Therefore, the surface potential change on the silicon oxide insulator of the a-Si:H ISFET, caused by a variation in pH, can be measured as a change in the gate voltage. The a-Si:H ISFET and the Ag/AgCl reference electrode are allowed to stand in 10 mM tris–HCl buffer solution at 18°C for 10–20 min. By comparison of the theoretical curve of the relationship between the surface potential and pH (Matsuo & Wise, 1974) to the pH characteristics of a-Si:H ISFET, It is assumed that the pK_a is 5 and the C_H (the capacity of the electric double layer per unit area) is $20\,\mu F/cm$. The linear V_g/pH characteristic of an a-Si:H ISFET is obtained over the pH range 5–10. The pH sensitivity is about 46 mV/pH at 18°C. The response times of the a-Si:H ISFET to pH change by the addition of acid or alkaline solution are very rapid, less than 30 s being

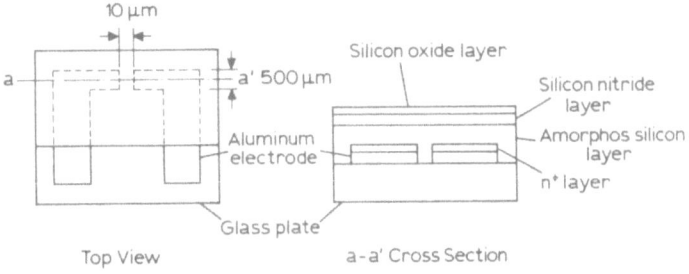

Fig. 3. Structure of amorphous silicon ISFET.

needed to reach a steady state value. Thus, a-Si:H ISFET was fabricated on glass by using a chemical vapor deposition technique.

3.1.5.2 CONSTRUCTION OF HYPOXANTHINE AND INOSINE SENSOR

Evaluation of freshness is important in the fish industry for the manufacture of high quality products. After fish die, the decomposition of ATP (adenosine-5'-triphosphate) in the fish meat sets in and ADP (adenosine-5'-diphosphate) and AMP (adenosine-5'-monophosphate) and related compounds are generated:

$$ATP \rightarrow ADP \rightarrow AMP \rightarrow IMP \rightarrow HxR \rightarrow Hx \rightarrow X \rightarrow U$$

where IMP is inosine-5'-monophosphate, HxR is inosine, Hx is hypoxanthine, X is xanthine and U is uric acid. Consequently, Hx accumulates with increase of storage time, and can be used as an indicator of fish freshness. It follows that simple and rapid methods for the determination of Hx and HxR are required in the food industry.

(*i*) *Hypoxanthine sensor* (*Tamiya* et al., 1988b): Xanthine oxidase is immobilized as follows: polyvinylbutyral and 1,8-diamino-4-aminomethyloctane are dissolved in dichloromethane. This polymer solution is dropped over the gate insulator of the a-Si:H ISFET which is then immersed in glutaraldehyde solution to promote the cross-linking reaction. Xanthine oxidase (XO) is thus immobilized on the membrane.

The enzyme FET and reference electrode are allowed to equilibrate in phosphate buffer. After the gate output voltage has reached a steady value, hypoxanthine solution is injected into the buffer, and the change in V_g is recorded. A linear relationship between the logarithm of hypoxanthine concentration and the rate of change of V_g per 1 min after injection was obtained in range 0·02–0·1 mM hypoxanthine.

(*ii*) *Inosine sensor* (*Gotoh* et al., 1988): An inosine sensor is made in the same way as a hypoxanthine sensor by using nucleoside phosphorylase and xanthine oxidase immobilized on a-Si:H ISFET simultaneously. After 90 s following injection of an inosine solution, the gate output voltage gradually increases and reaches a steady state for approximately 7 min. Xanthine formed by the decomposition of inosine catalyzed by nucleoside phosphorylase is subsequently oxidized by xanthine oxidase to uric acid. A linear relationship was obtained in the range of 0·02–0·1 mM by plotting the initial rate of the gate output voltage change against the logarithm of inosine concentration. The oxidation of hypoxanthine to uric acid by xanthine oxidase is initiated immediately after injection. The response to inosine, however, has a time lag of 90 s after injection which is attributed to the three step reaction. By using this time lag, the sensor can determine inosine and hypoxanthine simultaneously.

3.2 Micro-oxygen Electrode and its Application to Biosensor

3.2.1 Micro-oxygen Electrode

Clark-type oxygen sensing electrodes have been applied to various biosensors by immobilizing either enzymes or microorganisms which catalyze the oxidation of biochemical organic compounds. Demand has been increasing recently for miniaturized and integrated biosensors for use in clinical analysis. Several oxygen electrodes, based on conventional semiconductor technology, have been fabricated by several groups, but they have not yet reached the production line. One reason for this is that they contain a liquid electrolyte, making adhesion of the gas-permeable membrane to the substrate difficult, even if epoxy resin is used. Mass production of such a device is impossible. In this section, the development of a disposable oxygen electrode based on conventional semiconductor fabrication technology and the application of this electrode to biosensors are described. The key improvements include:

(1) Use of a porous material (agarose gel) to support the electrolyte solution.
(2) Use of a hydrophobic polymer (negative photoresist) as the gas-permeable membrane and cast directly over the porous material.

3.2.1.1 FABRICATION PROCESS OF MICRO-OXYGEN ELECTRODE (SUZUKI *et al.*, 1988*b*)

The oxygen electrode is fabricated in the following way. The basic procedure is similar to a conventional semiconductor fabrication process except for filling the U-shaped groove with agarose gel and not removing the photoresist in the final stage of the process.

(1) The silicon wafers (thickness $350\,\mu$m, diameter 5 cm) are washed in a boiled solution containing hydrogen peroxide, ammonium solution and water.
(2) The wafer is submitted to thermal oxidation. The oxidation temperature is $1000°$C. The thickness of the SiO_2 layer is $1\,\mu$m.
(3) The groove pattern is formed with a negative photoresist (Tokyo Oka, OMR-83), after which the other side is coated with the same photoresist. About 50–60 s of exposure is suitable when the MA-10 mask aligner is used.
(4) The SiO_2 layer is etched with a mixed solution consisting of 1 part 50% v/v hydrogen fluoride and 6 parts of 50% v/v ammonium fluoride. The remaining SiO_2 layer becomes the mask for the anisotropic silicon etching.
(5) The resist is removed in a 2/1 sulfuric acid/hydrogen peroxide solution at room temperature.
(6) The silicon is submitted to anisotropic etching in a 35% w/v potassium hydroxide solution. The temperature is maintained at $80°$C.
(7) The SiO_2 layer is removed with the same solution as used in (4).
(8) The silicon wafers are then washed with the same solution as used in (1).

(9) The wafers are submitted to thermal oxidation. The oxidation temperature is 1000°C. The SiO_2 layer is 500 nm thick.

(10) The resist pattern is formed for the gold electrodes by using the same photoresist as in (3).

(11) A 50-nm layer of chromium is deposited, followed by 1 μm of gold.

(12) The photoresist is then removed in warmed sulfuric acid.

(13) The U-shaped groove is filled with heated agarose gel containing a 0·1 M KCl solution using a microsyringe. Then the gel is cooled. The agarose gel concentration is 1% w/v.

(14) A hydrophobic polymer is applied, and this is followed by a photochemical reaction. In this case, the same photoresist (OMR-83) as was used to fabricate the oxygen electrode is used as the gas-permeable membrane and for insulation. The photoresist is spin-coated at 500 rpm for 5 s and 1500 rpm for 20 s, followed by exposure to ultraviolet light.

A linear relationship is obtained for an oxygen concentration between about 1 and 7·9 ppm (saturated) by using a 2-mm wide electrode when the terminal voltage between the two gold electrodes is 0·8 V. Similar calibration curves are obtained by using other oxygen electrodes.

3.2.2 *Glucose Sensor and CO_2 Sensor Using Micro-oxygen Electrodes*

The glucose sensor is fabricated by immobilizing glucose oxidase (GOD) on a sensitive part of the oxygen electrode by cross-linking with bovine serum albumin (BSA) and glutaraldehyde (GA). The enzyme-immobilized membrane is formed by dropping the sensitive part into a mixture containing 2 mg of GOD, 20 μl of 10% w/v BSA solution and 10 μl of 25% w/v GA solution. The glucose sensor responded as soon as the glucose solution was injected into the buffer solution, and stabilized for 5–10 min after the injection. The sensor responded almost linearly for glucose concentrations between 0·2 and 2 mM, which is comparable to conventional glucose sensors.

A microbial CO_2 sensor using this oxygen electrode was also constructed (Suzuki *et al.*, 1990). An autotrophic bacterium named S-17, which can use carbonate as the sole source of carbon, was obtained from The Fermentation Research Institute, Japan. Bacterial whole cells were immobilized on the micro-oxygen electrode as follows:

(1) The sensitive area of the oxygen electrode was immersed in 0·2% sodium alginate solution containing S-17 whole cells and then it was immediately immersed in a 5% (w/v) $CaCl_2$ solution to make a bacterial-immobilized calcium alginate gel.

(2) The negative photoresist as the gas-permeable membrane was formed over the bacteria-immobilized gel. The photoresist was only exposed to UV light for a short time. The response time of the sensor ranged from 2–3 min. The CO_2 was supplied by acidification of $NaHCO_3$. A linear relationship was obtained between the current decrease and $NaHCO_3$ concentration in the range 0·5–3·5 mM. The lowest detection limit was 0·5 mM $NaHCO_3$ within the margin of the noise amplitude. Above 3·5 mM, there was no significant increase in response.

3.3 Glucose Sensor Based on Mediator Measurement

In the type of biosensors based on oxygen consumption, for example, the glucose sensor and the alcohol sensor using glucose oxidase (GOD) or alcohol oxidase (AOD), sensor signal and dynamic range are greatly affected by the dissolved oxygen concentration. Variation of dissolved oxygen concentration may cause fluctuations on the electrode response, and the dynamic range of glucose determination is decreased by the lack of dissolved oxygen. If the redox reaction of an enzyme can be measured with an electrode directly, a biosensor which is not affected by dissolved oxygen concentration and has a wide dynamic range may be constructed. But generally redox enzymes do not show direct electron transfer at an electrode. Therefore, a mediator is necessary for the electrochemical oxidation of a redox enzyme. For instance, Foulds & Lowe, (1988) reported glucose oxidase immobilization in ferrocene-modified pyrrole polymer on a Pt electrode and used this system as a glucose sensor . Polypyrrole (PPy)-modified electrodes are prepared by electrooxidation of pyrrole. Electrodeposition of ferrocene-modified polypyrrole is effected by cycling the electrode between 0 and $+1\cdot0$ V in aqueous perchlorate containing pyrrole and [(Ferrocenyl) amidopropyl]pyrrole (FAPP). Entrapment of GOD in the redox copolymer (FAPP/pyrrole) is effected by cycling the working electrode between 0 and $+0\cdot8$ V. This sensor responds to glucose in the range of 1–100 mM. Ikeda *et al.* (1986) reported a glucose sensor composed of GOD and a carbon paste electrode (CPE) with mix-in *p*-benzoquinone (BQ) as a mediator, covered with a nitrocellulose film. This sensor detected glucose in the concentration range 10–150 mmol/dm^3. But the response of these sensors is dependent on the dissolved oxygen concentration. To overcome this problem, a glucose sensor using $NAD(P)^+$-dependent glucose dehydrogenase (GDH) was developed. Turner *et al.* (1987) reported an amperometric glucose sensor using quinoprotein, GDH, which was purified from *Acinetobacter calcoaceticus*. GDH was immobilized on a $1,1'$-dimethylferrocene-modified graphite foil electrode. The current density of this GDH electrode produced more than twice that of a graphite based-GOD electrode at 4 mM glucose. The calibration curve showed that this sensor could detect glucose in the range of $0\cdot5$–4 mM. When this sensor was treated with more GDH in the presence of glutaraldehyde, the range of response was increased up to 15 mM. But this treatment reduced the current output. The response of this electrode was unaffected by changes between an anaerobic environment and one of 100% oxygen saturation.

3.4 Glucose Sensor Based on Micro-Au Electrode with Mediator (Karube *et al.*, 1988)

A micro-glucose sensor was constructed with GOD and a modified electrode to which direct electron transfer can take place. It has been reported that GOD is trapped in a PPy membrane by electropolymerizing a polypyrrolemonomer in the presence of GOD. This method, however, has some disadvantages in clinical applications; it takes a long time to obtain a steady current in amperometric determination of glucose, and its sensitivity usually decreases after a week or so. The response of this electrode saturates at 15 mM glucose (37°C, pH 7·0). Moreover, there is a possibility of deterioration of PPy membrane by H_2O_2 produced by the enzyme reaction. The use of an electron mediator was examined to reduce the applied potential and to remove H_2O_2 directly. This led to stabilization and a rapid response of the glucose sensor. Considering the lowest redox potential (+0·1 V versus SCE), sufficient reaction rate for the GOD(red.) oxidation and its insolubility in water 1,1'-dimethylferrocene (DMFe) was found to be the best electron transfer mediator between the reduced form of GOD and the PPy-modified electrode. The reaction scheme is shown as follows:

$$\text{Glucose} + \text{GOD(ox.)} \rightarrow \text{Gluconolactone} + \text{GOD(red.)}$$

$$\text{GOD(red.)} + 2\text{DMFe}^+ \rightarrow \text{GOD(ox.)} + 2\text{DMFe} + 2\text{H}^+$$

$$2\text{DMFe(red.)} \rightarrow 2\text{DMFe(ox.)} + 2e^-$$

The micro-Au electrode utilized in this sensor has two working electrodes and one counter electrode on material with 1·6 mm width. The electrode is made as follows. First, titanium is sputtered onto a sapphire substrate to improve gold adhesion. Then, a gold layer (c. 1 μm thick) is deposited onto the titanium layer by sputtering. The gold layer is patterned by etching. The gold electrodes are covered with a negative photoresist for insulation except the sensitive area and the pad parts. PPy-modified electrodes are prepared by electropolymerization of pyrrole in aqueous solution. Immobilization of GOD is performed by adsorption of GOD onto PPy-modified electrode as follows: the PPy-modified electrode is dipped into a GOD solution (50 mg/ml) and is left overnight at 4°C; following this, the electrode is rinsed and dried. The GOD-adsorbed electrode is dipped into a dichloromethane solution containing 1% DMFe and 2% polyvinylbutyral (PVB) for 10 s and dried. DMFe is trapped in the PVB matrix. This DMFe immobilized-GOD/PPy electrode is characterized at an electrode potential of +0·1 V under a nitrogen gas atmosphere. This glucose sensor has a wide dynamic range, and the response saturation is not observed below a glucose concentration of 30 mM. This sensor is therefore suitable for direct measurement of glucose in blood.

4 INTEGRATED MULTIBIOSENSOR

In the field of clinical analysis, about 20 constituent elements are analyzed at the same time. When various substrates in a small amount of sample solution must be detected simultaneously, it is important and necessary to develop a very small multibiosensor. Recently, various kinds of integrated biosensors based on pH-selective ion-sensitive field effect transistors (pH-ISFET) and microelectrodes have been reported. In these biosensors, the pH-ISFETs and electrodes are coated with an enzyme-immobilized membrane. This necessitates an enzyme-immobilized membrane fabrication method that meets the following requirements; (1) the enzyme-immobilized membrane must be precisely deposited onto a transistor gate region or small working electrode; (2) a deposited membrane must not peel off the sensitive surface area during practical usage; (3) it must be possible to make different enzyme-immobilized membranes without mixing; (4) fabrication processes must be applicable to formation of wafers and compatible with the integrated circuit process. In this section, several integrated multibiosensors and fabrication methods are described.

4.1 Integrated Au Electrodes

4.1.1 Simultaneous Determination of Glucose and Galactose

In clinical analysis, simultaneous measurement of several kinds of biological substrates is important. In this section, integrated multibiosensors for the simultaneous measurement of glucose and galactose based on glucose dehydrogenase (GDH) and galactose oxidase (GAO), respectively, is discussed. GDH with pyrroloquinolinequinone (PQQ) as a co-enzyme catalyzes the oxidation of glucose as follows:

$$\text{Glucose} + \text{GDH(ox.)} \rightarrow \text{Gluconolactone} + \text{GDH(red.)}$$

$$2\text{Med(ox.)} + \text{GDH(red.)} \rightarrow 2\text{Med(red.)} + \text{GDH(ox.)}$$

where GDH(ox.) and GDH(red.) are the oxidized and reduced form of GDH, respectively, and Med(Ox.) and Med(red.) are similarly those of the mediators. Med(red.) is detected on the electrode by the reaction:

$$\text{Med(red.)} \rightarrow \text{Med(ox.)} + e^-$$

Because no electrons can transfer from PQQ to oxygen, this sensor is unaffected by the concentration of dissolved oxygen. On the other hand, GAO catalyzes galactose oxidation as follows:

$$\text{Galactose} + \text{GAO(ox.)} \rightarrow \text{Galactohexoaldose} + \text{GAO(red.)}$$

$$O_2 + \text{GAO(red.)} \rightarrow H_2O_2 + \text{GAO(ox.)}$$

It is possible therefore to measure the glucose and galactose concentration independently in the same solution even if GDH and GAO are immobilized on two separate electrodes in the same unit. Glucose can be detected by current increase during oxidation of the mediator, and galactose can be measured by current decrease during oxygen reduction.

Four gold working electrodes and a counter electrode were formed on Corning 7059 glass by vapor deposition. The area of the working electrode is $0.2\,\text{mm}^2$. One of the working electrodes is used as an immobilized GDH electrode and two as immobilized GDH and GAO electrodes. GDH was immobilized on the electrode as follows: 5 mg of GDH and 5 mg of BSA were dissolved in 100 μl of HEPES solution (pH 7·9) containing 10 mM MgSO$_4$ and 3 mM PQQ. The solution was spread on the electrode, and exposed to glutaraldehyde vapor at 30°C for 30 min. Enzyme solution containing 50 mg/ml GDH, 40 mg/ml GAO and 50 mg/ml BSA was used to prepare the immobilized GDH and GAO electrodes. The immobilization procedure was the same as that described above. The assay procedure was as follows: The enzyme electrode and the reference electrode were immersed in 10 mM HEPES solution (pH 7·4) containing ferrocene monocarboxylic acid (FCA) and stirred at 30°C. The potentials of the electrode for oxidizing mediators and the oxygen electrode were maintained at +350 and −300 mV, respectively. After the current output stabilized, glucose or galactose solution was injected into the buffer solution and the current change was measured, while air bubbled through the system. Glucose was measured in solutions containing various concentrations of galactose. The response toward glucose decreased with increasing galactose concentration. Because GAO reacts with both oxygen and mediators such as FCA, the response of this electrode to glucose depended on the galactose concentration. Simultaneous determination of both glucose and galactose can be performed as follows: firstly, galactose is measured by using an oxygen electrode, and then the glucose concentration is determined by using an appropriate calibration curve of glucose in the presence of galactose.

4.1.2 Simultaneous Determination of Glucose and Oxygen (Yokoyama et al., 1989)

Simultaneous determination of glucose and oxygen for in-vivo monitoring is very important not only because oxygen concentration monitoring in blood is necessary for understanding the respiration state of a patient, but also because the response of a glucose sensor using GOD is affected by oxygen concentration. In order to realize simultaneous monitoring of glucose and oxygen concentrations, an integrated multi-Au electrode is fabricated on a glass substrate and applied to glucose and oxygen sensors. GOD is immobilized on the electropolymerized polypyrrole film of a working electrode by using BSA and GA. 1,1'-dimethylferrocene (DMFe) is incorporated by dipping the GOD-immobilized part of the electrode into a DMFe-dissolved acetone solution. Another electrode is used as an oxygen electrode. An Ag/AgCl electrode is used as a reference electrode. The same counter and reference electrodes are used for both the glucose sensor and oxygen electrode. The terminal voltage of the glucose sensor and oxygen electrode are maintained at +0·1 V and −0·3 V versus Ag/AgCl electrode, respectively.

The characteristics of the glucose sensor and oxygen electrode are evaluated simultaneously in a 0·1 M phosphate buffer (pH 7·0, 30°C). The response time of this glucose sensor was about 1 min when 50 μl of 10 g/dl glucose sample solution was injected into 30 ml of phosphate buffer. The glucose sensor and the oxygen electrode responded independently and the oxygen electrode could be used to correct the glucose sensor.

4.2 Integrated ISFETs

Sibbald *et al.* (1984) reported the on-line, simultaneous measurement of blood K^+, Na^+, Ca^{2+} and pH by using a multi-functional ChemFET (chemical sensitive field effect transistor) attached to a flow-cell. But, the immobilization technique for a molecular recognition element in a very small area is not adequate when compared with the miniaturization technique of ISFET. In order to fabricate an integrated multibiosensor, it is necessary to develop an enzyme-immobilized membrane fabrication technology that satisfies the following requirements: (1) an enzyme-immobilized membrane must be deposited on a small sensitive region; (2) a deposited membrane must be resistant to detachment from the sensitive area in practical use; (3) fabrication processes must be compatible to the semiconductor fabrication processes.

Several methods for preparing enzyme-immobilized membranes of multibiosensors have been developed, such as a multibiosensor for glucose, urea and various ions, and based on ISFET. For example, a water-soluble photo-cross-linkable polymer solution containing an enzyme is spin-coated on an ISFET gate surface and enzyme is immobilized by polymerization using ultraviolet light irradiation (Hanazato *et al.*, 1986). This method is suitable for single-enzyme biosensor mass production if the enzyme-immobilized membrane can be patterned by the photolithographic technique. However excess enzyme is required since much of the enzyme solution does not contribute to the formation of the enzyme membrane when a spin-coating method is used to prepare enzyme-immobilized thin film. New membrane preparation methods which satisfy the above requirement to fabricate an integrated multibiosensor are described in the following section.

4.2.1 *Micropool Method (Kimura* et al., *1986)*

The micropool method employs making small wells on the ISFET gate region and pouring enzyme solution into each well. The fabrication of an integrated ISFET using silicon on sapphire (SOS) material and its application as a multibiosensor are described here and illustrated in Fig. 4.

(1) A silicon layer of the SOS wafer is anisotropically etched to form an 'island' of silicon.

(2) The source, drain and gate region are doped. The SiO_2 layer is then thermally oxidized. A polysilicon layer for the metal oxide semiconductor field effect transistor gates is deposited by chemical vapor deposition (CVD). Finally, to control the threshold voltage, phosphorous ions are implanted on the gate surface.

(3) An Si_3N_4 layer is deposited by CVD as a protectional membrane against saline water. It also works as a pH-sensitive membrane. Contact pads are created, and this is followed by deposition of aluminum. A gold layer is sputtered on the whole balk surface.

(4) A film resist photopolymer layer is adhered to the wafer surface and patterned as shown in Fig. 4 (Karube *et al.*, 1987). This photopolymer is used

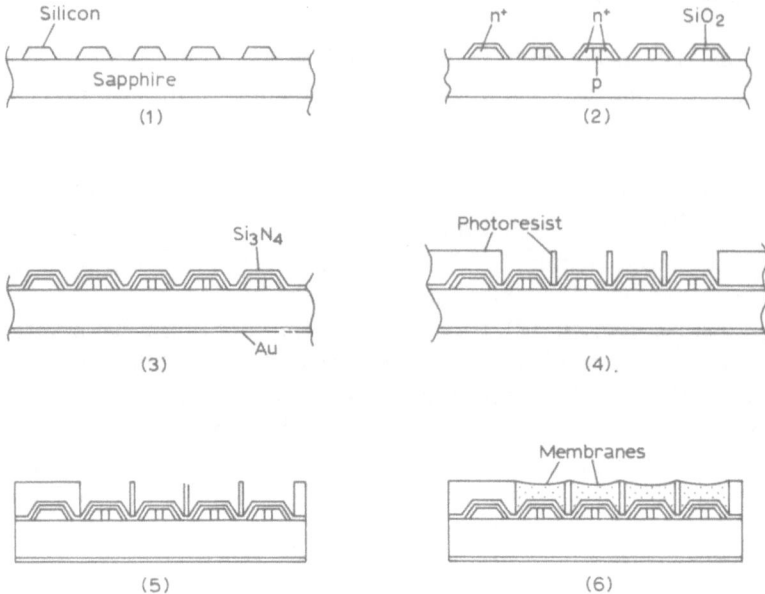

Fig. 4. Fabrication process by using micropool method.

to make micropools for four types of membrane. It also acts as a physical protectional membrane for the ISFET surface.

(5) The wafer is cut into individual sensor tips, which are placed on a flexible printed circuit board. Electric connections are made by aluminum bonding between the ISFET and the flexible print circuit board. The bonding area is then moulded by a heat hardening adhesive.

(6) Three kinds of membranes which are sensitive to potassium, glucose and urea, respectively, are formed:

   (i) Potassium-sensitive membrane. A potassium-sensitive membrane, which needs high temperature processing, is formed in the following way. Ten mg of valinomycin is dissolved in 120 μl of dioctyl adipate with 800 mg of negative photoresist polymer, OMR-83. After stirring, a small amount of the mixture is dropped into one of the sensing area pools using a microsyringe. The volume of one drop is about 0·03 μl. After prebaking at 80°C for 30 min one more drop is applied to the membrane surface. After a second prebaking and ultraviolet irradiation, the membrane is hardened by a final baking.

  (ii) Glucose-sensitive (urea-sensitive) membrane. 5 mg of GOD (or 1 mg urease) is dissolved in a 15% BSA solution. A small drop (0·05 μl) of this mixture is poured into one of the ISFET sensing areas by a microsyringe. After being dried at room temperature for about 10 min, 0·05 μl of 25% glutaraldehyde solution is placed on the dried enzyme, and the enzyme is immobilized. After 10 min the device is immersed in pure water.

(iii) Reference membrane. The role of this membrane is to protect the ISFET surface against physical destruction and to prevent bubble formation, while maintaining pH sensitivity. The process of making this membrane is the same as for enzyme membranes.

A gold layer works as a pseudo-reference electrode, and a potential of +1·6 V is added to the gold electrode. A urea-sensitive FET and potassium-sensitive FET shows a rapid response to urea and potassium, respectively. A urea-sensitive FET shows no response to glucose. A glucose-sensitive FET showed a gradual increase in response to glucose and a small response (less than 1 mV) to urea. The influence of diffusion of chemical species to the neighbouring electrode surface is smaller than expected. Glucose concentrations can be determined in the range 1–50 mg/dl. Urea concentrations can be determined from 1–100 mg/dl. The potassium differential output shows a linear relationship between 1×10^{-3} and 2×10^{-2} eq/liter and has a sensitivity of 60 mV/pK. This method makes it quite easy to create a membrane and to control the amount of enzyme immobilized. This procedure requires only small amounts of enzyme for immobilization, and hence it is an economical method. However, the major portion of the membrane has a tendency to be located towards the film resist side wall. As a result the membrane centre is rather thin. The thickness of the membrane must be uniform in order to obtain a large and rapid response. This is best achieved using an automatic microsyringe.

4.2.2 Lift-off Method (Kimura et al., 1988a; Nakamoto et al., 1988)

A lift-off method is one of the integrated circuit (IC) fabrication processes used to prepare a thin film on a small area with a large degree of precision. This method is therefore considered suitable for preparing enzyme-immobilized membranes on a small area in mass production of integrated multibiosensors. A fabrication process using the lift-off method for a multibiosensor is shown in Fig. 5. Individual processes to make a glucose/urea sensor based on ISFET is performed as follows:

(1) 1% γ-APTES aqueous solution is spin-coated on a wafer, and heated at 110°C for 5 min to modify the wafer surface amino groups by reacting with glutaraldehyde.

(2) A positive photoresist is spin-coated onto the wafer, then prebaked and treated with ultraviolet light through a photomask and developed to expose the gate region on which a GOD-immobilized membrane is to be deposited. The photoresist layer is 1·8 μm.

(3) 0·25 ml of BSA 28% (w/v) solution in 50 mM PIPES-NaOH (pH 6·8), 0·15 ml of GOD 23% (w/v) aqueous solution and 0·1 ml of 5% (v/v)

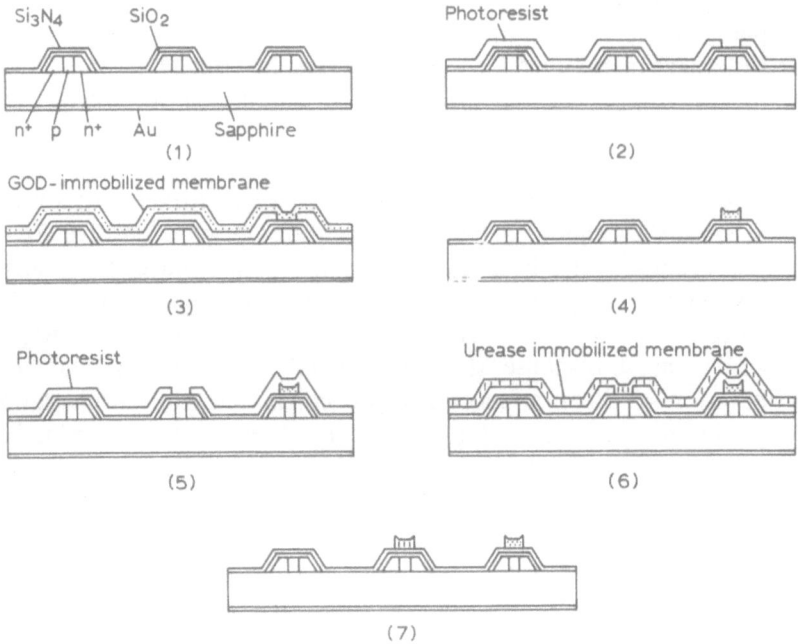

Fig. 5. Fabrication process by using lift-off method.

GA aqueous solution are mixed and then spin-coated on the wafer at 1500 rpm. The wafer is left at room temperature for 2 h to complete the cross-linking reaction. As a result, the wafer is covered with a GOD-immobilized membrane throughout (Fig. 5) (Ikeda *et al.*, 1986).

(4) Ultrasonic vibration is applied to the wafer in acetone. The photoresist between the GOD-immobilized membrane and the wafer dissolves in acetone. The GOD-immobilized membrane on the photoresist is thus lifted off in acetone. A precisely patterned GOD-immobilized membrane on the gate region is therefore obtained. The resulting wafer is washed in water and dried with nitrogen gas.

(5) The positive photoresist is spin-coated onto the wafer again. To prevent the enzyme inactivation from heat by baking, the photoresist is dried in an evacuated desiccator for 30 min instead of being baked, and patterned and developed as described in step (2) above to expose the gate region on which a urease-immobilized membrane is deposited.

(6) The urease-immobilized membrane is deposited onto the gate surface area as described in step (3). The enzyme solution containing 0·25 ml of BSA 28% (w/v), 0·15 ml of urease 5% (w/v) and 0·1 ml of 5% GA is spin-coated on the wafer.

(7) The urease-immobilized membrane is lifted off by the method described in (4) above.

By repeating the fabrication processes described above, the GOD-immobilized membrane and urease-immobilized membrane are precisely deposited on each gate surface in the wafer. The membranes are 100 μm × 400 μm. The thickness is about 1 μm. When the enzyme solution containing urease is spin-coated, the GOD-immobilized membrane is covered with photoresist. Therefore, contamination of the GOD-immobilized membrane by urease cannot take place. There is some fear that the photoresist could deteriorate the GOD activity. However the sensor shows the same response characteristics to glucose as a sensor prepared without treatment of the photoresist.

In the case of the glucose sensor, the linear relationship between differential output in steady state and glucose concentration was valid up to 90 mg/ml of sugar. The slope of the output began to decline at 90 mg/ml and a limit was reached at 200 mg/ml. In the case of the urea sensor, a linear relationship between the logarithm of the urea concentration and differential output was observed in the range from 1–100 mg/ml.

Another multibiosensor, in which a glucose sensor based on amperometry and a urea sensor based on potentiometry are integrated, has been realized by complementary use of ISFETs and micro-patterned planer gold electrodes. GOD-immobilized membrane, urease-immobilized membrane and BSA cross-linking membranes were deposited in appropriate positions by the lift-off method described above. The output of this sensor to glucose, 30 s after the addition of a concentrated solution, increases linearly with the logarithm of the glucose concentration from 1 mg/dl to 100 mg/dl. The output begins to decline at 100 mg/dl. The output to urea, 30 s after the addition of concentrated solution, rises sharply with the logarithm of the urea concentration, from 5 mg/dl to 100 mg/dl. The multibiosensor, which employs the complementary use of amperometric and potentiometric principle, can thus be effectively used to measure glucose and urea with high selectivity.

4.2.3 Ink Jet Nozzle Method (Kimura et al., 1988b)

In this section, an application of an ink jet nozzle, originally developed for printing equipment, is discussed as a tool for precise enzyme deposition onto an ISFET device. The ink jet nozzle construction is shown in Fig. 6. Two kinds of immobilization method were carried out using this equipment. The first method is gas phase immobilization and the fabrication process is as follows:

(1) Ten enzyme drops were emitted onto different positions of one ISFET gate area.

(2) To carry out preliminary immobilization, the sensor device was placed in a vapor chamber containing 25% (w/v) glutaraldehyde solution for 5 min.

(3) By repeating processes (1) and (2) five times, 50 drops of the enzyme solution were immobilized on one ISFET gate area.

(4) To complete the immobilization, the sensor device was immersed in 1% (w/v) glutaraldehyde solution for 5 min.

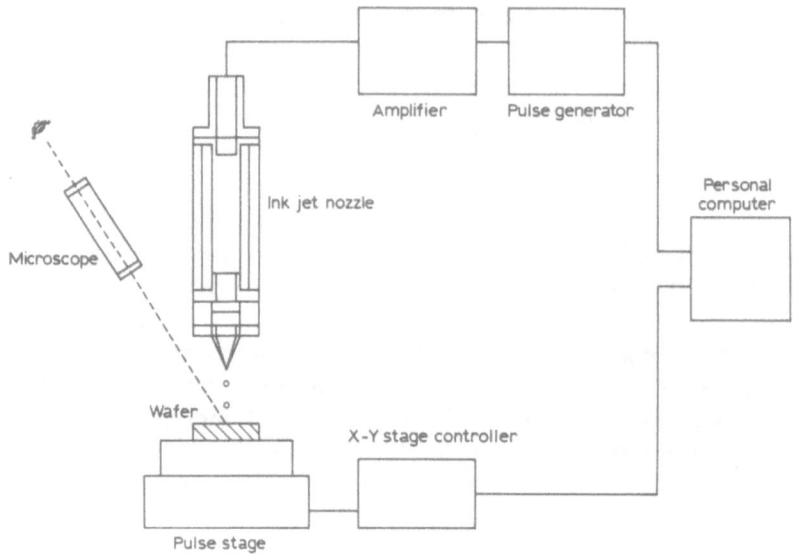

Fig. 6. Ink jet nozzle system.

By using this sensor, glucose was determined in the range of 2–100 mg/dl and urea was determined in the range 1–1000 mg/dl. In this method, only ten drops of enzyme solution cover a large part of the gate region. Enzyme position was precisely determined employing an *X-Y* stage controlled by a personal computer. The problem in using the gas phase immobilization method, however, is the length of time required. It takes about 1 h to emit and immobilize enzyme solution in one cycle.

The second method which may be used is liquid phase immobilization; this fabrication process is almost the same as the micropool method described in Section 4.2.1. The enzyme solution was emitted by the ink jet nozzle into micropools made by the photoresist. But, in this method, special skill was not needed to deposit an immobilized-membrane on a small sensing area. In the ink jet nozzle method, only a very small amount of enzyme is needed. So, highly expensive enzymes can be used efficiently for making a biosensor. A further merit of this method is that it is easy to select an enzyme among several enzymes and to determine the enzyme emission position by using a computer, so that many sorts of multibiosensors having different enzyme combinations could be realized on one wafer.

4.2.4 Direct Photopatterning Method (Hanazato et al., 1986)

In this section, the application of a direct photopatterning method to the immobilization of enzymes on ISFET is discussed. A water-soluble photo-cross-linkable polymer consisting of polyvinylpyrrolidone (PVP) and 2,5-bis(4'-azido-2'-sulfobenzal) cyclopentanone (BASC) sodium salt were used to immobilize GOD and lipase. The fabrication of the glucose sensor is as follows. The surface of an ISFET was treated by γ-APTES to improve the adhesive strength between the surface and the polymer membrane. A 0·1 ml aliquot of polymer solution (100 parts of water, 10 parts of PVP and 0·3 parts of BASC containing 10 mg of GOD and 10 mg of BSA) was spin-coated over the surface of the integrated ISFET having three ISFET elements at 2000 rpm for 2 min to make the thin film. A limited area of the membrane around the gate surface of one of the three ISFETs was exposed to UV-irradiation to polymerize the monomer and to entrap the enzyme. The ISFET was then immersed in water to form a GOD membrane. Thus, GOD membrane was formed on the limited area of gate surface. By repeating this procedure, a lipase membrane and BSA membrane were also formed on the other ISFET elements, respectively. The ISFETs were immersed in 3% (w/v) glutaraldehyde solution and then in 0·1 M glycine solution. The glucose sensor is useful for the determination of glucose in the concentration range 0–5 mM. When the lipid sensor is employed to measure triolein, it can be detected in the range 0–3 mM. According to cross-talk examination between glucose and triolein response of the integrated ISFET-biosensor, this multibiosensor can measure both glucose and triolein concentration simultaneously.

5 NOVEL BIOSENSORS BASED ON NEW TRANSDUCERS

Conventional biosensors have consisted of electrochemical devices. On the other hand, other transducers such as fiber optics, image sensors, piezoelectric devices, surface acoustic wave (SAW) devices, etc. are currently being investigated for use as biosensors. In the following section, novel biosensors utilizing novel transducers are discussed.

5.1 Image Sensors

Most clinical analyses are based on the determination of soluble substances in body fluids such as blood and urine. The direct analysis at cell or tissue level is very important in clinical diagnosis. In the case of cancer detection, highly sensitive and rapid detection methods for abnormal cells are required. Cell diagnosis is carried out mainly by visual inspection by trained experts or by using a flow cytometer.

Recently, much attention has been focused on image analyzing systems composed of an image sensor and a microcomputer system. Image sensors are classified into the X-Y address method and the charge transfer method. In the X-Y address method, optical signals of each address are read by switching on the corresponding circuit. The charge transfer method was initially demonstrated by using a bucket bridge device (BBD) and later, by the more advanced charge coupled device (CCD). Most image sensors in practical use are now of the CCD type. The CCD is an integrated semiconductor chip composed of photodiode arrays and charge transfer circuits. Electric charge

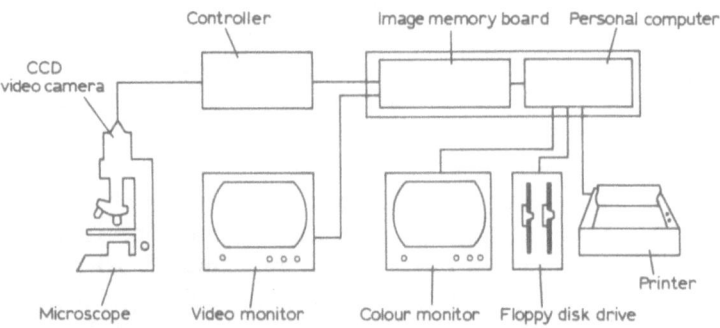

Fig. 7. Schematic diagram of the image sensor system.

accumulated at each photodiode is transferred systematically to the output terminal by controlling the electric potential in the chip. The output pulse height correlates with the brightness at the corresponding photodiode. Thus, a visual image focused on the CCD can be converted to a succession of analog pulses. Since the photodiodes are arranged with an approximate separation of 10 μm, the same degree of image resolution can be obtained. There are many advantages of solid state CCD image sensing compared with conventional vidicon. It is compact, highly sensitive, distortionless, no after-image is formed and it has low power consumption and a long operational life.

In the following section, the application of image analysis systems, consisting of a CCD image sensor and a personal computer, to cell and tissue diagnosis are discussed. Figure 7 shows a schematic diagram of the image sensor system. A phase-contrast transmitted light microscope was used to observe the cytolysis of tumor cells. A CCD video camera was mounted vertically onto the microscope. Video images of cells were displayed on a video monitor. The video display was fed into an image memory board connected to a 16 bit personal computer.

5.1.1 Detection of Tumor Cells (Suzuki et al., 1987a, b)
The specific detection of Line-10 hepatocarcinoma (L-10) cells is based on the complement-mediated cytotoxic reaction with monoclonal antibody 3C4. Cell lysis is followed by using a phase-contrast microscope. In the case of normal cells, the difference between intracellular and extracellular compartments produces a phase-lag of transmitted light because they cannot be lyzed. Therefore, normal cells look bright under the phase-contrast microscope, whereas damaged cells appear darker because they have lost membrane. Applying the decrease in brightness of lyzed cells, specific and non-specific cells are distinguished. The experimental procedure is as follows: 100 μl of cell suspension containing Line-10 cell and non-specific L2C cell, 100 μl of 3C4 antibody solution and 50 μl of rabbit serum (complement source) are mixed. After 30 min incubation at 37°C, the sample is transferred into a hemacyto-meter and analyzed by using the image sensor system. L-10 cells can be

detected quantitatively in the range 10–100%, by virtue of producing a linear relationship between the L-10 content and image area in this range. This process does not require a cell-washing step, making it possible to automate the whole system during the reaction stage. But, the detection limit of this system is not sufficient for cancer diagnosis. To overcome this problem, target cells were stained directly. By using the DNA (or RNA) specific fluorescent dye, propidium iodide, L-10 cells can be detected quantitatively when its content is above 2%.

5.1.2 Application to Latex Agglutination Test *(Matsuoka* et al., *1987; Tamiya* et al., *1988c)*

The turbidimetry of latex agglutination in immunoassay is already applied in practical uses, though the sensitivity is not always sufficient. It is known that specific and non-specific reactions differ in their initial rate of reaction. Therefore, a precise measurement of the turbidity change at an earlier stage of reaction would improve the sensitivity of turbidity measurements. Image sensing is a suitable device for such a purpose. In this section, the application of a linear (one-dimensional) image sensor to immunoassay based on latex agglutination is discussed. The experimental procedure is as follows. Human immunoglobulin G (IgG) solution and Ab-L (antibody-bound latex beads) suspension are mixed in a test tube and immediately transferred to a glass capilllary tube (0·8 mm dia. × 70 mm). The glass capillary containing human IgG and Ab-L is set on the sample holder of the image sensor which is placed in the dark box and illuminated with a fluorescent lamp through a slit. Light scattered by the capillary at a scattering angle of 120°, is passed through a lens and focused exactly on the sensing area of the image sensor. The optical intensity is adjusted with an iris and the progress of agglutination is followed using the image sensor. As Ab-L suspended in a glass capillary tube precipitated gradually, the linear image pattern changed. This linear image sensor system enabled human IgG to be detected in the range of 5×10^{-8} to 5×10^{-5} g/ml. In the latex agglutination immunoassay, an electric pulse accelerates the specific immunoreaction between Ab-L and antigen (Ag) by increasing the contact frequency of small particles. Such an acceleration improves the sensitivity of this assay method. By using the pulse immunoassay method, human IgG was determined in the range of $6·7 \times 10^{-8}$ to $6·7 \times 10^{-6}$ g/ml. It is thought that the combination of the image sensor system and the pulse immunoassay method can perform a rapid and precise immunoassay.

5.2 Quartz Crystal

It is well known that the resonant frequency of an oscillating piezoelectric crystal is affected by a change in mass at the crystal surface. This property has been applied to monitoring vapor deposition rates and for the chromatographic detection of gases. It has also been reported that modification of the crystal surface with organic compounds or enzymes that bind a particular gaseous substance can provide enhanced detection specificity. The properties of piezoelectric devices immersed in water or organic solvents have recently been studied by various researchers and several theoretical equations governing their behavior in liquids have been derived. Methods employing a coupled gravimetric/electrochemical assay have also been used for the determination of electrodeposited metal ions and anions and have been used for electrochemical analysis. Piezoelectric sensors are based on the measurement of small mass change on the surface of a piezoelectric crystal caused by the specific adsorption of molecules onto a specially modified surface. The relationship between surface mass change, Δm (g), resonant frequency, F (Hz) and frequency change (ΔF), is given by the Sauebrey equation:

$$\Delta F/F = -\Delta m/A\rho t,$$

where A is the crystal area covered by the adsorbed material (cm^2), ρ is the density of the quartz (g/cm^3) and t is the thickness of the uncoated crystal (cm).

5.2.1 Detection of Pathogenic Microorganisms

The first application of a quartz crystal to an immunosensor was reported by Shons *et al.* (1972). They measured anti-BSA antibody by using a BSA-immobilized quartz crystal. Leucocytes, immunoglobins, etc. have been also detected using quartz crystal. Muramatsu *et al.* (1986) reported the detection of *Candida albicans*. The determination of *C. albicans* has become important in clinical analysis. This yeast is found in the human body even under normal conditions, but an increase in cell population can induce infection and disease. *C. albicans* is conventionally assayed by visual inspection of antibody–antigen (*Candida* species) aggregate formation. The method, however, requires technical skill, is time-consuming and gives only a semiquantitative assessment of the cell concentration. Matsuoka *et al.* (1987) showed that a potentiometric method could be applied to a *C. albicans* immunoassay. In this method, the negatively charged cells are adsorbed onto an antibody-bound membrane, inducing a considerable change in the membrane potential, which is linear in the range 10^4–5×10^5 cells/cm^3. Another immunoassay based on an electric pulse technique was also proposed and applied to the *C. albicans* immunoassay and is linear in the range 10^7–6×10^7 cells/cm^3.

Piezoelectric crystals coated with immobilized anti-*Candida* antibody are applied to the immunoassay of *C. albicans*. This appears to be the first reported application of piezoelectric crystals for the detection of a microorganism. The piezoelectric crystals used were AT-cut quartz (8 mm × 8 mm × 0·18 mm), with a basic resonant frequency of 9 MHz. Silver electrodes were formed on the crystal by vacuum deposition and were plated with palladium. The palladium electrodes were treated by anodic oxidation. The electrodes were then treated by γ-APTES and GA, and then anti-*Candida* antibody was immobilized on the electrodes. The treated piezoelectric crystals were dipped in the microbial suspension for 30 min to allow reaction to occur between the immobilized antibody and the microbe. The crystals were then rinsed with 0·5 M NaCl and water and dried in air. The resonant frequency, F_2, was measured, and the frequency difference, $F = F_1 - F_2$, calculated. The correlation between the cell concentration and the resonant frequency shift resulting from binding of *C. albicans* was valid in the range 1×10^6–5×10^8 cells/cm^3. In another test, the number of cells adsorbed on the electrode was also measured by microscopy. These tests confirmed that the magnitude of F was related to the actual number of adsorbed microbes. The sensor showed no increase in response with microbe concentration for the yeast, *Saccharomyces cerevisiae*. The small constant increase in F is caused by non-specific adsorption.

5.2.2 Determination of IgG Subclass

Recently much attention has been directed towards the application of quartz crystals in a liquid. Muramatsu *et al.* (1987) reported application of AT-cut piezoelectric crystals to the determination of human IgG concentration and IgG subclasses in a solution. They immobilized protein A on the surface of a piezoelectric crystal as a recognition element for IgG, considering its high binding specificity toward IgGs. The experimental procedure is as follows. The crystal electrodes are modified with γ-APTES and glutaraldehyde, then Protein A is immobilized on the electrode via the surface aldehyde. After immobilization, the remaining unreacted aldehyde is blocked with glycine. The procedure for measurement of IgG concentration is as follows. The Protein A-modified piezoelectric crystal is placed inside the flow cell. First, the resonant frequency is monitored at a constant flow rate. The steady resonant frequency (F_1) is obtained. The water is drained off and the solution of human IgG is then injected into the cell. After incubation, the crystal is rinsed with 0·5 M NaCl solution to remove any non-specifically adsorbed IgG. This is followed by remeasurement of the steady resonant frequency (F_2) in constantly flowing water. IgG bound to Protein A is removed with glycine–HCl buffer, allowing the successive measurement of different human IgG concentrations. The determination of IgG subclass is as follows. Mouse γ-globulin or human γ-globulin solution is injected into the cell and the steady resonant frequencies, F_1 and F_2, measured as described above. The crystal is then rinsed with a stepped gradient of phosphate–citric acid buffer from pH 7 to pH 2·5, each pH change being 0·5 pH units. The cell is rinsed at each step, and the steady resonant frequency (F_{2pH}) is remeasured in flowing water.

The steady resonant frequency is decreased by the injection of γ-globulin solution. By rinsing with a stepped gradient of phosphate–citric acid buffer from pH 7 to pH 3, the steady resonant frequency is increased. The pattern of the overall resonant frequency shift during each of the separate rinsing steps corresponds to the result obtained by affinity chromatography. The peaks at pH 6·5, pH 5·5–4·5, and pH 3·5–3·0 indicate the presence of IgG$_1$, IgG$_{2a}$ and IgG$_{2b}$, respectively. The pattern of the overall resonant frequency shift during each of the separate rinsing steps in the case of bound human γ-globulins clearly shows a difference from that of mouse γ-globulins and corresponds to the results previously reported in which IgG_2 and IgG_4 were eluted at pH 4·7 and IgG_1 and IgG_4 were eluted at pH 4·3. This system can be applied to various analyses, through the application of protein–protein affinity reaction.

5.2.3 Determination of Pyrogen (Muramatsu et al., 1988)

Many of the applications of a piezoelectric crystal are based on its mass detector. On the other hand, in the studies of quartz crystals in contact with liquids, the resonant frequency shift and the electrical resistant included in the electric equivalent circuit of the quartz crystal were shown to be a function of the viscosity and density of the contacting liquid, as follows:

$$\Delta F = -F_0^{3/2}\left(\frac{\rho_L \eta_L}{\pi \rho_Q \eta_Q}\right)^{1/2}$$

where F_0 is the basic resonant frequency of the crystal, ρ_L and η_L are the absolute viscosity and density of the liquid, respectively, and ρ_Q and η_Q are the elastic modules and density of the quartz, respectively. Muramatsu *et al.* (1988) reported that the electrical resistance of quartz crystal has excellent linearity in relation to the viscosity of the liquid, and applied this to the measurement of endotoxin, which causes gelation of *Limulus* amebocyte lyzate. Endotoxin is a kind of fever-inducing pyrogen. It is produced by Gram-negative microbes and widely found in water samples. The determination of endotoxin is very important in medical products. Since the blood from a horseshoe crab is coagulated by the addition of endotoxin, the gelation of *Limulus* amebocyte lyzate (LAL) has been used for the determination of endotoxin, and named the *Limulus* test. The conventional technique is a manual inversion method, in which a mixture of LAL and sample solution is incubated in a test tube for 1 h at 37°C after mixing, and the decision that the concentration is lower than a certain value is made by observation of the settling gel. The conventional method is therefore only semiquantitative. The experimental procedure is as follows. The sample solution (0·2 ml) is added to solute lyophilized LAL; the mixture is poured into the well-type cell attached to one side of a quartz crystal, and the resistance and resonant frequency measured at 37°C. By monitoring the viscosity changes, endotoxin is determined in the range $1-10^5$ pg/ml.

REFERENCES

Bergveld, P. (1970). *IEEE Transactions on Biomedical Engineering*, **BEM-17**, 70.
Caras, S. & Janata, J. (1980). *Analytical Chemistry*, **52**, 1935.
Foulds, N. C. & Lowe, C. R. (1988). *Analytical Chemistry*, **60**, 2473.
Gotoh, M., Tamiya, E. & Karube, I. (1986). *Analytica Chimica Acta*, **187**, 287.
Gotoh, M., Tamiya, E., Momoi, M., Kagaya, Y. & Karube, I. (1987). *Analytical Letters*, **20**, 875.
Gotoh, M., Tamiya, E., Seki, A., Shimizu, I. & Karube, I. (1988). *Analytical Letters*, **21**, 1785.

Gotoh, M., Oda, S., Shimizu, I., Seki, A., Tamiya, E. & Karube, I. (1989). *Sensors and Actuators*, **16**, 55.

Hanazato, Y., Nakako, M., Maeda, M. & Shiono, S. (1986). *Proceedings of the Second International Meeting on Chemical Sensors*, p. 576.

Ikeda, T., Hamada, M. & Senda, M. (1986). *Agricultural and Biological Chemistry*, **50**, 883.

Karube, I., Tamiya, E., Dicks, J. M. & Gotoh, M. (1986). *Analytica Chimica Acta*, **185**, 195.

Karube, I., Wang, Y., Tamiya, E. & Kawarai, M. (1987). *Analytica Chimica Acta*, **199**, 93.

Karube, I., Tamiya, E., Sode, K., Yokoyama, K., Suzuki, H. & Hattori, S. (1988). *Annals of the New York Academy of Science*, **542** (*Enz. Eng.* **9**), 470.

Karube, I., Yokoyama, K., Sode, K. & Tamiya, E. (1989). *Analytical Letters*, **22**, 791.

Kimura, J., Kuriyama, T. & Kawana, Y. (1986). *Sensors and Actuators*, **9**, 373.

Kimura, J., Murakami, T., Kuriyama, T. & Karube, I. (1988*a*). *Sensors and Actuators*, **15**, 435.

Kimura, J., Kawana, Y. & Kuriyama, T. (1988*b*). *Biosensors*, **4**, 41.

Matsuo, T. & Wise, K. D. (1974). *IEEE Transactions on Biomedical Engineering*, **BEM-21**, 485.

Matsuoka, H., Tanioka, S. & Karube, I. (1987). *Analytical Letters*, **20**, 63.

Muramatsu, H., Kajiwara, K., Tamiya, E. & Karube, I. (1986). *Analytica Chimica Acta*, **188**, 257.

Muramatsu, H., Dicks, J. M., Tamiya, E. & Karube, I. (1987). *Analytical Chemistry*, **59**, 2760.

Muramatsu, H., Tamiya, E. & Karube, I. (1988). *Analytica Chimica Acta*, **215**, 91.

Nakamoto, S., Ito, N., Kuriyama, T. & Kimura, J. (1988). *Sensors and Actuators*, **13**, 1665.

Shons, A., Dorman, F. & Najarian, J. (1972). *Journal of Biomedical Materials Research*, **6**, 565.

Sibbald, A., Covington, A. K. & Carter, R. F. (1984). *Clinical Chemistry*, **30**, 135.

Suzuki, M., Tamiya, E., Kataoka, T., Tokunaga, T. & Karube, I. (1987*a*). *Clinical Chemistry*, **33**, 558.

Suzuki, M., Tamiya, E. & Karube, I. (1987*b*). *Analytical Letters*, **20**, 337.

Suzuki, H., Tamiya, E., Karube, I. & Oshima, T. (1988*a*). *Analytical Letters*, **21**, 1323.

Suzuki, H., Tamiya, E. & Karube, I. (1988*b*). *Analytical Chemistry*, **60**, 1078.

Suzuki, H., Kojima, N., Sugama, A., Takei, F., Ikegami, K., Tamiya, E. & Karube, I. (1989). *Electroanalysis*, **1**, 305.

Tamiya, E., Karube, I., Kitagawa, Y., Ameyama, M. & Nakashima, K. (1988*a*). *Analytica Chimica Acta*, **207**, 77.

Tamiya, E., Seki, A., Karube, I., Gotoh, M. & Shimizu, I. (1988*b*). *Analytica Chimica Acta*, **215**, 301.

Tamiya, E., Watanabe, N., Matsuoka, H. & Karube, I. (1988*c*). *Biosensors*, **3**, 139.

Turner, A. P. F., Karube, T. & Wilson, G. S. (eds) (1987). *Biosensors—Fundamentals and Applications*. Oxford University Press, Oxford.

Yokoyama, K., Sode, K., Tamiya, E. & Karube, I. (1989). *Analytica Chimica Acta*, **218**, 137.

INDEX

456

Index